MEDICAL RADIOLOGY

Diagnostic Imaging and Radiation Oncology

Interventional Radiation Therapy
Techniques – Brachytherapy

Contributors

G. Alth · H. Annweiler · H. W. Anton · J. M. Ardiet · J. J. Augsburger · K. Baier
F. Baillet · H. Bartelink · J. H. Borger · A. M. Borofsky · L. W. Brady · M. Busch
H. Busse · D. Chassagne · J. L. Chassard · C. T. Coughlin · J. L. Day · J. Dunst
B. Emami · R. Engenhart · H. Ernst · W. A. Fair · R. Fietkau · Z. Fuks · F. P. Gall
J. P. Gerard · B. J. Gerbi · F. H. Glaser · G. Grabenbauer · L. Grimard · J. Hammer
H.-P. Heilmann · B. S. Hilaris · H. Ikeda · T. Inoue · H. Junkermann · U. L. Karlsson
B. N. Kimmig · K.-H. Kloetzer · D. Kob · C. Koprowski · T. Kozuka · J. C. Kummermehr
K. Kuphal · C. K. Lee · S. H. Levitt · D. A. Lightfoot · P. K. Lommatzsch · K. Lutz
H.-B. Makoski · G. Marinello · A. M. Markoe · N. Masaki · R. E. Maxwell · J. J. Mazeron
J. F. Montbarbon · C. R. Moorthy · R. G. Müller · R.-P. Müller · F. Mundinger
D. Nori · C. G. Orton · C. B. Ostertag · S. Ozeki · J. Papillon · C. A. Perez · B. Pierquin
G. Pipard · R. A. Potish · R. Pötter · M. Riccabona · K. Rotte · R. Sauer · P. Schlag
G. Schlegel · M. H. Seegenschmiedt · J. A. Shields · J. Slanina · H. Sommerkamp
A. Sparenberg · W. Spitzer · V. Sturm · A. E. Tchelebi · N. Thesen
A. J. S. Tjokrowardojo · P. Touraine-Romestaing · K.-R. Trott · P. C. Veraguth
A. G. Visser · D. von Fournier · M. Wannenmacher · M. Weidenbecher · K. Weigel
W. F. Whitmore · F. Wilson · G. Wolf · N. Wolf · R. Woodleigh

Edited by

Rolf Sauer

Foreword by

Luther W. Brady and Hans-Peter Heilmann

Springer-Verlag
Berlin Heidelberg New York
London Paris Tokyo
Hong Kong Barcelona

Professor Dr. ROLF SAUER
Strahlentherapeutische Universitätsklinik
der Universität Erlangen-Nürnberg
Universitätsstr. 27
8520 Erlangen, Germany

MEDICAL RADIOLOGY · Diagnostic Imaging and Radiation Oncology

Continuation of
Handbuch der medizinischen Radiologie
Encyclopedia of Medical Radiology

With 193 Figures

ISBN 3-540-52465-7 Springer-Verlag Berlin Heidelberg New York
ISBN 0-387-52465-7 Springer-Verlag New York Berlin Heidelberg

Library of Congress Cataloging-in-Publication Data. Interventional radiation therapy techniques – brachytherapy / contributors, G. Alth ... [et al.]; edited by Rolf Sauer; foreword by Luther W. Brady and H.-P. Heilmann. p. cm – (Medical radiology) Includes bibliographical references. Includes index.
ISBN 3-540-52465-7 (alk. paper)
ISBN 0-387-52465-7 (alk. paper)
1. Radioisotope brachytherapy. 2. Cancer–Interventional radiology. I. Alth, Gerhart. II. Sauer, Rolf. III. Series. [DNLM: 1. Brachytherapy. 2. Neoplasms–radiotherapy. QZ 269 I628] RC271.R27I64 1991 616.99'406424–dc20
DNLM/DLC

Typesetting: Best-set Typesetter Ltd., Hong Kong
Printing: Appl, Wemding.Bookbinding: J. Schäffer GmbH & Co. K.G., Grünstadt
2113/3130-543210 – Printed on acid-free paper

Foreword

Brachytherapy is examined at length and in depth in this book, edited by ROLF SAUER, professor at the University of Erlangen. This treatment is basically the therapeutic utilization of encapsulated radionuclides placed close to or within a tumor. The entire technique involves either the insertion of molds or moulages in the proximity of a tumor or the interstitial application of radionuclide sources.

From a historical point of view, the first utilization of brachytherapy in treatment was in 1901, three years after the discovery of radium by Marie Sklodowska Curie. The episode was initiated when Pierre Curie gave a small amount of radium in a tube to Dr. HENRI DANLOS, suggesting that it be inserted within a tumor. Some two years later Dr. ALEXANDER BELL made the same suggestion in a letter to the editor of the *Archives of Roentgen Ray*.

Progress in the application of brachytherapy was rapid thereafter, and improvements in the technique allowed many different types of tumors to be treated by brachytherapy programs. Radon was used in the form of tiny glass tubes inserted within the tumor as a permanent implantation. Reports from various investigators demonstrating the usefulness of the technique began to appear in the literature.

In the early 1920s, CLAUDIUS REGAUD concluded that certain types of cancer could successfully be treated by radium needles of low intensity applied for periods of 6–10 days. Regaud's technique met with excellent results in intraoral cancer and was widely accepted.

In the 1930s, PATTERSON and PARKER developed a system of rules for the utilization of radionuclide sources allowing for uniform distribution of the radiation dose within the implanted region. Although complex, they became popular and widely accepted. At about the same time, EDITH QUIMBY developed a set of rules and tables for "the unsophisticated" radiotherapist who was deterred from using the Patterson and Parker system. This system provided for the designation of an average dose within the volume implanted. The sources were uniformly spaced but the best delivery was inhomogeneous in character.

In the late 1950s, various radium substitutes became available including iodine-125, iridium-192, and cesium-135, as well as other radionuclides. The widespread use of these radionuclides in various techniques are fully explored in depth and in a most scientific fashion by the contributors to this volume. This book represents an important and significant update on the utilization of this treatment technique in cancer.

From the data presented by the contributors to this volume, it is obvious that brachytherapy will continue to play a crucial role in the treatment of cancer. With afterloading devices, the more precise dosimetry for treatment programs as well as the innovations in the methods of application will contribute immeasurably to improving the potential for cure in cancer patients and at the same time minimizing the complications resulting from the treatment program.

LUTHER W. BRADY, M.D.
HANS-PETER HEILMANN, M.D.

Preface

Interventional radiotherapy embraces various techniques of operative radiation therapy: interstitial and intracavitary brachytherapy, interstitial hyperthermia and intraoperative radiation therapy. In this field, the radiation oncologist cooperates with specialists from several other fields such as neurosurgery, ophthalmology, ENT, surgical oncology, gynecology, urology, medical physics and basic sciences. In this context, the interdisciplinary character of radiation oncology proves particularly valuable in forging links between the various oncological disciplines. The radiotherapist is required to have a knowledge of surgical technique and an understanding of the special anatomy in the field concerned, in addition to manual dexterity, and the surgeon must acquire a knowledge of radiobiology, radiophysics and radiation protection. The activity of both is based on the latest developments in radiophysics and advances in radiobiology.

In this volume we limit ourselves to brachytherapy. Intracavitary or interstitial radiotherapy offers the considerable advantage of delivering a high dose to a limited target volume within a short time without subjecting the surrounding healthy tissue to excessive exposure. Advanced radiobiological perceptions permit the use of the most favorable dose rate.

As early as 1901 PIERRE CURIE made a small amount of radium available to Dr. HENRI ALEXANDRE DANLOS, a dermatologist at the Hôpital St. Louis, Paris, who used it in several surface applicators for the treatment of skin lesions. This was the beginning of brachytherapy – the treatment of lesions with radionuclides at short distance, in contrast to percutaneous teletherapy. Since that time, brachytherapy has developed by leaps and bounds.

In 1905 ROBERT ABBE, of St Lukes Hospital, New York, introduced for the treatment of skin tumors radium capsules which were placed in celluloid tubes implanted previously. This was the first afterloading treatment. In the same year, FRICKE, in Manchester, developed the first gynecological applicators. In 1925 FAILLA, of Memorial Hospital, New York, discovered that β-rays, which were responsible for tissue necroses, could be filtered by encapsulating radium and radon in attenuating material such as gold. HEYMAN and others at Radiumhemmet, Stockholm, treated patients suffering from cervical cancer by using three equal applications of 20 to 24 h in duration with intervals of 1–3 weeks between applications. At the Institute Curie, Paris, C. REGAUD observed that low-intensity radium applied over several days was more effective than higher-intensity radium tubes and shorter treatment periods. In Manchester, PATERSON and PARKER developed an interstitial dosimetric system for radium sources with different linear activity, the so-called Manchester system.

In 1934, JOLIOT-CURIE, Paris, discovered continued artificial radioactivity. In 1953 ULRICH HENSCHKE, in New York, carried out the first interstitial implants, employing postoperative afterloading techniques. He then systematically carried on the development of afterloading therapy: in 1955 he replaced radium-226 with iridium-192 wires, and in 1960 he developed an afterloading technique for intracavitary irradiation of gynecological malignancies. In 1965 the first iodine-125 seeds were applied in New York in HENSCHKE and HILARIS.

In the 1960s and 1970s, the increasing use of cobalt-60 teletherapy units and linear accelerators with electron beam capabilities caused the popularity of brachytherapy to decline. Physicians also began to suffer the consequences of excessive radiation exposure resulting from the improper handling of hazdardous radioactive material.

In the last decade, however, brachytherapy experienced a dramatic revival in conjunction with other new modalities such as hyperthermia and chemotherapy. The introduction of new radioisotopes with favorable nuclear physical properties, computer-assisted dosimetry and in particular the advent of modified afterloading techniques set off this revolutionary development. Afterloading techniques especially permit increased flexibility of implant design and reduction or even complete elimination of radiation exposure for surgeons and nursing staff. This resulted in more accurate placement of radioactive sources and more convenient and better care of the patient.

Interstitial therapy alone is instituted for palliative and curative purposes in the case of brain tumors, tumors of the head and neck, skin tumors, prostate carcinoma and gynecological malignancies. However, curative treatment is only possible with well-defined and well-differentiated tumors, e.g. T1-2 G1-2 prostate carcinomas, or with T1 NO MO tumors of the oral cavity. It constitutes the exception rather than the rule. In the event of recurrences or lesions not responsive to other treatment modalities, brachytherapy alone may still frequently produce a palliative effect.

As a rule, brachytherapy is employed in conjunction with external beam irradiation for boosting purposes. The radiobiological rationale indicates that a higher dose is required with increasing tumor size to bring the malignancy under control. The main tumor mass is treated interstitially with a booster dose, the subclinical peripheral portions of the primary tumor and the lymphatic drainage area are irradiated percutaneously. This procedure is mainly employed in the head and neck and with breast cancer, anal canal carcinoma and gynecological malignomas.

It makes sense to combine interstitial brachytherapy with interstitial hyperthermia, because the in-dwelling tubes for brachytherapy can also used for antennae and thermocouples. The given geometric arrangement of the applicators permits a uniform temperature distribution which is better than that achieved with external hyperthermia.

The idea of preparing the present volume was conceived with the help of LUTHER W. BRADY on the occasion of the international symposium on "Interventional Radiotherapy" held in autumn 1987 at Rothenburg ob der Tauber, FRG. We decided to restrict the range of topics to the most common indications, such as brain tumors, choroidal melanomas, head and neck tumors, mammary carcinomas, carcinomas of the anal canal, prostate carcinoma and gynecological malignomas.

The most prominent authors from Europe and overseas have supplied contributions on the most frequent and attractive applications of brachytherapy. They discuss indications and techniques and describe their personal experience. Our aim in publishing this book is to promote scientific exchange between the countries of the so-called Old World and our friends and colleagues in the United States and elsewhere. At the same time we wish to stimulate the diffusion of brachytherapy in Germany.

ROLF SAUER

Contents

1 Basic Principles

1.1 Principles of Combining External Beam and Brachytherapy
ROGER A. POTISH and SEYMOUR H. LEVITT. (With 2 Figures) 3

1.2 Fundamentals in Radiobiology
JOHANN C. KUMMERMEHR and KLAUS-RÜDIGER TROTT. (With 13 Figures) . . 7

1.3 Fundamentals in Physics
GINETTE MARINELLO. (With 2 Figures) 25

1.4 Brachytherapy Techniques
REINHOLD G. MÜLLER. (With 6 Figures) 35

1.5 High and Low Dose Rate Remote Afterloading: A Critical Comparison
COLIN G. ORTON. (With 1 Figure) 53

Special Considerations

2 Brain Tumors

2.1 Basic Principles of Brachycurietherapy of Brain Tumors
SEYMOUR H. LEVITT, BRUCE J. GERBI, CHUNG K. LEE,
and ROBERT E. MAXWELL 61

2.2 Imaging-Stereotaxic Implantation of Radionuclides in Intracranial Tumors
(Curietherapy and Brachycurietherapy)
FRITZ MUNDINGER 67

2.3 Experimental Dose Effects After Permanent and Temporary Interstitial
Irradiation of the Brain
CHRISTOPH B. OSTERTAG. (With 3 Figures) 81

2.4 Fractionated Afterloading Therapy in the Treatment of Malignant,
Inoperable Brain Tumors
ANNE SPARENBERG, KLAUS WEIGEL, and HELMUT ERNST. (With 3 Figs.) . . 85

2.5 External Stereotactic Focal Irradiation
BERNHARD N. KIMMIG, VOLKER STURM, and RITA ENGENHART
(With 5 Figures) 89

2.6 External Stereotactic Focal Irradiation of Arteriovenous Malformations
by a Routinely Used Linear Accelerator
HANS-BRUNO MAKOSKI 95

3 Choroidal Melanoma

3.1 Results After Brachytherapy Using ^{106}Ru/^{106}Rh Plaques for Choroidal Melanomas
P. K. LOMMATZSCH. (With 6 Figures) 103

3.2 Choroidal Melanoma: Role of Brachytherapy in Management
LUTHER W. BRADY, JERRY A. SHIELDS, JAMES J. AUGSBURGER, JOHN L. DAY, REGINALD WOODLEIGH, ARNOLD M. MARKOE, and ULF L. KARLSSON (With 6 Figures) 113

3.3 Experience with High-Dose β-Irradiation (^{106}Ru/^{106}Rh) of Choroidal Melanomas
ROLF-P. MÜLLER, R. PÖTTER, and H. BUSSE. (With 3 Figures) 123

4 Head and Neck Tumors

4.1 Classification of Oral Cavity and Oropharynx Carcinomas
ROLF SAUER 131

4.2 Interstitial Radiotherapy of Oral Cavity and Oropharynx Carcinomas (Paris Technique)
BERNARD PIERQUIN, JEAN JACQUES MAZERON, and LAVAL GRIMARD (With 10 Figures) 133

4.3 Interstitial Brachytherapy in Head and Neck Tumors
BASIL S. HILARIS, ANCA E. TCHELEBI, and CHITTI R. MOORTHY (With 2 Figures) 145

4.4 Manual Afterloading in Brachytherapy (Endocurietherapy with ^{192}Ir Wires and Needles)
PETER C. VERAGUTH. (With 7 Figures) 153

4.5 Temporary and Permanent Brachytherapy in Advanced Head and Neck Cancer – The Erlangen Experience
RAINER FIETKAU, MANFRED WEIDENBECHER, WOLFGANG SPITZER, and ROLF SAUER. (With 6 Figures) 159

4.6 ^{198}Au Implantation of Carcinoma of the Mobile Tongue
JOACHIM SLANINA, KLAUS KUPHAL, and MICHAEL WANNENMACHER (With 5 Figures) 169

5 Breast Cancer

5.1 Brachycurietherapy in Breast Cancer
FRANÇOIS BAILLET. (With 3 Figures) 179

5.2 Interstitial Boost Irradiation: Indications, Technique, Complications
BERNARD PIERQUIN, FRANK WILSON, and DANIEL CHASSAGNE (With 3 Figures) 187

5.3 Interstitial Boost Irradiation Technique and Dosage
KARSTEN ROTTE and KURT BAIER. (With 11 Figures) 191

5.4 Iridium-192 Afterloading for Boosting the Tumor Bed
J. H. BORGER and H. BARTELINK. (With 2 Figures) 199

6 Anal Canal Cancer

6.1 Carcinomas of the Anal Canal: An Introduction
ROLF SAUER and JÜRGEN DUNST 207

6.2 Role of Combined Radiochemotherapy: The Lyon Experience
J. PAPILLON, J. F. MONTBARBON, J. P. GERARD, J. L. CHASSARD, J. M. ARDIET, and P. TOURAINE-ROMESTAING. (With 1 Figure) 209

6.3 Combination Therapy of Anal Canal Cancer: A Report on External Irradiation With or Without Chemotherapy Followed by Interstitial Iridium 192
GUDRUN PIPARD. (With 1 Figure) 215

6.4 Radiochemotherapy of Anal Canal Cancer With or Without Interstitial Implants
JÜRGEN DUNST, GERHARD GRABENBAUER, NORBERT WOLF, FRANE PAUL GALL, and ROLF SAUER. (With 4 Figures) 221

6.5 Conservative Treatment of Anal Carcinoma: The Surgeon's View
P. SCHLAG. (With 1 Figure) 229

7 Prostatic Cancer

7.1 Interstitial Irradiation in Prostatic Cancer: Report of 10-Year Results
BASIL S. HILARIS, ZVI FUKS, DATTATREYUDU NORI, WILLIAM A. FAIR, and WILLET F. WHITMORE. (With 9 Figures) 235

7.2 Interstitial Radiotherapy: The Freiburg Experience
H. SOMMERKAMP and MICHAEL WANNENMACHER. (With 3 Figures) 241

7.3 Cancer of the Prostate: Brachytherapy Techniques
LUTHER W. BRADY, CHRISTOPHER KOPROWSKI, ANNE MARIE BOROFSKY, and DAVID A. LIGHTFOOT. (With 4 Figures) 245

7.4 Percutaneous Iodine-125 Implantation Guided by Ultrasound
JOSEF HAMMER and MARKUS RICCABONA. (With 6 Figures) 257

7.5 Transperineal ^{125}I Implantation Guided by Ultrasound: Preliminary Results of 150 Cases
GERHARD SCHLEGEL and KLAUS LUTZ. (With 3 Figures) 363

8 Gynecological Malignancies

8.1 Interventional Radiation Therapy Techniques in Gynecology
HANS-PETER HEILMANN. (With 12 Figures) 271

8.2 Technical Aspects of Bladder Dosimetry in Intracavitary Irradiation of Carcinoma of the Cervix
A. J. SUBANDONO TJOKROWARDOJO and A. G. VISSER. (With 6 Figures) 279

8.3 Intracavitary High Dose Rate Afterloading Therapy with Iridium 192: Basic Physical Measurements, Dosimetry, and Localization
NORFRIED THESEN. (With 10 Figures) 285

8.4 Present Status and Perspectives of High Dose Rate Afterloading in Gynecologic Malignancies
HEINER ANNWEILER and MANFRED BUSCH. (With 3 Figures) 291

8.5 Five- to Seven-Year Results in High Dose Rate Radiation of Cancer of the Cervix
GERHART ALTH. (With 4 Figures) 299

8.6 Prospective Clinical Trial Concerning High Dose Rate Afterloading Therapy in Cancer of the Cervix and Endometrium
DIETER KOB and KARL-HEINZ KLOETZER. (With 2 Figures) 307

8.7 Comparison Between High Dose Rate Afterloading and Conventional Radium Therapy
KARSTEN ROTTE. (With 9 Figures) 311

8.8 High Dose Rate Afterloading Compared with Conventional Brachytherapy
FELIX H. GLASER. (With 2 Figures) 319

8.9 Low and High Dose Rate Afterloading in Gynecological Malignancies
D. VON FOURNIER, H. W. ANTON, H. JUNKERMANN, and G. WOLF (With 7 Figures) 333

9 Interstitial Hyperthermia

9.1 Techniques and Clinical Experience of Interstitial Thermoradiotherapy
M. HEINRICH SEEGENSCHMIEDT, ROLF SAUER, LUTHER W. BRADY, and ULF L. KARLSSON. (With 1 Figure) 343

9.2 Interstitial and Intraoperative Hyperthermia
CHRISTOPHER T. COUGHLIN. (With 4 Figures) 359

9.3 Interstitial Thermoradiotherapy in the Treatment of Malignant Tumors
BAHMAN EMAMI and CARLOS A. PEREZ 365

9.4 Experiences With Interstitial Hyperthermia as a Sole Treatment Modality or Combined with Radiotherapy
TAKEHIRO INOUE, NORIE MASAKI, SHUZI OZEKI, HIROSHI IKEDA, and TAKAHIRO KOZUKA. (With 2 Figures) 371

Subject Index 375

List of Contributors 379

1 Basic Principles

1.1 Principles of Combining External Beam and Brachytherapy

ROGER A. POTISH and SEYMOUR H. LEVITT

CONTENTS

1 Introduction 3
2 Gy for Gy 3
3 The Linear Quadratic Model 3
3.1 Dose Rate Effects 3
3.2 Combination of External Beam and Brachytherapy 4
4 Nominal Standard Dose 5
5 Dose Specification and Biologic Effects 5
References 5

1 Introduction

External beam therapy and brachytherapy are customarily combined on a physical rather than on a biologic basis. This has been done both as a matter of tradition and because no other system has been unequivocally demonstrated to be superior. The purpose of the present chapter is to evaluate concordances and discordances between various biologic and physical prescription systems.

2 Gy for Gy

In a strictly physical sense, a Gy of external beam therapy is identical to a Gy of brachytherapy in that a specific amount of energy is absorbed within a defined volume. Although the simple addition of dose would appear logical, the biologic effects of external beam are, for a given dose, a function of numerous other parameters, ranging from energy to fractionation to volume. Even if these factors are held constant, the biologic effect of dose is not simply additive: 5 Gy added to 15 Gy does not elicit the same biologic effect as 5 Gy added to 70 Gy.

Nevertheless, the simple addition of physical dose has been empirically verified in the management of a variety of malignancies throughout the body. However, it is important to emphasize that this addition has been established only in these areas in which a wide variety of clinical experience has been accumulated with standard volumes, fraction sizes, and dose rates. Within these limits, for example, the total dose tolerable is 70–80 Gy for vaginal cancer and 80–90 Gy for cervical cancer at point A, varying slightly depending on the proportions of brachytherapy and external beam therapy. A similar though somewhat lower range of doses is traditional in the prescription of head and neck malignancies, whether the boost be given with brachytherapy or with shrinking field external beam therapy. The customary dose prescribed for boosting patients with excisonal biopsies of breast cancer is virtually identical whether brachytherapy or external beam is used for boosting.

It is important to emphasize that excellent clinical results have been obtained over decades of experience within very specific limits of both brachytherapy and external beam therapy. External beam fraction size has ranged between 1.7 and 2.0 Gy, while dose rates have typically ranged from 0.30 to 1.00 Gy/h (BARKLEY and FLETCHER 1976). The volume of both external beam therapy and brachytherapy must also be kept within standard limits. When treatment volumes are excessive, increased complications occur.

3 The Linear Quadratic Model

3.1 Dose Rate Effects

The linear quadratic model furnishes a basis for understanding the addition of external beam and

ROGER A. POTISH, M.D., Associate Professor, Departments of Therapeutic Radiology and of Obstetrics and Gynecology.
SEYMOUR H. LEVITT, M.D., Professor and Head, Department of Therapeutic Radiology – Radiation Oncology, University of Minnesota Hospitals and Clinics, Harvard Street at East River Road, Minneapolis, MN 55455, USA

Table 1. Dose rate relative effectiveness (RE)

	alpha/beta = 2.5	alpha/beta = 10.0
RE(0.357)	1.29	1.07
RE(0.30)	1.24	1.06
RE(0.65)	1.52	1.13
RE(1.00)	1.80	1.20

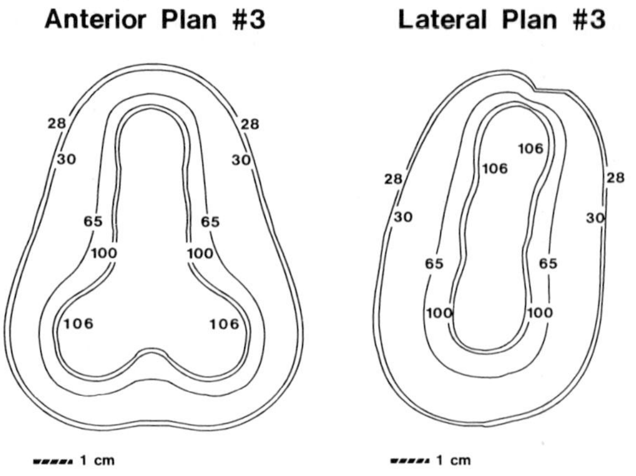

Fig. 2. Isodoses are presented in cGy/h around a standard Fletcher cesium application, with an alpha/beta ratio of 10.0

brachytherapy (Barendsen 1982; Thames et al. 1982; Withers 1985). The extrapolated tolerance dose (ETD) is defined as the tolerance dose for an infinite number of very small fractions (Barendsen 1982). The relative effectiveness (RE) of a prescribed dose rate is a function only of the mean half-life of repair of sublethal damage and of the alpha/beta ratio (Barendsen 1982). If a standard tolerance dose of 60 Gy at 0.357 Gy/h is utilized, the relative dose rate effectiveness can be calculated as a function of the alpha/beta ratio (Table 1). If, in turn, these values are normalized to a standard point A dose rate of 0.65 Gy/h, the isodoses of Figs. 1 and 2 are generated. As an example, the 1.00 Gy/h isodose is greater by 1.80/1.52, yielding a biologically greater effective dose rate of 1.18 Gy/h at an alpha/beta ratio of 2.5. Analogously, the effective dose rate at point B is 0.24 Gy/h rather than the physical dose rate of 0.30 Gy/h (i.e., 0.30 Gy/h × 1.24/1.52).

The implications of these effects can be considered for late responding tissues in Fig. 1, with an alpha/beta ratio of 2.5. The increase from 1.00 to 1.18 Gy/h results in shifting the isodose by only 2 mm. Thus, if two 50-h applications are utilized, an additional 1.8 Gy will be given to a minimally greater volume. This will be of no detectable clinical consequence to the relatively radioresistant adjacent vagina and uterus. The increase in effective dose more laterally will increase by 0.06 Gy/h, giving an additional 6 Gy to a volume that is 6 mm wider. Unless the total dose is close to tolerance, this increment would be very difficult to detect. Any bowel or bladder that is further away will have an even smaller difference in effective dose. Thus, the linear quadratic model implies that the normal tissues of the pelvis will be little affected by dose-rate effects from 0.30 to 1.00 Gy/h.

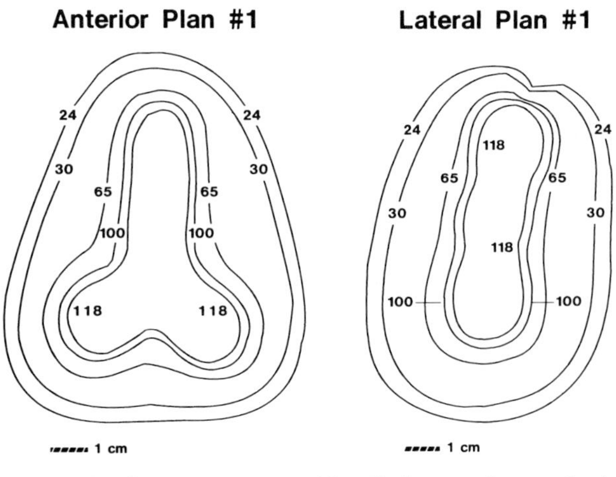

Fig. 1. Isodoses are presented in cGy/h around a standard Fletcher cesium application, with an alpha/beta ratio of 2.5

As shown in Fig. 2, there will be even smaller dose-rate effects for acute reactions and tumors, both of which have greater alpha/beta ratios. Two 50-h applications shift the 1.00 Gy/h isodose by less than a mm, and the effective dose increases by only 1.06. Further laterally, the 0.30 Gy/h isodose shifts by less than 2 mm and by only 0.05 Gy/h. Therefore, it is very unlikely that any clinically detectable effects would occur in either acute reactions or in tumors over the standard range of dose rates.

3.2 Combination of External Beam and Brachytherapy

The linear quadratic model can also be utilized to combine low dose rate brachytherapy with high dose fractionated external beam therapy. The following example assumes tolerance doses of 60 Gy (0.357 Gy/h) at point A with brachytherapy and 60 Gy at point B with external beam (2 Gy daily fractions). For an alpha/beta ratio of 2.5, the extrapolated tolerance dose is 77.1 Gy. Assuming the dose rate at point B to be one-quarter that at point A, the tolerance dose will be greater by a factor of 1.071. Thus, the physical dose of 15 Gy to point B by brachytherapy equals 0.21 of the extrapolated tolerance dose (15 Gy × 1.071/77.1 Gy). The remaining external beam dose to be delivered will be 0.79 of the extrapolated tumor dose, or 48 additional Gy. This results in a total dose of 63 Gy (15 from brachytherapy and 48 from external beam), differing only slightly from 60 Gy of external beam alone. Thus, for late radiation effects, there is little difference between physical dose and biologic dose.

A similar calculation with an alpha/beta ratio of 10.0 will demonstrate even smaller effects. With this dose rate, the extrapolated tolerance dose is 64.3 Gy for brachytherapy. Again, assuming the dose rate at point B to be one-quarter that at point A, the relative effectiveness is 1.018. The physical dose of 15 Gy to point B equals 0.24 of the extrapolated tolerance dose (0.15 Gy × 1.018/64.3 Gy). An additional 0.76 extrapolated tumor dose remains to be delivered – 46 Gy in 2 Gy fractions. The total is 61 Gy (15 from brachytherapy and 46 from external beam), virtually identical to the 60 Gy if administered solely by external beam.

4 Nominal Standard Dose

For several decades, it was hoped that a power formula could take treatment time and fraction size into account, both for external beam therapy and for brachytherapy. While a number of such systems have been developed, the most popular has been the Nominal Standard Dose (NSD) system (Ellis and Sorenson 1974). The validity of the NSD system has been challenged for a variety of reasons. Nevertheless, within generally accepted dose rates and fraction sizes, it may offer a number of useful predictions.

Time-dose fractionation (TDF) factors are assumed to be 100 for both 60 Gy of brachytherapy (0.357 Gy/h) and 60 Gy of external beam (30 fractions in 6 weeks) (Orton 1974). With the same radiation factors as in the above linear quadratic example, 15 Gy will be delivered to point B at 0.089 Gy/h, resulting in a TDF of 18. Thus, a TDF of 82 remains to be administered with external beam to achieve a biologically equivalent dose of 60 Gy. This will be 50 Gy of external beam, for a total physical dose of 65 Gy. This is somewhat greater than the dose suggested by the linear quadratic model, especially for late effects. If decay factors were utilized, then the additional external beam recommended by the NSD formulation would be even greater.

5 Dose Specification and Biologic Effects

The specification and prescription of brachytherapy dosage have remained problematic for the past 8 decades. A major physical problem is the rapid dose gradient around any brachytherapy application. Whatever dose rate is arbitrarily chosen, one can be certain that the dose rate will be quite different a few mm away. The choice of isodose for treatment prescription is arbitrary. If a prescription point is changed by a few mm, the dose rate may vary by tens of cGy/h. This effect is quite likely to overwhelm any biologic dose-response effects with standard treatment parameters. Both linear quadratic and power formulae suggest that a simple addition of brachytherapy and external beam Gy will be useful as long as standard dose rates and fraction sizes are prescribed. Any attempt to extend the addition of brachytherapy and external beam beyond the classic limits is frought with danger. In addition, high dose rate therapy awaits further empirical data before its late effects can be as well known as those of low dose rate therapy.

References

Barendsen GW (1982) Dose fractionation, dose rate and iso-effect relationships for normal tissue responses. Int J Radiat Oncol Biol Phys 8: 1981–1997

Barkley HT, Fletcher GH (1976) Volume and time factors in interstitial gamma-ray therapy. AJR 126: 163–170

Ellis F, Sorenson A (1974) A method of estimating biological effect of combined intracavitary low dose rate radiation with external radiation in carcinoma of the cervix uteri. Radiology 110: 681–686

Orton CG (1974) Time-dose factors (TDFs) in brachytherapy. Br J Radiol 47: 603–607

Thames HD, Withers HR, Peters LJ, Fletcher GH (1982) Changes in early and late radiation responses with altered dose fractionation: implications for dose-survival relationships. Int J Radiat Oncol Biol Phys 8: 219–226

Withers HR (1985) Biologic basis for altered fractionation schemes. Cancer 55: 2086–2095

1.2 Fundamentals in Radiobiology

JOHANN C. KUMMERMEHR and KLAUS-RÜDIGER TROTT

CONTENTS

1 Introduction . . . 7
2 Effects of Low Dose Rate Irradiation on Cells In Vitro . . . 8
2.1 Lethal and Sublethal Radiation Damage . . . 8
2.2 Differences in Dose Rate Response Between Cell Lines . . . 8
2.3 Repair of Sublethal Damage During CLDRI . . . 9
2.4 Influence of Cell Progression and Proliferation on the Dose Rate Effect . . . 9
3 Dose Rate Effects in Normal Tissues . . . 11
3.1 Bone Marrow . . . 11
3.2 Intestinal Mucosa . . . 12
3.3 Oral Mucosa . . . 13
3.4 Lung . . . 13
3.5 Central Nervous System . . . 14
3.6 Skin . . . 15
4 Effects on Experimental Tumours . . . 16
5 Mathematical Formulations of the Dose Rate Effect . . . 17
5.1 Empirical Time-Dose Formulae . . . 17
5.2 Extended Linear Quadratic Formalism . . . 18
5.3 Equating CLDRI and HDR Fractionation Protocols . . . 21
References . . . 22

1 Introduction

Apart from tumours of the skin, carcinoma of the cervix was one of the first cancers that could successfully be treated by radiotherapy alone. The eradication of bulky tumours without excessive damage to normal tissues had become possible by the introduction of brachytherapy. Among the factors that contributed to the effectiveness of this modality, better dose distribution unquestionably was of paramount importance, the improvements being due both to the favourable geometrical arrangement of the radiation source and the better penetration of the gamma-rays as compared with X-rays. Brachytherapy with radium sources was soon extended from cervical and endometrial cancer to other lesions such as cancer of the breast and tongue or lip (PIPARD 1989).

Another important aspect of radium brachytherapy is the effect of low dose rate. Radium treatment of pelvic tumours is associated with the exposure of large volumes of surrounding normal tissue at dose rates of 1 cGy/min or lower. The dependence of biological effectiveness on "radiation intensity" (LEA 1938) was recognized quite early. With the advent of cell culture techniques the biological principles of low dose rate sparing could be rationally studied and explained at the cellular level. It was soon realized, however, that the problems of clinical radiobiology required detailed in vivo studies to define possible differences between various normal tissues and tumours.

The need for such data became more pressing with the introduction of afterloading techniques and man-made radioisotopes providing dose rates that cover the range familiar from endocavitary radium treatment, e.g. 1 cGy/min, up into the region achieved by modern linear accelerators, e.g. 10 Gy/min. Within this range and particularly so over the two decades from 1 to 100 cGy/min the biological effectiveness of sparsely ionising radiation varies considerably with dose rate. The changes have been well identified at the cellular level. The overriding factor is undoubtedly repair of sublethal radiation damage, but low dose rate irradiation may also have consequences for the radiosensitivity of hypoxic cells, may affect the distribution of cells over the cycle, or in some cases even permit proliferation during exposure. A number of excellent reviews have been given on the cellular aspects (HALL 1972, 1981) that will briefly be summarized in this chapter. In addition, we will review the large body of data that has accrued from animal studies on normal tissues and transplantable tumours and outline the current mathematical model used to describe the dose rate effect in clinical and experimental radiobiology.

JOHANN C. KUMMERMEHR, Dr., Institut für Strahlenbiologie, Gesellschaft für Strahlen- und Umweltforschung (GSF), Ingolstädter Landstr. 1, 8042 Neuherberg, Germany

KLAUS-RÜDIGER TROTT, Professor Dr., Department of Radiation Biology, Medical College of St. Bartholomew's Hospital, Charterhouse Sqare, London EC1M 6BQ, UK

2 Effects of Low Dose Rate Irradiation on Cells In Vitro

2.1 Lethal and Sublethal Radiation Damage

Current concepts of cell inactivation by ionizing radiation assume that the loss of reproductive integrity is achieved through two different types of radiation damage. One part of the critical lesions is considered to be caused by primarily lethal events, irreparable by definition. As a consequence, their yield must be linearly proportional to the total dose and for the same reason be independent of dose rate. In addition to this type of damage, so-called sublethal lesions are imparted which only through their interaction become lethal. Before interaction and damage fixation have taken place, such lesions – by definition – may be healed or repaired at the molecular level. The simplest and most relevant case of interaction will be between two sublethal lesions closely related in space and time. The probability for this "overlapping" of two independent lesions to happen must necessarily be proportional to the dose squared. It is this part of the total radiation effect that can be modified by fractionation or protraction. In either situation sublethal damage from previous exposure will be partly or entirely repaired and no longer available for interaction, dependent on the dose per fraction and dose rate and also on the actual kinetics of the repair mechanism that becomes operative either between dose fractions or during low dose rate irradiation (LDRI).

The dual action of radiation is also reflected in single-dose survival curves of mammalian cells that to low LET (Linear Energy Transfer) radiation almost invariably display a "shoulder" or increasing curve slope. Strictly speaking, the "shoulder" expresses the increasing efficacy of interactive damage with rising dose. Evidence that this is equivalent to the capability of sublethal damage repair (SLDR) is derived from fractionated irradiation. It was first established by Elkind and Sutton (1960) that when a given dose is split into two fractions, the survival rate increases with increasing fraction interval to reach a maximum at about 3 h. In terms of the survival curve this led to a full restoration of the initial shoulder, demonstrable also on multiple fractionation. As a consequence, splitting the total dose into smaller and smaller fractions will result in a composite curve with ever decreasing slope that eventually approaches the initial slope of the single-dose curve. A similar situation arises in continuous low dose rate irradiation (CLDRI), because SLDR becomes possible *during* irradiation. Again, a limiting slope will be reached at very small dose rates when virtually all damage is due to irreparable lesions and all sublethal damage is effectively repaired before it has any chance of interaction.

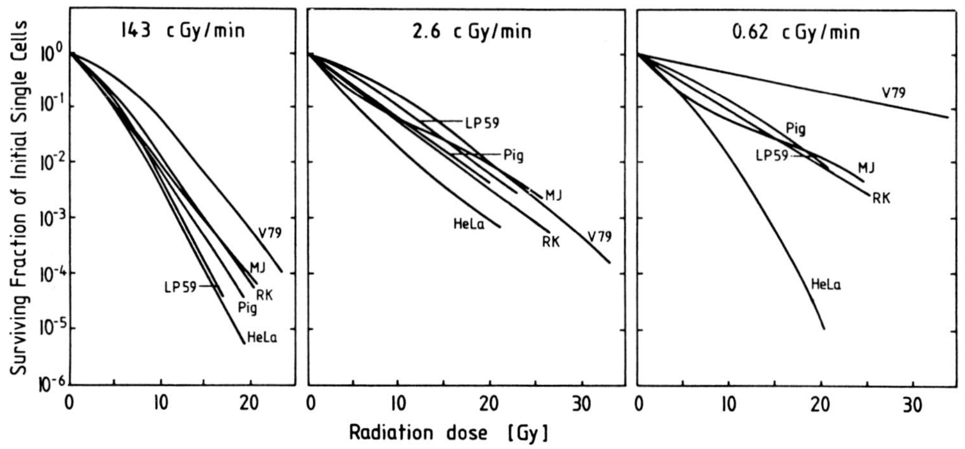

Fig. 1. Dose-survival curves of six different mammalian cell lines to radiation exposure at different dose rates. *Left*, 142.8 cGy/min; *middle*, 2.6 cGy/min; *right*, 0.62 cGy/min. Cell lines were mouse *LP59*, S3 *HeLa*, PK 15 *pig* kidney, rat kangaroo (*RK*), Indian muntjac (*MJ*) and Chinese hamster *V-79* cells. (Redrawn from Bedford et al. 1980)

2.2 Differences in Dose Rate Response Between Cell Lines

Experimental data on a large variety of cell lines in vitro have demonstrated the decreasing biological effectiveness of low LET radiation with fall-off in dose rate. As an example, the graphs in Fig. 1 show the response of six different cell lines to CLDRI at three selected dose rates. There is a basic trend towards flatter dose-response curves with lowered dose rate, but there are also remarkable differences between the cell lines. The V-79 cells benefit most, while in HeLa cells maximum sparing is already obtained at the intermediate dose rate. These differences are partly – but not entirely – explained by the differences present in the shoulder to single-dose irradiation; this shoulder is wide in V-79 cells, indicative of a large capacity of repair, while it is small in HeLa cells. However, a number of complicating factors exist that make prediction of LDR response from the acute survival curve unreliable. One such factor is *repair rate* which is not expressed in the shoulder and may differ between cell lines. One also has to realize that cellular radiosensitivity changes markedly as a cell progresses through the cycle, the more sensitive phases usually being G_2, mitosis and early S phase. In the acute survival curve this heterogeneity tends to minimize the shoulder (Fox and Nias 1971), while during CLDRI disturbances in cell progression may become relevant, resulting in an accumulation of cells in the more sensitive phases. Conversely, at very low dose rates cell division and multiplication may become possible in some lines but not in others. The superposition of these factors can create a complex situation, at least in rapidly proliferating cells in vitro, requiring a choice of suitable systems to assess the relative importance of repair, reassortment and proliferation.

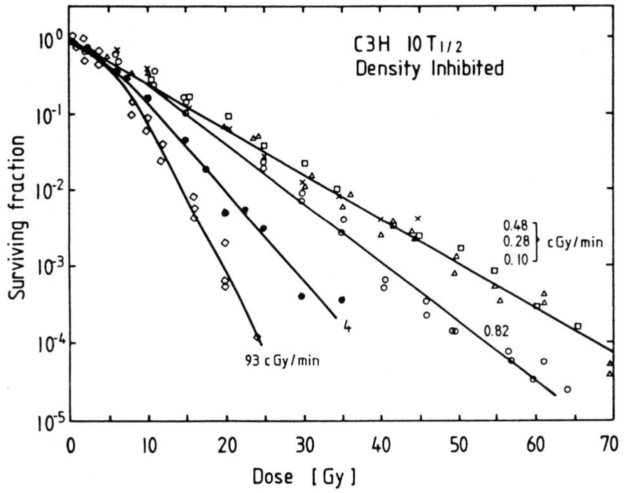

Fig. 2. Dose-survival curves of density-inhibited C3H $10T_{1/2}$ cells exposed to ^{137}Cs gamma-irradiation at the following dose rates: 93 cGy/min; 4.0 cGy/min; 0.82 cGy/min; △ 0.48 cGy/min; □ 0.28 cGy/min; × 0.10 cGy/min. (Redrawn from Wells and Bedford 1983)

2.3 Repair of Sublethal Damage During CLDRI

Most of the factors that confound the assessment of repair alone can be avoided by studying cell lines in plateau phase. Depending on culturing techniques, cells can be grown into a confluent state with minimal cell turnover. A very useful cell line in this respect is C3H 10T1/2, an embryonic mouse line that after confluency maintains a constant cell number with only slow proliferation and cell turnover.

In a careful study Wells and Bedford (1983) have followed cell survival over a wide range of dose rates down to a rate as small as 0.1 cGy/min. The resulting family of curves, shown in Fig. 2, reflects a sparing effect that is relatively small between 100 and 4 cGy/min, becomes extensive between 4 and 0.5 cGy/min, but cannot be extended below this dose rate. Full repair of SLD must therefore already have taken place at 0.5 cGy/min as an identical exponential curve is obtained both for 0.5 and 0.1 cGy/min. Moreover, when all the data were fitted to an extended linear quadratic model (see Sect. 5.2), the initial slope of the acute survival curve and the slope of the 0.5 cGy/min curve were fully identical. The cell kinetic studies done in parallel demonstrated that the slow cell turnover taking place in the controls came to a halt during CLDRI, creating extremely well-defined conditions. Similar data were obtained by Metting et al. (1985) in hamster ovary cells kept in stationary phase after confluency by nutritional deprivation, at dose rates between 100 and 0.3 cGy/min. The quantitative analysis gave clear evidence of two components of radiation damage, one being independent of dose rate while the other could be described by the accumulation and repair of sublethal damage.

2.4 Influence of Cell Progression and Proliferation on the Dose Rate Effect

While plateau-phase cultures have the advantage of permitting quantitative assessment of intracellular repair without disturbing factors, they are not a priori the most relevant in vitro models. The target cells of acutely responding tissues and of most tumours are in a state of active proliferation. When the dose rate falls below a critical level specific for a cell line or tissue, cell progression and division will set in. Apart from a possible direct influence of cell cycle phase on repair characteristics (THAMES and HENDRY 1987), this will alter the dose rate response. The problems have been addressed in a variety of studies using exponentially growing cell cultures. The diversity in response is well illustrated by the work of BEDFORD and MITCHELL (1973, 1977), MITCHELL and BEDFORD (1977) and MITCHELL et al. (1979a, b) that mainly contrasts the response of V-79 and HeLa cells. A basic difference between these lines is already apparent from Fig. 1, but is shown in greater detail in Fig. 3 where survival curves to a wider variety of dose rates are depicted. In this graph, V-79 cells display increased sparing with lowered dose rate, until at 0.6 cGy/min the curve becomes exponential and seemingly extrapolates the initial acute curve slope, suggesting that it is determined by irreparable damage alone. HeLa cells on the other hand, over a range of dose rates between 120 and 0.6 cGy/min, apart from less sparing in general display a less regular response. This becomes even more complex when the dose rates are closely staggered between 2.6 and 0.6 cGy/min (right panel in Fig. 3). While initially the dose responses are similar down to survival fractions of about 0.01, the curves subsequently separate in a paradoxical manner showing greater sensitivity with decreasing dose rate. The explanation for this *inverse dose-rate effect* is a gradual reassortment of cells into sensitive phases of the cycle that occur at a critical dose rate in HeLa cells. In V-79 cells, on the other hand, it turns out that the seemingly fully repaired curve at 0.6 cGy/min owes in fact some of its resistance to proliferation during CLDRI.

The underlying differences between the relatively resistant V-79 cells and the more sensitive HeLa cells have been elucidated by BEDFORD and MITCHELL (1973, 1977) and MITCHELL and BEDFORD (1977). Using synchronized cultures obtained by mitotic harvesting they demonstrated profound differences in progression between the cell lines during exposure to a fixed dose rate of 0.6 cGy/min; this dose rate was chosen as it is critical for HeLa cells, in the sense that their overall cell number (but not their survival) stays constant over a longer period of exposure. When studied by [^{3}H]-TdR (Thymidine) pulse label and mitotic count, cycle progression of the V-79 cells turned out to be little disturbed over several cycles, whereas HeLa cells were able to traverse a first S phase without delay but subsequently became blocked (Fig. 4). More detailed studies including observation of single cells by time-lapse

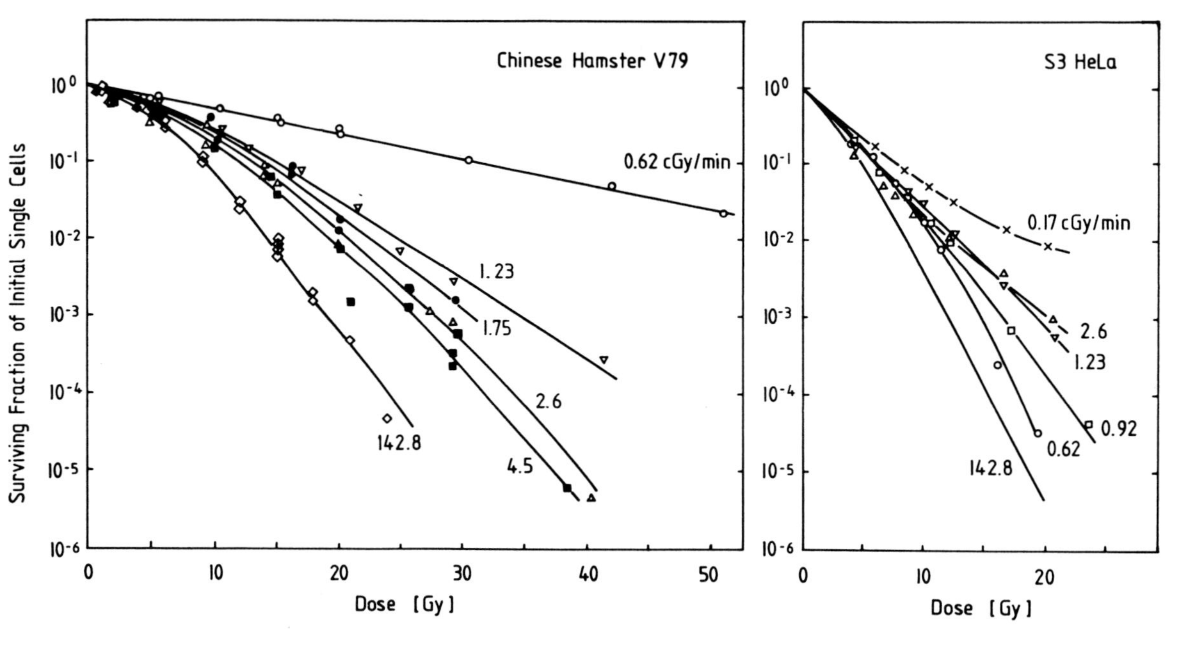

Fig. 3. Survival curves for exponentially growing V-79 cells (*left*) and S3 HeLa cells (*right*). Note the inversion of the dose rate effect in HeLa cells between 2.6 and 0.62 cGy/min. (Redrawn from MITCHELL et al. 1979)

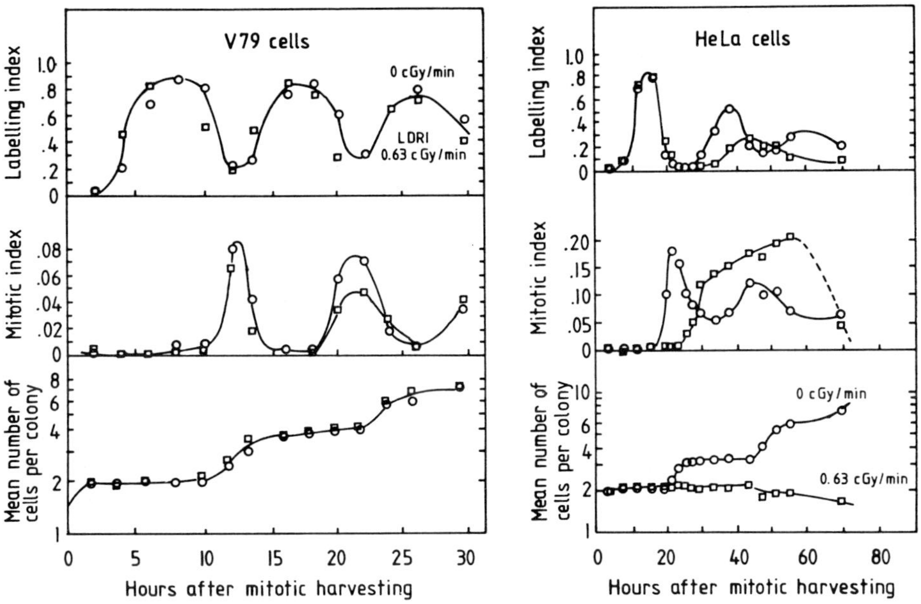

Fig. 4. Proliferation of synchronous V-79 (*left*) and HeLa (*right*) cells during continuous irradiation at 0.63 cGy/min. This dose rate effectively controls division of the HeLa cells but is nearly ineffective in V-79 cells over several cycles. From Bedford and Mitchell (1973)

cinematography revealed that they accumulated in G_2 for about 10 h and then entered mitosis but eventually failed to complete it (Bedford and Mitchell 1977).

As a consequence population growth, in terms of all morphologically intact cells, diverged greatly under CLDRI for the two cell lines. Growth of V-79 cells proceeded with little retardation up to total doses of 25 Gy, while HeLa cells ceased to multiply. For complete growth inhibition V-79 cells require a dose rate that is 7 times higher. Some but not all of this difference is due to accumulation in the G_2 and M phases for HeLa cells. An interesting finding in these studies was that it is the dose accumulated in G_2 that is crucial for the blocking action, as indicated by exposure of synchronized cells when they had migrated into G_2. Moreover, a very similar yield of G_2 accumulation and subsequent mitotic holding was achieved in HeLa cells by fractionated irradiation, provided the first dose hit the synchronized population in G_2 (Mitchell and Bedford 1977). Even asynchronous HeLa cell populations therefore become sensitized by cell cycle effects during extended exposure, while on the other hand such effects are not to be expected – and are not seen – after acute single doses.

In the search for an explanation of these differences a correlation was sought with various parameters, such as cycle time, DNA index, acute division delay and intrinsic cellular sensitivity (initial D_0). Taking growth arrest as the endpoint, responsiveness to low dose rate correlated best with the division delay imposed by acute single doses. In contrast, cell survival showed a marked correlation with the dose absorbed per cell cycle. In elegant experiments Szechter et al. (1978) were able to demonstrate this in V-79 cells whose cycle time was almost trebled by culturing under lowered temperature. A plausible explanation for this correlation was put forward earlier by Fox and Nias (1971). Assuming that under exposure to very low dose rate virtually no SLD accumulates and only irreparable events are effective, the probability of a cell producing two viable daughters will increase the shorter the cycle time. It must be kept in mind, however, that this aspect comes into bearing only when the dose rate is low enough to allow cell division. In the clinical situation in which cycle times of most tumour cells are at least 2–3 days, the doses accumulated per cycle will usually be too high to make this a major factor.

3 Dose Rate Effects in Normal Tissues

3.1 Bone Marrow

Bone marrow is not a critical organ in brachytherapy, but its response to LDRI has raised great experimental interest in conjunction with total body irradiation (TBI), as there is good evidence that leukaemic cells respond similarly to normal bone marrow stem cells. These data are useful in several ways. On the one hand, they illustrate the consistent correlation that exists between fractionation and dose rate effect. On the other hand, the studies have demonstrated that cell survival of the putative target cell (bone marrow stem cell) as measured directly by the endogenous or exogenous spleen colony assay, and tissue failure as defined by the $LD_{50/30}$ do not respond in an identical way to changes in dose rate.

In an early paper, BATEMAN et al. (1962) compiled the $LD_{50/30}$ data of small rodents after external TBI available at that time. Most data cover a range in dose rate from 10 to 1 cGy/min, and there is a significant rise in LD_{50} corresponding to a dose rate factor (ratio of isoeffective doses, DRF) of ca. 1.4. The authors chose an empirical fit that gave a linear relationship when LD_{50} was plotted against the inverse of dose rate cubed. The fit includes data points down to 0.1 cGy/min, but the associated exposure times of several days make proliferation of surviving cells during CLDRI a very likely factor of tolerance below 1 cGy/min. However, a significant dose rate effect also remains between 10 and 1 cGy/min, or even between 30 and 3 cGy/min (KREBS and JONES 1972); more recently this was confirmed by TRAVIS et al. (1985) who found a total DRF of 1.37 with most of the sparing taking place between 5 and 1 cGy/min.

In contrast to this, stem cell survival either when assayed in parallel to lethality (KREBS and JONES 1972; FRINDEL et al. 1972) or independently exhibited no significant dose rate effect in 7 of 8 studies compiled by GLASGOW et al. (1983). Exceptions are the data of PURO and CLARK (1972), that gave a DRF of 1.4 but are based on in vivo irradiation after cell transplantation into (already lethally irradiated) recipients, and the data by MARUYAMA et al. (1983) based on a study design in which a considerable acute top-up dose was given after CLDRI at 0.27 cGy/min to shorten the exposure time (and to minimize repopulation); as this protocol amounts to a two-fraction arm as compared with the acute single-dose irradiation, it is bound to overestimate a possible dose rate effect. The major evidence therefore points to the presence of a dose rate effect when lethality is the endpoint, whereas no such effect is detectable for the stem cells themselves. The biological reason of this divergence is unclear and could rely either on compromised functionality of the cellular offspring or on additional morbidity of other – more dose rate responsive – organs.

3.2 Intestinal Mucosa

A clinically more relevant target tissue in brachytherapy is intestinal mucosa. The main reason why jejunal epithelium has been frequently studied, however, is because it permits convenient and precise measurement of crypt cell survival in vivo by a microcolony formation assay (WITHERS and ELKIND 1970). The assay relies on the fact that after exceeding a threshold dose (e.g. 12 Gy HDR in the mouse) crypts with all clonogenic cells sterilized will degenerate within 2 or 3 days, while those in which one or more clonogenic cells have survived will form distinct foci of regeneration by 3.5 days, i.e. before the denudation of the villous surface has become fatal. The technique is simple as radiation can be delivered to the whole body without the need to immobilize the animal and surviving crypts are readily scored in histological sections. The derived dose-effect curves have a wide shoulder due to the multi-target nature of each crypt and can be followed downwards from about 140 crypts per circumference to less than 1. Moreover, cell survival can again be correlated with lethality, the lethal effect setting in at about 30–40 surviving crypts per circumference.

Several studies have been performed with the main emphasis on intracellular repair during CLDRI. In the study by FU et al. (1975) the survival curves flattened out between 2.74 Gy/min and 0.92 cGy/min, giving a steady increase in D_0 from about 1 Gy to 2.7 Gy, while at an even lower dose rate of 0.36 cGy/min it increased sharply up to approximately 7 Gy. This apparent resistance clearly reflects the onset of rapid proliferation, to be taken as a homeostatic response to the radiation-induced cell depletion. The enormous adaptive capacity of this response is best demonstrated by the ability of rat jejunum to maintain a constant, albeit somewhat reduced population size during exposure to 0.28 cGy/min over several days (LAMERTON and COURTENAY 1968; CAIRNIE 1967). As the latency time until accelerated repopulation starts is short in gut, a quantitative assessment of intracellular repair can only be done if the total treatment time is 2 days or less, tantamount to a minimum dose rate of ca. 1 cGy/min.

In more recent experiments HUCZKOWSKI and TROTT (1984, 1987) have shown an even greater increase in D_0 or isoeffective dose when varying the dose rate from 1.2 Gy/min to 2 cGy/min. The study contrasts continuous and fractionated irradiation at dose rates between 120 and 2 cGy/min

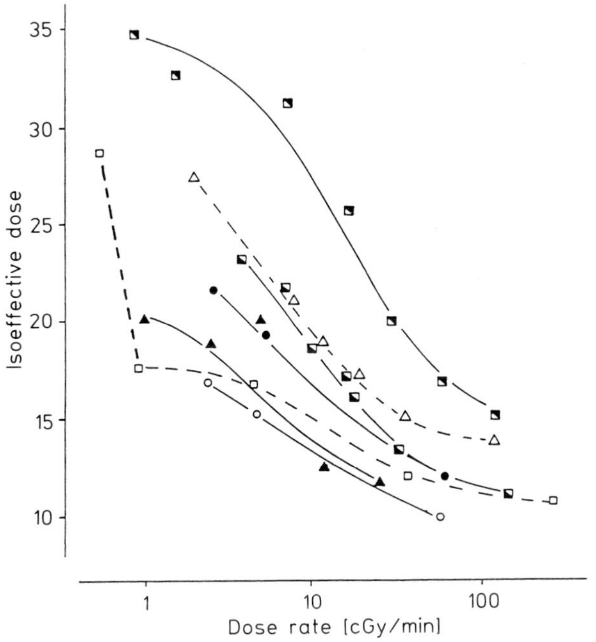

Fig. 5. Dose rate dependence of the isoeffective dose required for gastrointestinal death in 50% of the animals (*solid curves*) or for survival of 40 jejunal crypt stem cells per circumference (*dashed curves*) in the mouse. ◪; ◩ total body irradiation (TBI) ^{60}Co or X-rays, KREBS and LEONG (1970); ○; ●, TBI ^{60}Co or abdominal irradiation, WAMBERSIE et al. (1979); ▼, TBI ^{137}Cs, TRAVIS et al. (1985); □, TBI ^{137}Cs, FU et al. (1975); △, TBI ^{60}Co, HUCZKOWSKI and TROTT (1987)

and demonstrates in a quantitative way how the sparing effect of LDRI can be substituted by HDR fractionation and vice versa.

As with the hemopoietic syndrome it is interesting to compare the cellular response (crypt survival) and the clinical end-point (acute gastrointestinal syndrome). Data on the LD_{50} as a function of dose rate, as published by KREBS and LEONG (1970), WAMBERSIE et al. (1979) and TRAVIS et al. (1985), have been compiled in Fig. 5 along with the crypt survival data. Both the maximum extent of sparing and the most dynamic range of dose rate agree reasonably well, with maximum DRFs of almost 2 and the greatest impact of dose rate seen between 50 and 2 cGy/min.

While crypt survival has frequently been used as an assay because of its feasibility and precision, the clinically more relevant response of intestinal wall structures has been dealt with in only one study. KISZEL et al. (1985) treated a defined segment of rat rectum by an endocavitary afterloading technique using ^{192}Ir sources of different activity, resulting in dose rates to the submucosa of 200, 20 and 5 cGy/min. The endpoint chosen was fatal bowel obstruction within 200 days developing on the basis of a chronic ulcer. ED_{50} values (i.e. doses producing the effect in 50% of the animals) determined for the above dose rates were 27 Gy, 32.5 Gy and 51 Gy, respectively, rendering a $DRF_{5/200}$ of 1.9, comparable with that seen in many other tissues over this dose rate range.

3.3 Oral Mucosa

An early responding tissue that can limit the aggressiveness of treatment in teletherapy and brachytherapy is the mucous membrane of the oropharyngeal cavity. The difficulties are obvious in setting up a rodent model that does not imply an unacceptable burden to the animal (e.g. oropharyngeal death). An assay introduced in recent years quantitates the acute reaction of mouse lip following percutaneous irradiation of the whole snout.

SCALLIET et al. (1987) assessed the dose rate effect in this system over a range from 10.7 Gy/min down to 2.5 cGy/min. The data demonstrate that above 20 cGy/min the dose rate is hardly critical whereas between 20 and 2.5 cGy/min a distinct rise in tolerance takes place. The experiments were paralleled by fractionated HDR irradiation, and both data sets could well be fitted by a transformed linear quadratic dose relationship, giving estimates for the α/β ratio (7–8 Gy) and the half-time of repair (47 min).

3.4 Lung

Lung is an important late reacting tissue, and its response to LDRI has been investigated by several authors. Again these studies were elicited by the clinical problems of TBI where lung damage is the limiting factor. A remarkably high repair capacity has been established for lung in HDR fractionation experiments with complete repair, comparable only with CNS and kidney (THAMES and HENDRY 1987). The high repair capacity was also demonstrated in dose rate studies in which lung damage after thoracic irradiation was assessed in the mouse. This was done either in terms of LD_{50} (observation period starting after more than 30 days to exclude oesophageal death) or by the dose-dependent increase in breathing rate that develops

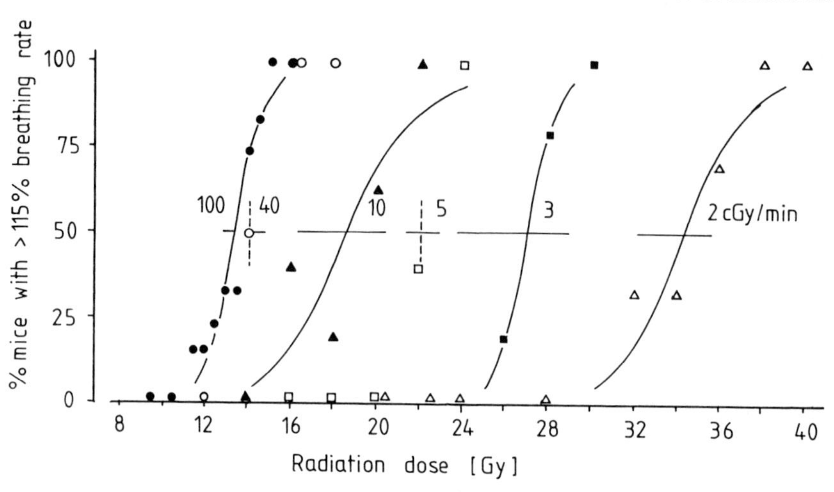

Fig. 6. Dose-response curves for mouse lung damage 14 weeks after thoracic irradiation. The frequency of animals with >115% breathing rate is plotted against dose. (Redrawn from Down et al. 1986)

with the onset of pneumonitis. An example is given in Fig. 6 which shows dose-response curves of breathing rate for dose rates between 1 Gy/min and 2 cGy/min, measured 14 weeks after radiation (Down et al. 1986). When replotted as isoeffective tolerance doses the data indicate a vast dose rate effect that should be further exploitable at even lower dose rates. This prediction is based on a data fit to the "incomplete-repair model" (Thames

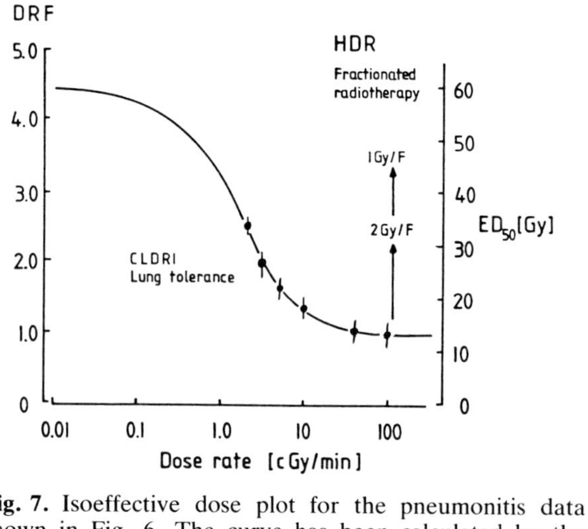

Fig. 7. Isoeffective dose plot for the pneumonitis data shown in Fig. 6. The curve has been calculated by the incomplete repair model of Thames (1985) as the best fit to the data. *Arrows* indicate the increase in ED_{50} by fractionated high dose rate (HDR) irradiation. *DRF*, ratio of isoeffective doses; *CLDRI*, continuous low dose rate irradiation. (Redrawn from Down et al. 1986)

1985) whose graphical representation is given in Fig. 7. It also demonstrates that LDRI at 2 cGy/min in this tissue is equivalent to conventional HDR fractionation with 2 Gy/fraction. Both in this paper and in a similar study by Hill (1983) there is some evidence that the dose rate effect is even slightly higher when the observation period is extended into a phase that is believed to represent lung fibrosis (40 weeks or more after irradiation). The DRF values reported for the later and earlier cut-off times, respectively, were 2.5 vs 2.15 (100 vs 2 cGy/min; Down et al. 1986) and 2.0 vs 1.8 (100 vs 5 cGy/min; Hill 1983). Using TBI and subsequent bone marrow rescue Travis et al. (1985) determined a DRF (25 vs 1 cGy/min) of 2.15 after 1 year of observation; the absolute LD_{50} values were lower than found after thoracic irradiation, but lung was considered to be the critical organ.

All studies thus show that lung benefits distinctly from LDRI, although *with the dose-rates actually employed* the increase in tolerance was not greater than in many other tissues.

3.5 Central Nervous System

Quite recently, data on the sensitivity of rat spinal cord to LDRI were communicated by Scalliet et al. (1989), based on the frequency of foreleg paraplegia up to 9 months after irradiating the rachis from C1 to T1 at dose rates between 178 and 3.33 cGy/min. The loss of animals from the extended anaesthesia (maximum 22.3 h) was considerable, and the lowest evaluable dose rate in terms of an ED_{50} was 6.5 cGy/min. Yet the available results could reasonably well be fitted in a comprehensive analysis that also included the response to HDR fractionation. The best combined estimate of α/β was 1.6 Gy in accordance with previous experiments, while the half-time of repair from the CLDRI data was about 1.5 h, again in very good agreement with previous estimates from HDR fractionation with short fraction intervals and incomplete repair. One must bear in mind, however, that the analysis of the CLDRI data did not comprise confidence limits and that these are often large.

Still, a half-time of 1.5 h may reasonably be assumed for the spinal cord. This is a good example of the consequences when a high SLD repair capacity (as reflected in the α/β ratio) combines with a relatively slow repair rate; as pointed out by the authors, the tissue might have a relatively small benefit until the dose rate is well below 2 cGy/min.

3.6 Skin

Most clinical studies into the time factor in general and more specifically into the biological effectiveness of LDRI have concentrated on skin. In particular, formulae to describe iso-effectiveness have been based on the vast material on skin tolerance under widely varying dose rates (see Sect. 5.1). Studies in rodent skin have mostly but not exclusively utilized the acute response. BAKER and LEITH (1977) after delivering a constant dose of 25 Gy of X-rays to the mouse leg at dose rates ranging from 367 cGy/min down to 4.2 cGy/min found a steep decrease in the severity of the acute response between 63 cGy/min and 4.2 cGy/min; in addition, clinical and histological examination for late damage 5 months after treatment again revealed a dose rate effect over the same range. Also in mouse foot, HENKELMAN et al. (1980) compared the effect of 1-, 2- and 10-fraction protocols, each administered at dose rates of 160, 16 and 6 cGy/min. Dose-rate dependent sparing was pronounced in the 1- and 2-fraction regimes but was hardly detectable in the 10-fraction arm in which the doses per session still ranged from 3 to more than 6 Gy. The formal analysis is based on a linear quadratic model and exponential repair kinetics; in addition, it introduces an interesting approach to utilize all data stimultaneously, across different damage levels, by means of a link function. While these data showed a consistent tolerance increase with falling dose rate, KAL and GAISER (1977) in rat foot skin unexpectedly found that radiation delivered in four daily fractions was more effective at a dose rate of 6.15 cGy/min than at >300 cGy/min. This result is difficult to interpret, but as a single top-up dose was applied 3 days after the fractionated treatment to bring the damage up to a measurable level the possibility was considered that the LDR schedule might have induced significant synchrony of the proliferating surviving cells. The study design represented the first half of a clinical protocol used by PIERQUIN et al. (1975), and the specific question was whether the therapeutic gain claimed by PIERQUIN for dose rates around 1.6 cGy/min could be reproduced by a somewhat higher and hence more practicable dose rate.

A different approach, addressing late skin response, was used in another study by KAL and SISSINGH (1974). Pre-irradiated grafts of rat dorsal skin were transplanted into normal recipients, and late graft contraction after 3–5 months was assessed. From the response to single X-ray doses (140 cGy/min) or to ^{137}Cs-CLDRI (1.4 cGy/min) a DRF of about 1.8 was deduced. Notably, this assay measures a late dermal (vascular?) reaction, different from epidermal damage.

The most important results on skin tolerance to LDRI were established by TURESSON and NOTTER (1979a, b, c) in experiments on pig skin, which supposedly is the most suitable animal model for human skin. The principal study design aimed at dose equivalence between CLDRI and HDR fractionation in order to test the validity of KIRK's formula (KIRK et al. 1975). In meticulously controlled experiments skin fields of 3-cm diameter (90% isodose) were irradiated by ^{137}Cs brachytherapy either at 150 or 2 cGy/min, and both the acute response (erythema and moist desquamation) and a subchronic skin reaction (discoloring after 3–4 months, indicative of imminent dermal necrosis) were scored. The DRF values derived were 1.97 and 2.55 for the acute and subchronic responses, respectively. This proves greater tolerance conferred by LDRI to the dermal tissue. The most striking result of the study, however, was that CLDRI tolerance doses did not match the fractionated HDR dose levels predicted as isoeffective by the KIRK formula (see Sect. 5.1). In fact, when changing from CLDRI to a hypofractionated HDR regime as usually employed in HDR afterloading, dermal tolerance would have been exceeded by at least 25%. Several biological factors apart from intracellular repair were discussed for their possible impact but were all rejected. For example, altered cell distribution during CLDRI if anything would be expected to make the cells more sensitive but is unlikely to play a role as porcine basal cells have cycle times distinctly longer than the CLDRI periods used (5 vs 1.5–3 days). The conclusion therefore was that the difference in *effective repair* associated with changing dose rate is incorrectly accounted for in the formula.

4 Effects on Experimental Tumours

In considering the possible biological advantages of brachytherapy it is also necessary to characterize the response of tumours. Although the geometrical dose distribution in most cases can be arranged to include the tumour bulk in high isodose contours, peripheral extensions and nests of tumour may be exposed to dose levels at which sterilization can also depend on the dose rate.

A number of studies have been done in transplantable rodent tumours to quantify cell survival after in vivo exposure over the relevant range of dose rates. In contrast to normal tissues, a principal difficulty in the interpretation of the results arises from the fact that virtually all rodent tumours contain a variable fraction of naturally hypoxic and hence radioresistant cells; this fraction will dominate the surviving fraction even at moderate dose levels. During CLDRI such cells may reoxygenate and disappear as a relevant subpopulation, but this is not easy to ascertain. One may therefore deal with a complex situation in which separate factors are difficult to delineate. A similar overall sensitivity may result when cells reoxygenate and resume repair activity or when resistant hypoxic cells persist. Chronically hypoxic cells seem to be unable to perform effective repair during CLDRI (Bedford and Hall 1966).

The study by Hill and Bush (1973) on the KHT mouse sarcoma is an illustrative example of the complexity that may underlie a seemingly simple response. Cell survival after in vivo exposure at 13, 3 and 0.7 cGy/min as measured by lung colony formation after i.v. injection of suspended cells followed a near exponential curve that was marginally more resistant than that of acutely hypoxic cells treated with HDR. When the oxygenation status of cells surviving 10 or 30 Gy of CLDRI was assessed, a surprisingly large fraction turned out to be oxic. One possible explanation is that the oxic cells initially present performed sufficient repair to survive preferentially while the hypoxic cells were unable to do so; this requires that the DRF be similar in size to the oxygen enhancement ratio. Alternatively and not unlikely as during HDR fractionation this tumour reoxygenates within hours, a large fraction of the hypoxic cells may have become oxic and resumed SLDR, thus cancelling out the loss of hypoxic protection.

A similar situation was faced in studies by Kal and Barendsen (1972) and Kal et al. (1975) on rat R1 rhabdomyosarcoma cells exposed in vivo to acute X-rays or ^{137}Cs LDRI at 1.5–2.5 cGy/min. Again, the responses to all three dose rates could be described by the same near exponential survival curve, with a D_0 slightly higher than that found for hypoxic cells irradiated at acute dose rates (DRF about 1.3). Again, cells tested after 20 h of CLDRI were found to be predominantly oxic and to harbour no detectable SLD. They must therefore have reoxygenated and performed SLDR. A slightly enhanced radiosensitivity to the acute test dose as

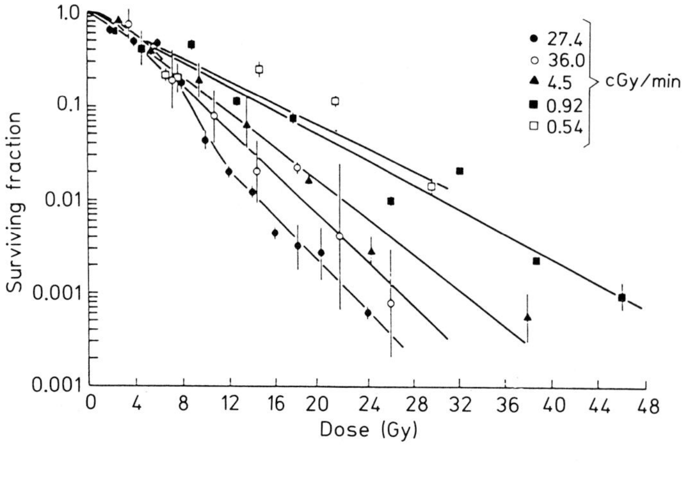

Fig. 8. Survival curves of EMT 6 tumour cells irradiated in vivo at various dose rates and subsequently plated in vitro. The biphasic curve shape to acute irradiation indicates the presence of hypoxic cells that dominate survival when a dose of 12 Gy is exceeded. (Redrawn from Fu et al. 1975)

compared with control cultures was ascribed to altered redistribution, but this effect was small.

Another extensive set of data comes from the study by Fu et al. (1975) on the EMT 6 tumour (Fig. 8), again based on in vivo irradiation and subsequent measurement of cell survival in vitro. When the dose rate was lowered from 274 cGy/min to 0.92 and 0.54 cGy/min, the biphasic survival curve flattened out to become exponential with an extrapolation number of 1.0 and a maximum D_0 of about 7 Gy. The oxygenation status of surviving cells along these curves was not assessed, but the initial response seen at the very low dose rates (i.e. a monophasic curve) clearly indicates that SLDR in oxic cells must have been the dominating factor.

It appears likely, therefore, that the overall response to LDRI in these animal tumours was governed by oxic cells, while the hypoxic population both due to its lack of SLDR and to reoxygenation did not represent a therapeutic problem. The controversy that remains can be pointed out in Fig. 9, taken from the review by Steel et al. (1986). If one assumes that the responses rely on oxic cells, then the increase in D_0 values by changing from acute to low dose rate and the resulting DRF values are unexpectedly high, contradicting the small repair capacity of most experimental tumours (Williams et al. 1984; Guttenberger et al. 1990). On the other hand, the saturation of

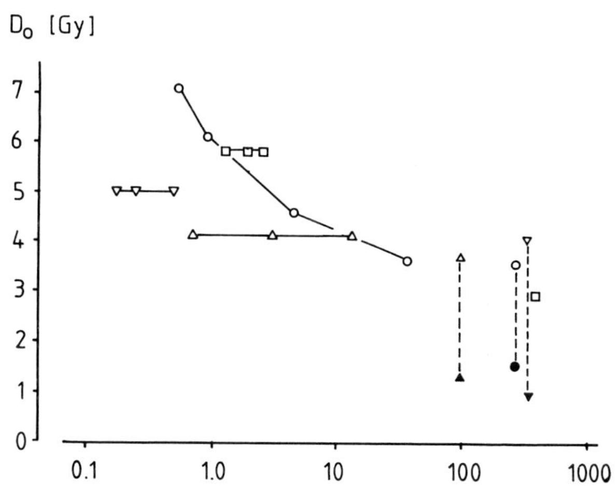

Fig. 9. Dose rate dependence of cellular radiosensitivity (*Do*) in four rodent tumours irradiated in vivo and assessed in vitro: □, rat rhabdomyosarcoma (Kal and Barendsen 1972); ▽, Lewis lung carcinoma (Shipley et al. 1983); △, KHT mouse sarcoma (Hill and Bush 1983); ○, EMT 6 carcinoma (Fu et al. 1975). The data pairs in the region 1–4 Gy/min are values for definitely oxic cells (*solid symbols*) or hypoxic cells (*open symbols*) irradiated in vitro. (From Steel et al. 1986)

the dose rate effect seen in all tumours except for EMT 6 could be partly explained by the small proportion of repairable damage and the relatively fast repair rates reported for rodent tumours (Williams et al. 1984; Rojas et al. 1989; Guttenberger et al. 1990) or human xenograft cell lines (Kelland et al. 1989). In summary, however, the reassuring message from Fig. 9 is that repair of tumour cells seems to saturate in a dose rate region in which normal tissue sensitivity still changes in a very dynamic way.

5 Mathematical Formulations of the Dose Rate Effect

5.1 Empirical Time-Dose Formulae

Mathematical formulae to describe isoeffective treatment for variation in dose rate or when changing from CLDRI to fractionated irradiation have been derived in several ways. A purely empirical approach similar to that underlying the NSD formula of Ellis (1969) was proposed by Kirk et al. (1972, 1975). The clinical database is formed by tolerance doses for acute skin response in patients treated at different dose rates and hence treatment times, as published by Mitchell (1960), Paterson (1963) and Ellis (1969). In addition, pig skin data, also relating to acute skin reactions, have been published by Atkins et al. (1972). In a double log plot of total isoeffective dose vs exposure time, a time exponent of between 0.24 and 0.30 was found to describe the data reasonably well for treatment durations between ca. 20 and 200 h, covering dose rates between 2.5 and 0.6 cGy/min.

The objections to this formula must be the same as to the NSD formula. A single exponent cannot sufficiently take into account tissue-specific repair, and as the power function is a merely empirical fit, extrapolation to dose rates outside the tested range can lead to grossly wrong tolerance predictions. The validity of the empirical approach becomes even more doubtful when used to calculate the equivalence of CLDRI and HDR fractionated treatment. For this purpose the following formula, based on the above time exponent of 0.29, has been given (Kirk et al. 1975):

$$CRE_c = k \cdot D \cdot T^{-0.29}$$

where D is total dose and T is total exposure time. The normalizing factor k was proposed to be 0.77–

0.80. In carefully controlled pig skin experiments using both early and subchronic skin reactions, TURESSON and NOTTER (1979a) determined a considerably smaller factor, i.e. 0.56–0.58. They conclude that the use of Kirk's factor when changing from CLDRI to HDR afterloading would have involved overdosage by at least 25%.

5.2 Extended Linear Quadratic Formalism

The current mathematical approaches to describe the dose rate effect have evolved from an entirely different concept. In contrast to the empirical models they are based on the underlying cellular dose-response relationship. In this respect the linear quadratic survival curve has become the most commonly used model since the mid-1970s. It is understood that in its present form it deals with the effect of repair alone and not with other biological processes that may play a role over very prolonged exposure times. However, in most cases repair is the overriding factor, and the effect of incomplete repair and dose rate are readily incorporated into the formula.

The salient assumption made in the model is that lethal events are produced either by primarily lethal (irreparable) hits or by the interaction of two sublethal lesions. The probability of total damage for single dose irradiation hence becomes

$$p = \alpha D + \beta D^2$$

with cell survival accordingly defined by

$$sf = e^{-\alpha D - \beta D^2}$$

$$\text{or} \quad -\ln(sf) = \alpha D + \beta D^2$$

When the dose is split into n fractions of size d given at intervals sufficient for complete repair, each fraction will equally contribute to the total effect E

$$E = -\ln(sf) = n\,(\alpha d + \beta d^2)$$

With LDRI the interactive term βd^2 becomes mathematically more complicated, but the underlying principle can readily be demonstrated (Fig. 10). When LDRI commences, SLD begins to build up according to the dose rate but at the same time decays according to the repair rate. The resulting curve that describes the accumulation of SLD with exposure time hence rises with an initial slope proportional to the dose rate but bends over continuously and eventually reaches a plateau when

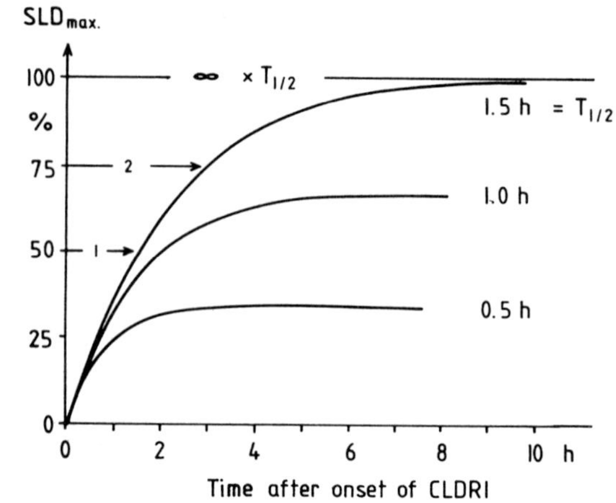

Fig. 10. Theoretical generation curves of sublethal damage (*SLD*) for a constant dose rate and three different half-times ($T_{1/2}$) of repair. *CLDRI*, continuous low dose rate irradiation

the generation of new SLD and its repair approach equilibrium. The 50% level of this plateau is attained after one repair half-time, the 75% level after two half-times etc., so that for practical purposes equilibrium can be assumed after 5 half-times. The plateau level depends linearly both on dose rate and repair half-time; doubling of either one therefore doubles the maximum SLD concentration attained at equilibrium. The contribution of the interactive β-term to the total effect at any time will be proportional to the product of SLD already accumulated and newly inflicted, and hence must be proportional to the dose rate squared.

Mathematically the accumulation of SLD integrates to the interactive term $\beta D^2\,F(\mu;T)$, where D is the total dose and $F(\mu;T)$ is a function of the exposure time T and the repair rate μ or the repair half-time $T_{1/2}$, as $\mu = (\ln 2)/T_{1/2}$). In its explicit form $F(\mu;T)$ reads

$$F(\mu T) = (2/\mu T)\,(1 - (1/\mu T)(1 - e^{-\mu T}))$$

which for $T \gg T_{1/2}$ *reduces to*

$$F(\mu;T) = (2/\mu T)\,(1 - 1/\mu T)$$

and eventually to $F(\mu;T) = 2/\mu T$

In experimental situations in which cell survival data can be obtained, the model can be employed to estimate the sensitivity parameters α and β as well as the repair half-time $T_{1/2}$. Extensive data sets for C3H $T10T_{1/2}$ cells irradiated in plateau phase with a great variety of dose rates have given excellent fits (METTING et al. 1985; WELLS and

Table 1. Repair half-times in tumour cell systems (Steel 1989)

Designation	Type	Half-time (h)
Human tumour cells		
HX 34	Melanoma	0.11
HX 118	Melanoma	0.16
HX 58	Pancreas	0.82
HX 156	Cervix	0.54
RT 112	Bladder	0.86
GCT 27	Testis	0.31
HX 142	Neuroblastoma	0.54
Murine tumour cells		
MT	Mammary tumour	0.094
Lewis	Lung tumour	0.092
B16	Melanoma	0.13

Bedford 1983; Steel et al. 1986). In addition, over the past few years a large number of human xenograft lines have been studied by Steel and his group (Steel 1989; Steel et al. 1987; Kelland et al. 1989). A selected list of the data is given in Table 1. The great variety which they reflect both in survival parameters and repair half-times may partly be due to the restricted number of dose rates tested (usually only two). Although differential sparing by LDRI in vitro could qualitatively be endorsed in vivo in two tumour lines by growth delay measurements (Kelland et al. 1989), the question remains as to how meaningful in vitro repair half-times of a few minutes may be for the in vivo situation.

The analysis of tissue responses in vivo requires a modification of the above formula. Gross tissue failure typically follows a sigmoid dose-response curve with a distinct threshold dose. For a given level of tissue response the underlying basic damage (cell survival) cannot be quantitated but is assumed to be equal, regardless of the radiation schedule by which it is produced. Analysis of the dose rate effect in this situation reduces to a comparison of isoeffective treatment. By dividing the common effect level E by β one obtains

$$E/\beta = \text{constant} = \alpha/\beta D_1 + D_1^2 F(\mu;T) = \alpha/\beta D_2 + D_2^2 F(\mu;T)$$

Both the "Incomplete Repair" model of Thames (1985) and its modification by Dale (1985) (see Sect. 5.3) have successfully been used to fit experimental animal data, resulting in parameter estimates for α/β and repair half-time. An illustrative example is the lung data by Down et al. (1986) based on changes in breathing rate recorded 14 weeks after CLDRI with dose rates between 100 and 2 cGy/min. The good agreement with the ICR model is obvious in Fig. 7, which also demonstrates the calculated extrapolation to lower dose rates than those actually used. In the case of lung tolerance it appears that at 2 cGy/min only half the repair capacity is exploited, and extensive further sparing should be possible by lowering the dose rate to <1 cGy/min. The model calculation also reveals that 2 and 1 cGy/min are the dose rates that are equivalent to fraction sizes of 2 and 1 Gy, respectively, in HDR fractionation protocols with complete repair.

In their analysis of oral mucosa (mouse lip), Scalliet et al. (1987) combined data from HDR fractionation and CLDRI over a wide range of fraction doses and dose rates. The α/β ratio derived was approximately 8 Gy as is typical for an acutely responding tissue; the best estimate of $T_{1/2}$ was 47 min. Similarly, in a simultaneous analysis of acute skin response in mouse foot including fractionated and single-dose HDR and LDR irradiation, Henkelman et al. (1980) demonstrated a most satisfactory fit for a common α/β of 7.5 Gy and a half-time of 1.3 h. These data indicate quite a range of tissue repair rates, of which jejunal crypts may mark the fast end. A mean half-time of 0.3 h was assessed for jejunal crypt stem cells from the experiments by Huczkowski and Trott (1984), which included continuous and fractionated LDRI over a wide range of dose rates. In a recent re-analysis of this work by Dale et al. (1988) evidence was presented that the apparent $T_{1/2}$ may depend on dose rate, with a maximum at 35 cGy/min ($T_{1/2}$ ca. 1 h) that decreases both with lower and higher dose rates ($T_{1/2}$ ca. 0.15 and 0.25 h, respectively).

It is important to discriminate between the different roles played by the two parameters, repair capacity (or α/β) and repair rate, for tissue tolerance with changing dose rate. This is visualized in the example given in Fig. 11: in the top panel the tolerance curves of a late responding model tissue are shown (α/β ratio 3 Gy) for repair half-times of 0.5, 1.0 and 2 h. Likewise in the bottom graph the tolerance curves of an early reacting tissue or of a tumour (α/β ratio 7.5 Gy) have been plotted. For simplicity, an identical acute tolerance dose (D_{ac}) of 16 Gy has been assumed for both tissues. The *maximum possible increase in tolerance dose* as dose-rate is minimized is clearly larger for the late responding tissue, amounting to a DRF_{max} of about 6 in this example, whereas for the early

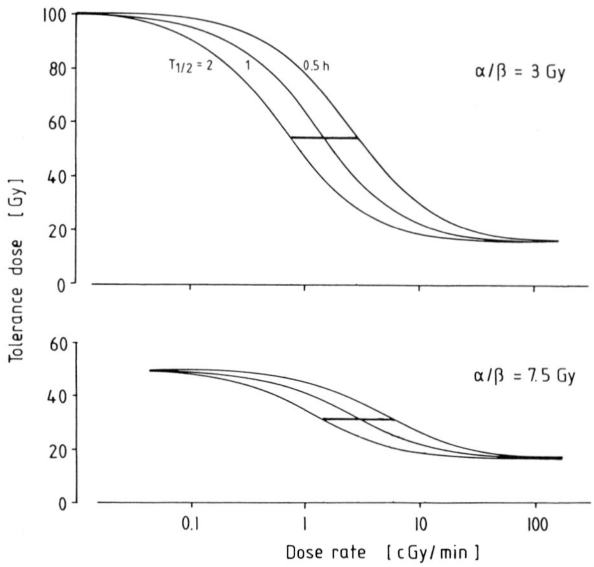

Fig. 11. Theoretical tolerance curves for late (α/β 3 Gy, *top*) and early reacting tissue (α/β 7.5 Gy, *bottom*) assuming a common acute tolerance dose of 16 Gy and repair half-times of 2, 1 and 0.5 h (from *left* to *right*)

responding tissue it is only about 3. The biological reason for this difference is the larger relative contribution of interactive damage in the late responding tissue at acute exposure. The mathematical term that describes the relation of repairable to irreparable damage in this situation is $D_{ac}/(\alpha/\beta)$. It can readily be shown that all tissues with identical ratio $D_{ac}/(\alpha/\beta)$ have *identical curve shape* and the same maximum tolerance increase, defined by $DRF_{max} = 1 + D_{ac}/(\alpha/\beta)$. There remains a difference between such curves, however, which regards their location on the dose rate axis. As they all have the same DRF at the inflection point (DRF_{infl}) and the inflection dose is defined by $DRF_{infl} \cdot D_{ac}$, curves that embody a higher absolute D_{ac} must proportionately be shifted towards higher dose rates.

The role of repair kinetics, on the other hand, is more straightforward. As seen in Fig. 11, the influence of repair rate is *exclusively on the location of the tolerance curve*, as expressed by their horizontal displacement with changing half-time in both panels. When the half-time is short, SLD is more effectively repaired during exposure and hence the tolerance curve rises – and saturates – already at relatively higher dose rates. As a consequence, the curves are horizontally displaced to the left with increasing half-time of repair. The factor by which the inflection dose rates are separated is necessarily identical to the ratio of repair half-times. As stated in section 4.2., the accumulation of SLD is proportional both to the dose rate and the repair half-time, and hence a doubling of half-time is exactly counteracted by a halving of dose rate.

Published tolerance curves from animal experiments usually cover only a section of the full dose rate range for the simple reason that very small dose rates require unduly long immobilization periods. An exception is TBI for the jejunal crypt survival assay, but full assessment of the dose rate effect in this system is thwarted by the onset of proliferation during exposure after a latency time of about 2 days.

A compilation of such curves for various tissue effects is given in Fig. 12, based on experiments that involved several dose rates. Tolerance increase becomes visible when the dose rate falls below 100 cGy/min, but seemingly in no case does it reach saturation at the lowest dose rates actually employed.

The relative increase in tolerance can better be demonstrated when all curves are normalized to the same tolerance dose for acute irradiation. This is shown in Fig. 13 for the data on lung (Down et al, 1986), jejunal crypts (Huczkowski and Trott 1987; Dale et al. 1988), skin (Henkelman et al. 1980), and mouse lip mucosa (Scalliet et al. 1989). The results are somewhat surprising in that over a wide range of dose rate (from 100 to 5 or 2 cGy/min) the DRF values of early and late damage are very little separated. Below 2 cGy/min, however, the extrapolated tolerance curve for lung as calculated by the model continues to increase steeply, while the early responses as exemplified

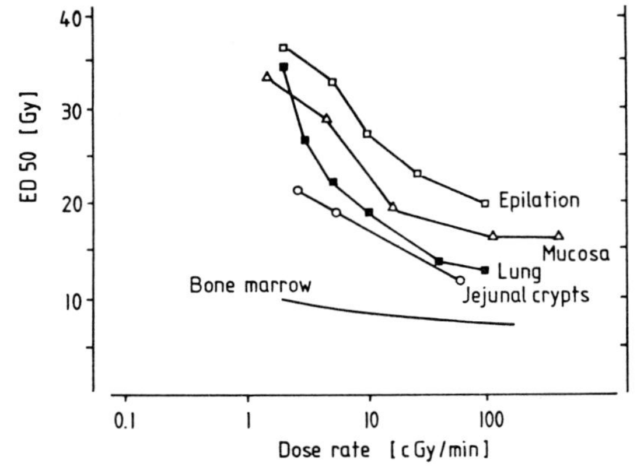

Fig. 12. Experimental curves describing dose rate dependence of normal tissue tolerance in the mouse. (Modified after Steel et al. 1986)

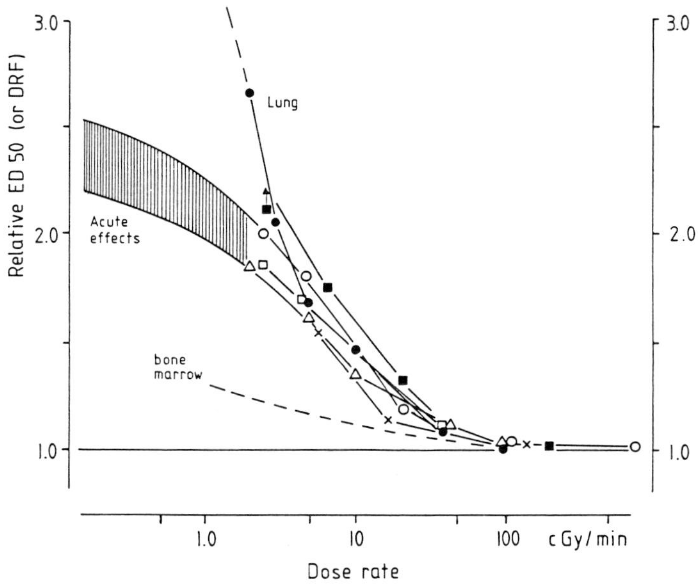

Fig. 13. Experimental tolerance curves of various normal tissues normalized to their acute ED_{50} values. *Shaded area* indicates extrapolation for acute effects. ●, Lung (Down et al. 1986); ■, spinal cord (Scalliet et al. 1989); □, jejunal crypts (Huczkowski and Trott 1987); ×, skin (Henkelman et al. 1980); ○, mucosa (Scalliet et al. 1987); △ epilation (Down, unpublished data). DRF, ratio of isoeffective doses

by oral and intestinal mucosa already tend to saturate at DRF_{max} values between 2 and 3. The great similarity in response of late and early effects between 100 and 5 cGy/min is somewhat surprising but does not negate the principles outlined above. The explanation must in fact be a circumstantial combination of factors. For example, the fast repair rate of jejunal cells and the lower α/β value of lung work in the same direction and make the curves of these tissues match at the intermediate dose rates. Mucosa and skin may have similar repair rates as lung, but their acute tolerance doses are definitely larger, minimizing the differences in $D_{ac}/(\alpha/\beta)$. The conclusion to be drawn from these animal data must be that the differences in dose rate response between tissues may actually be small at intermediate dose rates. Definitely low dose rates are required to effectively exploit the greater repair potential of late responding tissues.

5.3 Equating CLDRI and HDR Fractionation Protocols

The calculation of isoeffectiveness when changing from CLDRI of known tolerance to HDR afterloading treatment of a few large fractions, has long been a controversial issue. The linear-quadratic model offers a more biological and therefore safer approach than purely empirical formulae. Basically, the equations given above are already sufficient to do such calculations after minor transformation.

For example, it is readily shown that isoeffective treatments by HDR fractionation and CLDRI are related by

$$\frac{D_{HDR}}{D_{LDR}} = \frac{1 + (\beta/\alpha)D_{LDR} \cdot F(\mu;T)}{1 + (\beta/\alpha)d}$$

From a known tolerance dose D_{LDR} thus an isoeffective HDR treatment with a fraction size d can readily be computed, assuming a plausible repair halftime (e.g. 1.5 h) and the appropriate α/β for the tissue under study.

The above notation is identical with the concept of "Relative Effectiveness" (RE) as originally proposed by Barendsen (1982) and further elaborated by Dale (1985). The concept focuses on the effective slope of the cell survival curve, i.e. the slope of the chord that can be drawn from survival at dose zero down to the full effect E reached at dose D, or D/E. The largest conceivable isoeffective dose D, i.e. the ERD, is required when using an infinitely small dose-rate. The slope of this dose-response curve must be equal to α, the initial slope of the HDR survival curve. For finite dose-rates the corresponding chord slopes must all become steeper than α, the relative steepness over α being given by

$$RE = ERD/D = 1 + (\beta/\alpha)D_{LDR}\, F(\mu;T)$$

Similarly, for fractionated HDR treatment the relative chord slope is defined by

$$RE = 1 + (\beta/\alpha)d$$

When the total doses used in both modalities are the same, the RE's must be identical, and in this case the above equation reduces to

$$d = D\,F(\mu;T)$$

This equation indicates that to each dose per fraction (HDR) there exists a unique dose rate that has the same effectiveness. At first sight the relationship may also suggest that each doubling of d will be equivalent to a doubling of dose-rate, irrespective of α/β (FOWLER 1988). This is not strictly valid, as any change in d will necessarily alter the total dose, D, dependent on α/β. However, as for relatively long exposure time T (i.e., $T > 10T_{1/2}$) the function $F(\mu;T)$ practically reduces to the term $2/\mu T$, the equivalent dose-rate will also change over a wide range in a near linear manner dependent on μ but not on α/β. In a clinical situation where tolerance is defined by about 60 Gy delivered at 1 cGy/min or in 30 fractions of 2 Gy, the linearity is a good approximation up to dose fractions of about 8 Gy. The full meaning of this is that a fraction size of 4 Gy in HDR treatment implies the same decrease in repair for the above tissue as changing the dose-rate from 1 to 2 cGy/min, and going to 8 Gy per fraction encompasses the same loss in repair as using a dose-rate somewhat higher than 4 cGy/min. As has been shown earlier, the greatest selective benefit for late effects relative to acutely responding tissues or tumours is gained at very low dose-rates. HDR afterloading regimes using relatively large fraction sizes therefore unavoidably sacrifice a great deal of this selectivity.

The above isoeffect relationship derived from the extended linear-quadratic model has an interesting historical counterpart. Two decades ago, LIVERSAGE (1969) introduced a generalized formula that was devised to equate protracted and acute regimes of equal total dose. The formula defines the number of HDR fractions, N, into which the protracted dose must be split to be biologically isoeffective by

$$N = (uT/2) \cdot (1 - (1/uT)\,(1 - e^{-uT})$$

The identity with the foregoing equation is obvious when N is replaced by D/d. The agreement is by no means incidental. In deriving his formula, LIVERSAGE followed the concept of LAJTHA and OLIVER (1961) that accumulation of SLD is fully expressed in the shoulder of the acute cell survival curve while during protracted irradiation it is partly removed by an exponential repair process. The mathematical steps involved in equating acute and protracted irradiation therefore have much in common with the RE concept as they also rely on a comparison of curve slopes. In his elaboration LIVERSAGE gives a rather intuitive definition concerning the shape of the acute survival curve. His assumption, not based on real data or a model concept, is that the slope of the acute survival curve at dose D is equal to the effective curve slope between dose zero and 2D (see Appendix in his paper). As it turns out, this assumption fully meets the definition of a quadratic survival curve and hence makes his dose-response curve congruent with the one component in the linear-quadratic survival curve that can be affected by dose-rate.

References

Atkins HL, Fairchild RG, Robertson JS (1972) Dose-rate effects on RBE of Californium and Radium reactions of pig skin. Radiology 103: 439–442

Baker DG, Leith JT (1977) Effect of dose rate on production of early and late radiation damage in mouse skin. Int J Radiat Oncol Biol Phys 2: 69–77

Barendsen GW (1982) Dose fractionation, dose rate and iso-effect relationships for normal tissue responses. Int J Radiat Oncol Biol Phys 8: 1981–1997

Bateman JL, Bond VP, Robertson JS (1962) Dose-rate dependence of early radiation effects in small mammals. Radiology 79: 1008–1014

Bedford JS, Hall EJ (1966) Threshold hypoxia: its effect on the survival of mammalian cells irradiated at high and low dose-rates. Br J Radiol 39: 896–900

Bedford JS, Mitchell JB (1973) Dose-rate effects in synchronous mammalian cells in culture. – Observations using time-lapse cinemicrography. Radiat Res 54: 316–327

Bedford JS, Mitchell JB (1977) Mitotic accumulation of HeLa cells during continuous irradiation. Radiat Res 70: 173–186

Bedford JS, Mitchell JB, Fox MH (1980) Variations in responses of several mammalian cell lines to low dose-rate irradiation. In: Meyn RE, Withers HR (eds) Radiation biology in cancer research. Raven, New York, pp 251–262

Cairnie AB (1967) Studies in the intestinal epithelium of the rat; response to continuous irradiation. Rad Res 32: 240–264

Dale RG (1985) The application of the linear-quadratic dose-effect equation to fractionated and protracted radiotherapy. Br J Radiol 58: 515–528

Dale RG, Huczkowski J, Trott K-R (1988) Possible dose rate dependence of recovery kinetics as deduced from a preliminary analysis of the effects of fractionated irradiations at varying dose rates. Br J Radiol 61: 153–157

Down JD, Easton DF, Steel GG (1986) Repair in the mouse lung during low dose-rate irradiation. Radiother Oncol 6: 29–42

Elkind MM, Sutton H (1960) Radiation response of mammalian cells grown in culture. Radiat Res 13: 556–593

Ellis F (1969) Dose, time and fractionation: a clinical hypothesis. Clin Radiol 20: 1–7

Fowler JF (1989) Dose-rate effects in normal tissues. In: Mould RF (ed) Brachytherapy 2 – Proceedings of the 5th International Selectron Users' Meeting 1988. Nucletron International BV, Leersum, pp 26–40

Fox M, Nias AHW (1971) The influence of recovery from sublethal radiation damage on the response of cells to protracted irradiation at low dose rate. Curr Top Radiat Res Q 7: 71–103

Frindel E, Hahn GM, Robaglia D, Tubiana M (1972) Responses of bone marrow and tumor cells to acute and protracted irradiation. Cancer Res 32: 2096–2103

Fu KK, Phillips TL, Kane LJ, Smith V (1975) Tumor and normal tissue response to irradiation in vivo: variation with decreasing dose rates. Radiology 114: 709–716

Glasgow GP, Beetham KL, Mill WB (1983) Dose rate effects on the survival of normal hematopoietic stem cells of Balb/c mice. Int J Radiat Oncol Biol Phys 9: 557–563

Guttenberger R, Kummermehr J, Chmelevsky D (1990) Kinetics of recovery from sublethal radiation damage in four murine tumors. Radiother Oncol 18: 79–88

Hall EJ (1972) Radiation dose rate: a factor of importance in radiology and radiotherapy. Br J Radiol 45: 81–97

Hall EJ (1981) The radiobiological basis for low dose-rate brachytherapy. In: George FW (ed) Modern interstitial and intracavitary radiation management. Masson Publishing Corp., New York, pp 43–56

Henkelman RM, Lam GKY, Kornelsen RO, Eaves CJ (1980) Explanation of dose-rate and split-dose effects on mouse foot reactions using the same time factor. Radiat Res 84: 276–289

Hill RP (1983) Response of mouse lung to irradiation at different dose rates. Int J Radiat Oncol Biol Phys 9: 1043–1047

Hill RP, Bush RS (1973) The effect of continuous or fractionated irradiation on a murine sarcoma. Br J Radiol 46: 167–174

Huczkowski J, Trott KR (1984) Dose fractionation effects in low dose rate irradiation of jejunal crypt stem cells. Int J Radiat Biol 46: 293–298

Huczkowski J, Trott KR (1987) Jejunal crypt stem-cell survival after fractionated gamma-irradiation performed at different dose rates. Int J Radiat Biol 51: 131–137

Kal HB, Barendsen GW (1972) Effects of continuous irradiation at low dose rates on a rat rhabdomyosarcoma. Br J Radiol 45: 279–283

Kal HB, Sissingh HA (1974) Effectiveness of continuous low dose-rate gamma-irradiation on rat skin. Br J Radiol 47: 673–678

Kal HB, Barendsen GW, Bakker-van Hauwe R, Roelse H (1975) Increased radiosensitivity of rat rhabdomyosarcoma cells induced by protracted irradiation. Radiat Res 63: 521–530

Kelland LR, Steel GG (1989) Recovery from radiation damage in human squamous carcinoma of the cervix. Int J Radiat Biol 55: 119–127

Kelland LR, Tonkin KS, Steel GG (1989) A comparison of the in vivo and in vitro radiation response of three human cervix carcinomas. Radiother Oncol 16: 55–64

Kirk J, Gray WM, Watson R (1972) Cumulative radiation effect – II. Continuous radiation therapy – long-lived sources. Clin Radiol 23: 93–105

Kirk J, Gray WM, Watson R (1975) Cumulative radiation effect. IV: Normalization of fractionated and continuous therapy – area and volume correction factors. Clin Radiol 26: 77–88

Kiszel Z, Spietthoff A, Trott KR (1985) Chronische Strahlenfolgen am Enddarm der Ratte nach intrakavitärer Bestrahlung mit unterschiedlicher Dosisleistung. Strahlentherapie 161: 348–353

Krebs JS, Leong GF (1970) Effect of exposure rate on the gastrointestinal LD50 of mice exposed to ^{60}Co gamma rays or 250 kVp X-rays. Radiat Res 42: 601–613

Krebs JS, Jones DCL (1972) The LD50 and the survival of bone-marrow colony-forming cells in mice: effect of rate of exposure to ionizing radiation. Radiat Res 51: 374–380

Lajtha LG, Oliver R (1961) Some radiobiological considerations in radiotherapy. Br J Radiol 34: 252–257

Lamerton LF, Courtenay VD (1968) The steady state under continuous irradiation. In: Brown DG, Cragle RG, Noonan TR (eds) Dose rate in mammalian radiation biology. United States Atomic Energy Commission, conference 680410, pp 3.1–3.12

Lea DE (1938) A theory of the action of radiation on biological material capable of recovery: Part I. The time-intensity factor. Br J Radiol 11: 489–497

Liversage WE (1969) A general formula for equating protracted and acute regimes of radiation. Br J Radiol 42: 432–440

Maruyama Y, Nava C, Feola J, Beach JL, Hwang H-N, Williams A (1983) RBE of spleen CFU-s to low dose rate Cf-252 or photon radiation in vivo. Int J Radiat Oncol Biol Phys 9: 1049–1056

Metting NF, Roesch WC, Nelson JM (1985) Dose-rate evidence for two kinds of radiation damage in stationary-phase mammalian cells. Radiat Res 103: 204–218

Mitchell JS (1960) Studies in radiotherapeutics. Blackwell, Cambridge

Mitchell JB, Bedford JS (1977) Dose rate effects in synchronous mammalian cells in culture – II. A comparison of the life cycle of HeLa cells during continuous irradiation or multiple-dose fractionation. Radiat Res 71: 547–560

Mitchell JB, Bedford JS, Bailey SM (1979a) Dose rate effects in plateau-phase cultures of S3 HeLa and V79 cells. Radiat Res 79: 552–567

Mitchell JB, Bedford JS, Bailey SM (1979b) Dose rate effects in mammalian cells in culture – III. Comparison of cell killing and cell proliferation during continuous irradiation for six different cell lines. Radiat Res 79: 537–551

Paterson R (1963) The treatment of malignant disease by radiotherapy, 2nd edn. Arnold, London

Pierquin B, Baillet F, Brown CH (1975) Low dose rate irradiation in advanced tumors of head and neck. Acta Radiol 14: 497–504

Pipard G (1989) Historical aspects of interstitial brachytherapy and their implications for modern radiotherapy. In: Rotte K, Kiffer J (eds) Changes in brachytherapy. Wachholz, Nürnberg, pp 45–56

Puro EA, Clark GM (1972) The effect of exposure rate on animal lethality and spleen colony survival. Radiat Res 52: 115–129

Rojas A, Joiner MC, Johns H (1989) Recovery kinetics in mouse skin and CaNT tumours. Radiother Oncol 16: 211–220

Scalliet P, Landuyt W, van der Schueren E (1987) Effect of decreasing the dose rate of irradiation on the mouse lip mucosa. Comparison with fractionated irradiations. Radiother Oncol 10: 39–47

Scalliet P, Landuyt W, van der Schueren E (1989) Repair kinetics as a determining factor for late tolerance of central nervous system to low dose rate irradiation. Radiother Oncol 14: 345–353

Shipley WU, Peacock JH, Steel GG, Stephens TC (1983) Continuous irradiation of the Lewis lung carcinoma in vivo at clinically used "ultra" low-dose rates. Int J Radiat Oncol Biol Phys 9: 1647–1653

Steel GG (1989) Recovery kinetics deduced from continuous low dose rate experiments. Radiother Oncol 14: 337–343

Steel GG, Down JD, Peacock JH, Stephens TC (1986) Dose rate effects and the repair of radiation damage. Radiother Oncol 5: 321–331

Steel GG, Deacon JM, Duchesne GM, Horwich A, Kelland LR, Peacock JH (1987) The dose-rate effect in human tumour cells. Radiother Oncol 9: 299–310

Szechter A, Schwarz G, Barsa JM (1978) Continuous and fractionated irradiation of mammalian cells in culture – I. The effect of growth rate. Int J Radiat Oncol Biol Phys 4: 991–1000

Thames HD (1985) An "incomplete repair" model for survival after fractionated and continuous irradiations. Int J Radiat Biol 47: 319–339

Thames HD, Hendry JH (1987) Fractionation in radiotherapy. Taylor & Francis, London

Travis EL, Peters LJ, McNeill J, Thames HD, Karolis C (1985) Effect of dose-rate on total body irradiation: lethality and pathologic findings. Radiother Oncol 4: 341–351

Turesson I, Notter G (1979a) The response of pig skin to single and fractionated high dose rate and continuous low dose rate ^{137}Cs-irradiation – I. Experimental design and results. Int J Radiat Oncol Biol Phys 5: 835–844

Turesson I, Notter G (1979b) The response of pig skin to single and fractionated high dose rate and continuous low dose rate ^{137}Cs-irradiation – II. Theoretical consideration of the results. Int J Radiat Oncol Biol Phys 5: 955–963

Turesson I, Notter G (1979c) The response of pig skin to single and fractionated high dose rate and continuous low dose rate ^{137}Cs-irradiation – III. Re-evaluation of the CRE system and the TDF system according to the present findings. Int J Radiat Oncol Biol Phys 5: 1773–1779

Wambersie A, Stienon-Smoes M-R, Octave-Prignot M, Dutreix J (1979) Effect of dose rate on intestinal tolerance in mice. Implications in radiotherapy. Br J Radiol 52: 153–155

Wells RL, Bedford JS (1983) Dose-rate effects in mammalian cells – IV. Repairable and nonrepairable damage in noncycling C3H $10T_{1/2}$ cells. Radiat Res 94: 105–134

Williams MV, Denekamp J, Fowler JF (1985) A review of α/β ratios for experimental tumors: implications for clinical studies of altered fractionation. Int J Radiat Oncol Biol Phys 11: 87–96

Withers HR, Elkind MM (1970) Microcolony survival assay for cells of mouse intestinal mucosa exposed to radiation. Int J Radiat Biol 17: 261–267

1.3 Fundamentals in Physics

GINETTE MARINELLO

CONTENTS

1 Introduction 25
2 Physical Properties of Sources Used for Intracavitary and Interstitial Therapy 25
2.1 Production and Disintegration 25
2.2 Energy 26
2.3 Activity 27
2.4 Half-Life 28
2.5 "Average Life", T_a 30
3 Specification of Source Strength 30
4 Identification and Quality Control of Clinical Sources 31
4.1 Identification 31
4.2 Leakage and Surface Contamination Tests 31
4.3 Homogeneity Test 31
4.4 Calibration 32
5 Dose Distribution Around Sources 32
5.1 Absorbed Dose 32
5.2 Air Kerma Rate Constant 32
5.3 Absorbed Dose Calculation in a Medium 32
References 33

1 Introduction

Almost immediately after its discovery by Marie and Pierre Curie in 1898, radium was used to treat cancer. For years, plesiocurietherapy and interstitial therapy were performed using radium and its daughter element radon. With the advent of nuclear reactors, many new isotopes became available in the 1950s: needles of cobalt 60 (MYERS 1948) tantalum-182 wires, gold 198, and iridium 192 (SINCLAIR 1952; MYERS et al. 1953; HENSCHKE et al. 1953). They gradually replaced radium. Now, artificially produced radionuclides such as caesium 137, iridium 192 and more recently iodine 125 and ruthenium 106 have largely supplanted radium.

This chapter summarizes disintegration modes and pertinent physical characteristics of sources, the clinical applications of which are presented in other chapters. The different methods adopted to specify their strength and to control their quality are described. The calculation of absorbed dose is briefly presented.

GINETTE MARINELLO, Ph.D., Département de Carcinologie, Centre Hôpitalo Universitaire Henri Mondor, 51, Avenue de Maréchal de Lattre de Tassigny, 94010 Créteil, France

2 Physical Properties of Sources Used for Intracavitary and Interstitial Therapy

2.1 Production and Disintegration

Radionuclides can be found in nature or are produced artificially by fission product or irradiation of a stable isotope in a neutron flux. The amount of radioactive material produced is dependent upon the number of atoms being bombarded, the probability of a nucleus capturing a neutron, the intensity of neutron flux, and the time spent in the neutron flux (JOHNS and CUNNINGHAM 1983, GODDEN 1988).

Production and mode of decay of the most useful radionuclides are shown in Tables 1–7. The modes of decay which are of greatest interest in brachytherapy are beta and beta-gamma.

β-Emission consists of the ejection of a positive or a negative electron from the nucleus. It can lead to the ground state of the nuclide or to an excited state. The excess energy is then ejected very quickly by the emission of one or several γ-rays in cascade, and the radionuclide becomes a *β-γ-emitter*. For instance, phosphorus 32 and strontium 90 are β-emitters whereas caesium 137 and iridium 192 are β-γ-emitters.

α-Disintegration (a helium nucleus consisting of two protons and two neutrons is ejected) occurs mainly in heavy nuclei. Radium is a typical α-emitter which disintegrates to an excited state of radon, the excess energy being lost by the emission of a γ-ray of 0.18 MeV. Radon then decays by α-emission to form polonium 218, which itself emits α- and β-particles together with γ-rays with

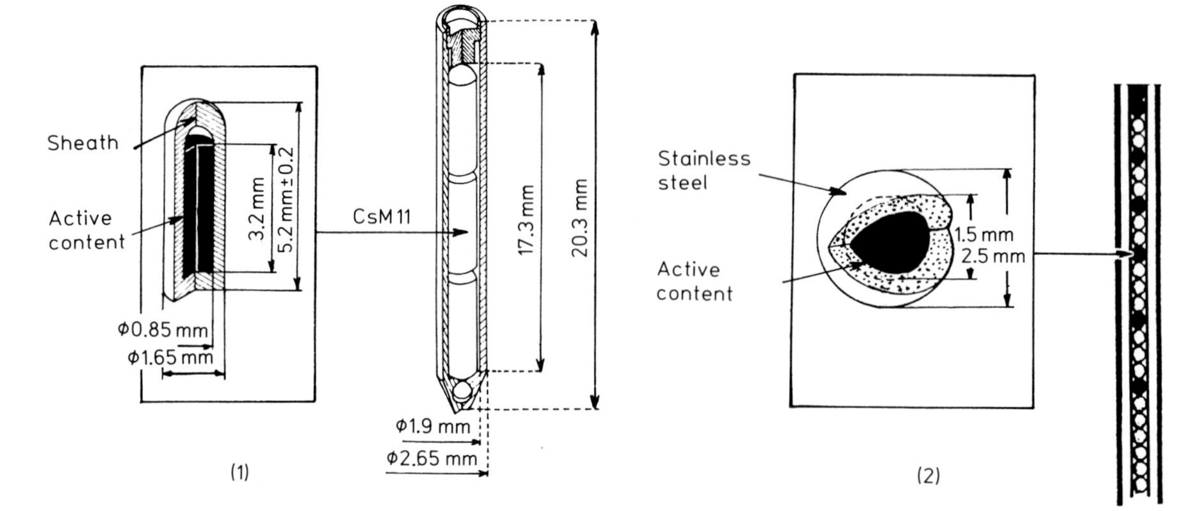

Fig. 1. Examples of caesium 137 sources as used in the Curietron–Oris–International CIS (*1*) and in the Selection – Nucletron Trading Ltd (*2*) afterloading systems

a total of 72 different energies (PAYNE and WAGGENER 1974).

Other modes of disintegration involve electron capture, electron conversion, isomeric transition, nuclear fission, etc. For detailed information the reader is referred to DUTREIX et al. (1982), JOHNS and CUNNINGHAM (1983), and GODDEN (1988).

2.2 Energy

The energy of the major radionuclides used in brachytherapy are presented in Tables 1–7. The *average energy* of the γ- or X-rays emitted is calculated taking into account the respective intensity of the different rays. The *maximum energy* of the associated β-radiation is indicated to allow an estimation of the inactive sheath required to absorb the β-rays (DUTREIX et al. 1982, CFMRI 1983).

Table 1. Recapitulative data for caesium 137

		Remarks
Production	Fission product from spent uranium fuel rods used in nuclear reactors	Small quantities of contaminant such as caesium 134 are often present (usually less than 1% by activity)
Mode of Decay	β and γ emissions to form the stable isotope barium 137: $^{137}_{55}\mathrm{Cs} \rightarrow {}^{137}_{56}\mathrm{Ba} + {}^{0}_{-1}\beta + \gamma$	
Energy	Average energy of γ-rays: 0.662 MeV Maximum energy of β-rays: 1.17 MeV	Beta radiation can be adequately filtered out by a sheath of platinum, 0.5 mm thick
Kerma rate constant	$\Gamma_\delta = 0.079\,\mu\mathrm{Gy}\cdot\mathrm{h}^{-1}\cdot\mathrm{MBq}^{-1}\cdot\mathrm{m}^2$	In old units, the exposure rate constant $\Gamma^*_\delta = 3.35\,\mathrm{R}\cdot\mathrm{h}^{-1}\cdot\mathrm{mCi}^{-1}\cdot\mathrm{cm}^2$
Half-life	30.15 years	Corresponding to a correction of 2% per year
Presentation	– Needles, tubes – Caesium 137 glass beads encapsulated in stainless steel and used to make source trains	
Clinical applications	– Needles and tubes for interstitial and intracavitary therapy – Source trains for manual or remote afterloading	

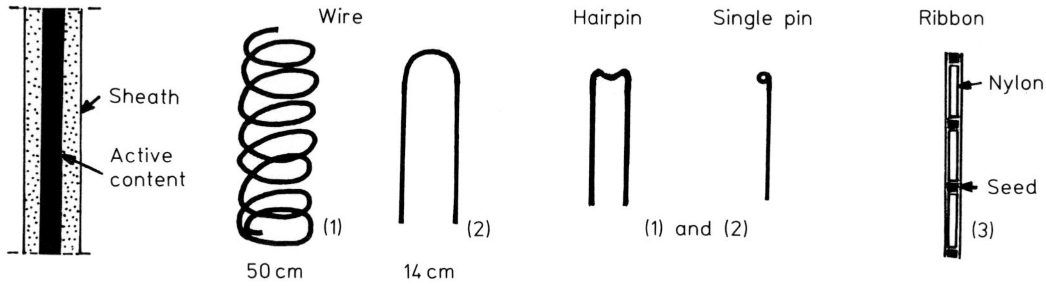

Fig. 2. Constructional details of iridium 192 wire and examples of sources available from Amersham International plc (*1*), Oris – International CIS (*2*), and Best Industries (*3*) or other manufacturers

The presented data (Tables 1–7) assume that the radionuclides are "bare" in air. In practice, brachytherapy sources are encapsulated, and thus consideration must be given to the absorption of radiation by the source capsule and self-absorption within the source. The average energy of γ- or X-rays varies according to the shape and the thickness of the source sheath. Likewise, the quantity of β-radiation decreases, or is even absent, if the sheath is thick enough. For instance, a sheath of platinum 0.1 mm thick is adequate to filter out the associated β-radiation of iridium 192. As all sources provided by the different manufacturers (Fig. 2) are sheathed in 0.1 mm of platinum, iridium 192 can be considered as a γ-emitter for brachytherapy purposes.

2.3 Activity

The amount of a radioactive material is specified in terms of *activity*. The definition given by ICRU (1980) is:

Table 2. Recapitulative data for iridium 192

		Remarks
Production	Produced by the n-γ reaction: $^{191}_{77}Ir$ (n-γ) $^{192}_{77}Ir$	Sources have to be supplied 8 days or more after irradiation in order to eliminate another isotope of short half-life produced at the same time: iridium 194
Mode of decay	β and γ emissions to form the stable isotope platinum 192: $^{192}_{77}Ir \rightarrow {}^{192}_{78}Pt + {}_{-1}^{0}\beta + \gamma$	
Energy	Average energy of γ-rays: 0.38 MeV Maximum energy of β-rays: 0.67 MeV	Sources are generally encased in a sheath of platinum, 0.1 mm thick, adequate to filter out the associated beta radiation
Kerma rate constant	$\Gamma_\delta = 0.1157\,\mu Gy \cdot h^{-1} \cdot MBq^{-1} \cdot m^2$	In old units, the exposure rate constant: $\Gamma_\delta^* = 4.9\,R \cdot h^{-1}\,mCi^{-1} \cdot cm^2$
Half-life	73.83 days	Neglecting source decay during clinical applications can lead to errors in the delivered dose of several percent (correction of about 1% per day)
Presentation	– Flexible wire 0.3, 0.5, or 0.6 mm in diameter – Single pin or hairpin, 0.5 and 0.6 mm in diameter – Seed sources preloaded in nylon ribbons	Wires and pins are easily cut to any desired length. They are internationally accepted as a 'sealed source' because they do not contaminate cutting instruments or release active particules
Clinical applications	Interstitial and intracavitary therapy (manual or remote afterloading)	

Table 3. Recapitulative data for cobalt 60

		Remarks
Production	Produced by the n-γ reaction: $^{59}_{27}Co\,(n-\gamma)\,^{60}_{27}Co$	
Mode of decay	β and γ emissions to form stable isotope nickel 60: $^{60}_{27}Co \rightarrow ^{60}_{28}Ni + ^{0}_{-1}\beta + \gamma$	
Energy	Average energy of γ-rays: 1.25 MeV Maximum energy of β-rays: 1.49 MeV	A thickness of 0.1 mm of platinum is adequate to absorb beta radiation, the dominant beta energy being 0.318 MeV
Kerma rate constant	$\Gamma_\delta = 0.309\,\mu Gy \cdot h^{-1} \cdot MBq^{-1} \cdot m^2$	In old units, the exposure rate constant: $\Gamma^*_\delta = 13.10\,R \cdot h^{-1} \cdot mCi^{-1} \cdot cm^2$
Half-life	5.27 years	Corresponding to a decay of about 1% per month
Presentation	– Source trains – Ophthalmic applicators	
Clinical applications	– Intracavitary remote afterloading systems (high dose rate) – Ophthalmology	

Table 4. Recapitulative data for gold 198

		Remarks
Production	Produced by the n-γ reaction: $^{197}_{79}Au\,(n\text{-}\gamma)\,^{198}_{79}Au$	
Mode of decay	β and γ emissions to form the stable isotope mercury 198: $^{198}_{79}Au \rightarrow ^{198}_{80}Hg + ^{0}_{-1}\beta + \gamma$	
Energy	Average energy of γ-rays: 0.420 MeV Maximum energy of β-rays: 0.966 MeV	A sheath of platinum 0.15 mm thick is adequate to filter out the majority of the β particles
Kerma rate constant	$\Gamma_\delta = 0.0548\,\mu Gy \cdot h^{-1} \cdot MBq^{-1} \cdot m^2$	In old units, the exposure rate constant $\Gamma^*_\delta = 2.32\,R \cdot h^{1} \cdot mCi^{-1} \cdot cm^2$
Half-life	64.68 h	
Presentation	Seeds in the form of small, square-ended cylinders with a core of gold of 0.5 mm in diameter encased in 0.15 mm of platinum	
Clinical applications	Permanent implantations: implanted directly into tissue using mechanical injectors	

The activity, A, of an amount of radioactive nuclide in a particular energy state at a given time is the quotient of dN by dt where dN is the expected value of the number of spontaneous nuclear transitions from that energy state in the time interval dt.

The SI unit for activity is the *becquerel* (Bq), which equals 1 disintegration per second.

Another unit, the curie (Ci) has been in use for many years. It is equal to 3.7×10^{10} Bq. Consequently, 1 Bq equals 2.703×10^{-12} Ci.

In practice, all these units can be altered by various factors of 10 through the use of appropriate prefixes. For instance, 10^{-12} = pico (p), 10^9 = giga (G), and one can write: 1 Ci = 37 GBp or 1 Bq = 2.703 pCi.

2.4 Half-Life

Radioactive decay occurs spontaneously, and there is no way to know when a given atom will

Table 5. Recapitulative data for radium 226 (in equilibrium with daughter products and filtered by 0.5 mm Pt)

		Remarks
Production	One of the members of the naturally occurring uranium series which starts with $^{238}_{92}U$ and ends with stable $^{206}_{82}Pb$	
Mode of decay	Radium disintegrates to form radon 222 (radioactive gas). If $^{222}_{86}Ra$ is prevented from escaping by being sealed in a capsule, there is a build-up of several daughter products that form the radium series, with emission of α and β particles together with γ-rays	
Energy	Average energy of γ-rays: 0.8 MeV Maximum energy of β-rays: 3.17 MeV	A thickness of 0.5 mm of platinum is adequate to absorb α particles and most of the β particles
Kerma rate constant	$\Gamma_\delta = 0.197\,\mu Gy \cdot h^{-1} \cdot MBq^{-1} \cdot m^2$	In old units, the exposure rate constant $\Gamma^*_\delta = 8.34\,R \cdot h^{-1} \cdot mCi^{-1} \cdot cm^2$ is equal to $8.25\,R \cdot h^{-1} \cdot mg^{-1} \cdot cm^2$
Half-life	1600 years	
Presentation	Needles and tubes generally formed by doubly encapsulating radium sulphate in platinum-iridium alloy	The double encapsulation minimises the risks of possible radium leak
Clinical applications	Interstitial and intracavitary therapy (direct loading)	

Table 6. Recapitulative data for iodine 125

		Remarks
Production	The isotope is produced as a daughter element of xenon 125, which is itself produced in a n-γ reaction: $^{124}_{54}Xe$ (n-γ) $^{125}_{54}Xe \rightarrow$ $^{125}_{53}I$ (e.c.)	
Mode of decay	Electron capture with emission of γ-rays, X-rays, and Auger electrons	
Energy	The principal photon emissions are: – a 35.48 keV γ-ray – 27.4 keV and 31.4 keV K_α and K_β characteristic X-rays	Because of their low energies, any electrons below 34.6 keV or L-characteristic X-rays below 3.7 keV are absorbed in the titanium jacket of the seed
Half-life	59.89 days	
Presentation	$^{125}_{53}I$ adsorbed onto the surface of a silver rod loaded into a hollow titanium tube 0.05 mm thick	
Clinical applications	– Permanent implantation directly into tissue using mechanical injectors or imbedded in sterile absorbable sutures and sewn into place – Temporary implants of high activity seeds for treatment of brain tumors	

disintegrate. However, on the average, one can predict that in a given time, called the half-life ($T_{1/2}$), half of the atoms will disintegrate (Tables 1–7). In the next half-life, one-half of the remaining atoms will decay, etc. Consequently, the number of atoms and the activity of any radionuclide vary exponentially with time.

The relationship between the activity and the elapsed time, t, is:

$$A = Ao \cdot \exp\left(-\frac{0.693 \cdot t}{T_{1/2}}\right)$$

where Ao is the initial activity.

Table 7. Recapitulative data for ruthenium 106

		Remarks
Production	Fission product	Possible impurity: ruthenium 103
Mode of decay	Essentially β emission: $^{106}_{44}Ru \rightarrow ^{106}_{45}Rh + ^{0}_{-1}\beta \rightarrow ^{106}_{46}Pd + ^{0}_{-1}\beta$	
Energy	Maximal energy of β-rays: 3.54 MeV	
Half-life	372.6 days	
Presentation	Ophthalmic applicators	Applicators are hemispherically shaped with their concave part coated with ruthenium 106 and their complex surface thick enough to absorb most of the beta radiation (radiation protection in handling)
Clinical applications	Intraocular tumors	

At a low dose rate (≤1 Gy/h), the duration of an application varies usually between 3 and 10 days. *For applications much shorter than the half-life of the radionuclide used*, the calculation of treatment time can be performed assuming that the source activity (or other units used to specify the source such as reference air kerma rate) is constant. This is always the case for radium 226 and caesium 137 sources. For caesium 137 sources, however, it is necessary to correct source activities for decay twice a year.

When source decay is short compared with the duration of the application, it is necessary to take into consideration the variation of activity during treatment time. For iridium 192 sources, correction for decay during treatment approximates 1% per day.

2.5 "Average Life", T_a

Sometimes sources are permanently implanted into patients. It is then necessary to know the total number of disintegrations that occur in an infinite period of time. It can be shown (Johns and Cunningham 1983), that a nuclide which decays exponentially over an infinite time with a half-life $T_{1/2}$ is equivalent to a nuclide which decays at a constant rate equal to its initial rate for a time, T_a, equal to:

$$T_a = \frac{T_{1/2}}{0.693} = 1.443\ T_{1/2}$$

and T_a is called the "*average life*" of the radionuclide. As a result of this, the total number of disintegrations for sources permanently implanted is equal to the initial activity, A_0, multiplied by the average life. The average life of iodine 125 and gold 198 commonly used for permanent implants is 86.4 days ad 93.3 hours, respectively.

3 Specification of Source Strength

Historically, source strength was defined in terms of the *mass of radium* encapsulated (in milligrams), together with the thickness of wall filtration (in millimeters of platinum). Total filtration was usually 0.5 mm Pt for radium needles and 1–2 mm Pt for radium tubes used in intracavitary therapy.

When sources of artificial radionuclides became available later, their strength was first specified in terms of *activity* (content activity) expressed in millicuries. But the activity of a source is not easy to measure and of little interest in practice because of the absorption by the source itself and its sheath. For this reason the term of equivalent or apparent activity was proposed by the IAEA (1967) and the ICRU (1970). *Apparent activity* is defined as the activity of a point source of the same nuclide which will give the same exposure rate in air at the same distance from the center of the source, this distance being large enough for the actual source to be considered as a point source. Apparent activity is expressed in millicuries.

To facilitate comparison between radium substitutes and radium, it was proposed to specify sources in *milligram radium equivalent (mg Ra eq)*. This is defined as that activity of a nuclide which delivers the same exposure rate at the same distance from the source as a point source of radium filtered by 0.5 mm of platinum. Such a

Table 8. Reference air kerma rate at 1 m for sources of apparent activity 1 mCi, of nominal exposure rate $1\,mR \cdot h^{-1} \cdot m^2$, or equivalent to 1 mg of radium filtered by 0.5 mm Pt

Strength of source	Reference air kerma rate ($\mu Gy \cdot h^{-1} \cdot m^2$)				
	$^{137}_{55}Cs$	$^{192}_{77}Ir$	$^{60}_{27}Co$	$^{198}_{79}Au$	$^{226}_{86}Ra$
1 mCi	2.92	4.28	11.43	2.03	7.29
$1\,mR \cdot h^{-1} \cdot m^2$	8.76	8.76	8.76	8.76	8.76
1 mg Ra eq (0.5 mm Pt)	7.23	7.23	7.23	7.23	7.23

method of source strength specification does not take into account source geometry, which severely limits its interest. For this reason, NCRP (1974) proposed that source strength should be specified in terms of the *exposure rate* at a distance of 1 m from the encapsulated source, perpendicular to the long axis of the source at its center.

As units used to express exposure rate do not belong to SI units, several organisations (CFMRI 1983, BCRU 1984, ICRU 1984) have recommended specification of source strength in terms of *reference air kerma rate*.

The reference air kerma rate of a source is the kerma rate to air, in air, at a reference distance of 1 m, corrected for air attenuation and scattering. It is expressed in micrograys per hour at 1 m ($\mu Gy \cdot h^{-1} \cdot m^2$). As a guide, the reference air kerma rate of sources of caesium 137, iridium 192, cobalt 60, gold 198, and radium 226, the strength of which is 1 mCi (apparent activity), $1\,mR \cdot h^{-1} \cdot m^2$ (exposure rate at 1 m), or 1 mg Ra eq, are shown in Table 8.

4 Identification and Quality Control of Clinical Sources

4.1 Identification

Some sources are provided with a means of identification, often in the form of a code number engraved on the source. This is the case for intestitial needles or intracavitary sources.

Iridium sources, because of their small size, carry no identification numbers. Nevertheless, it is necessary to be able to identify easily sources from different shipments. It is possible, while storing these sources in special containers including small compartments which can receive either individual wires or the branches of hairpins, to group the sources according to a specific linear reference air kerma rate (or linear activity). Standard model storage containers are sold commercially. Custom-made models meeting special needs are available from some manufacturers (Pierquin et al. 1987). When groups of sources having different reference air kerma rates are stored in a single safe, it is recommended to check the strength of sources quickly prior to clinical use. In very busy departments, it may be prudent to stock only a single source strength of any given size of wire at a time. In this situation the two sizes of iridium wires are easily differentiated by their diameter, and the containers can be of a much simpler design.

4.2 Leakage and Surface Contamination Tests

Although radiation sources are manufactured according to rigorous standards (numerous checks are obligatory prior to dispatch), sources must be tested for leakage and surface contamination prior to use. For example, the *swab test* can be used for most of the sources: the source is usually wiped with a swab of tissue moistened with ethanol or water. The swab is then checked for radioactivity and leakage is considered if the activity is more than 0.005 μCi (Jones 1988). These verifications should be performed periodically.

4.3 Homogeneity Test

The distribution of radioactivity within the source container should be checked. Except for iridium-192 sources, autoradiography is an appropriate technique (for details, refer to Johns and Cunningham 1983 or Jones 1988). This method can also be used to verify source position when sources are loaded into applicators. This check should be carried out when new sources or new applicators are received and then at regular intervals to insure that there is no movement of the radioactive material within the capsule and/or displacement of the sources within the applicator.

Control of the linear activity of iridium-192 wires or pins can easily be performed using a linear activimeter (Bernard et al. 1975; Dutreix et al. 1982). This consists of a radiation detector, in the form of three Geiger-Müller counters placed at the end of a lead collimator, which allows a length of 4 or 7 mm of wire to be measured at any one time. The wire is passed beneath the collimator and any variation in linear activity is detected. Of course, the same device can be used to verify other sources such as caesium-137 beads iridium-192 seed ribbons, etc.

One particular problem encountered with seed ribbons is the difficulty of producing ribbons with seeds of exactly the same reference air kerma rate (or activity) and spacing. Even though it is possible to take into account the variations in seed positions with the aid of specifically adapted computer programs, it is impossible in practice to account for the often quite significant variations in reference air kerma rate between seeds. Therefore, it is important to verify seed ribbons before use.

4.4 Calibration

The radioactive sources provided by the manufacturers are often accompanied by a certificate detailing their reference air kerma rate (or activity). Nevertheless, *it is strongly recommended that source strengths be verified locally prior to use.*

Total reference air kerma rate of the radioactive sources can be verified using a re-entrant ionisation chamber, previously calibrated with at least one source provided by a National Standard Laboratory. At the time of calibration, the re-entrant ion chamber has to be corrected to take into account the energy, the dimensions, and the position in the cavity of the radionuclide when these are different from the calibration source (Dutreix et al. 1982; Godden 1988).

Linear reference air kerma rate of iridium 192 wire or other linear sources can be easily checked using a linear activimeter calibrated under the same conditions as the re-entrant chamber. Homogeneity along the wire and absolute measurement of the linear reference air kerma rate (or activity) can be verified at the same time.

5 Dose Distribution Around Sources

5.1 Absorbed Dose

The *absorbed dose*, D, is defined by ICRU (1980) to be the quotient of dE by dm where dE is the mean energy imparted by ionising radiation to matter of mass dm, i.e.,

$$\mathrm{D} = \frac{\mathrm{d}E}{\mathrm{d}m}$$

The SI unit for absorbed dose is the *gray* (Gy) which is equal to 1 joule per kilogram. Another unit, the rad, has been used for many years. It is equal to 100 ergs per gram and related to the gray such that 1 gray = 100 rads.

5.2 Air Kerma Rate Constant

Introduced by ICRU (1980) as a replacement for exposure rate constant, Γ_δ^*, the air kerma rate constant, Γ_δ, is used to determine absorbed dose and is defined as:

$$\Gamma_\delta = \frac{\mathrm{D}^2}{\mathrm{A}} \cdot \left(\frac{\mathrm{d}K_{\mathrm{air}}}{\mathrm{d}t}\right)_\delta$$

where $\left(\frac{\mathrm{d}K_{\mathrm{air}}}{\mathrm{d}t}\right)_\delta$ is the air kerma rate due to the photons of energy greater than δ (expressed in keV) at a distance, D, from a point source of activity A.

The recommended unit for air kerma rate constant is $\mathrm{Gy} \cdot \mathrm{Bq}^{-1} \cdot \mathrm{s}^{-1} \cdot \mathrm{m}^2$. In practice, more convenient units are $\mu\mathrm{Gy} \cdot \mathrm{h}^{-1} \cdot \mathrm{MBq}^{-1}$ at one meter or $\mu\mathrm{Gy} \cdot \mathrm{h}^{-1} \cdot \mathrm{MBq}^{-1} \cdot \mathrm{m}^2$.

As a result, the air kerma rate from a point source ($\dot{\mathrm{K}}_\mathrm{a}$) is given by:

$$\frac{\mathrm{d}K_{\mathrm{air}}}{\mathrm{d}t} = \dot{\mathrm{K}}_\mathrm{a} = \frac{\mathrm{A} \cdot \Gamma_\delta}{\mathrm{D}^2}$$

where A is the activity in MBq.

This equation is the basis of brachytherapy dosimetry. Although many of the sources used in brachytherapy are not point sources, the air kerma rate from sources of particular geometry can be calculated by considering the source to be made of many point sources. Calculations of air kerma rate in air from sources of simple geometry like line, ring, circular disc, cylindrical, and spherical sources are developed in the book by Godden (1988).

5.3 Absorbed Dose Calculation in a Medium

The dose rate at a point near a given radioactive source depends upon several parameters. These include the distance to the source, the reference linear kerma rate of the source (or linear activity), the source shape, the composition and thickness of its metallic sheath, and the medium composition. Accurate mathematical calculation is possible but generally involves complex integrals which will not be described here (for details, see Shalek and

Stovall 1968, Stovall and Shalek 1968; Rosenwald and Dutreix 1970).

When a computer is available with adequate programs allowing corrections for autoabsorption and filtration of the γ-rays by the source material and its sheath, tissue attenuation, and source geometry, dose rates at any point could be calculated accurately, and isodose distributions around radioactive applications can be readily obtained.

When a computer is not available, various graphs and tables can be used to estimate the dose rates at particular points. The results obtained are generally correct for point or rectilinear sources arranged in simple patterns. Prior to calculations, it is necessary to check that graphs or tables correspond to sources identical to the ones being used and pertain to sources specified in the same units. Hand calculations are not accurate for loops, hairpins, and complex applications.

References

Bernard M, Guille B, Duvalet G (1975) Mesure du débit d'exposition linéique nominal des sources à une dimension utilisées en curiethérapie. J Radiol Electrol 56: 785–790

British Commission on Radiological Units (1984) Specification of brachytherapy sources. Br J Radiol 57: 941–942

Comité Francais de Mesure des Rayonnements Ionisants (1983) Recommandations pour la détermination des doses absorbées en curiethérapie. Bureau National de Metrologie, Paris (CFMRI report, no 1)

Dutreix A, Marinello G, Wambersie A (1982) Dosimétrie en curiethérapie. Masson, Paris

Godden TJ (1988) Physical aspects of brachytherapy. Hilger, Bristol (Medical physics handbooks, vol 19)

Henschke UK, James AG, Myers WG (1953) Radiogold seeds for cancer therapy. Nucleonics 11: 46–48

International Atomic Energy Agency (1967) Physical aspects of radio-isotope brachytherapy Technical report series, no 75, IAEA, Vienna

International Commission on Radiological Units and Measurements (1970) Specification of high activity gamma ray sources. ICRU, Washington (Report no 18)

International Commission on Radiological Units and Measurements (1980) Radiation quantities and units. ICRU, Washington (Report no 33)

International Commission on Radiological Units and Measurements (1984) Dose and volume specification for reporting intracavitary therapy in gynecology. ICRU, Washingon (Report no 38)

Johns HE, Cunningham JR (1983) The physics of radiology. Thomas, Springfield

Jones CH (1988) Quality assurance in gynecological brachytherapy. In: Dosimetry in radiotherapy, vol 1. pp 275–290. Edited by International Atomic Energy Agency, Vienna

Myers WG (1948) Application of artifically radioactive isotopes in therapy: cobalt 60. AJR 60: 816–823

Myers WG, Colmeny BH, McLellon WM (1953) Radioactive gold-198 for gamma radiation therapy. AJR 70: 258

National Commission on Radiological Protection (1974) Specification of gamma ray brachytherapy sources. NCRP, Washington (Report no 41).

Payne WH, Waggener RG (1974) A theorical calculation of the exposure rate constant for radium-226. Med Phys 1: 210–214

Pierquin B, Wilson JF, Chassagne D (1987) Modern brachytherapy. Masson, Paris

Rosenwald JC, Dutreix A (1970) Etude d'un programme sur ordinateur pour le calcul des doses en curiethérapie gynecologique. J Radiol Electrol 51: 651–654

Shalek R, Stovall M (1968) The M.D. Anderson method for computation of isodose curves around interstitial and intracavitary radiation sources. I. Dose from linear sources. AJR 102: 662–672

Sinclair WK (1952) Artificial radioactive sources for interstitial therapy. Br J Radiol 25: 417–419

Stovall M, Shalek RJ (1968) The M.D. Anderson method for the computation of isodose curves around interstitial and intracavitary sources. III. Roentgenograms for input data and the relation of isodose calculations to the Paterson Parker System. AJR 102: 677–687

1.4 Brachytherapy Techniques

REINHOLD G. MÜLLER

CONTENTS

1 Introduction . . . 35
2 Radioactive Sources . . . 35
2.1 Sealed Sources . . . 35
2.2 Unsealed Sources . . . 35
3 Techniques and Instrumentation . . . 36
3.1 Molds and External Applicators . . . 36
3.2 Intracavitary Applicators . . . 37
3.3 Interstitial Application . . . 37
3.4 Afterloading Techniques . . . 38
3.4.1 Manual Afterloading . . . 38
3.4.2 High Dose Rate Remote Afterloading . . . 39
3.4.3 Low Dose Rate Remote Afterloading . . . 39
4 Dose Prescription and Calculation . . . 39
4.1 Classic Prescription Rules . . . 39
4.2 Gynecologic Application Rules . . . 40
4.3 Recent Rules and Nomographs . . . 41
4.3.1 "Free Spaced" Interstitial Permanent Application . . . 41
4.3.2 Template and Multiplanar Temporary Application . . . 41
5 Source Localization and Three-Dimensional Reconstruction . . . 42
5.1 Radiography . . . 42
5.1.1 Orthogonal and Stereo Shift Projection . . . 42
5.1.2 Tomography and Different Techniques . . . 42
6 Calculation of the Dose Distribution . . . 43
6.1 Point Sources . . . 43
6.2 Line Sources . . . 43
6.3 Tubes and Others . . . 44
7 Reference Points and Reference Isodose . . . 44
8 Combination with Teletherapy . . . 45
8.1 Dose, Time, and Volume Dependent Response . . . 45
9 Experimental and Future Techniques . . . 46
9.1 New Isotopes and Sources . . . 46
9.2 Monoclonal Antibody Carrier . . . 46
9.3 Adjuvant Modalities . . . 46
9.3.1 Hyperthermia . . . 46
9.3.2 Chemical Modifiers . . . 46
References . . . 47

1 Introduction

"Radiotherapy is an evolving speciality, and therefore the practice of any particular time is liable to go out of date with surprising rapidity."

RALSTON PATERSON's statement from 1948 (PATERSON 1963) is as true as the fact that brachytherapy, a technique stemming from the very beginning of radiotherapy, is still a modern and developing method.

Historical considerations have to be kept aloof from this technical chapter. The reader is referred to the first chapter in *Modern Brachytherapy* by BERNARD PIERQUIN and coworkers (1987) and in *Renaissance of Interstitial Brachytherapy* by DEL REGATO (1978). Nevertheless, the historical view would be very instructive, showing brachytherapy as an intimate alliance between the clinician's endeavor, the physicist's basic research, the radiobiologist's knowledge, and the technician's potential. Sometimes in the past all this was focused together and represented in a single person.

2 Radioactive Sources

2.1 Sealed Sources

In a previous chapter the physical properties and constructional details are reviewed by Marinello for the most common sources. Even if no review of this topic can be complete, there is no need for a supplement in this chapter. For more information, especially about rarely used sources like ^{51}Cr, ^{182}Ta, and the neutron-emitting ^{252}Cf, and on the other hand the quite common β-emitting ^{90}Sr/^{90}Y and ^{106}Ru/^{106}Rh, the interested reader is referred to the literature (GLASGOW and PEREZ 1988; GODDEN 1988; HILARIS 1975; SHEARER 1981; TROTT 1987).

2.2 Unsealed Sources

The therapeutical application of unsealed sources such as colloid ^{198}Au in the past or the nuclei ^{131}I, ^{32}P, and ^{90}Y, which are always in use, is ranked

REINHOLD G. MÜLLER, Priv.-Doz. Dr. rer. nat., Institut für Radiologie, Krankenhausstr. 12, 8520 Erlangen, Germany with nuclear medicine and therefore not taken into consideration here (SPENCER 1978).

3 Techniques and Instrumentation

3.1 Molds and External Applicators

Molds of encapsulated radium for the therapy of superficial lesions were the first techniques in brachytherapy as well as in radiotherapy in general. WICKHAM and DEGRAIS (1910) from the Saint-Lazare Hospital in Paris developed the first rules of application. They recommended a uniform distribution of the sources over the superficial applicator. In the early 1920s more knowledge of the α- and β-components of the radium sources was available, and the pioneers started to carry out what nowadays is called dosimetry. It was EDITH QUIMBY who formed the rules of application to a system with variable area and calculated the source configuration by cutting up each source into a series of point sources (QUIMBY 1922, 1932, 1944; QUIMBY and CASTRO 1953). This is quite similar to the computation algorithms which are used in most planning systems for brachytherapy. PATERSON and PARKER from the Christie Hospital, Manchester, developed a refinement of Quimby's rules, dropping the stipulation of the uniform source application (PATERSON and PARKER 1934, 1938). They published a series of tables which promised a homogeneity in dose of $\pm 10\%$, provided the rules of application were observed. This formed the so-called Paterson-Parker or Manchester dosage system, which is based on a series of publications by PATERSON and PARKER (1934, 1938), PATERSON et al. (1936), TOD and MEREDITH (1938), and MEREDITH and STEPHENSON (1945a, b), and were edited by Meredith in 1947 (MEREDITH 1947, second edition 1967).

In the early 1960s PIERQUIN, DUTREIX and coworkers developed a dosage system for planar and biplanar application (PIERQUIN and FAYOS 1962, PIERQUIN and DUTREIX 1966). Considering the capabilities of the new artificial isotopes which were then available (HAHN 1956), they formulated a somewhat different predictive implant system, based geometrically on a triangular or alternatively rectangular mesh. In a series of papers this system evolved and was adapted to specific techniques, made possible due to the metallic isotope ^{192}Ir with its excellent physical properties, as can be seen in the chapter in this volume by GINETTE MARINELLO. While the Quimby and the Manchester systems were derived from the mold technique, the Paris system in principle was developed from experience in interstitial volume application.

Since the beginning of brachytherapy, mold techniques have lost significance because of modern surgery, interstitial therapy, or simple percutaneous radiotherapy with electrons. Nevertheless, there may sometimes be requests for molds. The principles of source distribution following the pattern of the Manchester system are reviewed elsewhere (GLASGOW and PEREZ 1988; GODDEN 1988) according to planar, circular, rectangular, and cylindrical molds as well as double plane applications.

For the treatment of choroidal and retinal tumors, β-emitting plaques or casual low-energy (^{125}I) and high-energy (^{60}Co) photon-emitting plaques have a certain interest. For shallow lesions ^{90}Sr/^{90}Y and ^{106}Ru/^{106}Rh are the isotopes of choice. The strontium applicators are commonly used for short-term fractionated irradiation of accessible lesions; the ruthenium applicators have to be fixed as long-term plaques (LOMMATZSCH 1983; Fig. 1). The radiation quality has to be chosen depending on the depth or height of the tumor. A current

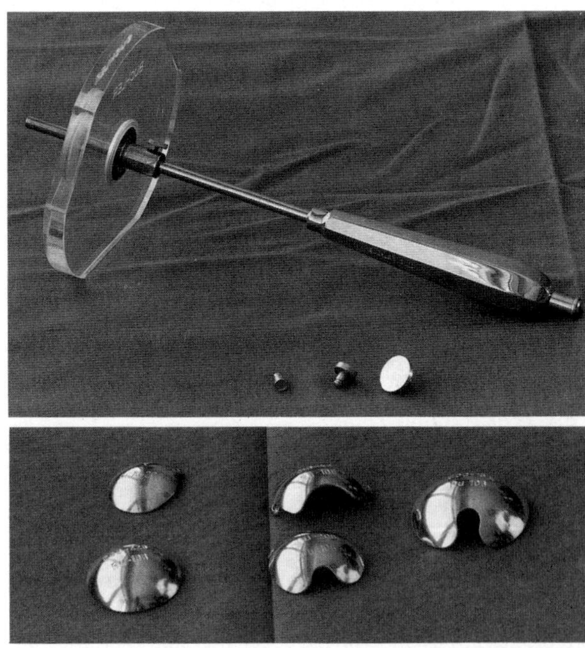

Fig. 1. a Plane and slightly concave ^{90}Sr applicators (diameters 6.2, 9.4 and 16 mm) with a straight forceps (Amersham Buchler, Braunschweig, FRG). **b** ^{106}Ru/^{106}Rh applicators after LAMMATZSCH (1977) (Akademie der Wissenschaften der DDR, Zentralinstitut für Isotopen- und Strahlenforschung, Berlin-Buch, DDR)

protocol (RTOG no. 84-04) recommends iodine plaques for lesions up to 3 mm, iridium up to 5 mm, and cobalt over 5 mm. There is an essential difference between β- and photon-emitting plaques. In contrast, the differences between photon-emitting isotopes seem to be nearly negligible, if any accompanying β-emission is filtered. The superficial dose levels are very high at the small surface of ophthalmological applicators, from 180 (^{60}Co) up to 500 Gy (^{90}Ru). This demands exact planning, reliable dosimetry, and skilful application. Individual plaque dosimetry is recommended but not feasible for most of the users. The exemplary papers dealing with dosimetry are of some importance (Magnus et al. 1968; Casebow 1971; Jones and Dermentzoglou 1971; Chan et al. 1972; Coffey et al. 1981). Loevinger described a method of calculating the dose distribution in the proximity of β-emitting plaques (Loevinger et al. 1956).

3.2 Intracavitary Applicators

The intracavitary application of radioactive sources (^{226}Ra) started nearly as early on as the mold technique. In principle, it could be called intracavitary mold application. For the most widespread gynecologic application there were two systems established in Europe, the Stockholm system from the Radiumhemmet (Heyman 1929, 1935; Heyman et al. 1941) and the Paris system from the Curie Foundation (Regaud 1929). The Paris system used a continuous application lasting approximately 120 h. In contrast, the Stockholm system used fractionated application with a time interval of at least 1 week between fractions. In both systems excellent clinical results were seen, and a basis was laid for the biological knowledge of the time-dose relationship. The Paris system was adapted by Tod and Meredith (1938) in Manchester. Fletcher et al. (Fletcher et al. 1953) later combined features of all three systems and introduced the first preloadable colpostat, which was modified for afterloading use by Henschke and Suit (Henschke 1960; Suit et al. 1963).

Through the years intracavitary application in gynecology persisted, developed, and led to good results because of the quite good radiotolerance of the vaginal mucosa, uterine corpus, cervix, and their surrounding tissue as well. However, the tolerance of the critical tissues, bladder and rectum, has to be watched carefully. Tamponing the structure (Stockholm) or screening the sources (Fletcher 1953) and direct measurement of the organ doses are the essential techniques. Since that time various detectors have been used in intracavitary dosimetry. As dose and dose rate at a certain point in the critical organ sensibly depend on the patient's anatomy and position (Jolesson and Backström 1969; Hunter; Thesen; Trokowar and Visser, this volume), a multiple detector system is strongly recommended.

Other intracavitary applications have been performed in all natural body cavities. The nasopharynx, esophagus, bronchus, bladder, anus, and rectum are treated this way. This will be completed by the intraluminal irradiation of, e.g., the bile duct, urethra,and probably in the future, the coronary arteries. For all these applications afterloading techniques are utilized.

3.3 Interstitial Application

Interstitial techniques are going to surpass intracavitary application. In relation to mold and intracavitary techniques, the interstitial application of radioisotopes guarantees better dose homogeneity in extended lesions, a reduced amount of equivalent activity and a steeper fall-off of dose outside the reference surface of the implant. In interstitial therapy a giant variety of needles, tubes, pins, hairpins, wires, grains, seeds, and ribbons is in use. It is nearly impossible to review all of them. Hence, the interested reader is referred to a number of textbooks (Fletcher 1980; George 1981; Godden 1988; Hilaris 1975; Marinello, this volume; Perez and Brady 1987; Pierquin et al. 1987; Shearer 1981; Vaeth 1978). The physical and geometrical properties of the radioisotopes obviously influence the technique of application (Trott 1987). The most important difference in these techniques results from permanent versus temporary application. ^{226}Ra, with a half-life of about 1600 years, the original radioisotope, forced the pioneers to develop removable applicators. The technique of permanent application requires a half-time value of the isotope between some days and a few weeks. First experiences with short half-life sources were collected by the two American physicists Duane (Duane 1915) and Failla (Failla 1926) with ^{222}R, a daughter isotope of ^{226}Ra. The artificial isotopes ^{198}Au and ^{125}I are appropriate and mostly used for permanent application. Considering radioprotection and radiobiology, ^{125}I becomes the radioisotope of choice for permanent application. It was Henschke who propagated this development at the Memorial Sloan-Kettering Cancer Center (HENSCHKE 1956; HENSCHKE et al. 1953; HILARIS et al. 1968; KIM and HILARIS 1975).

Direct handling of the radioactive isotope, e.g., the technique using the application pistol for gold grains (Amersham Buchler), is no longer up-to-date because of the serious exposure to the staff. Hence, permanent application has to utilize an afterloading technique just as like temporary applications. A temporary implant has a relatively high value of activity and should not be applied with the unprotected hand. That was the motive for the development of the innumerable afterloading techniques.

3.4 Afterloading Techniques

3.4.1 Manual Afterloading

Especially for interstitial therapy a variety of manual afterloading techniques has been developed (HENSCHKE et al. 1963; HILARIS 1975; PIERQUIN et al. 1987; SEYDEL 1977; SYED and FEDER 1977; VIKRAM and HILARIS 1981).

On the one hand there is the permanent implantation of ^{125}I. A modern reliable technique utilizes inactive hollow needles containing a mandrin, which are placed in an appropriate geometrical distribution inside the tumor. After calculation of the necessary amount of activity, a magazin holder and applicator (e.g., Mick applicator) is clamped on the needles. By drawing back the needles, the active seeds can be inserted at defined relative distances (Fig. 2). This technique was initiated by HENSCHKE and coworkers. It requires a compact volume to be implanted. Most of the possible applications are described in the handbook edited by HILARIS (1975).

However, for more complex target volumes, e.g., head and neck, anorectal region, urethra, bladder, extended gynecological application, and tumor bed applications in general, a carrying system is necessary for the active sources, which may be left during the complete treatment time. This can be provided by a stiff radioactive source itself, e.g., needle, pin, and hairpin, afterloaded through guide gutters (DELCLOS 1978; PIERQUIN et al. 1987). Iridium sources embeded in plastic ribbons and iridium wires have to be held in position by guide needles made of steel or plastic material. Numerous techniques are reviewed by DELCLOS (1978), HILARIS et al. (1968), HILARIS (1975), PEREZ and GLASGOW (1987), PIERQUIN et al. (1987), SEYDEL (1977), SYED and FEDER (1977), THIEL et al. (1987), VIKRAM and HILARIS (1981).

In order to obtain exact distances between the source carrier and to avoid crossing over and its resulting hot spots, different kinds of templates have been developed. There is the uniplanar or biplanar template designed after the Paris system application rule (PIERQUIN et al. 1987), which can be used for the fixation of double open-ended guide needles in the treatment of, e.g., breast,

a

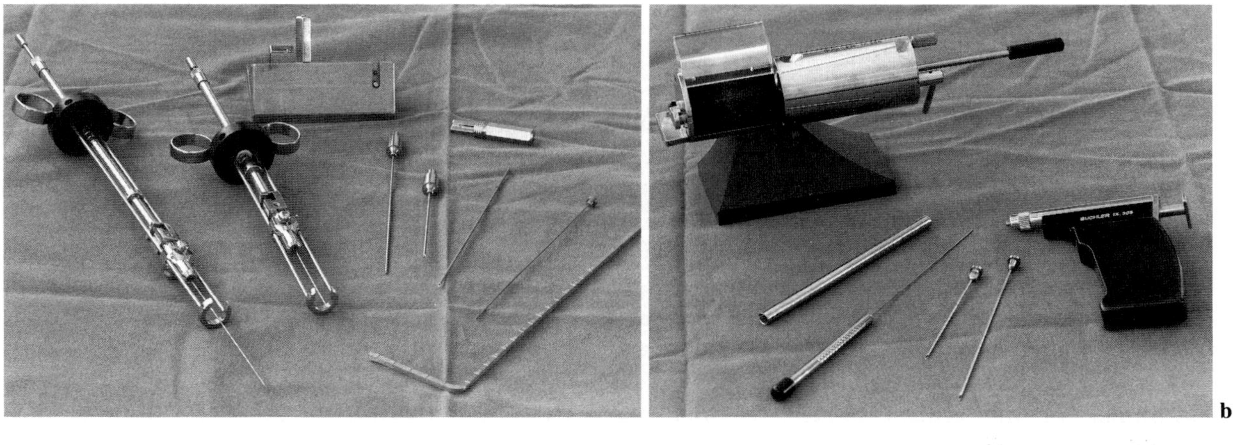

b

Fig. 2. a Mick applicator set, from *left* to *right*: long and short applicators, magazine holder, magazine, long and short needle with funnel for single seed implantation, needle with obturator (mandrin), and L-ruler (Mick Radio-Nuclear Instruments, Inc. Bronx, NY). **b** ^{198}Au application set, loading equipment, obturator needles, and pistol (Buchler GmbH, Braunschweig, FRG)

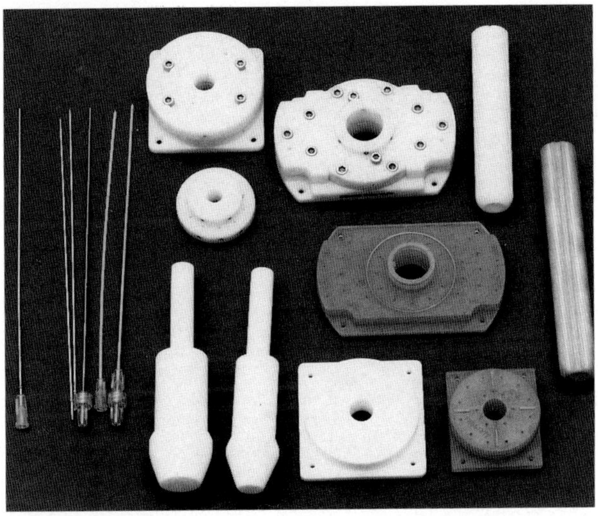

Fig. 3. Templates after SYED; needles, vaginal tubes, and anorectal obturators

skin, and soft tissue sarcomas. There are also one-ended templates mostly with a cylindrical symmetry designed for treatment of the anorectal region, female urethra, vagina, parametrium, prostate by perineal access, bladder, and similar applications in this region (Fig. 3) (GOFFINET et al. 1978; MARTINEZ et al. 1985; PUTHAWALA et al. 1985; SYED and FEDER 1977).

The template technique is also very beneficial in the treatment of intracranial tumors. With its multiple needles and sources, it renders exact definition and optimization of the dose distribution possible (ANDERSON 1985; FINDLAY et al. 1985; MCKAY et al. 1981; PEREZ and GLASGOW 1987).

3.4.2 High Dose Rate Remote Afterloading

By reason of radiation protection there has been increasing interest in changing the classic radium application techniques to remote afterloading (CHASSAGNE et al. 1969; HENSCHKE et al. 1966; NORI et al. 1983). Buchler's machine with a single oscillating source was the first commercial setup (available in the 1960s) which was mechanically programmable (ROTTE et al. 1973). Other manufacturers followed (SEAY et al. 1972). ANDERSON reviewed the different technical characteristics of eight systems (ANDERSON 1983a). The experience with some systems will be shown by ALTH et al. in this volume. Independent of the technical principle, the manufacturers and their users have endeavored to copy the classic radium techniques in their applicators and specific dose distributions. Everything can be achieved by remote afterloading including the Heyman packing method. In relation to the distribution of the physical dose there remain only marginal differences and problems. The main difficulty with high dose rate afterloading techniques is the adjustment of the fractionation regime, fraction size, and total dose for even tumor control and low rate of complications (ARAI 1978; ORTON 1981). In KUMMERMEHR's and ORTON's reviews in this volume these radiobiological problems are discussed in detail.

3.4.3 Low Dose Rate Remote Afterloading

The definition of low or high dose rate can be found in specific biological effects (ORTON, this volume). Reaching about 1 Gy/min and more at a certain point (e.g., point A in a gynecologic application) one can speak of a high dose rate, and below 1 Gy/h one should speak of a low dose rate. Between these two marks it should be called a medium dose rate. The biologic response to this intermediate dose rate is very complex, and any calculations about it are hardly possible.

Low dose rate remote afterloading furthers the clinical and radiobiological experience of classic radium application after the Quimby or Manchester system. In principle, this technique can replace all manual afterloading techniques. It is a further step to more safety and radiation protection. So far there have only been two commercial setups for interstitial application. One is the Micro selectron (Nucletron Trading, the Netherlands) with 15 independent channels; the other one is the Interpal C 38 (Interpal, Erlangen, FRG) with 38 channels (Fig. 4). Both machines can be loaded with either iridium ribbons or iridium wires. The number of applicators is as multifarious as for conventional manual afterloading (MEERTENS et al. 1988; THIEL et al. 1989). It may well be supposed that in the next decade the low dose rate remote afterloading will propagate similarly to the high dose rate remote technique.

4 Dose Prescription and Calculation

4.1 Classic Prescription Rules

The Quimby, Manchester, and Paris systems have already been mentioned above. The interested

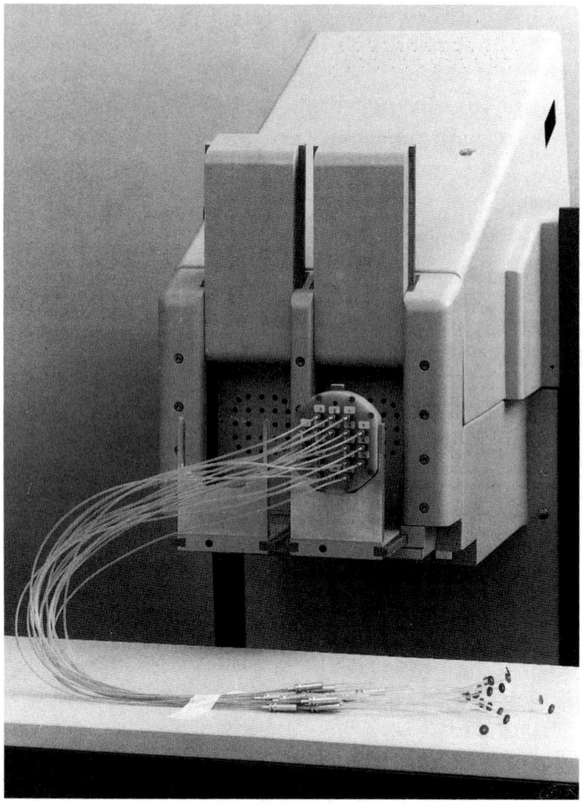

Fig. 4. Low dose rate afterloading machine Interpal C 38, with 2 × 19 channels and plastic application tubes

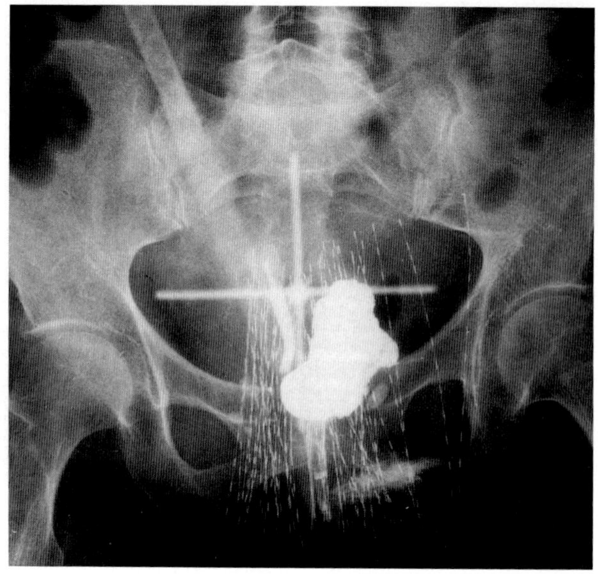

Fig. 5. Example of a gynecologic template application; the needles deviated because of the sturdiness of the tumor mass at the parametrium; even the divergent template (BERCHMANS et al. 1988) would fail in this case; underdosage at the tumor and hot spots in the middle are the consequences

reader is referred to the original papers and some textbooks (DUTREIX et al. 1982; GLASGOW and PEREZ 1987; GODDEN 1988; MEREDITH 1967; PIERQUIN et al. 1987). All of the distributions of sources inserted without a guiding technique show more or less significant deviations from the regular one, which may be planned after a certain system of rules. The same goes for flexible plastic guiding. Even the stiff steel guides arranged in a regular template may deviate seriously (Fig. 5).

This unsteadiness in source distribution requires a real spatial source reconstruction and dose calculation. Putting brachytherapy into practice, it is not possible to replace the calculation by any system of dosage tables. Since microcomputers are available in every radiotherapy department, a regular application in brachytherapy proceeds in a series of successive steps. One of them is the prospective planning of the number of sources, their spatial distribution, and amount of activity depending on tumor extension brachytherapy technique and additional therapy modalities. After application the retrospective planning supplies dose distributions and organ doses, making use of a real spatial source reconstruction. On the basis of these data the therapist determines the subsequent procedure as optimization by differential unloading for removable sources or, if necessary, additional application and defines the total treatment time in case of temporary application.

The method of dose prescription, calculation, and arguments for optimization depend on the technique of application. Generally speaking, there is a distinction between intracavitary, in particular gynecologic, application on the one hand and interstitial techniques on the other.

4.2 Gynecologic Application Rules

Speaking about dose prescription and absorbed doses in certain organs for intracavitary gynecologic applications in fact is speaking about the points A and B suggested by TOD and MEREDITH (1938; MEREDITH 1967). These points were created on anatomical structures but were related to the geometry of the position of the radium applicator itself. Nowadays, they are looked at in a slightly different manner (HORIOT, this volume; CHASSAGNE and HORIOT 1977; ICRU 1984; PEREZ et al. 1985).

4.3 Recent Rules and Nomographs

The rules which are briefly discussed here belong to interstitial application exclusively. The recently developed sources containing artificial nuclides, the different kinds of application, and the intraoperative insertion of a large number of sources make simplified rules necessary to be used at once in spite of computer simulation and retrospective dose calculation. Hence, especially the "free spaced" permanent application of, e.g., ^{125}I seeds and ^{198}Au grains requires such easy-to-use rules.

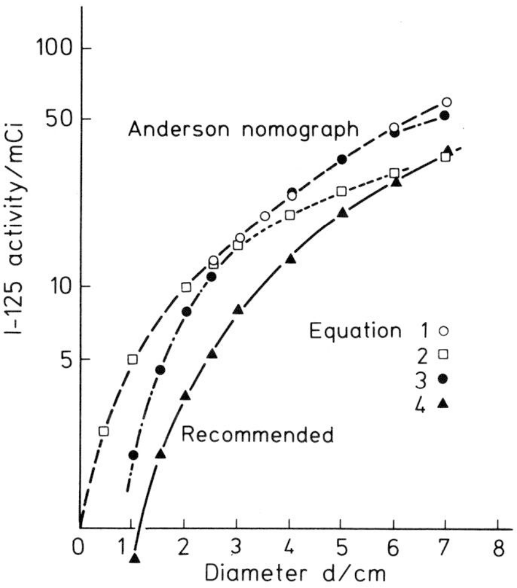

Fig. 6. Averaging dimension method for ^{125}I seeds with permanent implantation. *Upper curve* represents Anderson's nomograph resulting from three different equations (1a–c):

$$A = 5\ \text{mCi/cm} \cdot d \qquad d \leqslant 2.4\ \text{cm} \qquad (1a)$$
$$A = 3.87\ \text{mCi} \cdot \{d/\text{cm}\}^{1.293} \qquad 2.4\ \text{cm} < d \leqslant 3.24\ \text{cm} \qquad (1b)$$
$$A = 2.76\ \text{mCi} \cdot \{d/\text{cm}\}^{1.581} \qquad d > 3.24\ \text{cm} \qquad (1c)$$

Equation 2 is the simplified averaging method equal to Eqn. (1a).
Equation 4 is derived from the diffusion healing model (MÜLLER 1984) with

$$A = b\ \{d/2\ \text{cm}\}^{s} \cdot \text{MPD and} \qquad (4)$$
$$\text{MPD} = (D_0 - D_\infty)\ e^{-kd} + D^{\infty}$$
$$A = 1.4\ \{d/2\ \text{cm}\}^{2.4}\ (3e^{-0.33\text{cm}^{-1} \cdot d} + 1)\ \text{mCi}$$

Equation 3 is an approximation of ANDERSON's nomograph with the model of Eqn. (4) and $D_0/D_\infty = 10$. A, activity of ^{125}I; *d*, average diameter of target volume (cm); b, parameter to be fitted, approximately by the value $1.4\,\text{Gy}^{-1}$; MPD, minimal peripheral dose; D_0, tolerance dose for "zero-volume" approximately four times D_∞; D_∞, tolerance dose for "large volume" approximately of the value 60 Gy; k, parameter to be fitted, approximately by the value 0.33

4.3.1 *"Free Spaced" Interstitial Permanent Application*

In contrast to all classic radium application rules a prescription rule has been established for volume application with ^{198}Au grains (BUSCH 1966, 1977). Dose prescription rules in the form of nomographs have been developed at the Memorial Sloan Kettering Cancer Center in New York. ANDERSON and coworkers published these rules and their successive modifications (ANDERSON 1975, 1976; ANDERSON and AUBREY 1983). ANDERSON's nomographs started from a simple rule including averaging of the spatial dimension (see Eqn. 1, Fig. 6) (HENSCHKE and CEVEC 1968). The experience in Erlangen confirmed the conviction that the peripheral doses resulting from the nomographs are slightly overshooting, or rather, and with nearly the same consequence, the dose response to low dose rate irradiation – an iodine permanent implant gives an initial dose rate at the reference isodose from about 7 cGy/h – is underestimated. That is why an equation has been deduced for the activity to be implanted on the basis of a diffusion healing and tolerance model (Fig. 6, MÜLLER 1984).

In practice, there is no possibility of disentangling the radiobiological factors from the influence of physical and geometrical conditions. Hence, a prescription of dose is always the result of all these parameters. It is not out of the question that the rules may change in future. The fact that the rules are improved empirically by clinical experience makes them rather stable provided that defined modalities of application are adhered to.

4.3.2 *Template and Multiplanar Temporary Application*

The monoplanar application of removable sources is again the field for the classic application rules (Quimby, Manchester, Paris). But to tell the truth, monoplanar applications are a rarity reserved for the lip, vulva, skin, and very small breast. In general, tumors tend to grow in three dimensions, and they must be treated three dimensionally. Undoubtedly, the Paris system is favorable for volume implantation. The frequent comparison between the systems for monoplanar examples (GILLIN et al. 1984; GLASGOW and PEREZ 1987; PAUL et al. 1987, 1988) is of marginal interest. Recently, ANDERSON established spacing nomographs for mono- and multiplanar applicators of ^{192}Ir seeds in an approved manner (ANDERSON 1983b). They are as easy and quick to handle as his nomographs for iodine seeds.

The other class of prescription rules concerns specific source distributions of templates. In this context the Syed-Neblett Templates (see Fig. 3) have to be mentioned. Application techniques for these templates and the resulting dose distributions are described in a series of papers (ARISTIZABAL et al. 1985; KUMAR et al. 1986; MARTINEZ et al. 1985; PUTHAWALA et al. 1985; SYED et al. 1983). An optimization for the dose distribution can be achieved by differential unloading (SYED). Because of its flexibility the template application may be of interest for intracranial application, too (GUTIN et al. 1981, 1984; LEIBEL 1985). Chapter 2 of this volume contains some papers concerning this topic.

None of these prescription and spacing rules can replace a real source localization and dose calculation, not even with further refinement.

5 Source Localization and Three-Dimensional Reconstruction

5.1 Radiography

Radiography is the means for localization of sources inside an implant. The aim of localization in principle consists of a three-dimensional reconstruction of the position of the sources and the resulting dose distribution. In addition, it is necessary to know the spatial relation of the dose distribution to the organs and tumor. Conventional radiography allows the exact reconstruction of stiff applicators with a known geometry. For this reason orthogonal or stereo shift techniques are used. A "free spaced" multisource implant with the reduced spatial information of, e.g., orthogonal projections cannot be detected without doubt. Tomographic techniques achieve information which is more reliable, and mostly they allow the detection of organ contours. Unfortunately, the visual presentation of, e.g., CT information is dispersed in a series of different transverse sections.

5.1.1 Orthogonal and Stereo Shift Projection

It is standard for literature on the subject to deal with these localization techniques and algorithms (ANDERSON 1975, 1983a; GLASGOW and PEREZ 1987; GODDEN 1983). The nice trigonometric problem induced many physicists to create a computer code for localization and dose calculation. It would be superfluous to review details, however, it seems worthwhile to paraphrase some characteristics. The stereo shift techniques, first applied by MUSSEL (1956) and later by NUTTAL and SPIERS (1946), has turned out to be appropriate for planar implants or stiff applicator insertion. For real volume application the orthogonal projection technique is superior (EDMUNDSEN 1987; FITZGERALD and MANDERIL 1975; HUGHES 1956; LEETZ 1990; MOHAN 1981). The reliability of coordinate reconstruction can be improved by an isocentric multiplane reconstruction (ROSENTHAL and NATH 1983). As early as 1972 STOVALL and SHALEK reviewed the 20 computer codes developed at that time (STOVALL and SHALEK 1972). Most of them seemed to be acceptable (NEBLETT et al. 1978; POWERS et al. 1969). A volume implant may contain up to a hundred sources, and then it is comfortable and even better to leave it to a computer algorithm to decide on the coordinate matching. Hence, many of the innumerable programs utilize this feature (ALTSCHULER and FINDLY 1982; AMOLS and ROSEN 1981; BIGGS and KELLEY 1983; BULSKI and DADE 1986; EDMUNDSON 1987). Moreover, brachytherapy calculation sofware has to reconstruct not only the spatial distribution of the sources but also the dose distribution (see Sect. 6).

5.1.2 Tomography and Different Techniques

In principle, tomographic techniques collect the complete spatial information of a certain implant. At the time when no computed tomograph (CT) was available, PIERQUIN and FAYOS (1962) described a method of dose calculation using conventional transversal tomography (PIERQUIN et al. 1960, 1978, pp 68–77). BOISSERIE and MARINELLO developed a code using a micro computer, which was adapted to the Paris dosage system and line sources (BOISSERIE and MARINELLO 1979). This code is no longer up-to-date because it neglected the curvature of the wires. In 1983, WILKINSON et al. described a method of localization and dose calculation on CT image information for intracavitary treatment. Everyone will agree that CT must improve the accuracy of calculation, but up to now no such established method has been developed. This seems due to some practical reasons. For most radiotherapy departments it would be difficult, time-consuming, and expensive to obtain CT scans narrow enough (4 mm) with appropriate topograms (anterior and lateral) for further calculation. It seems not worthwhile achieving this accuracy. A different point is the ability of CT pictures to visualize organ contours and the tumor as well (MÜLLER et al. 1986).

Sonography, a further imaging technique, has been used as a guiding tool for, e.g., perineal access to the prostate for template application (LEE et al. 1988; NAG 1985). It seems to be unsuitable for postimplantation imaging, localization, and dose calculation. The same may also apply to the recently developed imaging technique without ionizing radiation, magnetic resonance imaging.

6 Calculation of the Dose Distribution

The basic data for calculation of two- or even three-dimensional dose distributions in principle are derived from those around the single sources. The measurement of these data is more complex than it seems at first view (WOOD 1981). Especially for a "free spaced" implant with sources of less than 1 mCi of ^{198}Ir (temporary) or ^{125}I (permanent) the mean distance between the sources is about 1 cm. Hence inside the implant, dose levels in the nearest proximity of the sources (under 0.5 cm) are of importance. It is hardly possible to measure these values (GEBHARDT and MÜLLER 1988, 1989; HINE and FRIEDMAN 1950). That is why up to now the best data have been obtained by calculation, calibrated to a certain point distant from the center point of the source. An isotropic source of the dimension of a mathematical point can be calculated most easily.

6.1 Point Sources

A simplified equation for the isotropic distribution of dose rate around such a hypothetical source has the following form:

$$\dot{D}(r) = A\Gamma_\delta r^{-2}\, g(r,M)$$

where $\dot{D}$ is the dose rate at the distance r; A is the source activity; Γ_δ is the air kerma rate constant, for all photons greater than a certain energy δ; r is distance; $g(r,M)$ is the build up and attenuation function inside medium M.

All these complex effects taken together are put into the function $g(r,M)$ for the complex photon emission spectrum. This $g(r,M)$ is usually described by a power expansion in r (MEISBERGER et al. 1968). Many models have been developed to account for these effects concerning a source inside tissue (BATHO and YOUNG 1964, 1967; HALE 1958; SOMOCOVITIS et al. 1967). Some of them were reviewed by GLASGOW and PEREZ (1987, p 222). Recent Monte Carlo calculations by DALE and WILLIAMSON (DALE 1982, 1983, 1986; WILLIAMSON 1986; WILLIAMSON et al. 1983b) assembled improved Meisberger coefficients for a number of sources in diverse tissues. For most isotopes and especially for ^{192}Ir these correction terms have a minimal influence on the dose distribution up to a distance of 5 cm or more, because of compensating effects of scattering and attenuation. Other influences such as the unisotropic source design, filtration, and source calibration are much more important.

6.2 Line Sources

Wires made of ^{192}Ir are most common, and there is also some interest in ^{182}Ta. HALL et al. worked out dosimetry for these sources (HALL et al. 1966). DUTREIX and colleagues investigated ^{192}Ir wires as used in the Paris system of application (DUTREIX et al. 1982; DUTREIX and MARINELLO 1987). All the output graphs, tables, and the Paris "escargot curve" of dose distribution apply to planes perpendicular to line sources extending only in one linear dimension (WELSH et al. 1983). The real localization of such line sources, including their curvature, is considered to be the state of the art (MAYLES et al. 1985). The calculation of the resulting dose distribution either utilizes the approximation of a chain of point sources (DUTREIX and WAMBERSIE 1968) or makes use of a dissection of a curvy line source into a polygon of straight parts (MAYLES et al. 1985). The latter is the common way. Thus the problem is reduced to a special case in the next section.

6.3 Tubes and Others

The sources first used in brachytherapy were sealed tubes of glass, gold, and platinum (filtration of the α- and β-components of radiation). In the year 1921 SIEVERT published an analytical solution to the problem of a cylindrical radium container. He also took the attenuation in the container wall into account and recognized the problem of the complex gamma ray spectrum emitted by ^{226}Ra, by separating the spectrum into three weighted parts of appropriate energy. In principle, SIEVERT solved the problem of unisotropic source design. While the point source calculation (Sect. 6.1) can be reduced to a 1-dimensional problem depending on the distance r only, SIEVERT's solution is 2-dimensional. For a point under calculation it contains the distance from the source axis b and a secant integral for the angle under which the source becomes effective in cylinder coordinates. SIEVERT distinguished four calculating zones. The zone nearly perpendicular to the axis of the source shows a modest integral form. The others are more complex. In general the dose rate formula from Sect. 6.1 can be replaced by:

$$\dot{D}(b,\theta) = A \cdot \Gamma_{\delta}(l \cdot b)^{-1} \, g(b,\theta,M_1,M_2)$$

where l is the length of active source; b is the distance of the calculation point to the source axis; θ is the angle under which active material appears.

In addition, the complex build-up and attenuation function g contains the azimuth angle θ (Sievert's integrals) and the attenuation in the wall material (M_1) and tissue (M_2) as well. Sievert's analytical solution is the basis of most dose calculation systems even though it was modified and refined by numerous authors. The reader interested in this topic is referred to some original papers (e.g., BATHO and YOUNG 1964; EVANS 1968; MEISBERGER et al. 1968; SHALEK and STOVALL 1968; SIEVERT 1921; WILLIAMSON et al. 1983a) and to WOOD's GODDEN's and SHALEK and STOVALL's excellent reviews (GODDEN 1988 pp 41–75; SHALEK and STOVALL 1990; WOOD 1981, pp 51–65).

Tubes other than of standardized ^{226}Ra are in principle accessible for the analogous calculations described above. For ^{137}Cs-encapsulated sources KRISHNASWAMY and BREITMAN calculated dose distributions dividing the active length into a sum of point sources (KRISHNASWAMY 1972a; BREITMAN 1974; WAGGENER et al. 1989).

DUTREIX and WAMBERSIE (1968) considered the dose distribution around so-called source trains and simulated them under certain conditions by using line sources. Following CASSELL's method of calculation, Amersham International published a variety of source train dose distributions (CASSELL 1983).

^{125}I seeds present serious unisotropy because of the low energy of the emitted photons and of the inhomogeneous design (LING et al. 1979, 1983; HARTMANN et al. 1983). Despite this fact, ANDERSON recommends taking iodine seeds as point sources because of practical considerations (ANDERSON et al. 1981a), provided that proper dose rate and attenuation factors are used (ANDERSON et al. 1981b; KRISHNASWAMY 1978; LING et al. 1983; MOHAN and ANDERSON 1978; MÜLLER 1984).

The dose calculation for ^{252}Cf seeds is complicated because of the relation between neutron dose and photon dose depending on the distance. In the early 1970s KRISHNASWAMY and BLOCH published calculated data (Monte Carlo) for point sources (KRISHNASWAMY 1971; Bloch et al. 1972). ANDERSON and SHAPIRO substantiated the dosimetry of this isotope (ANDERSON 1973, 1975; SHAPIRO et al. 1976).

Even if there is no appropriate algorithm for a certain source design, dose distributions for an implant can be calculated if measured dose patterns of single sources are utilized. Some programs support this alternative. Single source dose distribution can be stored as a dose matrix which can be used for a three-dimensional superposition of a real implant.

Last but not least, Monte Carlo techniques have to be mentioned. For dose calculations of an individual implant this method is still too time consuming. However, they are well established for basic calculations of the dose distribution of point sources (WEBB and FOX 1979; DALE 1982, 1983) and also for real source design (WILLIAMSON 1986, 1988; BURNS and REASIDE 1987, 1988).

7 Reference Points and Reference Isodose

The dose distribution of an implant shows steep gradients at the periphery and outside that space in which the radioactive sources are situated. This space of interest is not easy to define, and in any case the dose distribution inside is extremely inhomogeneous. The situation is wholly in contrast to the well-known conditions for percutaneous irradiation. Hence, it is of the utmost importance to establish a system of dose specification adequate for each class of application technique. For intracavitary application this has been approximately achieved with ICRU 38 (1984). For a "free spaced" application in interstitial therapy a different form of dose specification becomes necessary.

The dose specification using special points such as those common in gynecologic intracavitary application (Quimby, Manchester, Paris) is convenient from a practical point of view, but it is highly insufficient for defining the tumor dose and equally the dose and risk for critical organs. Therefore, a dose-volume analysis should be applied in every case. This dose-volume analysis has not yet been completely developed (ANDREW et al. 1985; ANDERSON 1986; PAUL et al. 1987, 1988; MECHTEL and MÜLLER 1988; NEBLETT et al. 1985; SAW and SUNTHARALINGAM 1988; WU et al. 1988). HORIOT's contribution in this volume discusses this problem in detail and deals with specific conditions of intracavitary application.

8 Combination with Teletherapy

Brachytherapy is an effective kind of radiotherapy, locally restricted to a volume from several cubic centimeters to a few hundred. This makes it suitable as a local boost for radiotherapy. On the other hand, brachytherapy applications in most cases demand supplemental percutaneous radiotherapy. The resulting combined treatment modalities raise a lot of questions. A suitable pattern for the spatial and chronological combination has to be determined. The decision for a certain treatment regime is highly influenced by clinical and radiobiological facts and considerations. As brachytherapy produces very different patterns of dose distribution and time course, the influence of these parameters on the dose response has to be considered and investigated.

8.1 Dose, Time, and Volume Dependent Response

This topic is discussed in principle and in detail by KUMMERMEHR (this volume), which is why, in this chapter, only a short statement will be made. Brachytherapy typically offers a wide range of dose rate and volume. In fact, the dose response of tumors and critical organs does correlate sensitively with these factors. Since the first application of radioactive sources in radiotherapy, that is the beginning of brachytherapy, this radiation response has been under investigation. In particular the comparison and combination of brachy- and teletherapy has been of interest. In the early 1920s SEITZ and WINTZ from Erlangen published rules for the biological dosimetry (SEITZ and WINTZ 1920). This was followed by a series of bioeffect systems. There is the area-volume-time nomogram from PATERSON (PATERSON 1963, p 35), which was evolved and cited by ELLIS (ELLIS 1971; ELLIS and SORENSEN 1974) and modified for use in brachytherapy by ORTON (1974, 1980) and KIRK et al. (1972). A somewhat different model was derived by DALE (1985) on the basis of the linear quadratic cellular dose response equation, which had been initiated by DOUGLAS and FOWLER (1976). The controversy among the scientific adherents to either system will amuse future generations. The systems are congruent for a certain interval of dose per fraction as KELLERER (1984) showed. It must be confessed that at present there does not exist any weighing system which would be able to allow us to superimpose all the different time regimes which are used in practice. What is more, all the established weighing systems should be applied with caution or not at all. Each additional correction term bears a suspicious fact. Hence, it might prove that the corrections for repair kinetics and cell proliferation, which depend on the dose rate, led the wrong way (LIVERSAGE 1969; KELLERER 1984). PIERQUIN and coworkers (1973, 1985) and INOUE and colleagues (1978) showed that there is no need for a dose rate correction in the range from about 0.3 to 1 Gy/h. This clinical experience could be better understood on the basis of a generalized linear quadratic model which has recently been proposed by MÜLLER (1985, 1989).

Besides these essential questions mentioned above, the geometrical matching of the dose distributions of the brachy- or the tele- part of the therapy regime seems rather simple. Nevertheless, the textbooks usually dedicate considerable space to this problem, especially as far as gynecologic applications are concerned (GODDEN 1988, pp 219–228; PEREZ 1987, pp 919–965; FLETCHER 1980, pp 720–828).

9 Experimental and Future Techniques

9.1 New Isotopes and Sources

Brachytherapy in this cosntext is thought to include all the techniques in radiotherapy in which a radiation source is brought into intimate contact with the target volume. Ignoring intraoperative and contact treatment with technically generated X-ray or particle beams, brachytherapy is restricted to the application of radionuclides. It can hardly be anticipated that anything new will occur in radioisotopes. The full scale of emitted particles, beam qualities, and half-life values suitable for radiotherapy has been tested. ^{252}Cf with a neutron component has been in use since the early 1970s (Castro et al. 1973). Modern source design containing, e.g., the isotopes ^{241}Am, ^{103}Pd, ^{75}Se, ^{145}Sm, or ^{169}Yb will hardly achieve a breakthrough. For further information the reader is referred to the supplement edited by Trott (1987).

9.2 Monoclonal Antibody Carrier

The idea to utilize the specific interaction of some antibodies with a group of tumor-associated antigens seems to be convincing. After a series of diagnostic tests had been established, first trials in tumor therapy were launched in the early 1980s (Hammersmith Oncology Group 1984; Leichner et al. 1981, 1984; Order 1981, 1984). Even if remissions were registered, there is one fact which can hardly be overcome: the low specific activity absorbed in the tissue. It is too low by some orders of magnitude (Leichner et al. 1984, 1990; Langmuir and Sutherland 1988). Thus radioimmunotherapy will not replace brachytherapy in the near future nor will it ever. By the way, a radioactive antibody is an unsealed source belonging to nuclear medicine as stated above (Sect. 2.2). For more information in detail the reader is referred to the recently published proceedings of the Second Conference on Radioimmunodetection and Radioimmunotherapy of Cancer (Goldenberg 1990). Particularly the contributions of Wessels (1990), Order et al. (1990), Williams et al. (1990), and Siegel et al. (1990) are of interest in this context.

9.3 Adjuvant Modalities

All these tumor entities which are not resectable in toto (R0) but are candidates for brachytherapy and even accessible by a certain technique, are only curable by irradiation if at all. All the adjuvant modalities to be mentioned here only slightly modulate the local tumor control probability, unless the gain of the adjuvant may be expected a priori in the systemic efficacy. The consequence of this statement, to which at least the radiotherapist should agree, is: There does not exist any compromise in dose for curative therapy even if an adjuvant is used. Exactly this essential will be suspended for palliation.

9.3.1 Hyperthermia

Hyperthermia and especially IHT are considered and reviewed in detail in Chapter 9 of this volume. There is no need to extend this topic here.

9.3.2 Chemical Modifiers

"The possibilities for interaction between chemotherapy drugs and radiation are limitless and it is not reasonable to discuss specific examples in this chapter. The reader may wish to refer to the proceedings of conferences devoted to the subject", Rodney Withers (1987), who made mention of some proceedings in the *Int J Radiat Oncol Biol Phys* (1982, 1984, 1986).

It is assumed that brachytherapy offers no additional aspects concerning sequential radiochemotherapy. In contrast, the simultaneous application of a chemical modifier during a long-term course of low dose rate brachytherapy may be thought to be an encouraging idea. Interest has been focussed on the radiosensitizing and radioprotecting drugs. Most experience comes from the so-called hypoxic cell radiosensitizers. These have been collected in clinical trials initiated by the Radiation Therapy Oncology Group (RTOG) and International Clinical Trials in Radiation Oncology (ICTRO) (Rubin et al. 1979, 1988). The results achieved in the trials, always combining fractionated percutaneous radiotherapy with modifiers like misonidazole or its analogues, have been completely discouraging. Up to now no substance has been found which would satisfy the therapist's expectations. It must be conceded indeed that the most intensively tested drug, misonidazole, was likely to be one of those less suitable.

In the same way, no valid notice exists of a therapeutical benefit following from the application of radioprotecting drugs. For a more extensive discussion of this topic the reader is referred to the excellent survey of WASSERMAN and KLIGERMAN (1987) and a series of suitable papers edited by HILL and BELLAMY (1990).

References

Altschuler MD, Findley PA (1982) Rapid accurate three-dimensional location of multiple seeds in implant radiotherapy. Med Phys 9: 612

Amols HI, Rosen II (1981) A three-film technique for reconstruction of radioactive seed implants. Med Phys 8: 210–214

Anderson LL (1973) Status and dosimetry for Cf-252 medical neutron sources. Phys Med Biol 18: 779–799

Anderson LL (1975) Dosimetry for interstitial radiation therapy. In: Hilaris BS (ed) Handbook of interstitial brachytherapy. Memorial Sloan Kettering Cancer Center, Publishing Science Group, Acton, pp 87–115

Anderson LL (1976) Spacing nomograph for interstitial implants of 125-I seeds. Med Phys 3: 48–51

Anderson LL (1983a) Remote afterloading in cancer management, part I. Afterloading design and optimization potential. In: Hilaris BS, Batata MA (eds) Brachytherapy oncology. Memorial Sloan Kettering Cancer Center, New York, pp 93–100

Anderson LL (1983b) Experiences with Ir-192. In: Wright AE, Boyer AL (eds) Advances in radiation therapy treatment planning. American Institute of Physics, New York

Anderson LL (1985) Physical optimization of afterloading techniques. Strahlentherapie 161: 264–269

Anderson LL (1986) A "natural" volume-dose histrogram for brachytherapy. Med Phys 13: 898–903

Anderson LL, Aubrey A (1983) Computerized dosimetry for I-125 prostate implants. In: Hilaris BS, Batata MA (eds) Brachytherapy oncology 1983. Memorial Sloan Kettering Cancer Center, New York, pp 57–63

Anderson LL, Wagner LK, Schauer TH (1981a) Memorial Hospital methods of dose calculation for 192-Ir. In: George FW (ed) Modern interstitial and intracavitary radiation cancer management. Masson, New York, pp 1–7

Anderson LL, Kuan HM, Ding IY (1981b) Clinical dosimetry with I-125. In: George FW (ed) Modern interstitial and intracavitary radiation cancer management, Masson, New York, pp 8–15

Andrew W, Zwicker RD, Sernick ES (1985) Tumor dose specification of I-125 seed implants. Med Phys 12: 27–31

Arai T (1978) Relationship between total isoeffect dose and number of fractions for the treatment of uterine cervical carcinoma by high dose rate intracavitary irradiation. Working party on the use of radionuclides and afterloading techniques in the treatment of cancer of the uterus. High Dose Workshop, London

Aristizabal SA, Valencia A, Ocampo G, Surwit E (1985) Interstitial parametrial irradiation in cancer of the cervix stage IIB–IIIB. Endocurie Hyperthermia Oncol 1: 41–48

Burns GS, Raeside DE (1987) Monte Carlo simulation of the dose distribution around 125-I seeds. Med Phys 14: 420–424

Burns GS, Raeside DE (1988) Two-dimensional dose distribution around a commercial 125-I seed. Med Phys 15: 56–60

Batho HF, Young MEJ (1964) Tissue absorption corrections for linear radium sources. Br J Radiol 37: 689–692

Batho HF, Young MEJ (1967) A revised table of tissue correction factors for linear radium sources. Br J Radiol 40: 785

Berchmans J, Scarbrough EC, Nguyen PD, Antich PP (1988) A diverging gynecological template for radioactive interstitial/intracavitary implants of the cervix. Int J Radiat Oncol Biol Phys 11: 461–465

Biggs PJ, Kelly DM (1983) Geometric reconstruction of seed implants using a three-film technique. Med Phys 10: 701–704

Bloch P, Krishnaswamy V, Hale J (1972) Dose tables for californium-252 implants. Am J Roentgenol 115: 822–833

Boisserie G, Marinello G (1979) Calcul automatique de la dose de base dans le Système de Paris, J Radiol 60: 327–332

Breitman K (1974) Dose rate tables for clinical Cs-137 sources sheated in platinum. Br J Radiol 47: 657–664

Bulski W, Dade M (1986) Treatment planning software for afterloading brachytherapy. Radiother Oncol 5: 59–64

Busch M (1966) Ein Dosierungsschema für die interstitielle Gammatherapie. Strahlentherapie [Sonderb] 64: 213–218

Busch M (1977) Dosierung bei interstitieller Therapie mit umschlossenen Gammastrahlern. Strahlentherapie 153: 589–593

Casebow MP (1971) The calculation and measurement of exposure distribution from Co-60 ophthalmic applicators. Br J Radiol 44: 618–624

Cassell KJ (1983) A fundamental approach to the design of a dose-rate calculation program for use in brachytherapy planning. Br J Radiol 56: 113–119

Castro JR, Oliver GD, Withers HR, Almond PR (1973) Experience with Californium-252 in clinical radiotherapy. Am J Roentgenol 117: 182–194

Chan B, Rotman M, Randall GJ (1972) Computerized dosimetry of Co-60 ophthalmic applicators. Radiology 103: 705–707

Chassagne D, Delouche G, Rocoplan JA, Pierquin B, Gest J (1969) Description et premieres essais du Curietron. J Radiol Electrol 50: 910–913

Chassagne D, Horiot JC (1977) Positions pour une définition commune des points de référence en curiethérapie gynécolographique. J Radiol Electrol 58: 371–375

Coffrey C, Sayeg J, Beach L, Song S, Landis C, Connor A (1981) Calibration of surface dose rate for a Sr-90 beta applicator: comparison of external, theoretical, and biological methods. Med Phys 8: 558

Dale RG (1982) A Monte Carlo derivation of parameters for use in the tissue dosimetry of medium and low energy nuclides. Br J Radiol 55: 748–757

Dale RG (1983) Some theoretical derivations relating to the tissue dosimetry of brachytherapy nuclides, with particular reference to iodine 125. Med Phys 10: 176–183

Dale RG (1985) The application of the linear-quadratic dose-effect equating to fractionated and protracted radiotherapy. Br J Radiol 58: 515–528

Dale RG (1986) Revisions to radial dose function data for 125-I and 131-Cs. Med Phys 13: 963–964

Del Regato JA (1978) Brachytherapy. In: Vaeth JM (ed) Renaissance of interstitial brachytherapy. Karger, Basel, pp 5–12

Delclos L (1978) Are interstitial radium applications passe? Front Radiat Ther Oncol 12: 42–56

Douglas BG, Fowler JF (1976) The effect of multiple small doses of X-rays on skin reactions in the mouse and a basic interpretation. Radiat Res 66: 401–426

Duane W (1915) On the extraction and purification of

radium emanation. Phys Rev 5: 311–314
Dutreix A, Marinello G (1987) Source localization and dose calculation methods. In: Pierquin B, Wilson JF, Chassagne D (eds) Modern brachytherapy. Masson, New York, pp 17–24
Dutreix A, Wambersie A (1968) Étade de la reparation des doses autour de sources poncuelles alignées application en curietherapie gynécologiques. Acta Radiol 7: 389–400
Dutreix A, Marinello G, Wamberie A (1982) Dosimétrie en curiethérapie. Masson, Paris
Edmundson CK (1987) Requirements for and quality assurance of computer treatment planning systems for brachytherapy. In: Kereiakes JG, Elson HR, Bom CG (eds) Radiation oncology physics 1986. American Institute of Physics, New York, pp 700–713 (Medical physics monograph no 15)
Ellis F (1971) Nominal standard dose and the ret. Br J Radiol 44: 101–108
Ellis F, Sorensen A (1974) A method of estimating biological effect of combined intracavitary low dose rate radiation with external radiation in carcinoma of the cervix uteri. Radiology 110: 681–686
Evans RD (1968) X-ray and γ-ray interactions. In: Attix FH, Roesch WC (eds) Radiation dosimetry, 2nd edn, vol I. Academic Paris, pp 94–155
Failla G (1926) The development of filtered radon implants. Am J Roentgenol 16: 507–526
Findlay PA, Wright DC, Rosenow U, Harrington FS, Miller RW (1985) 125-I interstitial brachytherapy for primary malignant brain tumors: technical aspects of treatment planning and implantation methods. Int J Radiat Oncol Biol Phys 11: 2021–2026
Fitzgerald LT, Mauderli W (1975) Analysis of errors in three-dimensional reconstruction of radium implants from stereo radiographs. Radiology 115: 455–458
Fletcher GH (1953) Cervical radium applicators with screening in the direction of bladder and rectum. Radiology 60: 77–83
Fletcher GH (ed) (1980) Textbook of radiotherapy, 3rd edn. Lea and Febiger, Philadelphia
Fletcher GH, Shalek KJ, Cole A (1953) Cervical radium applicators with screening in the direction of bladder and rectum. Physical Study. Radiology 60: 77–84
Gebhardt E, Müller RG (1988) Relative Dosismessungen im Nahfeld von Jod-125- und Ir-192-Seeds. In: Nüsslin F (ed) Medizinische Physik 1988. Deutsche Gesellschaft für Medizinische Physik, pp 446–500
Gebhardt E, Müller RG (1989) Dosisverteilung in der Umgebung radioaktiver Strahler der interstitiellen und intrakavitären Therapie. In: Leetz, HK (ed) Medizinische Physik 1989, Deutsche Gesellschaft für Medizinische Physik, pp 377–381
George FW (ed) (1981) Modern interstitial and intracavitary radiation management. Masson, New York
Gillin MT, Kline RW, Wilson JF (1984) Single and double plane implants: a comparison of the Manchester system with the Paris system. Int J Radiat Oncol Biol Phys, 10: 921–925
Glasgow GP, Perez CA (1987) Physics in Brachytherapy. In: Perez CA, Brady LU (eds) Principles and practice of radiation oncology. Lippincott, Philadelphia
Godden TJ (1988) Physical aspects of brachytherapy. Hilger, Bristol (Medical physics handbooks, vol 19)
Goffinet DR, Martinez A, Pooler D, Palos B (1978) Perineal brachytherapy. Renaissance of interstitial brachytherapy. Front Radiat Ther Oncol 12: 119–135
Goldenberg DM (ed) (1990) Second conference on radioimmunodetection and radioimmunotherapy of cancer, Princeton 1988. Cancer Res 50 [Suppl 3]: 1
Gutin PH, Phillips TL, Hosobuchi Y et al. (1981) Permanent and removable implants for the brachytherapy of brain tumors. Int J Radiat Oncol Biol Phys 7: 1371–1381
Gutin PH, Phillips TL, Wara WM, Leibel SA, Hosobuchi Y, Levin VA, Weaver KA, Lamb S (1984) Brachytherapy of recurrent malignant brain tumors with removable high-activity iodine-125 sources. J Neurosurg 60: 61–68
Hahn PF (ed) (1956) Therapeutic use of artificial radioisotopes. Wiley, New York
Hale J (1958) The use of interstitial radium dose rate tables for other radioactive isotopes. Am J Roentgenol 79: 49–53
Hall EJ, Oliver R, Shepstone BJ (1966) Routine dosimetry with tantalum-182 and iridium-192 wires. Acta Radiol 4: 155–160
Hammersmith Oncology Group and Imperial Cancer Research Fund (1984) Antibody-guided irradiation of malignant lesions: three cases illustrating a new method of treatment. Lancet: 1441
Hartmann GH, Schlegel W, Scharfenberg H (1983) The three-dimensional dose distribution of 125-I seeds in tissue. Phys Med Biol 28: 693–699
Henschke UK (1956) Interstitial implantation with radioisotopes. In: Hahn PF (ed) Therapeutic use of artificial radioisotopes. Wiley, New York, pp 375–397
Henschke UK (1960) "Afterloading" applicator for radiation therapy of the carcinoma of the uterus. Radiology 74: 834
Henschke UK, Cevec P (1968) Dimension averaging a simple method of dosimetry of interstitial implants. Radiobiol Radiother 9: 287–298
Henschke UK, James AG, Myers WG (1953) Radiogold seeds for cancer therapy. Nucleonics 11: 46–48
Henschke UK, Hilaris B, Mahan GD (1963) Afterloading in interstitial and intracavitary radiation therapy. Am J Roentgenol Radium Ther Nucl Med 90: 386–395
Henschke UK, Hilaris BS, Mahan GD (1966) Intracavitary radiation therapy in cancer of the uterine cervix by remote afterloading with cycling sources. Am J Roentgenol 96: 45–51
Heyman J (1929) The technique in the treatment of cancer uteri at Radiumhemmet. Acta Radiol X: 49–64
Heyman J (1935) The so-called Stockholm method and the results of treatment of uterine cancer at Radiumhemmet. Acta Radiol XVI: 129–148
Heyman J, Reuterwall O, Benner S (1941) The Radiumhemmet experience with radiotherapy in cancer of the corpus of the uterus: classification method of treatment and results. Acta Radiol 22: 11–98
Hilaris BS (ed) (1975) Handbook of Interstitial Brachytherapy. Publishing Sciences Group, Acton
Hilaris BS, Henschke UK, Holt JG (1968) Clinical experience with long half-life and low energy encapsulated radioactive sources in cancer radiation therapy. Radiology 91: 1163–1167
Hill BT, Bellamy AS (eds) (1990) Antitumor drug-radiation interactions. CRC Press, Boca Raton
Hine GJ, Friedman M (1950) Isodose measurements of linear radium sources in air and water by means of an

automatic isodose recorder. Am J Roentgenol 64: 989–998

Hughes HA (1956) Accuracy of foreign body localization from "Tube-Shift" radiographs. Br J Radiol 29: 116–119

ICRU, Report 38 (1984) Dose and volume specification for reporting intracavitary therapy. International Commission on Radiological Units and Measurements, Bethesda

Inoue T, Hori S, Miyata Y, Shigematsu Y, Fuchihata H, Tanaka Y (1978) Dose and dose rate in Ir-192 interstitial irradiation for carcinoma of the tongue. Acta Radiol [Oncol] 17: 27–32

Int J Radiat Oncol Biol Phys Chemical modifiers of cancer treatment. Proceedings (1982) 8: 323–815 (1984) 10: 1161–1814 (1986) 12: 1019–1545

Joleson I, Bäckström A (1969) Dose rate measurement in bladder and rectum. Acta Radiol 8: 343–359

Jones CH, Dermentzoglou F (1971) Practical aspects of Sr-90 ophthalmic applicator dosimetry. Br J Radiol 44: 203–210

Kellerer AM (1984) Verallgemeinerung des NSD-Konzepts auf Multifraktionierung sowie intracavitäre und interstitielle Therapie. In: Schmidt T (ed) Medizinische Physik pp 97–108

Kim JH, Hilaris B (1975) Iodine 125 sources in interstitial tumor therapy – clinical and biologic considerations. Am J Roentgenol 123: 163–169

Kirk J, Gray WM, Watson ER (1972) Commulative radiation effect. II. Continuous radiation therapy – Long-lifed sources. Clin Radiol 23: 93–105

Krishnaswamy V (1972a) Dose distributions about ^{137}Cs sources in tissues. Radiology 105: 181–184

Krishnaswamy V (1972b) Calculated debth does tables for californium-252 sources in tissue. Phys Med Biol 17: 56–63

Krishnaswamy V (1978) Dose distribution around an I-125 seed source in tissue. Radiology 126: 489–491

Kumar PP, Good RR, Hussian MB, Bartone FF (1986) Simple, accurate, safe and cost-effective percutaneous transperineal template technique for permanent 125-Iodine interstitial brachytherapy of prostate cancer. Strahlenther Onkol 162: 713–719

Langmuir VK, Sutherland RM (1988) Dosimetry models for radioimmunotherapy. Med Phys 15: 867–873

Lee F, Torp-Pedersen S, Meiselman L, Siders DB, Littrup P, Dorr RP, Pauly F (1988) Transrectal ultrasound in the diagnosis and staging of local disease after I-125 seed implantation for prostate cancer. Int J Radiat Oncol Biol Phys 15: 1453–1459

Leetz HK, Vogelsang U (1989) Unsicherheiten bei der Strahlenquellenlokalisation in der Brachytherapie. Strahlenther Onkol 165: 807–812

Leibel SA (1985) Interstitial implantation for the treatment of malignant brain tumors. Astro refresher course no. 402

Leichner PK, Klein JL, Garrison JB et al. (1981) A model for radioimmunoglobulin dosimetry. Int Radiat Oncol Biol Phys 7: 323–333

Leichner PK, Klein JL, Fishman EK et al. (1984) Comparative tumor dose from I-131 labeled polyclonal antiferritin, anti-AFP, and anti-CEA inprimary liver cancer. Cancer Drug Deliv 1: 321–328

Leichner PK, Yang NC, Wessels BW, Hawkins WG, Order SE, Klein JL (1990) Dosimetry and treatment planning in radioimmunotherapy. Front Radiat Ther Oncol 24: 109–120

Ling CC, Anderson LL, Shipley WU (1979) Dose inhomogeneity in interstitial implants using 125-I seeds. Int J Radiat Oncol Biol Phys 5: 419–425

Ling CC, Yorke ED, Spiro IJ, Kubiatowicz D, Bennett D (1983) Physical dosimetry of I-125 seeds of a new design for interstitial implant. Int J Radiat Oncol Biol Phys 9: 1747

Liversage WE (1969) A general formula for equating protracted and acute regimes of radiation. Br J Radiol 42: 432–440

Loevinger R, Japha EM, Browell GL (1956) Discrete radioisotope sources. In: Hine GJ, Brownell (eds) Radiation Dosimetry. Academic, New York

Lommatzsch P (1977) Die theapeutische Anwendung von ionisierenden Strahlen in der Augenheilkunde. Thieme, Leipzig

Lommatzsch PK (1983) Beta irradiation with Ru 106/Rh 106. Applicators of choroidal melanomas: sixteen years experience. In: Lommatzsch PK, Blodi LFC (eds) Intraocular Tumours. Springer-Verlag Berlin pp 355–363

Magnus L, Gobbeler T, Strotges W (1968) Tiefendosisberechnung für die Co-60-Augenapplikatoren CKA 5-11 (nach Stallard). Strahlentherapie 136: 170–177

Martinez A, Edmundson GK, Cox RS, Gunderson LL, Howes AE (1985) Combination of external beam irradiation and multiple-site perineal applicator (MUPIT) for treatment of locally advanced or recurrent prostatic, anorectal, and gynecologic malignancies. Int J Radiat Oncol Biol Phys 11: 391–398

Mayles WPM, Mayles HMO, Turner PC (1985) Physical aspects of interstitial therapy using flexible iridium-192 wire. Br J Radiol 58: 529–535

Mc Kay A, Gutin P, Hosobuchi et al. (1981) CT-stereo taxis and interstitial radiation for brain tumour. In: Moss AA, Goldberg HI (eds) International radiologic techniques: computerized tomography and ultrasonography University of California Printing Department, Berkley, pp 93–99

Mechtel M, Müller RG (1988) Rechnerische Kriterien zur Beurteilung von Quellenverteilungen in der interstitiellen Therapie. In: Nüsslin F (ed) Medizinische Physik. Deutsche Gesellschaft für Medizinische Physik, pp 441–445

Meertens H, Bartelink H, Minderhoud T (1988) First clinical experience with a remote afterloading system for low dose rate interstitial breast implants. Radiother Oncol 11: 387–393

Meisberger LL, Keller RJ, Shalek RJ (1968) The effective attenuation in water of the gamma-rays of gold-198, iridium-192, caesium-137, radium-226 and cobalt-60. Radiology 90: 953–957

Meredith WJ (ed) (1967) The Manchester system, 2nd edn. Livingston, Edinburgh

Meredith WJ, Stephenson SK (1945a) The calculation of dosage and an additional distribution rule of cylindrical "volume" implantations with Radium. Br J Radiol XVIII: 45–47

Meredith WJ, Stephenson SK (1945b) The use of radiographs for dosage control in interstitial gamma-ray therapy. Br J Radiol XVIII: 86–91

Mohan R, Anderson LL (1978) In: BRACHY II, Interstitial and intracavitary dose computation program user's guide. Memorial Sloan-Kettering Cancer Center, New York

Mohan R (1981) Computers in brachytherapy dose computation – the Memorial System. In: Shearer DR (ed) Recent advances in brachytherapy physics. Medical

Physics Monograph no 7, American Association Institute of Physics, New York pp 134–143
Müller RG (1984) Bestrahlungsplanung bei der interstitiellen Therapie. In: Schmidt T (ed) Medizinische Physik. Deutsche Gesellschaft für Medizinische Physik, pp 87–96
Müller RG (1985) A cell-kinetic model for dose response to low dose rate and fractionated irradiation. Radiat Prot Dosim 13: 185–189
Müller RG (1989) A generalization of the L-Q concept with respect to the cell kinetics. Proceedings of the 5th varian European clinac users meeting, San Remo (in press)
Müller RG, Thiel HJ, Düring A (1986) Die Computertomographie als Grundlage für die Bestrahlungsplanung und Dosisberechnung bei der Kombination von interstitieller und perkutaner Strahlentherapie. In: Frommhold W, Hübner KH (eds) Computertomographie in der Strahlentherapie. Thime, Stuttgart
Mussel LE (1956) The rapid reconstruction of radium implants: a new technique. Br J Radiol 29: 402–4087
Nag S (1985) Transperineal Iodine-125 implantation of the prostate under transrectal ultrasound and fluoroscopic control. Endocurie Hypertherm Oncol 1: 207–211
Neblett DL, King CJ, Schaeflein JW, Haymond HR (1978) Computerized dose distribution estimation system. Front Radiat Ther Oncol 12: 35–41
Neblett, DL, Syed AMN, Puthawala AA, Harrop R, Fray HS, Hogan SE (1985) An interstitial implant technique evaluated by contiguous volume analysis. Endocurie Hypertherm Oncol 1: 213–222
Nori D, Hilaris BS, Batata MA, Moorthy CR, Hopfan S (1983) Remote afterloading in cancer management Part II. Clinical applications of remote afterloaders. In: Hilaris BS, Batata MA (eds) Brachytherapy oncology 1983. Memorial Sloan Kettering Cancer Center, New York, pp 101–118
Nuttal JR, Spiers FW (1946) Dosage control in interstitial radium therapy. The general infirmary at Leeds. Br J Radiol 19: 135–142
Order SE (1981) Monoclonal antibody: potential role in radiation therapy and oncology. Int J Radiat Oncol Biol Phys 8: 1193–1201
Order SE (1984) Radioimmunoglobulin therapy of cancer. Compr Ther 10: 9
Order SE, Sleeper AM, Stillwagon GB, Klein JL, Leichner PK (1990) Radiolabeled antibodies: results and potential in cancer therapy. Cancer Res [Suppl] 50: 1011–1013
Orton CG (1974) Time dose factors (TDFs) in brachytherapy. Br J Radiol 47: 603–607
Orton CG (1980) Re-assessment of normalization between fractionated and continuous radiotherapy for the CRE and TDF equation. Br J Radiol 53: 374–375
Orton CG (1981) Radiological dose rate considerations with remote afterloading. In: Shearer DR (ed) Recent advances in brachytherapy physics. American Institute of Physics, New York, pp 190–200
Paterson R (1963) The treatment of malignant disease by radiotherapy, 2nd edn. Anrnold, London
Paterson R, Parker HM (1934) A dosage system for gamma-ray therapy. Br J Radiol VII: 592
Paterson R, Parker HM (1938) A dosage system for interstitial radium therapy. Br J Radiol XI: 252
Paterson R, Parker HM, Spiers FW (1936) A system of dosage for cylindrical distributions of Radium. Br J Radiol IX: 487
Paul MJ, Koch RF, Philip PC, Kahn FR (1987) Comparison of brachytherapy dosimetry systems: biplanar implant with equal and unequal areas. Endocurie Hypertherm Oncol 3: 55–66
Paul MJ, Koch RF, Philip PC (1988) Uniform analysis of dose distribution in interstitial brachytherapy dosimetry systems. Eur J Radiother Oncol (accepted)
Perez, CA, Brady LW (1987) Principles and practice of radiation oncology. Lippincott, Philadelphia
Perez CA, Glasgow GP (1987) Clinical applications of brachytherapy. In: Perez CA, Brady LW (eds) Principles and practice of radiation oncology. Lippincott, Philadelphia, pp 252–290
Perez CA, Kuske R, Glasgow GP (1985) Review of brachytherapy for gynecologic tumors. Endocuriether Hyperthermia Oncol 1: 153–175
Pierquin B, Dutreix A (1966) Pour une nouvelle méthodologie en curiethérapie; le Système de Paris (endo et plésio – radiothérapie avec préparation non radioactive). Note préliminare. Ann Radiol 9: 757–760
Pierquin B, Fayos JV (1962) Dosimetry by tomography in interstitial curietherapy: point technique. J Roentgenol 87: 585–592
Pierquin B, Chassagne D, Gasiorowski M (1960) Technique de dosimétrie en curiethérapie interstitielle par tomographie transversalle. Acta Radiol 53: 314–320
Pierquin B, Chassagne D, Baillet F, Paine CH (1973) Clinical observations on time-factor in interstitial radiotherapy using iridium-192. Clin Radiol 24: 506–509
Pierquin B, Chassagne DJ, Chahbazian ChM, Wilson JF (1978) Brachytherapy. Warren H. Green, St. Luis
Pierquin B, Calitchi E, Mazeron JJ, le Bourgeois JP, Leung S (1985) A comparison between low dose rate radiotherapy and conventionally fractionated irradiation in moderately extensive cancers of the oropharynx. Int J Radiat Oncol Biol Phys 11: 431–439
Pierquin B, Wilson JF, Chassagne D (1987) Modern brachytherapy. Masson, New York
Powers WE, Schneider AK, Schumate K, Fotenos H, Gallagher T (1969) Evaluation of methods of computer estimation of interstitial and intracavitary dosimetry. Am J Roentgenol Rad Ther Nucl Med 96: 59–65
Puthawala AA, Syed AM, Tannsey LA, Shanberg A, Austin PA, McNamara CS (1985) Temporary iridium-192 implant in the management of carcinoma of prostate. Endocuriether Hyperthermia Oncol 1: 25–34
Quimby EH (1922) The effect of the size of radium applicators on skin doses. Am J Roentgenol 9: 671–683
Quimby EH (1932) The grouping of radium tubes in packs or plaques to produce the desired distribution of radiation. Am J Roentgenol 27: 18–39
Quimby EH (1944) Dosage tables for linear radium sources. Radiology 43: 572–577
Quimby EH (1947) Radium Dosage in Radium Therapy. Am J Roentgenol 57: 622–627
Quimby EH, Castro V (1953) The calculation of dosage in interstitial radium therapy. J Roentgenol 70: 739–749
Regaud C (1929) Radium therapy of cancer at the radium institute of Paris. Am J Roentgenol 21: 1–24
Rosenthal MS, Nath R (1983) An automatic seed identification technique for interstitial implants using thee isocentric radiographs. Med Phys 10: 475–479
Rotte K, Linka F, Felder KD (1973) Intracavitäre Bestrahlung des Uteruskarzinoms durch ein Afterloading-Gerät mit punktförmiger Iridium-192-Quelle. Strahlentherapie 145: 523–528
Rubin P, Cowen RB, Rubin DJ (eds) (1979) The radiation

oncology research program: Recommended research proposals. Int J Radiat Oncol Biol Phys 5: 595

Rubin P, Tubiana M, Brady L (eds) (1988) International clinical trials in radiation oncology ICTRO. Int J Radiat Oncol Biol Phys 14 Suppl: S1–S214

Saw CB, Suntharalingam N (1988) Reference dose rates for single- and double-plane 192-Ir implants. Med Phys 15: 391–396

Seay DG, Hilbert JW, Moeller J, Alderman SJ, von Essen CF (1972) Therapy using a new remote-controlled high-intensity afterloading device. Radiology 105: 709–711

Seitz L, Wintz H (1920) Die kombinierte Röntgen-Radiumbehandlung im Rahmen der biologischen Dosierung. Zentralbl Gynakol 44: 529–536

Seydel HG (1977) Interstitial implantation in head and neck cancer. Semin Oncol 4: 399–406

Shalek RJ, Stovall M (1968) The M.D. Anderson method for the computation of isodose curves around interstitial and intracavitary radiation sources. I. Dose from linear sources. Am J Roentgenol, Radium ther Nucl Med 102: 662–672

Shalek RJ, Stovall M (1990) Brachytherapy Dosimetry. In: Kase KR, Bjärngard BE, Attix FH (eds) The dosimetry of ionizing radiation Vol III, Academic Press, San Diego, pp 259 321

Shapiro A, Schwatz B, Windham JP, Kereiakes JG (1976) Calculated neutron dose rates and flux densities from implantable Californium-252 point and line sources. Med Phys 3: 241–247

Shearer DR (ed) (1981) Recent advances in brachytherapy physics. Medical Physics Monograph no 7, American Association of Physicists in Medicin. American Institute of Physics, New York

Siegel JA, Pawlyk DA, Lee RE, Sasso NL, Horowitz JA, Sharkey RM, Goldenberg DM (1990) Tumor, red marrow, and organ dosimetry for 131-I-labeled anti-carcinoembryonic antigen monoclonal antibody. Cancer Res [Suppl] 50: 1039–1042

Sievert RM (1921) Die Intensitätsverteilung der primären – Strahlung in der Nähe medizinischer Radiumpräparate. Acta Radiol 1: 89–128

Somocovitis D, Young MEJ, Batho HF (1967) Apparent absorption of the gamma rays of radium in water. Br J Radiol 40: 771–777

Spencer RP (ed) (1978) Therapy in nuclear medicine. Grune and Stratton, New York

Stovall M, Shalek RJ (1972) A review of computer techniques for dosimetry of interstitial and intracavitary radiotherapy. Comput Programs Biomed 1: 125–136

Suit HD, Moore EB, Fletcher GH, Wornsnop R (1963) Modifications for Fletcher ovoid system of afterloading using standard-sized radium tubes (milligram and microgram). Radiology 81: 126–131

Syed AMN, Feder BH (1977) Technique of afterloading interstitial implants. Radiol Clin 46: 458–475

Syed AMN (1983) Temporary iridium-192 implantation in the management of carcinoma of the prostate. In: Hilaris BS, Batata MA (eds) Brachytherapy Oncology – 1983 pp 83–91, Memorial Sloan-Kettering Cancer Center, New York

Thiel HJ, Müller RG, Weidenbecher M, Sauer R (1987) Interstitielle Brachycurietherapie von HNO-Tumoren. In: Sauer R, Schwab W (eds) Kombinationstherapie der Oropharynx- und Hypopharynxkarzinome. Urban and Schwarzenberg, Munich

Thiel HJ, Herbst M, Fietkau R, Sauer R, Müller RG, Müller W, Puthawala A, Stauner J (1989) Ein neues vielkanaliges System (Inter-Pal-C-38). Strahlenther Onkol 165: 802–806

Tod MC, Meredith WJ (1938) A dosage system for use in the treatment of cancer of the uterine cervix. Br J Radiol 11: 809

Trott NG (ed) (1987) Radionuclides in brachytherapy: radium and after. Br J Radiol Suppl 21

Vaeth JM (ed) (1978) Renaissance of interstitial brachytherapy. Karger, Basel

Vikram B, Hilaris BS (1981) A non-looping afterloading technique for interstitial implants of the base of the tongue. Int J Radiat Oncol Biol Phys 7: 419–422

Waggener R, Lange J, Feldmeier J, Eagan P, Martin S (1989) 137-Cs dosimetry table for asymmetric source. Med Phys 16: 305–308

Wasserman TH, Kligerman M (1987) Chemical modifiers of radiation effects. In: Perez CA, Brady LW (eds) Principles and practice of radiation oncology. Lippincott, Philadelphia, pp 360–376

Webb S, Fox RA (1979) The dose in water surrounding point isotropic gamma-ray emitters. Br J Radiol 52: 482–484

Welsh AD, Dixon-Brown A, Stedeford JBH (1983) Calculation of dose distribution for Iridium-192 implants. Acta Radiol 22: 331–336

Wessels BW (1990) Current status of animal radioimmunotherapy. Cancer Res [Suppl] 50: 970–973

Whithers HR (1987) Biologic basis of radiation therapy. In: Perez CA, Brady LW (eds) Principles and practice of radiation oncology. Lippincott, Philadelphia, pp 67–98

Wickham L, Degrais P (1910) Radiumtherapy. Funk and Wagnalls, New York

Wilkinson JM, Moore CJ, Notley HM, Hunter RD (1983) The use of Selectron afterloading equipment to stimulate and extend the Manchester system for intracavitary therapy of the cervix uteri. Br J Radiol 56: 409–414

Williams LE, Beatty BG, Beatty JD, Wong JYC, Paxton RJ, Shievly JE (1990) Estimation of monoclonal antibody-associated 90-Y activity needed to achieve certain tumor radiation doses in colorectal cancer patients. Cancer Res [Suppl] 50: 1029–1030

Williamson JF (1986) The acuracy of the line and point source approximations in Ir-192 dosimetry. Int J Radiat Oncol Biol Phys 12: 409–414

Williamson JF (1988) Monte Carlo evaluation of specific dose constants in water for 125-I seeds. Med Phys 15: 686–694

Williamson JF, Morin RL, Kahn FM (1983A) Monte Carlo evaluation of the Sievert integral for brachytherapy dosimetry. Phys Med Biol 28: 1021–1032

Williamson JF, Morin RL, Kahn FM (1983B) Dose calibrator response to brachytherapy sources; a Monte Carlo and analytic evaluation. Med Phys 10: 135–140

Wood RG (1981) Computers in radiotherapy planning. Research Studies Press, Chichester

Wu A, Ulin K, Sternick ES (1988) A dose homogeneity index for evaluating 192-Ir interstitial breast implants. Med Phys 15: 104–107

1.5 High and Low Dose Rate Remote Afterloading: A Critical Comparison

COLIN G. ORTON

CONTENTS

1	Introduction	53
2	Physical Considerations	53
2.1	Mechanical Aspects	53
2.2	Convenience	54
2.3	Cost Effectiveness	54
2.4	Radiation Safety	54
3	Radiobiological Aspects	54
3.1	Redistribution	54
3.2	Repopulation	55
3.3	Reoxygenation	55
3.4	Repair	55
3.4.1	Problem	56
3.4.2	Solutions	56
4	Discussion	57
	References	57

1 Introduction

The recent surge in interest in high dose rate (HDR) remote afterloading and the availability of HDR equipment make it important to compare HDR techniques critically with the low dose rate (LDR) treatments they are replacing. Even though the HDR modality has been used for many years to treat a variety of diseases, far less experience has been gained with these techniques compared with conventional LDR brachytherapy. Hence, there is a reluctance on the part of many radiotherapists to even consider adoption of this new technology unless convincing evidence can be given to show that it represents an improvement.

The ultimate tests, and probably the only truly valid ones, are controlled clinical trials designed to compare these regimes. Such trials have been conducted for gynecological malignancies only and, even for these, have involved too few patients to allow a good statistical analysis. (WARD et al. 1974; HIMMELMANN et al. 1985). Lacking solid clinical evidence, an attempt will be made in this paper to compare these modalities solely from the viewpoint of physical and radiobiological considerations.

COLIN G. ORTON, Ph.D., Director, Medical Physics, Gershenson Radiation Oncology Center, Harper-Grace Hospitals and Wayne State University, 3990 John R. Street, Detroit, MI 48201, USA

2 Physical Considerations

The physical comparison of HDR and LDR remote afterloading techniques can conveniently be divided into four components: mechanical aspects, convenience, cost effectiveness, and radiation safety.

2.1 Mechanical Aspects

The hardware available for HDR and LDR treatments is in most respects similar. For example, almost all the applicators which are available for LDR remote afterloading are now also available for HDR therapy. Furthermore, the sizes of sources and catheters are similar. We will, therefore, only consider the few mechanical differences between the two systems, all of which appear to be to the advantage of HDR due to its increased ability to achieve a more stable positioning of applicators and sources in the patient.

With LDR implants, it is usually assumed that the sources, once localized, remain in their fixed positions with respect to the tumor and normal tissues for several days. However, at least with gynecological applications, this has been shown to be untrue. Catheters move, applicators move, and normal anatomy moves, especially when it is artificially packed away from the applicators, such as in treatment for carcinoma of the cervix (JOELSSON and BACKSTROM 1969). The patient's movements cannot be sufficiently constrained and packing materials cannot be expected to retain their original size and shape for long periods of time. This is not the case for HDR treatments, which last only a few minutes. Also, free use can be made of anchoring techniques to fix the hardware firmly in position during treatment, such as table clamps for gynecological applicators, clamped templates for interstitial implants, or balloons for endobronchial catheters. These mechanical constraints mean that geometrical precision is more readily assured, and this ensures more accurate dosimetry for both tumor and normal tissues, and improved packing and retraction of organs at risk of injury. In addition, if the high-dose regions from the implants are to be shielded when external beam treatments are applied, such as with midline blocks for whole-pelvis fields, the high-dose region is more precisely defined with the HDR method. The overall reduction in dose to normal structures and precision of placement of the sources is a major advantage of the HDR technique.

2.2 Convenience

In terms of convenience, the comparison is not so definitive. LDR treatments have the advantage that relatively few visits to the clinic are required, often only one. This is not necessarily the case with HDR (see Sect. 3). Also, the patient may prefer to be treated on an inpatient basis, with full anesthesia on implantation, and with the full nursing care possible with LDR remote afterloading equipment. On the other hand, many patients would rather not be hospitalized, especially with the bed restrictions usually required of brachytherapy inpatients. Similarly, many radiotherapy professionals prefer to treat their patients on an outpatient basis, since the entire procedure can be conducted in the radiotherapy facility. The radiotherapists and physicists do not have to leave the department, and no scheduling of shielded inpatient rooms is required.

2.3 Cost Effectiveness

The cost effectiveness of HDR compared with LDR techniques is highly controversial. If cost containment requires that hospitalization of patients be reduced, then HDR treatment may have an advantage. For example, in the USA, reimbusement mechanisms based upon disease related group (DRG) criteria appear to make such cost containment measures attractive. However, HDR remote afterloading equipment is more expensive than the LDR equivalent, and, unless a shielded room is already available (e.g., an old ^{60}Co teletherapy room), the cost of building a heavily shielded room can be considerable. The cost of shielding an LDR remote afterloading room, although not insignificant, is less. Furthermore, in order to anesthetize and treat patients in the radiotherapy clinic, it is often necessary to purchase expensive monitoring equipment. On the other hand, the HDR technique gives a clear advantage when the number of patients that can be treated on one machine is considered. Finally, one must compare the expenses associated with several hospital stays and associated care, with the costs of multiple visits to the radiotherapy facility.

2.4 Radiation Safety

Remote afterloading in general exhibits major advantages over conventional techniques as far as radiation safety is concerned. With the ability to return sources to the shielded safe whenever entering the room of a patient undergoing LDR treatments, and the extensive shielding by the thick walls of HDR rooms, doses to radiotherapy staff, nurses, other patients, and visitors are readily maintained well within regulatory limits for both modalities. The HDR technique does exhibit a slight advantage over LDR in that the doses to staff and other patients are essentially zero, not just low. Also, with HDR the sources are confined to the radiotherapy facility, so the HDR method has the advantage that the sources are used in a more readily controlled environment. Such sources are less likely to be mishandled or lost.

3 Radiobiological Aspects

It is convenient to simplify the biological comparison of HDR and LDR treatment regimes by considering the potential impact of the 4 R's of radiotherapy, namely redistribution, repopulation, reoxygenation, and repair.

3.1 Redistribution

The potential therapeutic advantage of radiation-induced redistribution of cells into synchronous phases of the cell cycle and the concomitant change in radiation sensitivity is so dependent upon the specific tumors and normal tissues at risk that it is not possible at this time to justify using this as a criterion to demonstrate preference for either technique. Consequently, any advantage (or disadvantage) attributable to redistribution will be disregarded.

3.2 Repopulation

Repopulation is a time-dependent phenomenon. Since a course of radiotherapy (fractionated external treatments plus brachytherapy) typically is spread over 6–8 weeks, and since the application of HDR or LDR implants will have little influence upon this overall treatment time, it is unlikely that repopulation will have a major impact upon the relative benefits of HDR and LDR regimes. Also, even if it is possible to effect a small reduction in overall treatment time (and hence tumor cell repopulation) by employing one technique or the other, this may be offset by a reduction in the repopulation of normal tissue cells.

3.3 Reoxygenation

As was the case with repopulation, reoxygenation is a time-dependent phenomenon although, instead of several weeks, it probably takes only a few days to manifest itself. Again, if the overall treatment time of a course of radiotherapy is the same, it is unlikely that reoxygenation will be significantly different for these two modalities. An advantage with HDR might be achieved if fractions are spaced a few days apart, since little reoxygenation could be expected *during* an LDR treatment. However, this might be offset by a reduced oxygen enhancement ratio (OER) at low dose rate. Overall, differences in reoxygenation are likely to be minor.

3.4 Repair

In contrast, repair has a major impact. Repair of sublethal damage is a continual process, which progresses for several hours after cells are first damaged by radiation. Since repair takes time but is usually completed within 24 h after exposure, unless prevented by further radiation damage, it is to be expected that negligible repair takes place *during* an HDR exposure lasting just a few minutes, whereas complete repair occurs *between* fractions. In contrast, with LDR treatments lasting many hours, repair is both continually occurring and continually being inhibited by further irradiation.

Hence, repair is inhibited for both fractionated treatments at high dose per fraction and high dose rate continuous treatments. For each type of cell there will be a specific dose per fraction of HDR treatment that will produce the same overall repair as a specific dose rate LDR treatment for the same total dose, i.e., it is possible to match HDR and LDR regimes *for a specific cell type*. Then, if the dose per fraction for an equivalent HDR technique is found for a specific normal tissue at risk of damage, further reduction in the dose per fraction, keeping the total dose constant, will allow more repair of sublethal damage. The HDR method would then *appear* to have an advantage over the LDR regime (see Fig. 1). However, this is not necessarily so, since this analysis has not accounted for the concomitant increase in repair of tumor cells. Fortunately, there is some evidence that for low doses (and dose rates), tumor cells are more sensitive to radiation than the cells of the late-responding tissues which are at risk (THAMES et al. 1983). Therefore, the reduction in dose per

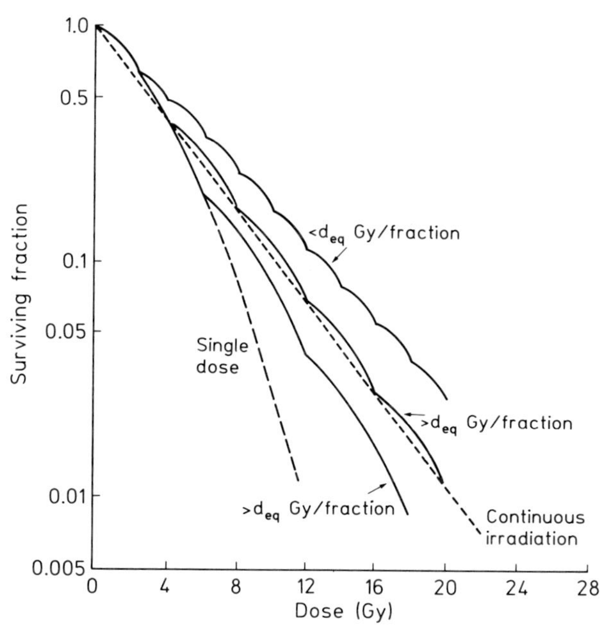

Fig. 1. Hypothetical cell surviving fraction curves illustrating the effect of changing the dose per fraction of high dose rate (HDR) treatments. At dose per fraction d_{eq} the HDR cell survival curve is equivalent to that of an (unspecified) LDR continuous exposure

fraction of the HDR treatments discussed above ought to protect tumor cells less than it protects the normal late-reacting tissues, i.e., there should be a "therapeutic advantage."

In summary, for any specific LDR technique, it ought to be possible to replace it with an HDR regime with a large enough number of treatments. With more than this number of fractions, a therapeutic gain should be achieved.

These qualitative arguments can be represented mathematically by the use of an appropriate bioeffect dose model. Such models, in order to be accurate, need to be highly complicated and account for a wide variety of biological effects. However, as shown by the previous discussions, it is probable that only repair has a major impact upon the relative benefits of HDR and LDR modalities. A useful mathematical representation of repair is given by the linear-quadratic (L-Q) dose-effect model. This is used below to determine how best to replace a low dose rate regime with a high dose rate one and to illustrate the influence of the number of fractions upon the relative effectiveness of the HDR treatments.

3.4.1 Problem

A radiotherapist wishes to replace a conventional low dose rate boost implant of 25 Gy at 0.5 Gy/h by a high dose rate technique delivered in about the same overall treatment time. How can this best be accomplished?

Firstly, let us assume that the radiotherapist would like the HDR treatments to result in equivalent, or even improved, biological effectiveness with respect to local control of the tumor, whilst maintaining the same risk of late normal tissue damage. Mathematically, if biological effectiveness is represented by the L-Q bioeffect doses (ERDs) to tumor (t) and late-responding tissues (l), then (Dale 1985):

$$(ERD_t)_{HDR} \geqslant (ERD_t)_{LDR}$$

and

$$(ERD_l)_{HDR} = (ERD_l)_{LDR}$$

where

$$(ERD)_{HDR} = Nd\left[1 + \frac{d}{\alpha/\beta}\right]$$

and

$$(ERD)_{LDR} = RT\left[1 + \frac{2R}{\mu(\alpha/\beta)}\left(1 - \frac{1}{\mu T}\right)\right]$$

where N = number of HDR fractions of dose d (Gy); T = treatment time (in hours) for an LDR regime at dose rate R (Gy/h); and α/β and μ are L-Q model parameters. For the purposes of this problem, the following representative values of α/β and μ will be assumed (Dale 1985); $(\alpha/\beta)_t = 10\,Gy$; $\mu_t = 1.4\,h^{-1}$. $(\alpha/\beta)_l = 2.5\,Gy$; $\mu_l = 0.46\,h^{-1}$

3.4.2 Solutions

If 3 fractions are to be delivered at 1 fraction/day, thus keeping the overall elapsed treatment time at 2 days, then substitution in the ERD equations gives:

$d = 4.80$ Gy

$(ERD_t)_{HDR} = 21.3$; $(ERD_t)_{LDR} = 26.8$

$(ERD_l)_{HDR} = 46.6$; $(ERD_l)_{LDR} = 46.6$

Clearly, the biological effectiveness o such an HDR regime is reduced as far as the tumor is concerned. This does not meet the radiotherapist's requirements.

In order to improve the effectiveness on the tumor, a hyperfractionated technique is required (Thames, et al, 1983). If it is assumed that 6 fractions are to be delivered at 2 fractions/day, allowing enough time between fractions for complete repair of sublethal damage, then the ERD equations give:

$d = 3.33$ Gy

and $(ERD_t)_{HDR} = 26.6$
(all other ERDs remaining as before)

These regimes are thus essential equivalent [since previously $(ERD_t)_{LDR} = 26.8$].

A therapeutic advantage could be gained by treating with 9 fractions at 3 fractions/day. Substitution in the ERD equations yields:

$d = 2.65$ Gy

and $(ERD_t)_{HDR} = 28.9$

The biological effectiveness for the tumor is thus improved. This solution needs to be slightly modified because a typical 3 fractions/day regime allows insufficient time between fractions for complete repair. For example, if 5 h are allowed between the 3 daily fractions, a more complex ERD equation needs to be used (Dale et al. 1988), and it can be shown that this yields:

$d = 2.45$ Gy

$(ERD_t)_{HDR} = 27.5$

This still represents an improvement over the LDR method.

4 Discussion

In terms of physical comparisons of HDR and LDR modalities, the HDR technique has one obvious advantage, namely the ability to fix the applicators and sources in a known geometry in the patient without risk of their moving with respect to the tumor and normal tissues during the treatment. Other physical differences such as convenience, cost effectiveness, or radiation safety are minor in comparison.

With reference to radiobiological considerations, it is the repair of sublethal cellular damage that should be of significant importance when comparing HDR and LDR treatments. Application of basic radiobiological models suggests that it ought to be possible to replace a conventional LDR regime with an HDR schedule which will maintain comparable probabilities of late tissue reaction and local tumor control. Also, with further fractionation of the HDR treatments, a therapeutic advantage should be realized.

References

Dale RG (1985) The application of the linear-quadratic dose-effect equation to fractionated and protracted radiotherapy. Br J Radiol 58: 515–528

Dale RG, Huczkowski J, Trott KR (1988) Possible dose rate dependence of recovery kinetics as deduced from a preliminary analysis of the effects of fractionated irradiations at varying dose rates. Br J Radiol 61: 153–157

Himmelmann A, Holmberg E, Oden A, Skogsberg K (1985) Intracavitary irradiation of carcinoma of the cervix stage IB and IIA. Acta Radiol [Oncol]. 24: 139–144

Joelsson I, Backstrom A (1969) Dose rate measurements in bladder and rectum. Acta Radiol [Ther] [Stockh] 8: 343–359

Thames HD, Peters LJ, Withers HR, Fletcher GH (1983) Accelerated fractionation vs hyperfractionation: rationales for several treatments per day. Int J Radiat Oncol Biol Phys 9:127–138

Ward AJ, Stubbs B, Dixon B (1974) Carcinoma of the cervix: establishment of a hyperbaric oxygen trial associated with the use of the Cathetron. Br J Radiol 47: 319–325

Special Considerations

2 Brain Tumors

2.1 Basic Pinciples of Brachycurietherapy of Brain Tumors

Seymour H. Levitt, Bruce J. Gerbi, Chung K. Lee, and Robert E. Maxwell

CONTENTS

1 Clinical Rational . . . 61
2 Implantation Techniques . . . 62
3 Selection of Radionuclide . . . 62
4 Patient Selection Criteria . . . 63
5 Dose Specification . . . 63
6 Brain Implantation Technique . . . 63
6.1 Preimplantation Planning and Patient Work-up . . . 64
6.2 Presurgical CT Scans . . . 64
6.3 Surgical Implantation . . . 64
6.4 Verification CT Scans . . . 64
6.5 Dosimetry . . . 65
7 Clinical Results . . . 65
8 Conclusions . . . 65
References . . . 65

1 Clinical Rational

Interstitial brain irradiation for primary malignant brain tumors is not a new technique but has only recently attained clinical significance due to the newly developed ability to perform stereotactic localization and biopsies (Mundinger et al. 1978; Szikla et al. 1979; Gutin et al. 1981). Primary cerebral neoplasms are well suited to a local strategy using radiation for tumor control since (1) extraneural metastases are uncommon, (2) the majority of the lesions that recur do so within 2 cm of their site of origin, (3) more than 90% of anaplastic astrocytomas and glioblastomas are localized to one side of the brain (Hochberg and Pruitt 1980), (4) tumors are not always amenable to open surgical decompression or extensive cytoreduction because of their anatomic location, (5) when surgery cannot be performed, the most effective adjunct has clearly been proven to be radiation, (6) both chemotherapy and chemical modifiers (sensitizers/protectors) have been shown to be ineffective (Walker et al. 1978), (7) it is still too early to identify the clinical efficacy of hyperthermia in the treatment of these tumors, and (8) the results of novel fractionation schemes are still to be determined.

Since brain tumors can be well localized using computerized tomography (CT) and magnetic resonance imaging (MRI) scans, the ability to deliver a high radiation dose to a limited region offers many advantages over conventional methods of radiation dose delivery. With modern, CT-guided, stereotactic implantation techniques, a high degree of positional accuracy is possible using a limited surgical technique (Findlay et al. 1985; Gutin et al. 1981; Gutin et al. 1984; Mundinger et al. 1978; Szikla et al. 1979; Thiel et al. 1983; Weaver 1986). This allows for a high dose to be directly deposited in the primary lesion while greatly sparing the surrounding unaffected brain tissue. This is important since destroyed normal neural tissue never regenerates, but partial functional recovery is possible following limited injury. It must be kept in mind that the radiosensitivity of normal brain parenchyma is nearly equivalent to that of the majority of the primary tumors requiring irradiation. This is especially true for low grade astrocytomas. Consideration of these factors clarifies the advantage of a technique which delivers a high dosage of irradiation to a limited amount of normal brain tissue.

Irradiation methods must take into account the complex anatomical and functional organization of the normal brain. Of particular concern are intracranial structures such as the visual tracts, motor strip, hypothalamus, etc. in the vicinity of the tumor and the radiation dose they will receive. A final caveat is that the definition of the target volume is determined mostly by indirect, non-stereotactic means from a synthesis of the clinical data, neurological findings (angiograms, ventri-

Seymour H. Levitt, M.D., Professor, Bruce J. Gerbi, Ph.D., Chung K. Lee, M.D., Department of Therapeutic Radiology, Radiation Oncology.

Robert E. Maxwell, M.D., Department of Neurosurgery, University of Minnesota Hospitals and Clinics, Harvard Street at East River Road, Minneapolis, MN 55455, USA culograms, CT scans), and neurophysiologic and neuropathologic factors (SZIKLA 1987).

Present data appear to indicate that MRI is more accurate in defining the true extent of a tumor than CT or any other presently available diagnostic modality. Thus, some of the failures in treatment may be related to an inadequate estimate of the true extent of the tumor (KELLY et al. 1987). In the future, MRI will have to be used to ensure adequate definition of the treatment volume.

2 Implantation Techniques

Various methods of performing brain implantation have been described in the literature (MUNDINGER et al. 1978; GUTIN et al. 1981; FINDLEY et al. 1985; ABRATH et al. 1986). These techniques may be categorized as (1) "radiosurgical" or "focal spot" irradiation which involves the implantation of one or only a few sources into the center of the target volume (MUNDINGER et al. 1978; GUTIN et al. 1981; THIEL et al. 1983). This technique requires high activity sources and results in such a high-dose gradient from the center of the target volume to its periphery that radiation necrosis, which necessitates an additional surgical procedure, is expected in every case (GUTIN et al. 1981, 1984). (2) A "Quimby style" technique involves spacing sources of equal activity uniformly throughout the target volume. This necessitates the implantation of more catheters but results in increased uniformity of radiation dosage throughout the target volume. However, the central region still experiences a higher dose rate than the periphery of the volume. (3) A technique was developed at the National Cancer Institute (NCI) whose objective is to keep the dose homogeneous throughout the target volume while using the fewest number of catheters as possible (FINDLAY et al. 1985). This technique is similar to the Paterson-Parker method for volume implants with two main differences: The NCI system avoids implantation of active seeds into healthy tissue beyond the target volume, and it limits the implantation of seeds in the center of large tumors to the central source plus additional sources only at the ends of the catheter line (FINDLEY et al. 1985).

The technique developed at the University of Minnesota essentially follows the Quimby philosophy. It was designed to provide a fairly uniform dose distribution (20%–30%) throughout the target volume, while keeping the required number of catheters to a minimum. Also, finding a limited surgical technique that could be done under local anesthesia and that required a minimum of time in the CT scanner room was a primary concern.

3 Selection of Radionuclide

The most commonly used radionuclides for this procedure are iodine 125 and iridium 192. Iridium 192 has been used as a brachytherapy source for many years and has the following advantages: (1) Source strength calibration techniques are well established at the national level (LOFTUS 1980), and institutional calibration techniques are also well documented (WILLIAMSON et al. 1982); (2) the exposure rate constant has been extensively investigated and is well known (GLASGOW and DILLMAN 1979); (3) the dose distribution around an individual seed is only slightly anisotropic and can be approximated in routine planning by considering a seed to be a point source, to an accuracy of 1%–2% (WILLIAMSON 1986); (4) the individual sources are extremely durable with little chance of rupture and are currently available for the modest cost of about $3.00 per seed; and (5) the decrease in dose with distance from the source is essentially the same as radium 226, allowing the same distribution rules that apply for radium to be used with iridium.

The major difference between iodine 125 and any other radionuclide used for brachytherapy is the low energy (27–35 keV) of the emitted photons. These low-energy photons are rapidly attenuated in both the tissues of the body and by minimal amounts of shielding (HILARIS et al. 1976; GUTIN et al. 1984). In fact, the radiation protection aspects of this radionuclide are the primary reason for its popularity. There are reports indicating that iodine 125 has a greater radiobiological effectiveness in the treatment of malignancies than other isotopes (FREEMAN et al. 1982; KIM and HILARIS 1975), but this has been disputed (DA SILVA et al. 1984). There are some serious disadvantages associated with the use of iodine 125. The most important of these is the anisotropic dose distribution around individual seeds caused by the attenuation of the low-energy photons in the endweld of the source (LING et al. 1983). Additionally, the cost of individual seeds depends on their activity and can range from $10.00 for a low-activity seed to over $300.00 for high-activity seeds. Because of the high cost of the high-activity seeds, institutions are strongly motivated to use the seeds for several implants. During the reuse of one of these high-activity seeds, the source encapsulation ruptured, leading to a large dose to the thyroid of the patient and exposure of approximately 60 hospital workers. Similar incidents have also been reported for the low-activity seeds used for permanent implants (USNRC 1986). Accurate iodine-125 source calibration is also more difficult to perform. Most departments use deep-well ionization chambers for their calibration checks, and it has been shown that the response of these units is highly dependent on the source position in the well. Also, in-house dose calibrator standardizations that are performed for one model of iodine-125 seed are not valid for other models of iodine seed (Weaver 1986). Finally, the dose distribution near iodine 125 is more greatly affected by tissue inhomogeneities such as bone, air cavities, or calcifications when these are in close proximity to the seeds (Weaver 1986).

The dose around an individual iodine-125 seed decreases more rapidly in surrounding tissue than for other radionuclides (Krishnaswamy 1978; Ling et al. 1983). The benefit of this rapid dose fall-off is that the dose to normal tissue beyond the target volume decreases more rapidly than with other radionuclides. This has been shown to be true for large, spherical tumor masses greater than 5 cm in diameter. However, for masses of 2 cm in diameter, there is essentially no difference in the dose fall-off away from the treated region between iridium 192 and iodine 125 (Rosenow et al. 1987).

Iridium 192 is currently the radionuclide of choice at this institution because of the advantages listed above. However, at other institutions where the ease of radiation shielding have outweighed other considerations, iodine 125 is often preferred.

4 Patient Selection Criteria

We feel that for patients to be eligible for this procedure they should be of sufficient age, ambulatory, and conscious so as to take care of themselves during the implant without close supervision by the nursing staff; the size of the primary lesion should be less than 5 cm in its greatest dimension; the histopathology must be newly diagnosed or recurrent glioblastoma, grade III astrocytoma, or recurrent grade I or II astrocytoma; the tumor must be located supratentorially and must be confined to one hemisphere of the brain; and the patient must also be free of active infectious process. These selection criteria are similar to those of institutions using iodine 125 (Findlay et al. 1985) and differ only in the requirement that the patient be able to care for himself. In our experience, this particular requirement has not restricted our use of this technique if the patient met the other selection criteria.

5 Dose Specification

For primary lesions, 45–50 Gy of external beam therapy is delivered to the whole brain in 1.8–2.0 Gy fractions, giving five fractions per week. An additional 25–30 Gy is delivered to a more confined region using iridium 192 brain implantation. For recurrent tumors, iridium 192 brain implantation is used to deliver 35–50 Gy to the involved area with the total dose from all treatments not to exceed 90–95 Gy in the treated region. The target volume for implantation includes the area of enhanced uptake as observed on the CT scan plus an additional margin of 0.5 cm around that region.

6 Brain Implantation Technique

The primary technical challenge in brain implantation is the accurate placement of the radioactive sources in the target volume. To accomplish this goal, the following sequence of steps should be undertaken: Preimplantation planning and patient work-up; presurgical CT scanning and implantation system attachment; surgical implantation of catheters or radioactive material; postsurgical catheter location verification by radiotherapy simulator (orthogonal radiographs for seed localization and dosimetry) or CT scan verification to image catheters in relation to the target volume; and dosimetry (calculation of isodose distributions around target volume).

6.1 Preimplantation Planning and Patient Work-up

Before the patient is admitted for the procedure, the treatment volume to be implanted is determined using all available diagnostic information. Additional diagnostic procedures are also requested at this time if necessary. MRI has been beneficial in this regard, and it has recently been shown that MRI studies more accurately describe the true region of involvement for low-grade cerebral neoplasms. For these lesions, the CT contrast-enhanced area was found to underestimate the size of the lesion in comparison with brain biopsy results. For high-grade lesions, the CT contrast-enhanced region accurately defined the tumor tissue volume (Kelly et al. 1987; Burger 1987).

6.2 Presurgical CT Scans

The purpose of this step is to provide the input data for the actual stereotactic brain implantation procedure. In essence, the target volume is localized in terms of the reference frame provided by the stereotactic device to be used for the procedure. Various instruments have been tried to establish a frame of reference for stereotactic implantation of the target volume. At our institution, the apparatus designed by Brown–Roberts–Wells (BRW) (Heilbrun et al. 1983) and supplied by Radionics, Inc. is currently in use. There are three basic components to the BRW stereotactic system: the head ring, the localizer ring, and the arc drilling system. The head ring is firmly attached, using four screws, to the patient's skull and serves as the foundation for the localizer ring and the arc drilling assembly. The localizer assembly is in place on the head ring during preimplantation CT scans and appears on the scans as an arrangement of fiducial points. These fiducial points establish a frame of reference between the drilling assembly and the target volume to be implanted. Contrast-enhanced localization CT scans are taken at 3-mm intervals through the entire extent of the target volume. Parasagittal reconstructions are also recommended to document the target extent in planes perpendicular to the transverse scan. From these scans the most superior and inferior borders of the target volume are identified, and an acceptable external entry point is chosen and marked with a radiopaque marker. The x, y coordinates of the fiducial points from the localizer assembly are essential input data to a dedicated Epson computer that calculates the coordinates to be set on the drilling assembly.

6.3 Surgical Implantation

The arc drilling assembly replaces the localizer ring when in the operating room and is also attached to the head ring. Of the many different techniques used in the operating room for the placement of catheters, we have found it most convenient to use a limited surgical procedure that requires only local anesthesia. The stereotactic coordinates as determined from the computer program are set on the arc drilling assembly, local anesthetic is injected, and a small burr hole is made for the insertion of a plastic catheter. The end of the catheter is placed 0.5 cm beyond the contrast-enhanced region as seen on the CT scan. Additional catheters are inserted parallel to the central source as determined by the radiotherapist. The actual arrangement of sources is dependent on the two drilling templates that are also commercially available from Radionics. Of the two templates (0.7 and 1.0 cm center-to-center spacing), we prefer 0.7 cm spacing and place the catheters in every other position to achieve a 1.4 cm spacing. With this arrangement of catheters it is possible to implant our maximum volume of 5 cm diameter using 19 catheters. The dose homogeneity in this situation is also 20%, not including the region close to the sources.

After all the catheters have been inserted, they are glued to a silastic patch that was initially sutured to the scalp of the patient. The protruding catheters are protected using a foam covering. Following this, the patient is moved to intensive care for observation and finally into a private room. When the placement of the catheters has been verified by post-surgical CT scans, the radioactive material is afterloaded into the hollow catheters in the patient's room.

6.4 Verification CT Scans

Since the location of the target volume in relation to the catheters can only be seen using CT scans, an additional contrast-enhanced CT series is performed after surgery. This provides actual visual verification that the catheters and the radioactive material are properly placed.

6.5 Dosimetry

In order to perform the dosimetry for the arrangement of radioactive material in the brain, orthogonal radiographs with dummy sources in place are taken using a radiotherapy simulator. The dosimetry is done by identifying each seed uniquely on both the anterior and lateral radiographs of the orthogonal film pair. The seeds are separately digitized into a treatment planning computer which calculates the dose distribution for the implant. The duration of the implant is determined from these isodose curves. At our institution, typical spacing between the seeds in the catheters is 1.0 cm center-to-center while the seed activity is between 0.40 and 0.50 mg Ra eq.

7 Clinical Results

The University of Minnesota Hospital has conducted a phase 1 clinical study of CT-guided stereotactic implantation for the treatment of malignant astrocytomas and recurrent brain tumors using iridium 192. We have treated 16 patients using this method, and 14 patients are evaluable. Nine patients (5 grade III patients and 4 grade IV patients) were treated previously in conjunction with external beam irradiation, and 5 patients were treated for recurrent brain tumors.

The median age was 49 years (range 24–66 years) for patients with primary brain tumor and 39 years (range 15–55) for patients with recurrent brain tumor. Tumor size was limited to 5 cm.

Two of the patients treated primarily are alive without evidence of disease at 38 and 32 months, and 3 patients are alive with radiographic evidence of disease by CT and/or MRI scan at 14, 17, and 26 months. One patient underwent a posttreatment debulking procedure. Four patients died of disease at 9, 11, 13, and 14 months. Patients who were treated for recurrent disease died of disease at 2, 6, 6, 15, and 28 months.

Complications have been minimal; two patients developed scalp infection, and one patient had a history of four previous operations. One patient was eliminated from the study group because of bleeding during the procedure which led to an emergency craniotomy.

8 Conclusions

Interstitial brain irradiation for primary malignant tumors is an important new approach to the treatment of a very difficult problem. For many years there has been difficulty controlling primary brain tumors despite the fact that they remain relatively localized and do not metastasize. The basic problem has been the inability to eradicate the neoplasm without destroying the vital parts of the brain including normal tissue. This in many instances has led to either underdosage or overdosage when using external radiation therapy of any sort. There are certain instances in which it is appropriate to use interstitial isotopic brachytherapy in the treatment of brain tumors with the current development of more accurate diagnostic techniques to determine the true extent of neoplastic malignancies. There are a number of approaches to the use of interstitial therapy, one of which recommends the placement of intense sources locally and accepts the problem of subsequent necrosis. Our approach has been to attempt to deliver a uniform dose throughout the tumor-bearing area with the hope that intensive radiation will destroy the local tumor without producing too much damage to the surrounding normal tissue or vital structures. As noted in the following chapters, the approach of interstitial brachytherapy in the treatment of brain tumors has had a beneficial effect in a number of institutions in which the appropriate neurosurgical, neuropathologic, neuroradiologic, and radiation physics and computer personnel are available and willing to partake in this particular procedure. The inherent risks involved require complete dedication to this meticulous, complex, and involved approach.

References

Abrath FG, Henderson SD, Simpson JR, Moran CJ, Marchosky JA (1986) Dosimetry of CT-guided volumetric Ir-192 brain implant. Int J Radiat Oncol Biol Phys 12: 359–363

Anderson LL (1976) Spacing nomograph for interstitial implants of ^{125}I seeds. Med Phys 3: 48–51

Burger PC (1987) The anatomy of astrocytomas. Mayo Clin Proc 62: 527–529

Da Silva VF, Gutin PH, Deen DF, Weaver KA (1984) Relative biological effectiveness of ^{125}I sources in a murine brachytherapy model. Int J Radiat Oncol Biol Phys 10: 2109–2111

Findlay PA, Wright DC, Rosenow U, Harrington FS, Miller RW (1985) ^{125}I interstitial brachytherapy for primary malignant brain tumors: technical aspects of treatment planning and implantation methods. Int J Radiat Oncol Biol Phys 11: 2021–2026

Freeman ML, Goldhagen P, Sierra E, Hall EJ (1982) Studies with encapsulated ^{125}I sources. II. Determination of the relative biological effectiveness using cultured mammalian cells. Int J Radiat Oncol Biol Phys 8: 1355–1361

Glasgow GP, Dillman LT (1979) Specific -ray constant and exposure rate constant for ^{192}Ir. Med Phys 6: 49–52

Gutin PH, Phillips TL, Hosobuchi Y, Wara WM, Mackay AL, Weaver KA, Lamb S, Hurst S (1981) Permanent and removable implants for the brachytherapy of brain tumors. Int J Radiat Oncol Biol Phys 7: 1371–1381

Gutin PH, Phillips TL, Wara WM, Leibel SA, Hosobuchi Y, Levin VA, Weaver KA, Lamb S (1984) Brachytherapy of recurrent malignant brain tumors with removable high-activity iodine-125 sources. J Neurosurg 60: 61–68

Heilbrun MP, Roberts TS, Apuzzo MLJ, Wells TH, Sabshin JK (1983) Preliminary experience with the Brown–Roberts–Wells (BRW) computerized tomographic stereotaxic guidance system. J Neurosurg 59: 217–222

Hilaris BS, Holt GJ, St German J (1976) The use of iodine-125 for interstitial implants. Department of Health, Education, and Welfare, Washington (DHEW Publication (FDA) 76-8022)

Hochberg FH, Pruitt A (1980) Assumptions in the radiotherapy of glioblastoma. Neurology (NY) 30: 907–911

Kelly PJ, Daumas-Duport C, Scheithauer BW, Kall BA, Kispert DB (1987) Stereotactic histologic correlations of computed tomography- and magnetic resonance imaging-defined abnormalities in patients with glial neoplasms. Mayo Clin Proc 62: 450–459

Kim JH, Hilaris BS (1975) Iodine 125 source in interstitial tumor therapy. AJR 123: 163–169

Krishnaswamy V (1978) Dose distribution around a ^{125}I seed source in tissue. Radiology 126: 489–491

Ling CC, Yorke ED, Spiro IJ, Kubiatowicz D, Bennett D (1983) Physical dosimetry of ^{125}I seeds of a new design for interstitial implant. Int J Radiat Oncol Biol Phys 9: 1747–1752

Loftus TP (1980) Standardization of iridium-192 gamma ray sources in terms of exposure. J Res Natl Bur Stans (US) [Phys Chem] 85A: 19–25

Mundinger F, Weigel K (1984) Long-term results of stereotactic interstitial curietherapy. Acta Neurochir [Suppl] (Wien) 33: 367–371

Mundinger F, Birg W, Ostertag CB (1978) Treatment of small cerebral gliomas with CT-aided stereotaxic curietherapy. Neuroradiology 16: 564–567

Rosenow UF, Findlay PA, Wright DC (1987) The NCI atlas of dose distributions for regular ^{125}I brain implants. Radiother Oncol 10: 127–139

Szikla G (1987) Central nervous system. In: Pierquin B, Wilson JF, Chassagne D (eds) Modern brachytherapy. Masson, Paris, pp 287–299

Szikla G, Schlienger M, Betti O, Talairach J, Cohadon F, Rougier A, Pigneux J, Benabid AL, Vrousos P, Chiroussel J, Sedan R, Peragut JC, Farnarier P, Pecker J, Scarabin JM, Vallee B, Gilbert P (1979) Combined interstitial and external irradiation of gliomas. A progress report. In: Szikla G (ed) Stereotactic cerebral irradiation. Elsevier, New York, pp 329–338

Thiel HI, Huk WI, Müller R, Sauer R (1983) Die computertomographisch geleitete, stereotaktische interstitielle Therapie von Hirnfumoren mittels temporäreroder permanenter Implantation von 125-I-seeds. Fortschr Geb Rontgenstr 138: 348–355

US Nuclear Regulatory Commission (1986) IE Information Notice No 86–84

Walker JD, Alexander E, Hunt WE, MacCarty CS, Mahaley MS, Mealey J, Norrell HA, Owens G, Ransohoff J, Wilson CB, Gehan EA, Strike TA (1978) Evaluation of BCNU and/or radiotherapy in the treatment of anaplastic gliomas. J Neurosurg 49: 333–343

Weaver K (1986) Dosimetry of I-125 sources with emphasis on brain implants and eye plaques. Med Phys 13:626

Williamson JF (1986) The accuracy of the line and point source approximations in Ir-192 dosimetry. Int J Radiat Oncol Biol Phys 12: 409–414

Williamson JF, Khan FM, Sharma SC, Fullerton GD (1982) Methods in routine calibration of brachytherapy sources. Radiology 142: 511–516

2.2 Imaging-Stereotaxic Implantation of Radionuclides in Intracranial Tumors (Curietherapy and Brachycurietherapy)

FRITZ MUNDINGER

CONTENTS

1	Introduction	67
2	Methods	67
2.1	Curietherapy	67
2.2	Brachycurietherapy	70
3	Material and Results	70
3.1	Pituitary Adenomas	70
3.2	Skull Base and Parasellar Tumors	71
3.3	Nonresectable Low-Grade Gliomas (WHO I, II)	71
3.3.1	Thalamus Tumors	72
3.3.2	Diencephalic Region	72
3.3.3	Pineal Region	73
3.3.4	Mesencephalic Brainstem and Pons Region	73
3.3.5	Summary	73
4	Discussion	74
4.1	Pituitary Tumors	74
4.2	Parasellar Tumors	74
4.3	Intracerebral Tumors	75
4.4	Gliomas of the Midline	75
	References	76

1 Introduction

Curietherapy, using the natural radionuclide radium, was very popular from the 1920s to the 1940s, when it was replaced by artificial radioactive nuclides such as cobalt 60, cesium 157, and iridium 192. Intracranial curietherapy and brachycurietherapy with artificial radioactive isotopes did not come into use until the 1940s. This method, which we have been employing in Freiburg for the treatment of intracranial tumors since 1951, is described in detail elsewhere (MUNDINGER 1966, 1970a).

Over the past 25 years, the colloid application of yttrium 90 (occasionally phosphorus 32, gold 198, or rhenium 186) has been chosen chiefly for the intracavitary curietherapy of such lesions as craniopharyngiomas, tumor cysts, and arachnoid cysts (KNÜFERMANN et al. 1981; MUNDINGER 1956, 1958, 1969a; MUNDINGER and RIECHERT 1969; MUSOLINO et al. 1985; POLLACK et al. 1988; SCHAUB et al. 1979; STURM et al. 1988; SZIKLA et al. 1984a). For interstitial curietherapy and brachycurietherapy, iridium 192 and iodine 125, which I introduced for intracranial therapy in 1959 and 1979, respectively, have been preferred (MUNDINGER 1969b, 1970c, 1975, 1979).

2 Methods

In 1970, I presented an overview of the implantation techniques and the long-term results of 504 tumor patients, along with the rationale behind deciding on the method (Table 1) (MUNDINGER 1970a). Since 1951, intracranial implantation of the nuclides has been carried out with the stereotaxic device which Riechert and I developed[1] (MUNDINGER and RIECHERT 1969; RIECHERT and MUNDINGER 1956, 1959). Major technical advancements included the computerization of the stereotaxic method (BIRG and MUNDINGER 1973) as well as the direct integration of computerized tomography (CT) in the stereotaxic operation for localization and dosimetry in 1981 (BIRG and MUNDINGER 1982; MUNDINGER and BIRG 1988) and of magnetic resonance imaging (MRI) in 1984 (BIRG et al. 1985; MUNDINGER 1988b), referred to today as imaging stereotaxy.

2.1 Curietherapy

The technique of curietherapy used in our department and our partial results have been published elsewhere (MUNDINGER 1963, 1966, 1970a, 1981,

FRITZ MUNDINGER, Prof. Dr. med., Universität Freiburg, St.-Josefs-Krankenhaus, Hermann-Herder-Str. 1, 7800 Freiburg, Germany (former Medical Director of the Dpt. of stereotaxy and neuronuclear medicine)

[1] Manufactured by F.L. Fischer MET, 7800 Freiburg, Germany. In the USA distributed by Leibinger and Fischer, Dallas, 8350 Sterling Street, Irving/TX 75063.

Table 1. Postoperative survival times of patients with brain tumors following radiation with interstitially permanent implanted radioisotopes and combinations. (From MUNDINGER 1970a)

Tumor classification	Radioisotope	$t_{\bar{x}} \pm s$ (months)	t_{max} (months)	t_{min} (months)	Surviving patients (months)
Glioblastoma	^{198}Au + roentgen	8.5 ± 4.9	12.0	5.0	
	^{182}Ta		6.4		
	$^{182}Ta + {}^{60}Co$	14.3 ± 2.0	15.8	12.7	
Astrocytoma (dedifferentiated)	^{198}Au	6.5 ± 3.0			
	$^{198}Au + {}^{60}Co$		13.0		
	^{182}Ta	26.7 ± 19.1	45.9	7.5	
	$^{182}Ta + {}^{60}Co$		14.8		
	$^{182}Ta + {}^{60}Co$ + roentgen		30.1		
Oligodendroglioma (dedifferentiated)	^{198}Au + roentgen	17.4 ± 2.0	28.7	16.0	
	$^{198}Au + {}^{182}Ta$		43.8		
	$^{198}Au + {}^{60}Co$		33.0		
	^{182}Ta		6.9		
Ependymoma	^{198}Au		17.7		
Malignant meningioma	^{182}Ta				85.9
	$^{182}Ta + {}^{60}Co$		30.5		
	^{182}Ta + roentgen		56.5		

Table 2. Bioptically confirmed histology and number of interstitial irradiation treatments (1952 – 31.8.1988)

	n
Astrocytoma	997
Oligodendroglioma	153
Glioblastoma	504
Ependymoma	60
PNET	76
Meningioma	32
Metastases	162
Sarcoma	15
Pituitary adenomas	264
Craniopharyngioma	81
Other lesions	126
Hypophysectomy	57
Pallidotomy	21
Total	2548

1982a, b; MUNDINGER et al. 1979; MUNDINGER and WEIGEL 1984b, 1988b). This report presents the up to 10 year results of 649 low grade gliomas taken from a series of 2548 (August 1988) bioptically confirmed cases (Table 2) with 2154 radionuclide implantations (Table 3).

All of the tumors in the series had been biopsied immediately prior to the implantation. The intraoperative investigation of the specimens in smear preparation after staining with methylene blue is absolutely necessary to plan an optimum subsequent therapy and to assess the survival chances, as well to decide on any additional treatments such as interstitial or percutaneous irradiation or chemotherapy.

The immediate histological examination in the smear preparation makes it possible to arrive at a decision before the operation has been ended. In addition, paraffin embedding is always carried out by means of conventional methods, as well as immunochemical and electron microscopic evaluations, through which a reliable diagnosis with accurate grading can be achieved in 97% of patients (APUZZO and SABSHIN 1983; BRAUS and MUNDINGER 1988; KIESSLING et al. 1984; KLEIHUES et al. 1984; LUNSFORD 1988; MUNDINGER 1982c, 1985b, 1986a, 1987a). Directly following the tumor biopsy, curietherapy or brachycurietherapy or a

Table 3. Radionuclides and number of stereotactic interstitial irradiation procedures (1952 – 31.8.1988)

	n
^{32}P	32
^{60}Co	179
^{90}Y	44
^{182}Ta	21
^{198}Au	129
^{125}I	684
^{125}I Brachycurie	40
^{192}Ir GammaMed	365
^{192}Ir	660
Total	2154

Table 4. Database of stereotactic biopsy and curietherapy

Tumor type	Biopsy only only	Curietherapy $^{125}I^{a}$	$^{196}Ir^{b}$	*n*
Gliomas (June 30, 1985)				
Astrocytoma I	80	67	59	206
Astrocytoma II	203	106	91	400
Astrocytoma III	156	44	23	223
Astrocytoma IV	144	20	14	178
Oligodendroglioma II	21	11	21	53
Oligodendroglioma III	7	2	8	17
Total	611	250	216	1077
Nonglial tumors				
Ependymoma	6	12	6	24
Primitive neuroectodermal tumor	18	9	4	31
Germinoma	19	10	7	36
Teratoma	9	1	3	13
Metastasis	78	14	3	95
Unclassified tumor	7	1	3	11
Total	137	47	26	210

[a] ^{125}I from 1979 to 1985
[b] ^{192}Ir from 1965 to 1985

combination of both can be performed with the cannula or catheter that is already in place. The risk is minimal, and with modern computer techniques the operation can be kept very short.

We now have a database of records from 1287 patients in whom we performed biopsies and interstitial irradiation treatment between 1965 and June 30, 1985 (Table 4). This database was set up and evaluated together with Birg, Huber-Stentrup, and Weigel. Most of the patients were followed up regularly at 3-month intervals and later every 6 months to 1 year. Follow-up included neurological examination and CT or MRI control. The closing date of the follow-up was June 30, 1985.

Table 4 gives a listing of tumors according to therapy: those treated by curietherapy with either iridium 192 or iodine 125 and those treated by biopsy alone. This last group, however, includes patients who, after biopsy, were also treated by operation alone, operation with external irradiation, or external irradiation alone. The complications that occurred in 2125 patients (1976–1985) are listed in Table 5 (Mohadjer et al. 1987).

Today, we use exclusively iridium 192 and iodine 125 for curietherapy. The peripheral tumor accumulation dose of protracted long-term irradiation is ordinarily 120 Gy for iridium 192. When the nuclide is implanted with a thin catheter, the last 15% of the dose accumulation can be interrupted by removing the catheter, which results in a clinical accumulated radiation dose corresponding to the 2.75 times the half-life of the nuclide

Table 5. Complications after stereotactic biopsy with or without subsequent implantation of radionuclides (1.1.1976–31.13.1985) ($n = 2125$)

Operative mortality 1.08%	
Hemorrhage with permanent neurological deficits	0.94%
Edema, spasm of arteries, etc., with transient neurological deficits	0.10%

(Birg et al. 1979; Mundinger 1970a; Mundinger and Riechert 1969).

For *midline lesions* such as those in the diencephalon, pineal region, and brainstem-pons region, the peripheral accumulated radiation dose was reduced (corresponding to the tumor surface) to 90 Gy for iridium 192. This was also the case for larger tumors and after previously applied interstitial or percutaneous irradiation. We use iridium 192 for small midline lesions and for fibrillary astrocytomas and other benign tumors (WHO I, II). These usually prove to have grown into the tissue several millimeters beyond the point at which they can be visualized by CT or MRI and are already infiltrated with cell nests. Iodine 125 is sufficient for sharply defined foci. The peripheral tumor accumulated dose is 20% lower with iodine 125, corresponding to its 14-day shorter half-life. The accumulated dose is 100 Gy or 80 Gy, respectively. If the dose is reduced any further, recurrences will appear relatively soon, and the long-term results will not be satisfactory. Likewise, overdosing with more than 120 and 100 Gy, respectively, results in poor survival times. (see also Table 12).

2.2 Brachycurietherapy

We use intraoperative, high dose rate, one-time irradiation with GammaMed, the afterloading iridium-192-contact radiation device which Sauerwein and I developed in 1963[2]. The newer version of the device is called GammaMed II (MUNDINGER 1969a, 1985a; MUNDINGER and SAUERWEIN 1966). The doses are delivered by afterloading GammaMed irradiation applied once or twice with 30–40 Gy (total 100 Gy) over a period of 3–5 months. If a boost of only 20 Gy is given, iodine-125 curietherapy via catheter with 70 Gy or additional percutaneous irradiation with 45–55 Gy is carried out.

For some time now we have stopped using brachycurietherapy over several days with a high-dose iridium-192 or iodine-125 catheter (ABRATH et al. 1986; BERNSTEIN and GUTIN 1981; DYCK 1983; BOUZAGLOU et al. 1985; CHAVAUDRA et al. 1979; DAVIS et al. 1983; GUTIN et al. 1981, 1984, KELLY et al. 1978; MÜLLER et al. 1982; MURRAY et al. 1984; ROSSMANN et al. 1985; SALCMAN et al. 1984; SAUER 1988; SZIKLA et al. 1981; THIEL et al. 1984, 1988; WILLIS et al. 1988), and the very involved technique of fractionated 74-day iridium-192 afterloading device irradiation with daily doses delivered through a cannula inserted into the tumor and secured on the bone does not offer any advantages (PANNEK et al. 1988; WEIGEL et al. 1987).

Apart from the problems associated with radiation protection with patients receiving several days' irradiation, the results of patients irradiated over several days were not better than those of the intraoperative one-time afterloading GammaMed irradiation, and the hospitalization stay was longer. This can be seen by comparing the three methods in 75 patients with malignomas, which was drawn up together with Weigel (Tables 6, 7). For a look at further results achieved with brachycurietherapy, the reader is advised to refer to earlier publications (MUNDINGER 1966, 1969a, 1970a, 1982b, 1984, 1986a; THIEL et al. 1984).

[2] Manufactured by Isotopen-Technik Dr. Sauerwein GmbH, 5657 Haan, Germany.

Table 6. Mean values of survival, radius, and peripheral tumor dose after one-time high dose rate brachycurietherapy with ^{192}Ir (GammaMed) or several days' high dose rate brachycurietherapy with ^{125}I

Iridium 192 (GammaMed)	*n*	Gy	mm	months
Anaplastic glioma III	6	27	15.1	7.3
Glioblastoma IV	16	30	15.6	10.3
Metastasis	2	29	18.0	18.5
Iodine-125 Brachycurie				
Anaplastic glioma III	10	38	15.1	11.5
Glioblastoma IV	9	40	18.5	5.0
Metastasis	5	40	13.2	12.0

Table 7. Mean survival and peripheral tumor dose after fractionated high dose rate brachycurietherapy with ^{192}Ir afterloading device (14 days)

	n	Gy	months
Anaplastic glioma III	10	24–30	10.0
Glioblastoma IV	17	24–30	12.0

3 Material and Results

3.1 Pituitary Adenomas

In our monograph published in 1967 Riechert and I reported (MUNDINGER and RIECHERT 1969) on the clinic report, pathology, treatment methods, and results in patients with pituitary adenoma, including our own results with several hundred patients.

This chapter focuses first of all only on the recurrences of extracerebral pituitary adenomas, which in most patients were observed after repeated open surgery or after the combination of surgery and curietherapy. Most of them were implanted with iridium 192 via a transcranial and transnasal/transsphenoid stereotaxic approach and some with iodine 125 (MUNDINGER 1970c, 1975, 1982b, 1987b; TALAIRACH et al. 1962).

Table 8 lists the 12-year follow-up results of tumors treated primarily or secondarily with iridium 192 after open surgery (MUNDINGER and BUSAM 1979). The comparison of the adenomas, which today are generally resected by microneurosurgery, as FAHLBUSCH and BUCHFELDER (1986) reported from a large series, indicates that apart from the necessary operation for recurrences, there is a trend to expose these tumors to additional percutaneous irradiation. However, according to

Table 8. Survival of pituitary adenomas after stereotactic iridium 192 curietherapy at 12-years' follow-up

	n	1	3	5	8 (%)	10	12 years
Chromophobe adenoma	39	95	92	89	89	85	85
					intrasellar 87		
Eosinophilic adenoma	34	100	94	92	83	83	83

Long-term results of visual disturbances after stereotactic iridium 192 curietherapy at 12-years' follow-up

	n	Amelioration	Unchanged	Worse
		(%)		
Chromophobe adenoma	39	55.2	44.8	–
Eosinophilic adenoma	34	23.5	64.7	11.8

Long-term results of working capacity after stereotactic iridium 192 curietherapy at 12-years' follow-up

	n	Full	Limited	Incapacity
		(%)		
Chromophobe Adenoma	39	50.0	34.6	15.4
Eosinophilic Adenoma	34	48.4	38.7	12.9

Table 9. Mortality and complications of 86 patients with pituitary adenomas after stereotactic ^{192}Ir curietherapy at 12-years' follow-up

	Percentage
Immediate operative death	0
General mortality up to 12 years	15.5
(as a result of operation 2.9%)	
(recurrence of adenoma 2.9%)	
Complications	
Delayed radiation damage (ophthalmologic changes, hypopituitarism)	2.0
Empty sella	2.0
Seizure	1.2

Table 10. Indications for radiotherapy of pituitary adenoma (in the event that microsurgical technique is no longer sufficient or is contraindicated)

Interstitial curietherapy with ^{125}I or ^{192}Ir for simple-shaped, well-defined tumors that are directly accessible to CT/MRI stereotaxic puncture

Percutaneous radiotherapy for complex-shaped, cryptogenic tumors that are not directly accessible

studies conducted by SNYDER et al. (1986), hypothalamopituitary insufficiency results in 55%–67% some 4.2 years after percutaneous irradiation.

Although this percentage is lower among patients treated by surgery alone (13%), it is also very low in our series (2%) and proves to be favorable in view of the 12-year follow-up (Table 9). Our experience has shown that for nonfunctional recurrent pituitary adenomas which have infiltrated the interbrain and are no longer resectable, implantation with iodine 125 is more effective and causes fewer side effects (especially considering that the interbrain has already been damaged by the adenomas itself) than additional percutaneous irradiation with 40–46 Gy.

Here it should be taken into account that in these differentiated tumors with WHO grading I, percutaneous irradiation in the hormone-inactive adenoma effects no regressive change in size, as opposed to low-dose rate interstitial curietherapy, which acts on the center of the adenoma. This is the criterion for the indication (Table 10).

3.2 Skull Base and Parasellar Tumors

Skull base and parasellar, nonresectable tumors are another indication for interstitial curietherapy, usually with a combination of iridium 192 and iodine 125. Nasopharyngeal carcinomas that have penetrated the cranial base can be approached transfrontally or from the facial part of the skull. In most cases percutaneous irradiation has already been performed so that these patients are considered "treated out". We have found that curietherapy has an additional palliative effect (ETOU et al. 1988). This indication for interstitial curietherapy, despite its therapeutic relevance, has always been used reluctantly.

3.3 Nonresectable Low-Grade Gliomas (WHO I, II)

In the following, 598 nonresectable gliomas are evaluated. Table 11 shows that in comparison with biopsy, in which other treatment forms are included (MUNDINGER et al. 1988a; MUNDINGER and WEIGEL 1988a), the patients treated interstitially with iridium 192 have the longest survival times, followed by iodine 125. The Karnofsky

Table 11. Life expectancy of patients with low-grade gliomas who underwent interstitial irradiation compared with those who had biopsies only (as of June 1985)

	Tumor type	n	Percentage of patients surviving at			
			1 year	3 years	5 years	10 years
Iodine 125	Astrocytoma I	67	95	70	55	–
Iridium 192	Astrocytoma I	59	94	86	78	61
Biopsy only	Astrocytoma I	79	73	49	45	21
Iodine 125	Astrocytoma II	106	87	31	28	–
Iridium 192	Astrocytoma II	91	86	45	31	26
Biopsy only	Astrocytoma II	196	86	17	6	–

Table 12. Life expectancy of low-grade gliomas after curietherapy (1.1.1965–30.6.1985) Dependency on peripheral accumulated tumor dose and volume of tumors

Tumor dose (cGy)	n	Mean life expectancy (in months)	2 y (%)	3 y (%)	5 y (%)	10 y (%)
Iridium 192						
12000–18200	7	41.2	42.5	15.5	–	–
12000	104	68.6	67.5	55.0	37.5	25.0
9000–11000	54	72.3	80.0	70.0	45.0	27.5
4000– 9000	6	21.1	57.5	–	–	–
Total	171					
Iodine 125						
Tumor dose (cGy)						
12000	83	46.4	62.5	47.5	42.5	–
9000–11000	59	41.8	62.5	55.0	–	–
4000– 9000	42	39.1	65.0	–	–	–
Total	184					
Volume of tumors (cm^3)						
0.9– 40	298	66.1	76.9	51.8	43.0	29.4
40–120	257	38.8	72.2	41.3	27.3	–
$\geqslant$120	45	26.1	62.5	44.0	–	–

scale is satisfactory to good over the entire follow-up period, as is shown in more detail another publications (HUBER-STENTRUP 1988; MUNDINGER 1988a, c, MUNDINGER and WEIGEL 1984a, 1985). Table 12 demonstrates the importance of the accumulated peripheral tumor dose and tumor size for life expectany.

3.3.1 Thalamus Tumors

Out of a series of 275 biopsied patients with thalamus tumors 100 low-grade gliomas along with the patients' survival times were reported in 1984 (MUNDINGER and WEIGEL 1984a). Patients treated with iridium 192 and iodine 125 have clearly better survival chances than those undergoing biopsy alone.

3.3.2 Diencephalic Region

We have also seen that in 303 patients with tumors of the diencephalon region, irradiation with iridium 192 followed by iodine 125 produces better results throughout all tumor gradings than other treatment methods (MUNDINGER 1970b, 1987c; MUNDINGER et al. 1979; MUNDINGER and HOEFER 1974; MUNDINGER and METZEL 1970). Only in rare cases, for instance, if the tumor has penetrated the ventricular system, is surgical reduction carried out. Resecting smaller tumors in this region results in a less favorable outcome, as it causes damage to the midline structures, something every neurosurgeon has experienced. In our 303 patients, surgery shortened their lives, as did percutaneous irradiation, to which these differentiated gliomas show little or no response (GREENBERGER et al. 1977; GREENWOOD 1973; KALFF et al. 1988; KELLY

1988). For this reason, the literature lacks long-term results of resection of gliomas of the diencephalon, even when performed with very sophisticated techniques (Apuzzo and Sabshin 1983; Beks et al. 1987; Kelly et al. 1978; Müller and Pötter 1988).

3.3.3 *Pineal Region*

The results of curietherapy for tumors in the pineal region were published in 1988 (Mundinger and Weigel 1988b). Of the 163 patients with tumors of the pineal region which we biopsied, 80 were exposed to interstitial irradiation.

The risk is low, with a mortality of 0.6% (caused by hemorrhage) and morbidity of 1.8%. The 5-year follow-up indicates that iodine 125 should be given preferentially for treating tumors in this region. After 5 years, 76% of the patients with a WHO I astrocytoma are still alive, as opposed to only 21% after iridium-192 implantation. Of the patients with a WHO II astrocytoma, 24% of those treated with iodine 125 are alive after 5 years, as opposed to 11% of those treated with iridium 192. Better survival results were achieved with smaller tumors and with a smaller dose (70 Gy). It is also possible that the better results are due to the reduced perifocal reaction of the mesencephalon after irradiation with iodine 125.

3.3.4 *Mesencephalic Brainstem Pons Region*

Our series of 186 patients with tumors in the mesencephalon brainstem pons region was reported in 1988 (Mundinger et al. 1988a). They are divided up according to the histologically confirmed WHO classification and Karnofsky scale. Although mortality and complications are tolerable (Table 13), the indication for curietherapy in this region must be made very meticulously and is dependent on the clinical picture. On the other hand, biopsy is indispensable, especially since in 12% (and this also applies to other locations) the lesions diagnosed as tumors by imaging methods turned out not to be tumors when biopsied (hematoma cysts, necrotizing abscesses, etc.) and thus have a completely different prognosis. Nontumors accounted for 12% of our total biopsy series (Mundinger et al. 1986; Mundinger 1986b). The decision to implant iridium 192 or iodine 125 will have to made much more strictly when dealing with lesions in the mesencephalon brainstem pons region. In nondelineated, high-grade tumors – biopsy mandatory – percutaneous irradiation as a palliative treatment is the only possible measure left.

Table 13. Complications and mortality after 186 CT-stereotaxic biopsies of the midbrain/brainstem/pons region without and with curietherapy

Complications	*n*	%	Mortality *n*	Mortality %
			3	1.6
Hemorrhage	4	2.2	All with malignant tumors died after 1 day due to herniation (edema); no hemorrhage	
Infarction	1	0.5		
Paresis	3	1.6	1	0.5
Seizures	4	2.2	Patient with glioblastoma died after 8 days due to herniation	

3.3.5 *Summary*

In general, the indication for curietherapy requires that the nonresectable low-grade lesion or recurrence be small, the neurological deficits show only slight damage, and the gliomas be well defined by means of neuroimaging methods in combination with biopsy (Tables 13, 14). In the case of infiltrating fibrillary astrocytomas WHO II, treatment with iridium 192 should be given

Table 14. Indicators for CT-stereostactic, one-time *high dose rate* ^{192}Ir or several days ^{125}I brachycurietherapy (biopsy obligatory)

Intracranial malignant tumors	
Primarily brachy-curietherapy	– Small volume hemispheric tumors – Tumors in functionally important region (central, temporal, parietal) Nonresectable, deep-seated white matter tumors
Secondary brachy-curietherapy	– After operation and external irradiation – Recurrences after external irradiation
Not indicated in processes around midline structures	
Nonresectable, deep-seated malignant tumors around cerebral/cerebellar midline structures	
WHO grading III	– ^{192}Ir or ^{125}I curietherapy (permanent implantation)
WHO grading IV	– External beam irradiation only
Brachycurietherapy not indicated	

Table 15. Indications for CT-stereotactic, *low dose rate* ^{192}Ir or ^{125}I curietherapy (permanent implantation) (biopsy obligatory)

Primarily	– Nonresectable, low-grade tumors (WHO I, II) around the cerebral/cerebellar midline – Nonresectable tumors (WHO II–III, III) followed by external irradiation
Secondary	– Resting low-grade tumors (WHO I, II) after partially resection – High-grade tumor recurrences after operation and external irradiation – After external irradiation only
Brachycurietherapy not indicated	

preference whenever possible, unlike in pilocytic astrocytomas WHO I, which usually can be so well visualized that the volume can be determined and dosimetry calculated accurately. In this case low-energy iodine 125 is the nuclide of choice.

For higher grade gliomas additive percutaneous irradiation of 45–55 Gy is indicated. The results of this tumor treatment were presented in 1986 (Mundinger 1986a; see also Nicholas et al. 1988; Pierquin 1964, 1976; Schlienger et al. 1979; Szikla et al. 1979, 1984b).

4 Discussion

4.1 Pituitary Tumors

Interstitial curietherapy for pituitary tumors was applied very early on (Mundinger 1969b, 1970c; Mundinger and Riechert 1969; Talairach et al. 1962). In the preimaging era, the widened hypophyseal fossa and cisternography with air or gas for intrasellar and suprasellar spread were the localizing tools used for the intracranial, later transsphenoid, stereotaxic approach, and the implantation of phosphorus-32, gold-198, and yttrium-90 seeds and iridium 192. The numerous contributions on the effectiveness of this method were compiled in our publications. The long-term results presented in this article, as well as the projected ones, demonstrate the success of interstitial curietherapy in the treatment of functional and non-functional pituitary adenomas and for hypophysectomy. It used to be indicated both primarily as well as secondarily following surgical resection or decompression. The indications are listed in Table 10.

This underwent some change after Hardy, who adopted the transnasal/transsphernoid approach from Hirsch back at the beginning of the century, introduced the microsurgical transsphenoid operation (Hardy 1969; Hirsch 1910).

Interstitial curietherapy is now indicated again for recurrent tumors that can no longer be resected, for instance, after repeated attempts at open surgical resection. On the one hand, there seems to be evidence that microsurgery alone is not always sufficient, particularly in the case of functional pituitary adenomas. On the other hand, external radiotherapy, which has undergone a revival especially in the USA, is laden with a high risk of delayed radiation damage. Hypothalamic insufficiency has been observed in 55%–67% of patients after surgery and external radiotherapy, and in 50%–55% after follow-up irradiation alone, after an average of 4.2 years (Snyder et al. 1986). The patients will be seriously restricted in terms of their ability to work, and their quality of life will be diminished (Mundinger 1987b). This is because the hypothalamus in the vicinity of an adenoma that has spread suprasellarly will be exposed to almost the same dose as the tumor itself (40–45 Gy), even under external pendulum irradiation or pendulum – convergent external beam irradiation. The functioning of the hypothalamus cannot tolerate this.

In contrast, with curietherapy the rapid dose fall-off and the protraction of the very low dose rate irradiation reduce the exposure of the tumor's surroundings. Furthermore, it is possible to discontinue the irradiation if necessary by removing the catheter before the calculated accumulated radiation dose has been completely absorbed.

4.2 Parasellar Tumors

Further indications for imaging-stereotaxic curietherapy are parasellar tumors and tumors of the cranial base, such as nasopharyngeal carcinomas which have infiltrated the cranial base (Etou et al. 1988). In these patients the first measures taken are surgery and external irradiation therapy. Should these measures cease to have any effect, or if the tumors are recurrences that have been "treated out", interstitial curietherapy is able to achieve a palliative result, if necessary, through repeated reimplantations. This indication should be given more attention in the future.

4.3 Intracerebral Tumors

In recent years, various study groups have reported on intracerebral tumors treated with high-dose rate (brachycurietherapy) application and dosimetry – usually with iodine 125 and iridium 192 – and have provided postoperative results. Most of these reports, however, deal with anaplastic gliomas (WHO III), glioblastomas, or malignant tumors (WHO IV). The results are encouraging and are consistent with the results we have published. This also applies to recurrences of malignant tumors which are considered no longer treatable by common therapeutic methods. Our experience has shown that with the intraoperative, single, very high dose irradiation with afterloading instruments – in our case with iridium-192 GammaMed – a similar palliative result can be achieved, or in the case of glioblastomas, even a better result than with the continuous high dose rate spread over several days or the high dose rate afterloading brachycurietherapy fractionated over 14 days.

The major advantage of intraoperative, single, very high dose irradiation is that the dose is administered within just a few minutes, so that unlike with implants which continue to deliver radiation doses for several days, there are no problems later on in terms of radiation protection.

For treatment of high-grade tumors in the past few years, the combined application of intraoperative very high dose rate irradiation with the very low dose rate irradiation with iodine 125 (in very small tumors with iridium 192) seems to be the superior method. With this technique, the intraoperative boost bridges the time needed for a biologically effective dose to be accumulated for iridium 192 or iodine 125 applied with low activities as a protracted long-term irradiation. Also, percutaneous irradiation with 40–45 Gy and a larger field can be employed subsequently. Our investigation, which is currently being evaluated, will show to what degree this combined irradiation treatment improves survival by achieving the required Karnofsky scale of 6–8. The trend seems to be going in that direction (JELLINGER et al. 1981).

A further advantage is that biopsy, which always preceeds interstitial irradiation with high and very low dose rate curietherapy, is incorporated into the stereotaxic operation, and the patient can be either allowed to return home or referred to further therapy very quickly (for our patients the average hospital stay is 3.8 days). High dose rate and low dose rate curietherapy can be combined in one operation, even in the case of metastatic tumors. In 88% of the metastases it was not until imaging-stereotaxic biopsy was performed that we could start to look for the primary tumor (WEIGEL et al. 1984).

Very low dose rate curietherapy, that is, protracted long-term irradiation or permanent irradiation, which I developed and introduced into the treatment of intracranial brain tumors, has also proved to be effective in tumors of other organs, such as prostatic, tongue, or bronchial cancer (HILARIS 1975; KNÜFERMANN et al. 1981).

This therapy, which I initially applied with tantalum 182 (MUNDINGER 1958), then with iridium 192 and iodine 125 for intracranial tumors, for a long time was not accepted as a treatment modality for low-grade astrocytomas. The 10-year results presented here, along with radiobiological experimental findings like those compiled by BERNSTEIN and GUTIN (1981), BOUZAGLOU et al. (1985), CHAVAUDRA et al. (1979), HILARIS (1975), MUNDINGER (1970a, 1979), demonstrate its radiotherapeutic efficacy. Patients who later underwent surgery alone or in combination with percutaneous irradiation, or percutaneous irradiation alone, are subsumed under the heading "biopsied only". Most of these patients had gliomas in the diencephalon, mesencephalon, and brainstem-pons region, for which surgery, even microsurgery, was not or did not appear to be indicated. In these patients clearly better survival times were achieved with curietherapy.

4.4 Gliomas of the Midline

For nonresectable WHO I and II low-grade gliomas of the midline, curietherapy as the sole therapeutic method is the most effective one. These tumors are highly resistant to externally administered radiation (50–65 Gy) (BEKS et al. 1987; GONZALES GONZALES and VAN DIJK 1988; GREENBERGER et al. 1977; HUBER-STENTRUP 1988; KALFF et al. 1988; LEIBEL and SHELINE 1987; MUNDINGER 1963). In other words, they show no or only few radioregressive changes, no or only little shrinkage or necrosis (GONZALES GONZALES and VAN DIJK 1988). They manifest only small cystic degenerations in the sense of inner decompression of the perifocal region. This we were able to confirm through re-biopsies.

With curietherapy, the low-grade tumors can be

effectively treated by administering the maximum dose to the tumor center itself with a rapid dose fall-off towards the periphery and with long-term protraction of the dose delivery over 5–8 months until a level of between 70 and 120 Gy, depending on the location, has been reached. This, however, requires that the tumor volume be below 120 mm^3 (ideal is less than 40 mm^3) (see Table 12) and that the lesion be well defined by imaging methods so that an exact dosimetry can be calculated. Reimplantations can be successful in the case of recurrent tumors or in regions which have been underdosed, something we have learned from many examples in our patient population, which now numbers 2200. A detailed presentation of the data does not fit into the scope of a summary such as this.

We do not know of any other long-term results in low-grade gliomas of the midline treated with very low dose rate curietherapy. The reason for this is that curietherapy is just beginning to make headway. Another reason is that a valid 10-year evaluation requires a minimum of 15 years' follow-up. The lack of data on the external irradiation of low-grade gliomas that are confirmed by biopsy is due to the fact that imaging-stereotaxic biopsy has only recently become employed on a wider scale. The more recently published results by Beks et al. (1987) and by Bernstein et al. (1984) based on a few thalamic tumors are disappointing. Moreover, in patients with space-occupying lesions in the brainstem and pons, radiotherapy (and cerebrospinal fluid shunting methods in the case of occlusion hydrocephalus) was initiated on the basis of the clinical and imaging diagnosis, often without obtaining biopsy confirmation (Greenberger et al. 1977; Leibel and Sheline 1987; Ryoo et al. 1979; Sauer 1988).

As we now know from experience, what appear to be space-occupying lesions turn out in 12%–18% of patients to be cysts, necroses, inflammatory lesions, multiple sclerosis (MS), and glioses. In these patients the result of irradiation therapy could be misinterpreted as being effective, as such lesions – apart from MS – show no clinical progression.

The results in the compiled statistics of German neurosurgeons with over 1500 patients (Wüllenweber et al. 1973) of low-grade, resected or debulked, and externally irradiated gliomas are not significantly better than those achieved by surgery alone. A multinational, randomized, double arm study on the combination of surgery/external radiation and radiation alone has now been started in the Benelux countries and in Germany. The study is to run a minimum of 10 years. This means that we cannot expect to have statistically supported answers earlier than 15 years from now (Krauseneck 1987). Our results already provide justification for preference being given to very low dose rate curietherapy alone after biopsy to treat low-grade gliomas and other low-grade tumors.

References

Abrath FG, Henderson SD, Simpson JR, Moran CJ, Marchosky JA (1986) Dosimetry of CT-guided volumetric Ir-192 brain implant. Int J Radiat Oncol Biol Phys 12: 359–363

Apuzzo MLJ, Sabshin JK (1983) Computed tomographic guidance stereotaxis in the management of intracranial mass lesions. Neurosurgery 12: 277–284

Beks JWF, Bouma GJ, Journée HL (1987) Tumours of the thalamic region. A retrospective study of 27 cases. Acta Neurochir (Wien) 85: 125–127

Bernstein M, Gutin PH (1981) Interstitial irradiation of brain tumors. A review. Neurosurgery 9: 741–750

Bernstein M, Hoffmann HJ, Halliday WC (1984) Thalamic tumors in children. Long-term follow-up and treatment guidelines. J Neurosurg 61: 649–656

Birg W, Mundinger F (1973) Computer calculations of target parameters for a stereotactic apparatus. Acta Neurochir (Wien) 29: 123–129

Birg W, Mundinger F (1982) Direct target point determination for stereotactic brain operations from CT data and the calculation of setting parameters for polar-coordinate stereotactic devices. Appl Neurophysiol 45: 387–395

Birg W, Schneider J, Bauer S, et al. (1979) An interactive program system for the stereotactic interstitial implantation of radionuclides in brain tumors. INSERM Symp 12: 77–80

Birg W, Mundinger F, Mohadjer M, et al. (1985) X-ray and MR-stereotaxy for functional and nonfunctional neurosurgery. Appl Neurophysiol 48: 22–29

Bouzaglou A, Dyck P, Solt-Bochmann LG, Gruskin P (1985) Stereotactic interstitial implantation of brain tumors. Endocuriether Hyperthermia Oncol 1: 99–112

Braus DF, Mundinger F (1988) Neuroimaging Fehldiagnosen bei MS. 61st Annual Meeting of the Deutschen Gesellschaft für Neurologie, Sept 22–24, Frankfort

Chavaudra J, Schlienger M, Szikla G (1979) Some considerations on the physical and clinical aspects of stereotactic cerebral irradiation. INSERM Symp 12: 177–184

Davis RL, Barger GR, Gutin PH, Phillips TL (1983) Response of human malignant gliomas and CNS tissue to 125-I brachytherapy: a study of seven autopsied cases. Acta Neurochir [Suppl] (Wien) 33: 301–305

Dyck P (1983) Stereotactic biopsy and brachytherapy of brain tumors. University Park Press, Baltimore

Etou A, Mohadjer M, Mundinger F (1988) Curietherapy of reoccuring nasopharyngeal carcinoma invading the skull base. Skull Base Study Group. IV. International Congress: Surgery of the sella region and paranasal sinuses, June 3–6, Hannover (in press)

Fahlbusch R, Buchfelder M (1986) Transsphenoidale Chi-

rurgie der sellären und parasellären raumfordernden Prozesse. In: Walter W, Krenkel (eds) Jahrbuch der Neurochirurgie 1988. Regensberg and Biermann, Münster, pp 53–72
Gonzales Gonzales D, van Dijk JDP (1988) Ergebnisse der Strahlentherapie bei Gliomen niedrigen Malignitätsgrades. In: Bamberg M, Sack H (eds) Therapie primärer Hirntumoren. Zuckschwerdt, Münich pp 196–200
Greenberger JS, Cassady JR, Levene MB (1977) Radiation therapy of thalamic, midbrain and brain stem gliomas. Radiology 122: 463–468
Greenwood J Jr (1973) Radical surgery of tumors of the thalamus, hypothalamus and third ventricle area. Surg Neurol 1: 29–33
Gutin PH, Phillips TL, Hosobuchi Y, Wara WM, et al. (1981) Permanent and removable implants for the brachytherapy of brain tumors. Int J Radiat Oncol Biol Phys 7: 1371–1381
Gutin PH, Phillips TL, Wara WM, Leibel SA, Hosobuchi Y, Levin VA, Weaver KA, Lamb S (1984) Brachytherapy of recurrent malignant brain tumors with removable high-activity iodine-125 sources. J Neurosurg 60: 61–68
Hardy J (1969) Transsphenoidal microsurgery of the normal and pathological pituitary. Clin Neurosurg 16: 185–217
Hilaris BS (1975) Handbook of interstitial brachytherapy. Publishing Sciences Group, Acton
Hirsch O (1910) Endonasal method of removal of hypophyseal tumors. With report of two successfull cases. JAMA 55: 772–774
Huber-Stentrup M (1988) Stereotaktische interstitielle Curie-Therapie mit Iridium-192 und Jod-125 bei nicht resezierbaren niedergradigen Gliomen des Gehirns. Verlaufsbeobachtung mit 10jährigen Langzeitergebnissen bei 659 Patienten. Dissertation, University of Freiburg
Jellinger K, Volc D, Podreka I, et al. (1981) Ergebnisse der Kombinationsbehandlung maligner Gliome. Nervenarzt 52: 41–50
Kalff R, Pospiech J, Roosen K, et al. (1988) Therapie der supratentoriellen Astrozytome niedrigen Malignitätsgrades. In: Bamberg M, Sack H (eds) Therapie primärer Hirntumoren. Zuckschwerdt, Münich, pp 210–213
Kelly PJ (1988) Volumetric stereotaxis and computer-assisted stereotactic resection of subcortical lesions. In: Lunsford LD (ed) Modern stereotactic neurosurgery. Nijhoff, Boston, pp 169–184
Kelly PJ, Olson MH, Wright AG (1978) Stereotactic implantation of iridium-192 into CNS neoplasms. Surg Neurol 10: 349–354
Kiessling M, Kleihues P, Gessaga E, et al. (1984) Morphology of intracranial tumors and adjacent brain structures following interstitial iodine-125 radiotherapy. Acta Neurochir [Suppl] (Wien) 33: 281–289
Kleihues P, Volk B, Anagnostopoulos J, et al. (1984) Morphologic evaluation of stereotactic brain tumour biopsies. Acta Neurochir [Suppl] (Wien) 33: 171–181
Knüfermann H, Bruggmoser G, Wannenmacher M (1981) Die interstitielle Strahlentherapie in der Behandlung des Prostatakarzinoms. In: Wannenmacher M, Schreiber HW, Gauwerky F (eds) Kombinierte chirurgische und radiologische Behandlung maligner Tumoren. Urban and Schwarzenberg, Münich, pp 63–69
Krauseneck P (1988) "Therapie primärer Hirntumoren", In: Bamberg M, Sack H (eds) Therapie primärer Hirntumoren. Zuckschwerdt, Münich, pp 410–415
Leibel SA, Sheline GE (1987) Radiation therapy for neoplasms of the brain. J. Neurosurg 66(1): 1–22
Lunsford LD (1988) Diagnosis and treatment of mass lesions using the Leksell stereotactic system. In: Lunsford LD (ed) Modern stereotactic neurosurgery. Nijhoff, Boston, pp 145–168
Mohadjer M, Milios E, Mundinger F (1987) Risiken der CT- (MRI-) stereotaktischen intracraniellen Biopsie. 8th Congress of the Gesamtverband Deutscher Nervenärzte, May 28–30, Kiel
Müller RP, Pötter R (1988) Behandlungsergebnisse nach Strahlentherapie von zerebralen und zerebellären Astrozytomen im Kindesalter. In: Bamberg M, Sack H (eds) Therapie primärer Hirntumoren. Zuckschwerdt, Münich, pp 214–217
Müller R, Thiel HJ, Herbst M (1982) Dose planning in interstitial therapy with 125-I-seeds. 1st Annual Meeting of the European Society for Therapeutic Radiology and Oncology, London
Mundinger F (1956) Eine einfache Methode der lokalisierten Bestrahlung von Großhirngeschwülsten mit radioaktivem Gold. MMW 89: 23–25
Mundinger F (1958) Beitrag zur Dosimetrie and Applikation von Radio-Tantal (^{182}Ta) zur Langzeitbestrahlung von Hirngeschwülsten Fortschr Geb Röntgenstr 89: 86–91
Mundinger F (1963) Die interstitielle Radio-Isotopen-Bestrahlung von Hirntumoren mit vergleichenden Langzeitergebnissen zur Röntgentiefentherapie. Acta Neurochir (Wien) 9: 89–109
Mundinger F (1966) Treatment of brain tumors with radioisotopes. Prog Neurolog Surg 1: 202–257
Mundinger F (1969a) Erfahrungen mit der stereotaktischen interstitiellen Brachytherapie mit Ir-192-GammaMed bei infiltrierenden Hirntumoren. Fortschr Geb Röntgenstr 110: 254–261
Mundinger F (1969b) Die intraselläre protrahierte Langzeitbestrahlung von Hypophysenadenomen mittels stereotaktischer Implantation von Iridium-192. Acta Radiol (Stockh) 8: 55–62
Mundinger F (1970a) The treatment of brain tumors with interstitially applied radioactive isotopes. In: Wang Y, Paoletti P (eds) Radionculide applications in neurology and neurosurgery. Thomas, Springfield, pp 199–265
Mundinger F (1970b) Interstitial radioisotope therapy of intractable diencephalic tumors by the stereotaxic permanent implantation of iridium-192, including bioptic control. Confin Neurol 32: 195–203
Mundinger F (1970c) Intraselläre Iridium-192 Permanent-Implantation bei Hypophysenadenomen. In: Busche KA (ed) Fortschritte auf dem Gebiet der Neurochirurgie. Hippokrates, Stuttgart, pp 83–87
Mundinger F (1975) Interstitial curietherapy in the treatment of pituitary adenomas and for hypophysectomy. Prog Neurol Surg 6: 326–379
Mundinger F (1979) Rationale and methods for interstitial iridium-192 brachy-curietherapy and iridium-192 or iodine-125 protracted long-term irradiation. INSERM Symp 12: 101–116
Mundinger F (1981) Die stereotaktische interstitielle Therapie nicht resezierbarer intracranieller Tumoren mit Ir-192 und I-125. In: Wannenmacher M, Schreiber HW, Gauwerky F (eds) Kombinierte chirurgische und radiologische Behandlung maligner Tumoren. Urban and Schwarzenberg, Münich, pp 90–112
Mundinger F (1982a) Stereotactic interstitial therapy of

nonresectable intracranial tumours with iridium-192 and iodine-125. In: Kärcher KH, et al. (eds) Progress in radio-oncology II. Raven, New York, pp 371–380

Mundinger F (1982b) Implantation of radioisotopes (curietherapy). In: Schaltenbrand G, Walker AE (eds) Textbook of stereotaxy of the human brain. Thieme, Stuttgart, pp 410–435

Mundinger F (1982c) CT-stereotactic biopsy of brain tumors. In: Voth D, Gutjahr P, Langmaid C (eds) Tumours of the central nervous system in infancy and childhood. Springer, Berlin Heidelberg New York, pp 234–246

Mundinger F (1984) Stereotaktische intrakranielle Bestrahlung von Tumoren mit Radioisotopen (Curie-Therapie). In: Dietz H, Umbach W, Wüllenweber R (eds) Klinische Neurochirurgie, vol 2. Thieme, Stuttgart, pp 519–565

Mundinger F (1985a) Technik und Ergebnisse der interstitiellen Hirntumorbestrahlung. In: Heilmann HP (ed) Spezielle Strahlentherapie maligner Tumoren. Springer, Berlin Heidelberg New York, pp 179–214 (Handbuch der medizimischen Radiologie, vol 19/4)

Mundinger F (1985b) CT-stereotactic biopsy for optimizing the therapy of intracranial processes. Acta Neurochir (Wien) 35: 70–74

Mundinger F (1986a) Stereotactic biopsy and technique of implantation (instillation) of radionuclids. In: Jellinger K (ed) Therapy of malignant brain tumors. Springer, Berlin Heidelberg New York, pp 134–194

Mundinger F (1986b) Die CT-stereotaktische interstitielle Strahlenbehandlung bei nicht resezierbaren Hirntumoren mit Radionukliden. Fortschritte in der interdisziplinären Onkologie, Sept 19–20, Hamburg

Mundinger F (1987a) Stereotactic biopsy and technique of implantation (instillation) of radionuclei. In: Kärcher KH, Kogelnik HD, et al. (eds) Progress in radio-oncology II. Raven, New York, pp 371–380

Mundinger F (1987b) Die Strahlenbehandlung der Hypophysen-Adenome. 41st Österrischer Ärztekongreß, Oct 26–31, Vienna

Mundinger F (1987c) Indikation und Ergebnisse der interstitiellen Curietherapie nicht resezierbarer niedriggradiger Tumoren des Zwischenhirns. Deutsche Gesellschaft für Neurochirurgie 2. Arbeitstagung der Arbeitsgemeinschaft Neuro-Onkologie, Nov 5–7, Tübingen

Mundinger F (1988a) Interstitielle Curietherapie bei Gliomen niedrigen Malignitätsgrades. In: Bamberg M, Sack H (eds) Therapie primärer Hirntumoren. Zuckschwerdt, Münich, pp 285–290

Mundinger F (1988b) Stereotactic biopsy and implantation of radionuclides guided by computed tomography or magnetic resonance imaging for therapy of brain tumors. In: Schmidek HH, Sweet WH (eds) Operative neurosurgical techniques, vol 1. Grune and Stratton, Orlando, pp 491–514

Mundinger F (1988c) CT- and MR-guided stereotaxis interstitial implantation of radioisotopes into non-resectable intracranial tumors: long term results of 10 years experience. Annual Meeting of the American Association of Neurological Surgeons, Apr 24–28, Toronto

Mundinger F, Birg W (1988) The imaging-compatible Riechert-Mundinger-system. In: Lunsford LD (ed) Modern stereotactic neurosurgery. Nijhoff, Boston, pp 13–25

Mundinger F, Busam B (1979) Stereotactic interstitial iridium-192 permanent implantation of pituitary adenomas. INSERM Symp 12: 187–197

Mundinger F, Hoefer T (1974) Protracted long-term irradiation of inoperable midline tumors by stereotactic curietherapy using iridium-192. Acta Neurochir [Suppl] (Wien) 21: 93–100

Mundinger F, Metzel E (1970) Interstitial radioisotope therapy of intractable diencephalic tumors by the stereotactic permanent implantation of iridium-192, including bioptic control. Confin Neurol 32: 195–202

Mundinger F, Riechert T (1969) Hypophysentumoren, Hypophysektomie. Klinik, Therapie, Ergebnisse. Thieme, Stuttgart

Mundinger F, Sauerwein (1966) "GammaMed", ein neues Gerät zur interstitiellen, nur einige Minuten dauernden Bestrahlung von Hirngeschwülsten mit Radioisotopen, auch intraoperativ anwendbar. Acta Radiol (Stockh) 5: 48–52

Mundinger F, Weigel K (1984a) Long-term results of stereotactic interstitial Curietherapy. Acta Neurochir [Suppl] (Wien) 33: 367–371

Mundinger F, Weigel K (1984b) Stereotactic curietherapy of thalamic tumours. J Neurooncol 2: 264–278

Mundinger F, Weigel K (1985) CT-stereotactic interstitial irradiation therapy of nonresectable and recurrent intracranial tumors in children and adolescents. In: Voth D, Krauseneck P (eds) Chemotherapy of gliomas: basic research, experiences and results. de Gruyter, Berlin, pp 241–259

Mundinger F, Weigel K (1988a) Considerations in the usage and results of curietherapy. In: Lunsford LD (ed) Modern stereotactic neurosurgery. Nijhoff, Boston, pp 245–258

Mundinger F, Weigel K (1988b) Die stereotaktische Biopsie und interstitielle Therapie von Raumforderungen der Pinealisregion. In: Bamberg M, Sack H (eds) Therapie primärer Hirntumoren. Zuckschwerdt, Münich, pp 361–366

Mundinger F, Busam B, Birg W, et al. (1979) Results of interstitial iridium-192 brachy-curietherapy and iridium-192 protracted long-term irradiation. INSERM Symp 12: 303–320

Mundinger F, Weigel K, Fürmaier R, Volk B (1986) CT and MRI diagnoses of intracranial tumours compared with the results of stereotactic biopsy. In: Poeck K, Freund HJ, Gänshirt H (eds) Neurology. Springer, Berlin Heidelberg New York, pp 469–476

Mundinger F, Birg W, Etou A, Mohadjer M (1988a) CT/NMR-stereotaktische Diagnose und Curie-Therapie von Tumoren des Mittelhirn/Hirnstamm/Brücken/Bereichs. In: Walter W, Krenkel (eds) Jahrbuch der Neurochirurgie 1988. Regensberg and Biermann, Münster, pp 51–72

Mundinger F, Birg W, Huber-Stentrup M, et al. (1988b) 10 years follow up results after iridium-192- and iodine-125-permanent implantation of low grade astrocytomas. Annual Meeting of the American Association of Neurological Surgeons, Apr 24–28, Toronto

Murray KJ, Blumberg A, Strubler K, et al. (1984) Permanent radioactive iodine seed implants following radical resection in recurrent human malignant high-grade astrocytomas. J Neurooncol 2: 277–282

Musolino A, Munari C, Blond S, et al. (1985) Traitement stéréotaxique des kystes expansifs de craniopharyngiomes par irradiation endocavitaire beta (Re-186, Au-198, Y-90). Neurochirurgie 31: 169–178

Nicholas M, et al. (1988) Interstitial brachytherapy for malignant brain tumors: technique and results. In:

Lunsford LD (ed) Modern stereotactic neurosurgery. Nijhoff, Boston, pp 235–243

Pannek HW, Oppel F, Ernst H (1988) Die Behandlung maligner Gliome durch das fraktionierte Afterloading in Kombination mit perkutaner Bestrahlung. In: Bamberg M, Sack H (eds) Therapie primärer Hirntumoren. Zuckschwerdt, Münich, pp 297–301

Pierquin B (1964) Précis de curiethérapie. Masson, Paris

Pierquin B (1976) The destiny of brachytherapy in oncology. AJR 127: 495–499

Pollack IF, Lunsford LD, Slamovits TL, et al. (1988) Stereotaxic intracavitary irradiation for cystic craniopharyngiomas. J Neurosurg 68: 227–233

Riechert T, Mundinger F (1956) Beschreibung und Anwendung eines Zielgerätes für stereotaktische Hirnoperationen (II. Modell). Acta Neurochir (Wien) 3: 308–337

Riechert T, Mundinger F (1959) Stereotaxic instruments. In: Schaltenbrand G, Bailey P (eds) Introduction to stereotaxis with an atlas of the human brain. Thieme, Stuttgart, pp 437–471

Rossmann KJ, Shetter AG, Speiser BL, Nehls D (1985) Stereotactic afterloading iridium implants in treatment of high-grade astrocytomas. Endocuriether Hyperthermia Oncol 1: 49–57

Ryoo MC, Kind GA, Chung T, et al. (1979) Irradiation of primary brainstem tumors. Radiology 131: 503–507

Salcman M, Sewchand W, Amin P, et al. (1984) CT-guided stereotactic surgery and interstial irradiation for glial tumors. J Neurooncol 2: 279–282

Sauer R (1988) Gegenwärtiger Stand der Strahlenbehandlung der Gliome hohen Malignitätsgrades. In: Bamberg M, Sack H (eds) Therapie primärer Hirntumoren. Zuckschwerdt, Münich, pp 75–82

Schaub C, Bluet-Pajot MT, Videau-Lornet C, et al. (1979) Endocavitary beta irradiation of glioma cysts with colloidal rhenium-186. INSERM Symp 12: 293–302

Schlienger M, Bouhnik H, Missir O, Constans JP, Szikla G (1979) Association of temporary interstitial Ir-192 implantation and external radiotherapy in the management of supratentorial tumors. 12: 117–121

Snyder PJ, Fowble BF, Schatz NJ, et al. (1986) Hypopituitarism following radiation therapy of pituitary adenomas. Am J Med 81(3): 457–462

Sturm V, et al. (1988) Intracavitary irradiation of cystic craniopharyngiomas. In: Lunsford LD (ed) Modern stereotactic neurosurgery. Nijhoff, Boston, pp 229–233

Szikla G, Schlienger M, Betti O, et al. (1979) Combined interstitial and external irradiation of gliomas. A progressed report. INSERM Symp 12: 329–338

Szikla G, Betti O, Szenthe L, et al. (1981) L'expérience actuelle des irradiations stéréotaxiques dans le traitement des gliomes hémisphériques. Neurochirurgie 27: 295–298

Szikla G, Musolino A, Miyahara S, et al. (1984a) Colloidal rhenium-186 in endocavitary beta irradiation of cystic craniopharyngiomas and active gliomas cysts: long-term results, sideeffects and clinical dosimetry. Acta Neurochir [Suppl] (Wien) 33: 331–339

Szikla G, Schlienger M, Blond S, Daumas-Dupport C, et al. (1984b) Interstitial and combined interstitial and external irradiation of supratrentorial gliomas. Results in 61 cases treated 1973–1981. Acta Neurochir [Suppl] (Wien) 33: 355–362

Talairach J, Sizkla G, Bonis A, et al. (1962) Déstruction stéréotaxique de l'hypophyse non tumorale par les isotopes radioactifs. Presse Med 70: 1399–1402, 1449–1451

Thiel JH, Huk WJ, Müller R, Sauer R (1984) Die temporäre und permanente Brachycurietherapie inoperabler Hirntumoren mit Jod-125-Seeds. Klinikarzt 8: 736–750

Thiel HJ, Sauer R, Müller RG, et al. (1988) Kombinierte interstitielle und perkutane Bestrahlung von Hirngliomen: Probleme der Dosierung und der Dosisverteilung. In: Bamberg M, Sack H (eds) Therapie primärer Hirntumoren. Zuckschwerdt, Münich, pp 291–296

Weigel K, Mohadjer M, Mundinger F (1984) CT-stereotaxy for differential diagnosis and radiotherapy of intracranial metastases. Neurosurg 12: 87–93

Weigel K, Sparenberg A, Wehr M, Brock M, Ernst H, Mundinger F (1987) Brachy-Curie-Therapie: Fraktioniertes Afterloading. Aktuelle Therapieformen bei malignen intrakraniellen Tumoren. 38. Jahrestagung der Deutschen Gesellsch für Neurochirurgie, May 3–6

Willis BK, Heilbrunn MP, Sapozink MD, McDonald PR (1988) Stereotactic interstitial brachytherapy of malignant astrocytomas with remarks on postimplantation computed tomographic appearance. Neurosurgery 23(3): 348–354

Wüllenweber R, Kuhlendahl H, Miltz H (1973) Astrocytomas of the cerebral hemispheres. A review on 1.500 cases. In: Proccedings of the German Society of Neurosurgery, vol 3. Excerpta Medica, Amsterdam, pp 100–107

2.3 Experimental Dose Effects After Permanent and Temporary Interstitial Irradiation of the Brain

CHRISTOPH B. OSTERTAG

CONTENTS

1 Introduction 81
2 Materials and Methods 81
3 Results 81
4 Discussion 82
References 84

1 Introduction

Clinical experience has demonstrated that gliomas of the brain are relatively resistant to conventional irradiation (teletherapy) (BLOOM and WALSH 1975; BURGER et al. 1979; LEIBEL and SHELINE 1987). Thus, the dose-effect curve for tumor control and the tissue damage curve obtained in many clinical situations turn out to be almost identical. This has led to various attempts to increase the therapeutic spectrum of radiotherapy. Apart from the stereotaxic focusing of radiation with linear accelerators, brachytherapy has been receiving a considerable amount of attention again in the past few years (BERNSTEIN and GUTIN 1981; GUTIN and LEIBEL 1985; OSTERTAG 1986). The dose rates used for interstitial radiation vary from very high dose rates for different afterloading techniques to very low dose rates for permanent implants with low-energy emitters such as iodine 125. The dosimetry calculations for interstitial irradiation have been based largely on clinical experience (ANDERSON et al. 1981; HILARIS 1975). To devitalize a tumor with high local radiation doses no longer poses as serious a problem as the damage done to the tumor-free tissue. This damage is not so much the harm inflicted by direct radiation but rather that caused by secondary chronic vasogenic edema. In order to obtain confirmed biological data as a foundation on which to base the clinical dosimetry, we studied the effects of doses delivered through γ-emitters with different energy and dose rates.

CHRISTOPH B. OSTERTAG, Prof. Dr., Abteilung Stereotaktische Neurochirurgie, Neurochirurgische Universitätsklinik, 7800 Freiburg, Germany

2 Materials and Methods

Experimental dose effects on the brain (beagle) caused by the γ-emitters iodine 125, iridium 192, and gold 198 were examined over an exposure period of 5–365 days (Table 1). Permanent, low activity, low dose rate implants were compared over time with temporary, high activity, high dose rate implants, both delivering the same accumulated total radiation dose. Particular attention was paid to the morphological development, size, and composition of the radiation necrosis, as well as to changes in the blood-brain barrier. Disruptions of neurological functions were monitored by daily observation and by recordings of evoked potentials. The reference dose was taken as the cumulative dose of 26 000 cGy at a distance of 5 mm from the center of the seed (Fig. 1).

3 Results

Radionecroses were manifest as early as 10 days after implantation (Fig. 2). A marked glial reaction

Table 1. Principal physical properties of the isotopes used experimentally

Isotope	Half-life (days)	Energy		
		Alpha	Beta	Gamme (MeV)
Gold 198	2.7	–	(+)[a]	0.41
Iridium 192	74.2	–	(+)[a]	0.30 –0.61
Iodine 125	60.2	–	–	0.028–0.035

[a] β-energy shielded by platinum coating.

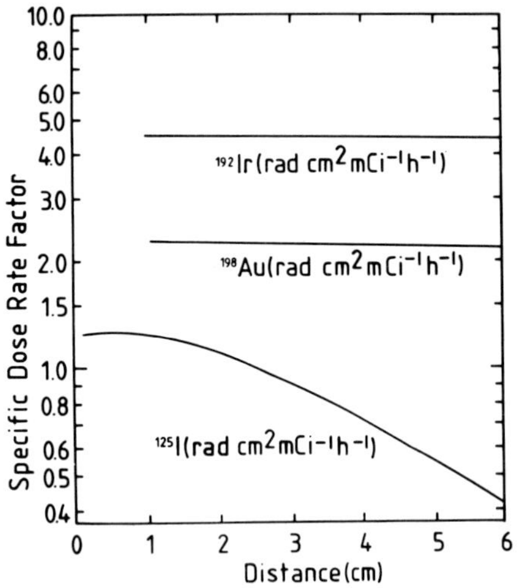

Fig. 1. Comparison of specific dose factors of the γ-emitters used. The γ-energy of iodine 125 is rapidly absorbed by the tissue (modified after Hilaris 1975)

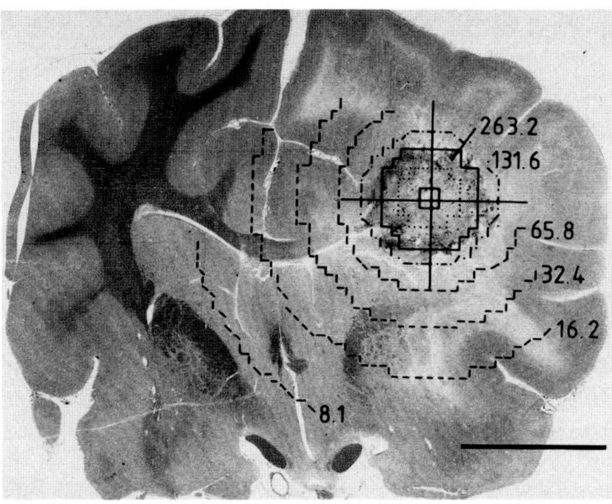

Fig. 2. Coronal brain section (Klüver-Barrera stain) with isodose overlay through an experimentally induced radionecrosis [temporary (10 days) gold-198 implant (31.8 mCi) after 46 days' survival]. A sharply demarcated, calcified necrotic zone (volume 0.9 ml) can be recognized. Note the considerable mass effect with midline shift due to vasogenic edema and the pallor of the ipsilateral white matter. *Horizontal bar* corresponds to 1 cm; *numbers* indicate isodoses in Gy

could be seen in the white matter adjacent to the radiation lesion. The neurons next to the radionecrosis but not incorporated, however, appeared to be morphologically intact, even at a distance of 0.2 mm. The vessels showed the typical hyaloid vessel wall degeneration and were then incorporated into the radionecrotic zone.

A remarkable factor is the time course of the lesion growth (Fig. 3). Whereas the size of the radiation lesion resulting from low dose rate, permanent implants had not reached its maximum until after 70 days, the radionecrosis after high dose rate, temporary implants showed the most growth after 25 days, after which the lesion decreased relatively quickly in comparison with the radionecrosis induced by permanent implants. A high dose delivered over a short period effected a larger volume of necrosis with the inevitably induced vasogenic edema. Under low doses and long-term exposure, small necroses were produced along with moderate but long-lasting vasogenic edema.

When temporary implants were used, lively macrophage activity set in after the emitter had been removed and the zone of necrotic tissue had decreased. The extent of the demyelination in the white substance was directly related to the intensity of the vasogenic edema. The glial reaction and demyelination clearly exceeded the physical penetration depth of the γ-radiation, at least that of iodine-125 gamme radiation and were therefore considered secondary effects (Groothuis et al. 1987; Janzer et al. 1986; Ostertag et al. 1982, 1983, 1984).

4 Discussion

With interstitial irradiation very high radiation doses can be applied locally. As with conventional radiotherapy, however, the risk of radiation-induced damage to the healthy tissue depends on the cumulative dose, volume irradiated, and dose rate. The effects of scattering and attenuation in dose distribution in the tissue vary considerably for the three isotopes used. The specific dose rate factor, i.e., the dose rate per unit activity, divided by the geometrical attenuation is not a constant with distance, particularly not for iodine 125 (Fig. 1).

How much radiation exposure the healthy brain can tolerate is relatively well-known from measurements taken after externally administered radiation (teletherapy) (Caveness 1980). The upper limit is generally considered to be 6000 cGy in standard fractionation (Bloom and Walsh 1975). Other values apply for interstitial irradiation. We found the volume of the radionecrosis and the size of the vasogenic edema to be highly

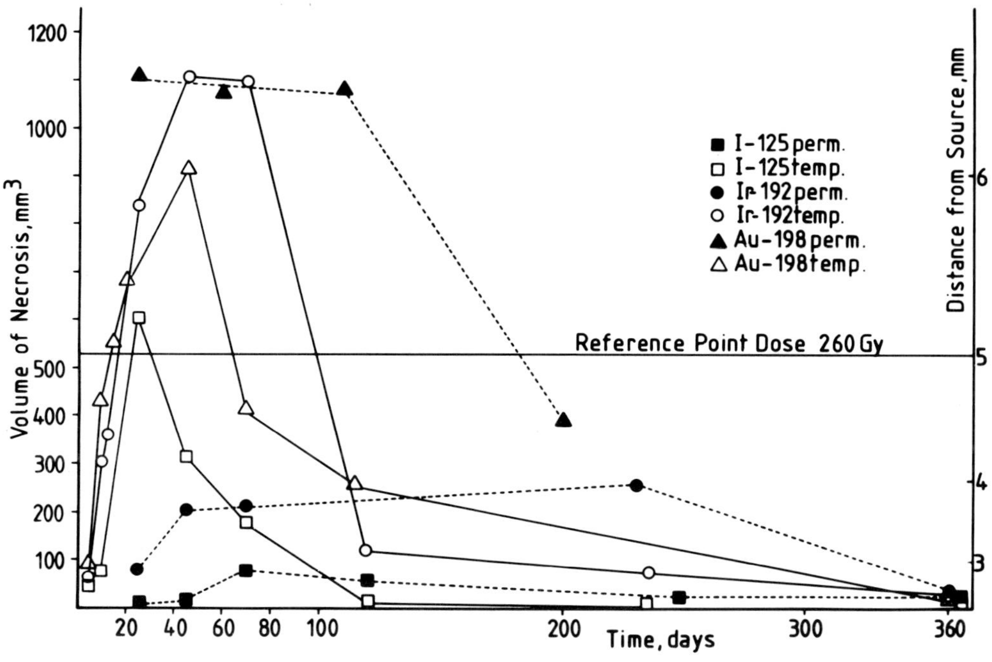

Fig. 3. Necrosis volumes in the time course for permanent, low activity, low dose rate implants in comparison with temporary, high activity, high dose rate implants. Each symbol represents one animal

influenced by the dose rate and the application time. Lower dose rates caused small radionecroses, which did not increase in size after 100 days. Since the dose is delivered continuously, typical repair mechanisms of the cells such as reproduction, redistribution, and repair of sublethal damage are obstructed. The continuous low dose rate, however, was effective enough to inhibit macrophage activity, which is why the radionecrosis had not decreased after an observation period of at least 1 year. We, too, observed a long-lasting edema as a side effect (Fike et al. 1985; Groothuis et al. 1984; Turowsky et al. 1986; Warnke et al. 1988). A high dose rate administered over a short time produced a larger necrosis accompanied by an extensive vasogenic edema. However, after rapid fall-off in tissue to a low dose, repair processes such as macrophage activity were able to take action, and the lesion almost completely disappeared. The temporary, geographically confined opening of the blood-brain barrier is considered a specific side-effect of interstitial irradiation in the brain. The extent and the time course of the blood-brain barrier disruption have been clarified with quantitative methods, which other authors have described qualitatively using imaging methods (Fike et al. 1985; Groothuis et al. 1984, 1987; Turowsky et al. 1986). It could be shown that in comparison with normal brain the irradiated brain tissue develops higher capillary permeability and maintains this for a period of several weeks after irradiation (Groothuis et al. 1987; Warnke et al. 1988).

Under clinical conditions using permanent implants the time factor cannot be changed, which means that no corrections can be made. This also applies to overdoses, so that consequently overdosing is frequently observed in clinical situations. Moderate underdosing often requires reimplantation or results in incomplete control of the tumor growth. Larger tumors present a problem because of the risk of long-lasting, chronic, vasogenic edema. The edema phase can be shortened by using temporary implants with a moderate dose rate, in which case repair processes are free to take place. For clinical application it was concluded that temporary implants with dose rates of 10 cGy/h might be more favorable. The results obtained so far demonstrate a positive tendency. Experimental data from investigations on healthy brain are not completely transferable to tumors. We are therefore making efforts to test these observations experimentally on a model of a virus-induced anaplastic glioma (Ostertag et al. 1984).

Acknowledgement. This study was supported by the Deutsche Forschungsgemeinschaft.

References

Anderson LL, Hsin MK, Ing-Yuan D (1981) Clinical dosimetry with 125-I. In: George FW (ed) Modern interstitial and intracavitary radiation cancer management. Masson, New York

Bernstein M, Gutin PH (1981) Interstitial irradiation of brain tumors: a review. Neurosurgery 6: 741–750

Bloom HJG, Walsh LS (1975) Tumors of the central nervous system. In: Blood HJG, Lemerle J, Neidhardt MK (eds), Cancer in children. Clinical management. Springer, Berlin Heidelberg New York

Burger PC, Mahaley MS, Dudka L, Vogel FS (1979) The morphologic effects of radiation administered therapeutically for intracranial gliomas. Cancer 44: 1256–1272

Caveness WF (1980) Experimental observations: delayed necrosis in normal monkey brain. In: Gilbert HA, Kagan AR (eds) Radiation damage to the nervous system. Raven, New York, pp 1–38

Fike JR, Cann CE, Phillips TL, Bernstein M, Gutin PH, Turowsky K, Weaver KA, Davis RL, Higgins RJ, DaSilva V (1985) Radiation brain damage induced by interstitial 125-I sources: a canine model evaluated by quantitative computed tomography. Neurosurgery 16: 530–537

Groothuis DR, Vriesendorp F, Mikhael M, Blasberg RG, Patlak C (1984) Quantitative measurement of brain tumor capillary permeability by CT: application in canine gliomas and implications for use in patients. Neurology [Suppl 1] (NY) 34: 185–186

Groothuis DR, Wright DC, Ostertag CB (1987) The effect of I-125 interstitial radiotherapy on blood-brain barrier function in normal canine brain. J Neurosurg 67: 895–902

Gutin PH, Leibel SA (1985) Stereotactic interstitial irradiation of malignant brain tumors. Neurol Clin 3: 883–893

Hall EJ (1978) Radiobiology for the radiologist. Harper and Row, Hagerstown

Hilaris BS (1975) Handbook of interstitial brachytherapy. Publishing Sciences Group, Acton

Janzer RC, Kleihues P, Ostertag CB (1986) Early and late effects on the normal dog brain of permanent interstitial Iridium-192 irradiation. Acta Neuropathol (Berl) 70: 91–102

Krishnaswamy V (1978) Dose distribution around an 125-I seed source in tissue. Radiology 126: 489–491

Leibel SA, Sheline GE (1987) Radiation therapy for neoplasms of the brain. J Neurosurg 66: 1–22

Ostertag CB (1986) Stereotactic brachytherapy of brain tumors. In: Gerosa MA (ed) Brain tumors biopathology and therapy. Pergamon, Oxford, pp 175–183

Ostertag CB, Hossmann KA, v.d. Kerckhoff W (1982) Radiation effects of iridium-192 implants in the cat brain. Nucl Med 21: 99–104

Ostertag CB, Weigel K, Warnke P, Lombeck G, Kleihues P (1983) Sequential morphological changes in the dog brain after interstitial Iodine-125 irradiation. Neurosurgery 13: 523–528

Ostertag CB, Warnke P, Kleihues P, Bigner D (1984) Iodine-125 interstitial irradiation of virally induced dog brain tumors. Neurol Res 6: 176–180

Sondhaus CA (1981) I-125: physical properties, photon dosimetry and effectiveness. In: George FW (ed) Modern interstitial and intracavitary radiation cancer management. Masson, New York

Turowsky K, Fike JR, Cann CE, Higgins RJ, Davis RL, Gutin PH, Phillips TL, Weaver KA (1986) Normal brain iodine-125 radiation damage: effect of dose and irradiated volume in a canine model. Radiology 158: 833–838

Warnke PC, Groothuis DR, Ostertag CB (1988) Quantitative Bestimmung der bidirektionalen kapillären Permeabilität im Verlauf bei interstitieller Bestrahlung. In: Bamberg M, Sack H (eds) Therapie primärer Hirntumoren. Zuckschwerdt, Münich, pp 63–66

2.4 Fractionated Afterloading Therapy in the Treatment of Malignant, Inoperable Brain Tumors

ANNE SPARENBERG, KLAUS WEIGEL, and HELMUT ERNST

CONTENTS

1 Introduction . . . 85
2 Methods . . . 85
3 Dosimetry . . . 86
4 Clinical Results . . . 86
5 Discussion . . . 87
5.1 Method-Related Limitations . . . 87
5.2 Recidivation . . . 87
5.3 Radionecrosis . . . 87
6 Conclusions . . . 87
References . . . 87

1 Introduction

Due to postoperative radiotherapy, the mean survival period of patients with malignant gliomas has been raised from 4.5 to 10.5 months (WALKER et al. 1979). A combination of most thorough surgical removal of the tumor and postoperative radiotherapy (JELSMA and BUCY 1967; SHELINE 1977) has become a generally recognized therapy concept. No long-term, statistically significant increase in the survival period could be achieved by raising the target dose from 60 to 80 Gy (SALAZAR et al. 1979), and this notion was abandoned in view of the possibility of complications (radionecroses) (CHIN et al. 1986).

By altering fractionation schemes (SIMPSON and PLATTS 1976), further attempts were made to influence positively the survival period by superfractionation (HINKELBEIN et al. 1984; SHIN et al. 1983) as well as by accelerated irradiation (KEIM et al. 1986). These therapy approaches have had just as little success in becoming established as the intraoperative electron irradiation approach (ABE and TAKAHASHI 1982; GOLDSON et al. 1984). On the contrary, interstitial radiotherapy, especially by Mundinger (MUNDINGER 1981, 1984; THIEL et al. 1983), has gained a solid position, particularly in the treatment of less malignant brain tumors. The results of the treatment of more highly malignant brain tumors could not, however, be significantly improved (GUTIN et al. 1984; MUNDINGER and WEIGEL 1986). The principle of dose fractionation in brachytherapy was not utilized in the past.

2 Methods

First under conventional stereotactic conditions, later under computerized tomography (CT) control, a guide tube with an outer diameter of 3 mm and closed on the brain side is implanted in the tumor. The target corresponds to the center of the caudal tumor extension (Fig. 1). The irradiation axis is also set at the focus of the tumor. Before the tube is inserted, 3–5 tissue samples are removed for histological classification. The guide tube is fixed to the skull cap by means of a special screw-bolt with a clamping screw at the top to prevent

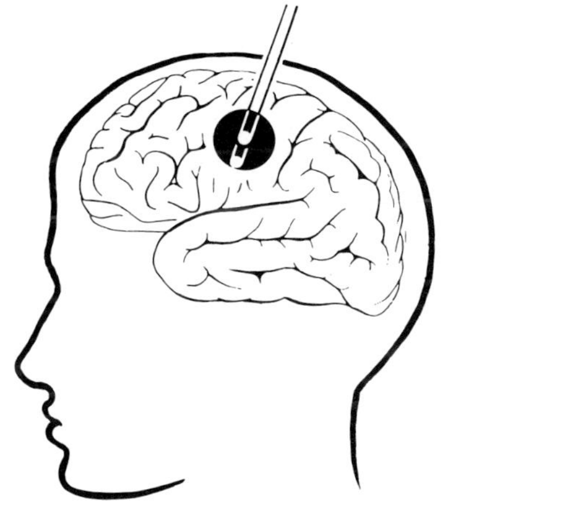

Fig. 1. Irradiation axis set at the tumor center to the caudal tumor extension (SPARENBERG et al. 1988)

ANNE SPARENBERG Dr. med., HELMUT ERNST Prof. Dr. med., Abteilung Radiologie und Strahlentherapie
KLAUS WEIGEL, Dr. med., Abteilung Neurochirurgie, Universitätsklinikum Steglitz, Freie Universität Berlin, Hindenburgdamm 30, 1000 Berlin 45, Germany

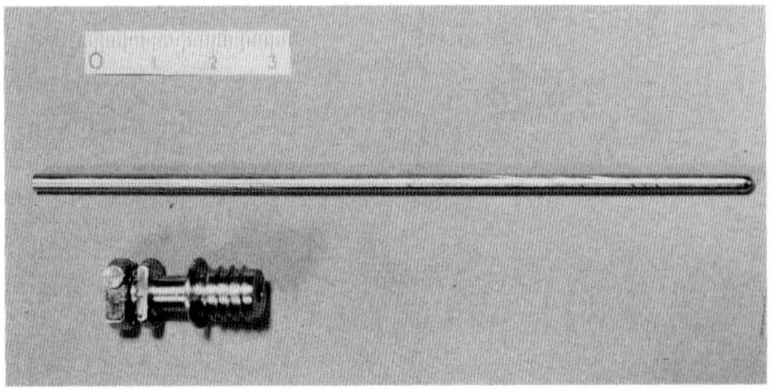

Fig. 2. Screw-bolt with clamping screw, guide tube

longitudinal shifting (Fig. 2). The intervention is terminated by closure of the skin around the spool-like upper part of the screw-bolt and sterile coverage of the implant (ERNST et al. 1986).

The afterloading is applied with the patient lying down. A radionuclide applicator with an outer diameter of 2 mm is inserted into the guide tube; at the other end, it is connected to the hose of the afterloading apparatus (System Buchler). Iridium 192 with an initial activity of 300 GBq and a diameter of 1 mm is used as the source of irradiation (SPARENBERG et al. 1988). It is possible to choose a great variety of dose profiles, which should be adjusted to the tumor volume as closely as possible.

3 Dosimetry

The afterloading therapy is performed twice a day at intervals of 7–8 hours with single doses of 2 Gy each, which ideally encompass the tumor volume. The target dose at the margin of the tumor is 30 Gy. Patients who had had no radiological pretreatment received percutaneous postirradiation by means of cobalt 60 or a particle accelerator. The entire cranium is thereby irradiated with an additional dose of 30–40 Gy four times a week; this results in a dose of 60 or 70 Gy in the border area between the tumor and healthy tissue.

4 Clinical Results

Within the last 2.75 years, we have treated 35 patients using the described treatment concept. Of these 20 belonged to the histological group of glioblastoma, 10 suffered from astrocytoma III, and 5 from astrocytoma II. Most of the brain tumors were localized in the temporoparietal area or in the thalamic region (Table 1). The dominating pretherapeutic symptoms were paresis, followed by seizure and cranial nerve dysfunctions (Table 2).

Table 1. Localization of the tumors

	n
Temporoparietal region	16
Thalamus region/basal ganglia	10
Frontotemporal region	7
Midbrain pons region	2

Table 2. Pretherapeutic symptoms

	n
Paresis	23
Attacts	18
Cranial nerve dysfunctions	12
Aphasia	5
Headache	5

Table 3. Mean survival time post afterloading treatment

	n	months
Total	35	8.7
Patients without pretreatment	13	10.9
Patients with pretreatment	22	7.9

Of the 35 patients, 23 died within a period of 2.75 years, which corresponds to a mean survival period of 8.7 months for the total patient group (Table 3). The patients who earlier had received no specific pretreatment survived 10.9 months (n = 13), in comparison with the 7.9 months for the patients who had received pretreatment or who had completed radiotherapy before they were sent to us.

In 13 cases, tumor progression was the main cause of death, which means that most of the patients died of tumor recurrence or of progression of the primary tumor (Table 4). It is unknown to us whether or not any deaths were due to radionecrosis, because we were not able to perform an autopsy on all patients.

According to the Kaplan-Meier method, the life expectancies estimated 6 and 12 months post afterloading therapy are 65% and 23%, respectively, for the entire patient group (n = 35), and 76% and 38%, respectively, for the group without any specific pretreatment (n = 13).

Table 4. Cause of death

	n
Tumor progression	13
Cardiovascular failure	4
Lung embolism	4
Pneumonia	2

5 Discussion

5.1 Method-Related Limitations

Due to the rotational symmetry of the isodoses, only absolutely symmetrical bodies can be irradiated. Should, for anatomic reasons, only an excentrical implantation of the guide tube be achieved, homogenic irradiation of the tumor volume would no longer be possible (Fig. 3). We therefore increased the actual measured tumor diameter by 2 cm, e.g., the diameter used for CT purposes was 2 cm larger.

5.2 Recidivation

The high recidivity of this type of tumor is doubtless a result of its biological characteristics (THIEL et al. 1983) that are responsible for its very unfavorable prognostic classification. Consequently, astrocytomas generally recidivate with a higher grade of malignancy. SALAZAR and RUBIN (1976) also point out that spreading – be it progressive or multicentric – is present in about 80% of all glioblastomas. According to JELLINGER (1978) the portion of multifocal occurrences is estimated at 2.5%–6%; SCHIEFER et al. (1978) cite values between 0.5% and 10%. The tendency for infiltration thus receives greater importance from a therapeutic point of view. In our patient collective, the high recidivity is doubtless partially due to the fact that exact CT-controlled placement of the afterloading tube could be performed only in the last three patients.

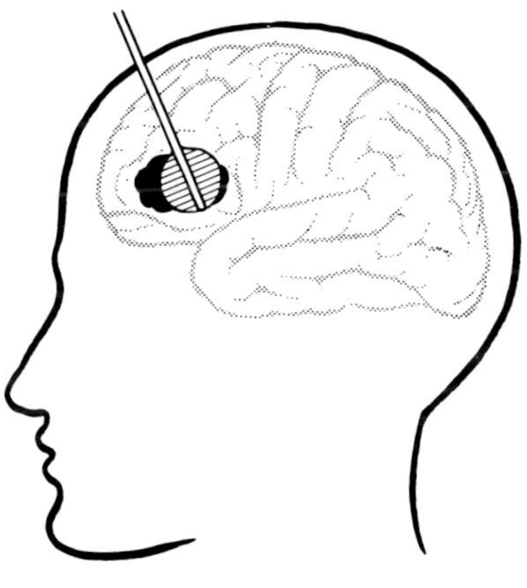

Fig. 3. Excentrical implantation of the guide tube

5.3 Radionecrosis

According to SHELINE et al. (1980), the general radiation reaction can be divided into three phases: (a) the acute reaction, (b) the early delayed reaction, and (c) the late delayed reaction.

The acute reaction, which is primarily of a vasogenic nature (CSANDA 1980), is occurs only rarely in the above-mentioned therapy. In the opinion of several authors, however, vasogenic edema is a signpost for the late delayed reaction (BURGER et al. 1979; HINKELBEIN and WANNENMACHER 1982; OSTERTAG et al. 1984).

In our patient collective, some suffered a temporary deterioration about 3 months after radiation therapy that correlated with an early delayed reaction. This deterioration was clinically reversible with high doses of cortisone.

6 Conclusions

This is the first time that the principle of dose fractionation in brachytherapy for malignant brain tumors is used, resulting in a beneficial effect on perifocal edema formation. With this method higher doses can be generated in the tumor center than with percutaneous irradiation. On the other hand, the surrounding healthy tissue can be protected. Patients who have already completed radiotherapy can receive palliative therapy without any isolation.

References

Abe M, Takahashi M (1982) Intraoperative Strahlentherapie. Strahlentherapie 158: 585–593

Burger PC, Mahaley MS, Dudka L, Vogel FS (1979) The morphologic effects of radiation administered therapeutically for intracranial gliomas. A postmortem study of 25 cases. Cancer 44: 1256–1272

Burger PC, Vogel FS, Green SB, Strike TA (1985) Glioblastoma multiforme and anaplastic astrocytoma. Pathology criteria and prognostic implications. Cancer 56: 1106–1111

Chin HW, Maruyama Y, Young B, Markesbery W, Tibbs P, Goldstein S (1986) A clinical study with brain brachytherapy for malignant gliomas. Strahlenther Onkol 162: 433–436

Csanda E (1980) Radiation brain edema. Adv Neurol 28: 125–146

Ernst H, Scheffler A, Oppel F, Brock M, Brust V, Bauer R, Pannek HW, Förster A (1986) Fraktionierte Afterloading-Bestrahlung als neues Verfahren zur Therapie inoperabler Hirntumoren. Strahlenther Onkol 162: 437–440

Goldson AL, Streeter OE, Ashayeri E, Collier-Manning J, Barber JB, Fan K-J (1984) Intraoperative radiotherapy for intracranial malignancies. Cancer 54: 2807–2813

Gutin PH, Phillips TL, Wara WM, Leibel SA, Hosobuchi Y, Levin VA, Weaver KA, Lamb S (1984) Brachytherapy of recurrent malignant brain tumors with removable high-activity iodine-125 sources. J Neurosurg 60: 61–68

Hinkelbein W, Wannenmacher M (1982) The radiosensitivity of the infant brain. In: Voth D, Gutjahr P, Langmaid C (eds) Tumours of the central nervous system in infancy and childhood. Springer, Berlin Heidelberg New York, pp 291–300

Hinkelbein W, Bruggmoser G, Schmidt M, Wannenmacher M (1984) Die Kurzzeitbestrahlung des Glioblastoms mit hohen Einzelfraktionen. Strahlentherapie 160: 301–308

Jellinger K (1978) Glioblastoma multiforme: morphology and biology Acta Neurochir (Wien) 42: 5–32

Jelsma R, Bucy PC (1967) The treatment of glioblastoma multiforme of the brain. J Neurosurg 27: 388–400

Keim H, Potthoff PC, Neiss A, Trott KR (1986) Lebensqualität und Überlebenszeit von Glioblastompatienten nach accelerierter Strahlentherapie. Zentralbl Radiol 132: 372

Mundinger F (1981) Stereotaktische Therapie nicht resezierbarer intracranieller Tumoren mit Ir^{192} und Jod^{125}. In: Wannenmacher M, Schreiber HW, Gauwerky F (eds) Kombinierte chirurgische und radiologische Behandlung maligner Tumoren. Urban and Schwarzenberg, pp 86–108

Mundinger F (1984) Stereotaktische intrakranielle Bestrahlung von Tumoren mit Radioisotopen (Curie-Therapie). In: Dietz H, Umbach W, Wüllenweber R (eds) Klinik und Therapie. Thieme, Stuttgart pp 519–565 (Klinische Neurochirurgie, vol 2)

Mundinger F, Weigel K (1986) Stereotactic curietherapy of thalamic tumours. In: Walker MD, Thomas DGT (eds) Biology of brain tumour. Nijhoff, Boston, pp 261–268

Ostertag CB, Groothuis D, Kleihues P (1984) Experimental data on early and late morphologic effects of permanently implanted gamma and beta sources (Ir-192, J-125 and Y-90) in the brain. Acta Neurochir [Suppl] (Wien) 33: 271–280

Salazar OM, Rubin P (1976) The spread of glioblastoma multiforme as a determining factor in the radiation treated volume. Int J Radiat Oncol Biol Phys 1: 627–637

Salazar OM, Rubin P, Feldstein MC, Pizzutiello R (1979) High dose radiation therapy in the treatment of malignant gliomas. Final report. Int J Radiat Oncol Biol Phys 5: 1733–1740

Schiefer W, Hasenbein B, Schmidt H (1978) Multicentric glioblastomas. Methods of diagnosis and treatment. Acta Neurochir (Wien) 42: 89–95

Sheline GE (1977) Radiation therapy of brain tumors. Cancer 39: 873–881

Sheline GE, Wara WM, Smith V (1980) Therapeutic irradiation and brain injury. Int J Radiat Oncol Biol Phys 6: 1215–1228

Shin KH, Muller PJ, Geggie PHS (1983) Superfractionation radiation therapy in the treatment of malignant astrocytoma. Cancer 52: 2040–2043

Simpson WJ, Platts ME (1976) Fractionation study in the treatment of glioblastoma multiforme. Int J Radiat Oncol Biol Phys 1: 639–644

Sparenberg A, Ernst H, Weigel K (1988) Fraktionierte Afterloading-therapie bei malignen, inoperablen Hirntumoren. In: Bamberg M, Sack H (eds) Therapie primärer Hirntumoren. Zuckschwerdt, Münich, pp 302–305

Thiel HJ, Huk WJ, Müller R, Sauer R (1983) Die computertomographisch geleitete stereotaktische interstitielle Therapie von Hirntumoren mittels temporärer oder permanenter Implantation von ^{125}Jod-Seeds. Fortschr Geb Rontgenstr 138(3): 348–355

Walker MD, Strike ThA, Sheline GE (1979) An analysis of dose-effect relationship in the radiotherapy of malignant gliomas. Int J Radiat Oncol Biol Phys 5: 1725–1731

2.5 External Stereotactic Focal Irradiation

BERNHARD N. KIMMIG, VOLKER STURM, and RITA ENGENHART

CONTENTS

1 Introduction 89
2 Irradiation Technique and Planning 89
3 Solitary Brain Metastases 91
4 Arteriovenous Malformations 92
References 93

1 Introduction

"Radiosurgery" is a somewhat provocative term for a special concept of radiotherapy (LEKSELL 1951). It describes a percutaneous, stereotactically guided irradiation delivering a single high dose with collimated narrow beams. The precise stereotactic localization of the target point and a steep dose gradient outside the target volume allow the administration of high doses to a lesion without damage of adjacent normal tissue.

Units for radiosurgery were designed at Stockholm using multiple external cobalt-60 γ-sources, at Boston operating with a high-energy proton beam of a cyclotron, and at Berkeley operating with helium ions accelerated by a synchrocyclotron (STEINER 1986; KJELLBERG et al. 1983a, b; FABRIKANT et al. 1984).

An attractive alternative to these complicated and expensive facilities is to modify a conventional linear accelerator for the purpose of radiosurgery. At the present time numerous therapy groups have developed or are developing radiosurgery units associated with commercial linear accelerators in a special moving field arrangement, as in Vincenza, Montreal, Buenos Aires, and Boston (COLOMBO et al. 1986; PODGORSAK et al. 1988; BETTI et al. 1983; LUTZ et al. 1988). At the German Cancer Research Center in Heidelberg such a system was started in 1982/1983 and has been available for the treatment of patients since then (HARTMANN et al. 1985; STURM et al. 1987; ENGENHART et al. 1989).

Classic indications for radiosurgery are cerebral arteriovenous malformations (AVMs) and benign, radioresistant, primary tumors of the brain. We have extended the indication to solitary brain metastases of radioresistant primary tumors (STURM et al. 1987). This review will concentrate on the technical aspects of stereotactic irradiation and on the treatment of brain metastases and arteriovenous malformations.

2 Irradiation Technique and Planning

For stereotactic, single, high-dose irradiation we use a system containing three components (HARTMANN et al. 1985): (a) a modified linear accelerator (Mevatron 77, Siemens AG, FRG), (b) a stereotactic localization and target positioning system (Fischer, FRG), and (c) a 3-dimensional planning system (SCHLEGEL et al. 1984).

Irradiation is carried out with a series of moving fields in noncoplanar plans. After every gantry motion the treatment table with the patient rotates to another position (Fig. 1). Additional circular tungsten collimators reduce the penumbra and permit adaptation to diverse field diameters. The superposition of the irradiation fields deliveres spherical dose distributions with steep dose gradients outside the target volume (7–15%/mm) comparable with the profiles of cobalt-60 units or 185-MeV protons used for radiosurgery (Fig. 2).

The stereotactic localization system consists of a Riechert-Mundinger head frame modified with a measuring phantom for artifact-free use in a computerized tomography (CT) scanner (STURM et

BERNHARD N. KIMMIG, Priv.-Doz. Dr. Dr., RITA ENGENHART, Dr. Radiologische Klinik der Universität Heidelberg, Abteilung Klinische Radiologie, Im Neuenheimer Feld 400, 6900 Heidelberg, Germany

VOLKER STURM Prof. Dr., Neurochirurgische Klinik der Universität Köln, Abteilung Stereotaxie und funktionelle Neurochirurgie, Joseph-Stelzmann-Str. 9, 5000 Köln 41, Germany

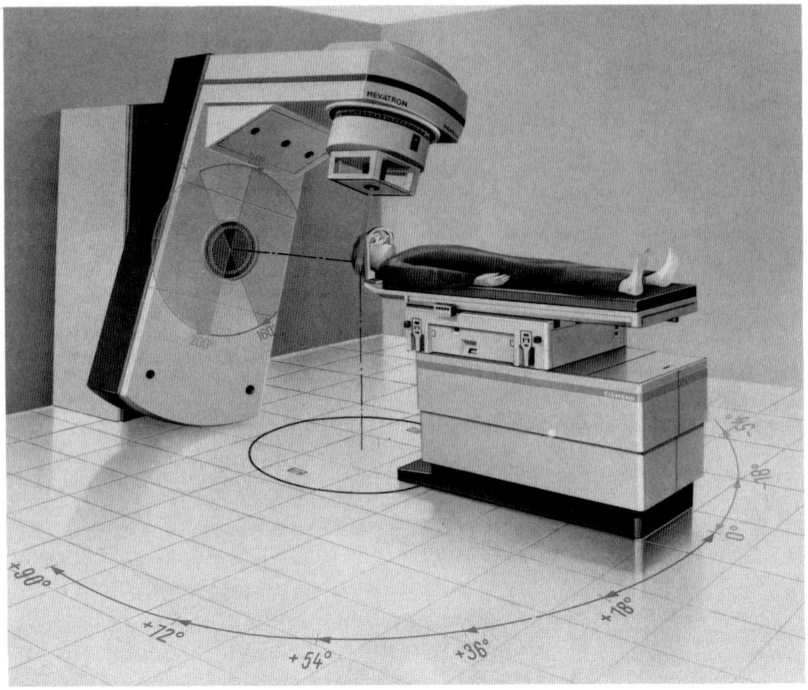

Fig. 1. Schematic illustration of gantry and patient's couch rotation around the isocenter during stereotactic focal irradiation

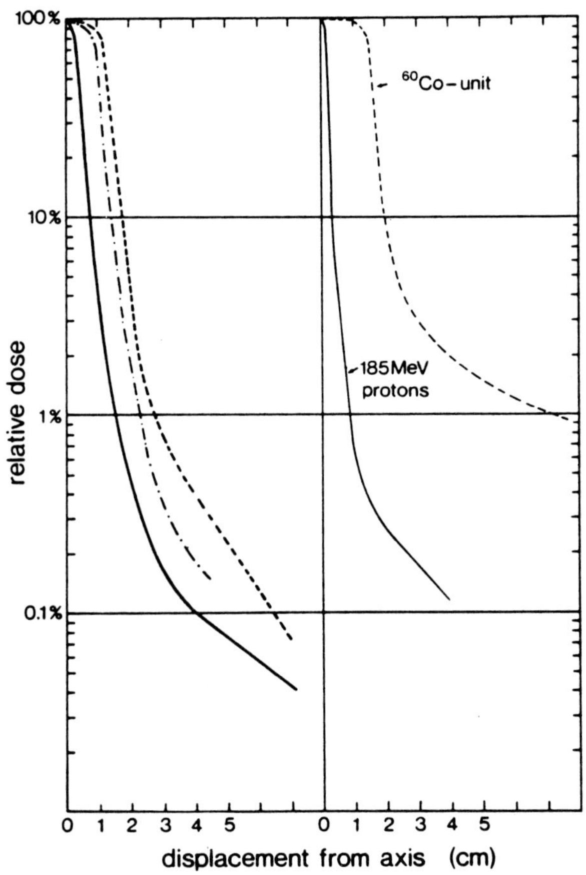

Fig. 2. Dose profiles of collimated narrow beams. *Left*, the 9, 21, and 26 mm beams from the linear accelerator using the described moving field technique; *right*, beams of other irradiation units for radiosurgery

al. 1983). Plexiglass squares with embedded steel wires allow direct measurement of the target point coordinates from the CT scan. A corresponding positioning system allows adjustment of the target point to the isocenter of the irradiation facility with stereotactic precision. Essential for this irradiation technique is an exact control of the different moving parts: the central beam of the irradiation field, the axis of the patient table, and the axis of the gantry have to meet precisely in one point: the isocenter.

An alternative system for patient fixation consists of an individual light cast mask, which can be attached to the CT scanner couch and the linear accelerator couch by a wooden-based stereotactic frame. This mask system is less accurate than the stereotactic head frame attached to the patient's skull; nevertheless, the accuracy of reproducibility is relatively high, with a mean deviation of ±1 mm and a maximum deviation of ±2 mm.

Computer programs have been developed for 3-dimensional treatment planning (Fig. 3). Field

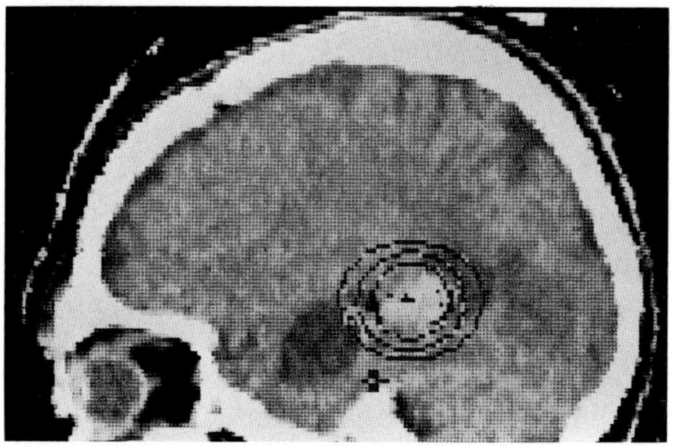

Fig. 3. Sagittal reconstruction of CT scans showing an arteriovenous malformation in the region of the basal ganglia. *Lines* represent the 80%, 50%, and 30% isodoses for focal irradiation

diameter and target point coordinates can be gained from CT scans of the measuring phantom or from selective cerebral arteriograms performed with the stereotactic head frame.

3 Solitary Brain Metastases

The treatment of choice for multiple brain metastases is radiotherapy (BORGELT et al. 1980). Surgery should be considered first for solitary metastases, which occur in 20%–30% (ZIMM et al. 1981). For patients with solitary metastases whose lesions are not accessible or in whom surgery is impossible without the risk of disabling neurologic deficit, radiotherapy is also indicated. Primary tumors with solitary brain metastases are generally adenocarcinomas of the kidney, colon, or lung.

Since 1984, 37 patients with solitary metastases were treated with single high doses under stereotactic conditions. Indications were inoperable solitary metastases of relatively radioresistant primary tumors (STURM et al. 1987). Kidney (13/17) and lung (11/37) cancer predominated as primary tumors. In 23 patients radiosurgery was carried out as a primary treatment procedure. Some 14 patients were treated for local recurrent metastases – 5 after previous surgery and 9 after conventional fractionated whole brain radiotherapy with 50 Gy.

On CT scans for the planning procedure we found more than one metastasis in 12 patients. As many as three separate metastases were accepted for treatment, receiving identical doses as if they were one field. A single dose of 10–40 Gy was applied, with respect to tumor volume, localization, and histology. The patients were hospitalized for 2–3 days and dismissed from hospital day after irradiation. The irradiation was carried out under steroid protection.

The median survival of the patients after irradiation was 5 months, with a mean follow-up of 6.7 months. Some 14 patients developed additional brain metastases outside the target volume 4–13 months after radiosurgery. Four of them had had previous conventional whole brain irradiation. Seven patients are still alive. A total of 28 patients died from progression from the underlying extracranial cancer, and 2 died of progression of their brain metastases. A significant clinical improvement of neurologic deficits was achieved in 26 patients (84%), 10 of them had complete clinical remissions, and 16 had partial clinical remissions. An improvement of the CT findings was achieved in 70% of patients, with 4 complete and 10 partial remissions. An example for a complete remission of a solitary metastasis after radiosurgery is shown in Fig. 4. Of 12 patients showing no change on CT study, 5 had no improvement of clinical deficits. Two patients demonstrated increased neurological deficits in conjunction with positive CT findings. They died of their progressive disease 2.5 and 4 months after irradiation.

A severe treatment-related complication occurred in one patient who died 15 h after irradiation from an inferior herniation attributed to an increase in preexisting edema. The 80% isodose curve surrounded a 42-mm cerebellar metastasis

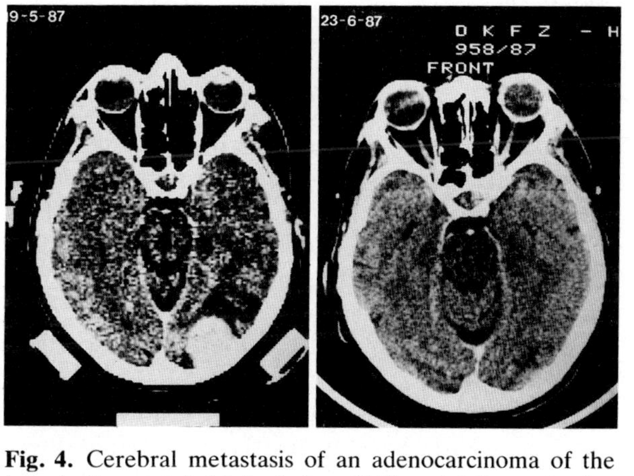

Fig. 4. Cerebral metastasis of an adenocarcinoma of the lung in the occipital lobe before (*left*) and 1.5 month after (*right*) stereotactic irradiation with 20 Gy

which received 2500 cGy. Three other patients developed a reversible perifocal edema 2–6 months after irradiation which responded well to steroids. One patient is suffering from an irreversible radiation-induced edema. He has received 4 mg steroids per day over the past 26 months and is now free of neurological symptoms.

The major advantage of radiosurgery in comparison with conventional fractionated radiotherapy is the short treatment time and the rapid onset of the treatment effect in patients with a short lifespan. The clinical responses usually begin within 1 week, in some cases within 2 days. Since this procedure is well tolerated and side effects or complications are rare, radiosurgery is a realistic indication for the treatment of relatively radioresistant, solitary brain metastases.

4 Arteriovenous Malformations

Cerebral AVMs are conceived as congenital vascular anomalies. Although they are not true neoplasms – because there is no evidence for an autonomous cellular proliferation – they can resemble tumors in their clinical features, with progressive growth producing a mass effect and destroying the adjacent brain tissue. The tumorlike growth is caused by arterial blood pressure, leading to a progressive dilatation of the affected vessels. A tremendous complication is massive intracranial bleeding which occurs in up to 25% of patients with AVMs. The annual risk rate of hemorrhage is about 2%–3%. Mortality for all patients with AVMs approaches 15%, that for patients with deep AVMs approaches 25% – which is an exceptionally high risk for a benign disease (Davis and Symon 1985; Drake 1983; Graf et al. 1983).

Surgical excision is the treatment of choice. An alternative is embolization. Radiation therapy is indicated for AMVs in vital or sensitive regions of the brain, which cannot be excised without risk of a disabling neurologic deficit, and for residual AVMs that remain after partial embolization or incomplete surgical removal. The most effective method of radiotherapy for AVMs is radiosurgery (Steiner 1986).

Since 1983, 55 patients (f: 24, m: 31) with cerebral AVMs have been treated in Heidelberg by stereotactic, single high dose irradiation. The age of the patients ranged from 8 to 64 years with a mean value of 29 years. Initial clinical symptoms were hemorrhage (50%), seizure (24%), and paresis (22%). Previous operative treatment without complete remission had been performed in 7 patients.

Radiation doses between 8 and 29 Gy were applied as related to the 80% isodose contour which enclosed the volume of pathological vessels. The mean field diameter was 40 mm (ranging from 12 to 54 mm). Integral doses to the whole brain were estimated to be under 1 kg Gy.

A total of 30 patients had a follow-up period longer than 18 months. In 14 of them, a complete remission of the malformation was found by angiography; in 5 others, partial obliteration of the pathologic vessels could be demonstrated. No change was seen in 11 patients, all of them with target doses below 12 Gy.

Acute complications have not been observed. Late side effects were found in 7 patients, occurring as reversible neurologic deficits after a latency of 6–12 months. Permanent radiogenic neurological deficit was seen in 1 patient, who developed a persistent, incomplete hemiparesis. Three patients suffered from acute hemorrhage 6, 9, and 10 months after radiation therapy. Because of the well-known long latency (1–2 years) between irradiation and obliteration of the angioma, these bleedings can be assumed to be spontaneous events with a natural risk of about 2% per year.

Conventional fractionated radiotherapy has been performed for cerebral AVMs by Johnson (1975), Glanzmann (1978), and Makoski et al. (1984, 1988) with contradictory results (Table 1). Despite some occasional success, fractionated irradiation must be considered ineffective and has now been abandoned (Steiner 1986).

Radiosurgery for AVMs has been performed by

Table 1. Results of radiation therapy for arteriovenous malformations

Reference	*n*	Therapeutic results	
		CO	PO
Johnson (1975)	20	45%	25%
Glanzmann (1978)	13	0%	0%
Makoski et al. (1988)	25	12%	48%
Kjellberg et al. (1983b) Kjellberg (1986)	439	22%	56%
Fabrikant (1988)	300	65%	25%
Steiner (1988)	600	>90%	3%–6%
Kimmig et al. (1991)	30	47%	17%

CO, complete obliteration; *PO*, partial obliteration

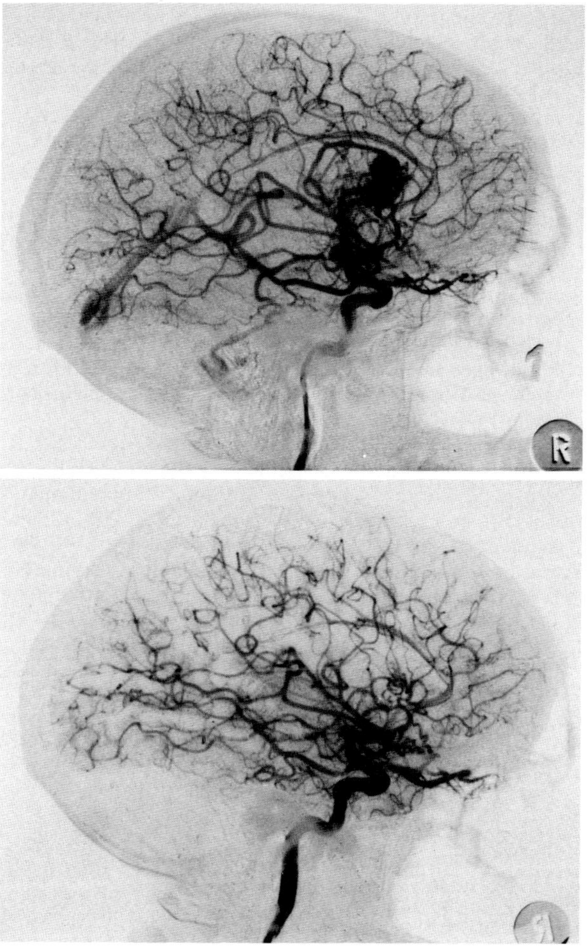

Fig. 5. Arteriovenous malformation in the region of the basal ganglia before (*above*) and 14 months after (*below*) stereotactic irradiation with 30 Gy

Kjellberg et al. (1983 a, b; Kjellberg 1986) with high energy protons, by Fabrikant et al. (1984) with helium ions, and by Steiner (1988) using a cobalt-60 unit with favorable results (Table 1). Our therapeutic results are preliminary with respect to the small number of patients and the short mean follow-up period till now. The technique does seem to be as effective as proton beam therapy at least. Differences from the excellent data reported by Steiner can be due to three causes:

1. The limit of 18 months for follow-up is relatively short regarding the latency of the radiation effect.
2. Most of the AVMs treated by Steiner were less than 2.5 cm in size. Against that, half of our patients suffered from tumors greater than this value, and the efficiency of radiotherapy is supposedly correlated with AVM volume.
3. In the 11 patients without therapeutic effect, tumor doses under 12 Gy were delivered to avoid damage to sensitive tissue like chiasma. Apparently doses under 12 Gy are not sufficient for the obliteration of AMVs.

Little is known about the dose effect curve for occluding angiomas and about the influence of tumor volume to this curve, but probably AVMs need high doses of about 20 Gy or more for complete obliteration (Fig. 5). Whether low doses without complete obliteration protect against hemorrhage is an open question as yet.

References

Betti O, Derechinsky V (1983) Irradiation stereotaxique multifaisceaux. Neurochirurgie 29: 295–298

Borgelt B, Gelber R, Kramer S, Brady LW, Chang CH, Davis LW, Perez CA, Hendrickson FR (1980) The palliation of brain metastases: final results of the first two studies by the radiation therapy oncology group. Int J Radiat Oncol Biol Phys 6: 1–9

Colombo F, Benedetti A, Pozza F, et al. (1986) Radiosurgery using a 4 MeV linear accelerator. Acta Radiol [Suppl] (Stockh) 369: 600–607

Davis C, Symon L (1985) The management of cerebral arteriovenous malformations. Acta Neurochir (Wien) 74: 4–11

Drake CG (1983) Arteriovenous malformations of the brain. The options for management. N Engl J Med 309: 308–310

Engenhart R, Kimmig B, Wowra B, Sturm V, Höver KH, Schneider S, Wannenmacher M (1989) Stereotaktische Einzeitbestrahlung cerebraler Angiome. Radiologe 29: 219–223

Fabrikant JI (1988) Radiosurgery workshop. National Institutes of Health, June 9–10, Washington

Fabrikant JI, Lymann JT, Hosobuchi Y (1984) Stereotactic heavy-ion Bragg peak radiosurgery for intracranial vascular disorders: method for treatment of deep arteriovenous malformations. Br J Radiol 57: 479–490

Glanzmann C (1978) Zerebrale arteriovenöse Mißbildungen: Verlauf bei 18 Fällen nach Radiotherapie. Strahlentherapie 154: 305–308

Graf CJ, Perret GE, Torner JC (1983) Bleeding from cerebral arteriovenous malformations as part of their natural history. J Neurosurg 58: 331–337

Hartmann GH, Schlegel W, Sturm V, Kober B, Pastyr O, Lorenz WJ (1985) Cerebral radiation surgery using moving field irradiation at a linear accelerator facility. Int J Radiat Oncol Biol Phys 11: 1185–1192

Johnson RT (1975) Radiotherapy of cerebral angiomas. With a note on some problems in diagnosis. In: Pia HW (ed) Cerebral angiomas. Springer, Berlin Heidelberg New York, pp 256–259

Kjellberg RN (1986) Stereotactic bragg peak proton beam radiosurgery for cerebral arteriovenous malformations. Ann Clin Res [Suppl] 47: 17–19

Kjellberg RN, Hanamura T, Davis KR, Lyons SL, Adams RD (1983a) Bragg peak proton-beam therapy for arteriovenous malformations of the brain. N Engl J Med 309: 269–274

Kjellberg RN, Davis KR, Lyons SL, Butler W, Adams RD (1983b) Bragg peak proton-beam therapy for arteriovenous malformations of the brain. Clin Neurosurg 31: 248–290

Leksell L (1951) The stereotaxic method and radiosurgery of the brain. Acta Chir Scand 102: 316–319

Lutz W, Winston KR, Maleki N (1988) A system for stereotactic radiosurgery with a linear accelerator. Int J Radiat Oncol Biol Phys 14: 373–381

Makoski HB, Nocken U, Fiebach BJO, Zeilstra DJ (1984) Die Radiotherapie arteroivenöser Malformationen des Hirns. Strahlentherapie 160: 159–165

Makoski HB, Zeilstra DJ, Nocken U (1988) Arteriovenöse Malformationen des Hirnschädels – Strahlentherapeutische Aspekte. In: Bamberg M, Sack H (eds) Therapie primärer Hirntumoren. Zuckschwerdt, Münich pp 245–249

Podgorsak EB, Olivier A, Pla M, Lefebvre PY, Hazel J (1988) Dynamic stereotactic radiosurgery. Int J Radiat Oncol Biol Phys 14: 115–126

Schlegel W, Scharfenberg H, Doll J, Hartmann G, Sturm V, Lorenz WJ, (1984) Three dimensional dose planning using tomographic data. In: IEEE (ed) Proceedings of the eight international conference on the use of computers in radiation therapy. IEEE, Silver Spring, pp 191–196

Steiner L (1986) Radiosurgery in cerebral arteriovenous malformation. In: Fein J, Flamm E (eds) Textbook of cerebro-vascular surgery, vol 4. Springer, Berlin Heidelberg New York, pp 1161–1215

Steiner L (1988) Radiosurgery in arterio-venous malformation. Meeting: Surgical Neuroangiography, May 2–6, New York

Sturm V, Pastyr O, Schlegel W, Scharfenberg H, Zabel HJ, Netzeband G, Schabbert S, Berberich W (1983) Stereotactic computer tomography with a modified Richert-Mundinger device qas the basis for integrated stereotactic neuroradiological investigations. Acta Neurochir (Wien) 64: 87–102

Sturm V, Kober B, Höver KH, Schlegel W, Boesecke R, Pastyr O, Hartmann G, Schabbert S, zum Winkel K, Kunze S, Lorenz WJ (1987) Stereotactic percutaneous single dose irradiation of brain metastases with a linear accelerator. Int J Radiat Oncol Biol Phys 13: 279–282

Zimm S, Wampler GL, Stablein D, Hazra T, Young HF (1981) Intracerebral metastases in solid – tumor patients: natural history and results of treatment. Cancer 48: 384–394

2.6 External Stereotactic Focal Irradiation of Arteriovenous Malformations by a Routinely Used Linear Accelerator

Hans-Bruno Makoski

CONTENTS

1 Introduction 95
2 Arteriovenous Malformations 95
3 Patients and Methods 96
4 Results 97
5 Discussion 97
6 Outlook 97
References 98

1 Introduction

The biological difference in radiosensitivity of cells within and outside the target volume plays an important part in *radiotherapy*, especially in malignant tumors. Fractionated doses are needed for the recovery of normal tissue. If the dose gradient is very steep at the edge of the target volume, a single dose sufficient to cause cell death may be given: *radiosurgery*. Stereotactic radiosurgery implies the application of well-collimated beams of ionizing radiation. It is of especial interest in the head for the treatment of malignant tumors or metastases (Sturm et al. 1987) or pituitary adenomas or vascular disorders like carotid-cavernous fistulas (Yasunaga et al. 1987) and arteriovenous malformations (Steiner et al. 1974, 1977).

Radiosurgery with gamma-units is only available in a few places in the world and restricted to lesions not exceeding 3.5 cm in diameter. Radiosurgery with heavy ions has the advantages of the Bragg peak, but the diameters of the beams do not exceed 5 cm. Worldwide, there are only three centers providing charged particle therapy. Irradiation with neutron beams through 9 noncoplanar ports is done in Seattle. As already proposed by Steiner (1982, 1985, personal communication), there is a need for using teletherapy equipment, which has the advantages of treating large lesions not accessible by radiosurgery and above all of being available in all radiotherapy centers.

Hans-Bruno Makoski, Prof. Dr. med., Strahlenklinik – Radioonkologie – Nuklearmedizin, Städtische Kliniken Duisburg, Zu den Rehwiesen 9/Kalkweg, 4100 Duisburg 1, Germany

2 Arteriovenous Malformations

Intracranial arteriovenous malformations (AVMs) are benign but potentially lethal abnormalities drained by regular arteries and veins which can be dilated. They account for from 1.9% up to 3% of all intracranial masses, only 7% of these lying in the posterior fossa. Intact AVMs rupture at an annual rate of 2%–3%, the rate of rebleeding being 6% in the 1 year after a hemorrhage. About 10% of patients die with the first bleeding, and 20% with each rebleeding (Graf et al. 1983). Bleeding may also come from concomitant aneurysms (1.5%–13%).

AVMs can be asymptomatic or present with symptoms from headache to bleeding. Due to noninvasive brain section tomography [computerized tomography (CT) and magnetic resonance imaging (MRI)] the detection rate of AVMs has been increased. Growing numbers of patients will be presenting for therapy. Radical surgical excision is the best therapy when it can be done without a great risk of a disabling neurologic deficit. When total resection is not feasible, there is a variety of methods of treatment left: embolization, balloon occlusion, cryosurgery, laser therapy, and radiotherapy. Radiotherapy of AVMs has been proposed as a last resort for very unfavorable cases, which influences the results (Drake 1983; Glanzmann 1978; Sutton 1984; Steiner et al. 1974). Failures were due to unsystematic or imprecise treatment.

Johnson (1975) was the first to report on follow-up angiograms, giving a cure rate by megavoltage therapy of 45% in 20 of 100 patients; all nine, small, deeply placed AVMs were absent in angiograms obtained more than 2 years after treatment. There were failures with large lesions (GLANZMANN 1978), so radiotherapy was thought to be ineffective. Stereotactic radiosurgery by the Stockholm technique (STEINER 1972) has become the "gold standard" for treatment results. A report was given for 135 AVMs: In 85 patients with AVMs smaller than 30 mm, 40% of the malformations had been obliterated by the end of 1 year, and in 67, 84% of the AVM had disappeared on follow-up angiograms 2 years after treatment (STEINER 1984). In a proton-treated series a complete angiographic response of the AVM at 2 years was obtained in 22% of 260 patients studied and nearly total obliteration in 29% (KJELLBERG 1986). MEHDORN and GROTE (1988) did follow-up angiograms on 30 patients 2 years after proton treatment: only in 2 was there complete occlusion; 21 patients did not show any change at all. Several centers were successful in developing linear-accelerator-based radiosurgical techniques, some with plane rotations and others with more sophisticated movements of converging arcs and couch. Single treatments are used up to 50 Gy with field sizes of 1-cm diameter, or two sessions of 20 Gy each with targets between 1 and 3 cm in diameter. (HARTMANN et al. 1985; LUTZ et al. 1988; PODGORSAK et al. 1988; SAUNDERS et al. 1987). Reference doses of the Heidelberg group are 8–28 Gy (KIMMIG et al. 1988). The latest data on 22 patients with follow-ups longer than 24 months of 61 patients are: complete obliteration in 10, partial obliteration in 6, and no change in 6 (KIMMIG et al. 1989, updated results).

Table 1. Duisburg protocol for irradiation of arteriovenous malformation

1. Clinical work-up
2. Round with neurologist, neurosurgeon, (neuropathologist), neuroradiologist, and radiation oncologist
3. Patient's consent
4. Individual fixation of the patient (Neofract) for all further procedures
5. Cerebral angiography
6. CT or MRI scanning
7. Treatment planning (SIDOS-U2)
8. Simulation
9. Localization at the linear accelerator (port films)
10. Radiotherapy (5 × 2 Gy/week up to 50 Gy, or 4 × 5 Gy/7 days up to 20 Gy)
11. CT scanning after radiotherapy
12. Follow-up with CT (or MRI) at half-yearly intervals
13. Angiography 12–24 months after radiotherapy
14. Further follow-up depending on the results obtained

Since 50 Gy is supposed to have the same relative biologic effectiveness whether delivered by protons, γ-rays or X-rays, conventional radiologic treatment should have a similar therapeutic effect if it can be concentrated on the AVM (DRAKE 1983). The recommendation for such a treatment with immobilization by casts (KINGSLEY et al. 1980), not with a stereotactic apparatus, was adopted by us in 1982 (Table 1) (MAKOSKI et al. 1984). Results of the so-called semi-stereotactic irradiation were documented in a controlled study by ZEILSTRA (1987; MAKOSKI et al. 1988, 1989).

3 Patients and Methods

From August 1982 until March 1989, 84 patients with intracranial AVM were treated with a linear accelerator in a semi-stereotactic fashion as described earlier (MAKOSKI et al. 1984). Each patient was to be seen by the same group of experienced specialists (neurologist, neuroradiologist, neurosurgeon, radiotherapist, and physicist). Radiotherapy was undertaken when the site of the AVM was considered unfavorable and associated with a high risk of operative morbidity, whenever embolization was not feasible, or when the patients objected to surgery.

There were 52 men (2–67 years, mean 27.33 years) and 32 women (9–71 years, mean 25.75 years). Bleeding had occurred in 39 patients, 33 had fits, in 21 patients there had been neurological deficits like hemiparesis or hemianopsia, and 10 patients complained of headache. Time, dose, and fractionation stayed constant for 53 consecutive patients: 50 Gy in 35 days, 5 times 2 Gy per week, giving 966 neuret[1] (SHELINE et al. 1980). Since June 1987 the regime has been changed for 20 patients under the auspices of comparable risk for the brain tissue: 20 Gy in 7 days, 4 times 5 Gy (978 neuret). Two weekly doses of from 4 Gy up to 48 Gy (1285 neuret) were tolerated by 3 patients without complaints, the same as 51 Gy, 3 times 3 Gy per week (1187 neuret) in 1.

To determine the localization of the AVM precisely and to prevent a geographical miss with fractionated radiotherapy, a system of a plexiglass frame and an individually molded polyurethane cast is used. Identical positioning for each pro-

[1] Neuret = $D \cdot N^{-.44} \cdot T^{-.06}$ (D = total dose in rads, N = number of fractions, T = treatment time in days)

cedure (deviation 2–3 mm) is warranted by an additional mouthpiece. The system is compatible with all our diagnostic and therapeutic units. The volume to be treated is measured on angiograms and tomograms. Optimized treatment plans are obtained using the tomographic data. The AVM has to receive 100% of the total dose (ICRU 29), the surrounding brain tissue not more than 40% thereof. Localization with the device is done with a simulator, and field controls are made at the linear accelerator before the first treatment and at weekly intervals (MAKOSKI et al. 1984). Photon treatment was done with a 10-MV linear accelerator with a 100-cm focus-to-axis distance: 60 times with single beam rotation, 11 times 2 ports, and 7 times 3 ports. One patient was treated with a four-field setup. Field sizes varied between 3.5 × 3.5 cm^2 and 7 × 9 cm^2.

Table 2. Results of treatment of arteriovenous malformation with dose of 50 Gy (5 times 2 Gy/week; n = 50)

	n	Percentage
Complete obliteration	9	18
Partial obliteration (≥50%)	9	18
Partial obliteration (≤50%)	11	22
No change	15	30
Deterioration	6	12

Table 3. Results of treatment of arteriovenous malformation with dose of 20 Gy (4 times 5 Gy/7 days; n = 17)

	n	Percentage
Complete obliteration	1	5.8
Partial obliteration (≥50%)	5	29.4
Partial obliteration (≤50%)	5	29.4
No change	6	35.3
Deterioration	0	0

4 Results

Data are available for all 84 patients for up to 7 years. All treatment modifications were shown to be feasible; no therapy-related complications could be seen (MAKOSKI et al. 1988; NOCKEN et al. 1983). Seven patients died after radiotherapy, three due to rebleeding 4, 7, and 15 months into the latency period. For the other four patients, no bleeding had been confirmed. A woman suffering from epileptic fits had a seizure 1 month after treatment while bathing and drowned; autopsy did not show any signs of haemorrhage nor damage to the brain tissue. A young girl had rebleeding 3 years after radiotherapy; embolization was tried without success, so a radical resection was done with subsequent neurological deficits.

A critical analysis has been made for the first 25 patients including angiographic studies up to 30 months after radiotherapy (ZEILSTRA 1987). Three of these patients died in the period at risk, i.e., before 12–24 months. The AVMs were classified into three groups as to size: (I) smaller than 8 cm^3 (diameter up to 25 mm; n = 9), (II) from 8 cm^3 to 65 cm^3 (diameter 25–50 mm; n = 15), and (III) larger than 65 cm^3 (diameter more than 50 mm; n = 1).

Three AVMs in 2 patients were totally obliterated; in 5 patients with six AVMs the lesions showed a reduction of 50% or more; in 6 patients the reduction was less than 50%; and in 11 patients there were no changes. Reduction of volume is clearly dependent on the size of the AVM: In group I in 6 patients with 7 AVM it was 75%; in group II with 5 patients, 38%. As to total volume the reduction was 56% in group I, 12% in group II, and none in group III (MAKOSKI et al. 1988).

The 1989 update gives results on 76 of 84 patients (7 died, 1 patient was treated only 3 months before the deadline). The two treatment schemes (50 Gy vs. 20 Gy) have been evaluated separately, indicating no major difference in response (Tables 2, 3). An improvement of obliteration rates in the 20-Gy group may be expected as the time of observation is much shorter than with the 50-Gy group. Critical examination of all patients at risk will be needed.

5 Discussion

Surgical removal of intracranial AVMs is the first line of treatment. If it is not feasible, radiotherapy may be considered as the alternative to conservative treatment (DRAKE 1983). To select the best therapy, the natural course of untreated AVMs has to be considered. Smaller AVMs are found to bear a higher risk of hemorrhage than larger ones, the overall mortality lying between 6% and 17% (GRAF et al. 1983). Radiotherapy had been described as ineffective (GLANZMANN 1978), but irradiation had not been carried out in a controlled study. Favorable reports of anecdotal character were given by Poulsen (1987), Sutton (1984), Tognetti et al. (1985), Wolkov and Bagshaw (1988), Yasunaga et al. (1987), stressing the usefulness of conventional radiotherapy. Controlled studies had been initiated by Johnson (1975), Steiner (1982), personal communication and Makoski et al. (1984, 1989). Preliminary data are available indicating the efficacy of this treatment modality, although the series is as yet incomplete. Surprisingly, there is not a striking difference from results of patients treated by stereotactic proton therapy (Mehdorn and Grote 1988). On the other hand, data presented by Steiner (1984) and the groups using heavy-ion therapy (Fabrikant 1989; Kjellberg 1986) could not be matched (Larson 1989). Preliminary data of much more sophisticated radiographic treatment (Kimmig et al. 1989; Saunders et al. 1987) with dynamic stereotactic radiosurgery are hopeful, but they cannot discredit any systematic conventional treatment as of yet (Fabrikant 1989).

6 Outlook

Obliteration of an AVM is a result of hyperplasia of the intima of the vessel wall. An interval of 12–24 months is generally regarded as a minimum period to elapse before these vascular changes occur. Any protection offered by radiotherapy will therefore be delayed for such a period. Methods of shortening or easily monitoring this latency period have to be looked for. A total dosage of 50 Gy has been adopted from radiosurgery (Steiner et al. 1974). Even doses up to 60 Gy have been recommended (Steiner 1982, personal communication), twice weekly of 5 Gy (1606 neuret) or 3.5–42 Gy (1124 neuret). More dose-finding studies will be necessary. That is why a series of patients was given 20 Gy in 4 fractions over 7 days (967 neuret) as compared with our large series of conventional fractionation up to 50 Gy (978 neuret).

Confirmation of dose distribution to irregular volumes requires more sophisticated techniques, e.g., dynamic stereotactic radiotherapy as a single treatment. It will be worthwhile to compare both methods in the very same institution.

References

Drake CG (1983) Arteriovenous malformations of the brain. The options for management. N Engl J Med 309: 308–310

Fabrikant JI (1989) Workshop on radiosurgery. ISR-Congress, Paris

Glanzmann C (1978) Zerebrale arteriovenöse Mißbildungen: Verlauf bei 18 Fällen nach Radiotherapie. Strahlentherapie 154: 305–308

Graf CJ, Perret GE, Torner JC (1983) Bleeding from cerebral arteriovenous malformations as part of their natural history. J Neurosurg 58: 331–337

Hartmann GH, Schlegel W, Sturm V, Kober B, Pastyr O, Lorenz WJ (1985) Cerebral radiation surgery using moving field irradiation at a linear accelerator facility. Int J Radiat Oncol Biol Phys 11: 1185–1192

Johnson RT (1975) Radiotherapy of cerebral angiomas. In: Pia HW, Gleave JRW, Grote E, Zierski J (eds) Cerebral angiomas. Springer, Berlin Heidelberg New York, pp 256–259

Kimmig B, Engenhart R, Wowra B, Marin-Grez M, Sturm V, Höver KH (1988) Stereotaktisch gezielte Strahlentherapie zerebraler arteriovenöser GefäßmiBbildungen (AVM). Zentralbl Radiol 136: 639

Kimmig B, Engenhart R, Höver KH, Wowra B, Schneider S, Wannenmacher M (1989) Radiosurgery for cerebral angiomas using a linear accelerator. Int J Radiat Oncol Biol Phys [Suppl 1] 17: 234–235

Kingsley DPE, Bergström M, Bergren B-M (1980) A critical evaluation of two methods of head fixation. Neuroradiology 19: 7–12

Kjellberg RN (1986) Stereotactic Bragg peak proton beam radiosurgery for cerebral arteriovenous malformations. Ann Clin Res [Suppl 47] 18: 17–19

Larson DA (1989) Radiosurgery. Refresher Course, ASTRO-Congress, San Francisco

Lutz W, Winston KR, Maleki N (1988) A system for stereotactic radiosurgery with a linear accelerator. Int J Radiat Oncol Biol Phys 14: 373–381

Makoski H-B, Nocken U, Fiebach BJO, Zeilstra DJ (1984) Die Radiotherapie arteriovenöser Malformation des Hirnes. Strahlentherapie 160: 159–165

Makoski H-B, Zeilstra DJ, Nocken U (1988) Arteriovenöse Malformationen des Hirnschädels – strahlentherapeutische Aspekte, In: Bamberg M, Sack H (eds) Therapie primärer Hirntumoren. Zuckschwerdt, Münich, pp 245–249

Makoski H-B, Zeilstra DJ, Bettag W (1989) Die Strahlenbehandlung intrakranieller arteriovaskulärer Malformationen – Erfahrungen mit semi-stereotaktischer Technik. Radiobiol Radiother 30(3): 213–220

Mehdorn HM, Grote W (1988) Zur Strahlentherapie inoperabler arteriovenöser Gefäßmißbildungen. In: Bamberg M, Sack H (eds) Therapie primärer Hirntumoren. Zuckschwerdt, Münich, pp 250–255

Nocken U, Ewen K, Makoski H-B (1983) Somatisches Strahlenrisiko bei Hirnbestrahlungen. Strahlentherapie 159: 548–550

Podgorsak EB, Olivier A, Pla M, Lefebvre P-Y, Hazel J (1988) Dynamic stereotactic radiosurgery. Int J Radiat Oncol Biol Phys 14: 115–126

Poulsen MG (1987) Arteriovenous malformations: a summary of 6 cases treated with radiation therapy. Int J Radiat Oncol Biol Phys 13: 1553–1557

Saunders WM, Winston KR, Siddon RL, Svensson GH,

Kijewski PK, Rice RK, Hansen JL, Barth NH (1988) Radiosurgery of arter ovenous malformations of the brain using a standard linear accelerator: rationale and technique. Int J Radiation Oncology Biol Phys 15: 441–447

Sheline GE, Wara WM, Smith V (1980) Therapeutic irradiation and brain injury. Int J Radiat Oncol Biol Phys 6: 1215–1228

Steiner L, Leksell L, Greitz T, Forster DMC, Backlund EO (1972) Stereotaxic radiosurgery for cerebral arteriovenous malformations. Acta Chir Scand 138(1972): 459–464

Steiner L (1984) Treatment of arteriovenous malformations by radiosurgery. In: Wilson CB, Stein BM (eds) Intracranial arteriovenous malformations. Williams and Wilkins, Baltimore, pp 295–314

Steiner L, Leksell L, Forster DMC, Greitz T, Backlund E-O (1974) Stereotactic radiosurgery in intracranial arterio-venous malformations. Acta Neurochir [Suppl] (Wien) 21: 195–209

Steiner L, Backlund E-O, Greitz T, Leksell L, Noren G, Rähn T (1977) Radiosurgery in intracranial arteriovenous malformations. Excerpta Med Int Congr Ser 433: 168–180

Sturm V, Kober B, Höver K-H, Schlegel W, Boesecke R, Pastyr O, Hartmann GH, Schabbert S, zum Winkel K, Kunze S, Lorenz WJ (1987) Stereotactic percutaneous single dose irradiation with a linear accelerator. Int J Radiat Oncol Biol Phys 13: 279–282

Sutton ML (1984) Adult central nervous system. In: Easson EC, Pointon RCS (eds) The radiotherapy of malignant disease. Springer, Berlin Heidelberg New York, pp 225–227

Tognetti F, Andreoli A, Cuscini A, Testa C (1985) Successful management of an intracranial arteriovenous malformation by conventional irradiation. J Neurosurg 63: 193–195

Wolkov HB, Bagshaw MA (1988) Conventional radiation therapy in the management of arteriovenous malformations of the central nervous system. Int J Radiat Oncol Biol Phys 15: 1461–1464

Yasunaga T, Takada C, Uozumi H, Saito Y, Uena S, Hatanaka Y, Baba Y, Takahashi M (1987) Radiotherapy of spontaneous carotid-cavernous sinus fistulas. Int J Radiat Oncol Biol Phys 13: 1909–1913

Zeilstra DJ, Bettag W, Makoski H-B (1986) Radiotherapy of intracranial AVM's. J Neurosurg 64: 525

Zeilstra DJ (1987) Semi-stereotaktische Bestrahlung inoperabler arteriovenöser Mißbildungen des Hirns. Thesis, University of Cologne

3 Choroidal Melanoma

3.1 Results After Brachytherapy Using $^{106}Ru/^{106}Rh$ Plaques for Choroidal Melanomas

P.K. LOMMATZSCH

CONTENTS

1 Introduction 103
2 Patients and Methods 103
3 Results 104
3.1 Tumor Regression 105
3.2 Visual Acuity 108
3.3 Tumor Size 108
3.4 β-Irradiation and Xenon Coagulation 108
3.5 Repeated Radiotherapy 108
3.6 Radiogenic Side Effects 109
3.7 Enucleation After β-Irradiation 109
3.8 Deaths After β-Irradiation 109
3.9 Survival Rate After $^{106}Ru/^{106}Rh$ Therapy 109
4 Discussion 109
References 111

1 Introduction

The choice of therapy in the management of uveal melanomas is still an unresolved issue. The ideal procedure does not exist which would destroy the tumor locally without inducing the development of metastases and without local side effects.

Enucleation is still indicated for patients with large tumors or painful or blind eyes. However, the treatment of melanomas in eyes with useful vision remains controversial (ZIMMERMAN et al. 1978; MANSCHOTT and VAN PEPERZEEL 1980). Since STALLARD (1966) published his first results obtained with cobalt-60 plaques, the preference for radiotherapy of uveal melanomas has increased (LOMMATZSCH 1983). Several irradiation procedures are available, but every method has its advantages, hazards, and side effects (CHAR et al. 1980; GRAGOUDAS et al. 1980).

$^{106}Ru/^{106}Rh$ plaque therapy was introduced in ophthalmology in 1964 (LOMMATZSCH 1974). The goal of this investigation is to assess the efficiency of β-irradiation of patients with posterior uveal melanoma based on long-term follow-up examinations.

2 Patients and Methods

Between 1964 and 1984, a total of 309 patients (168 men, 141 women) suffering from choroidal melanoma were treated with $^{106}Ru/^{106}Rh$ applicators after diagnosis had been confirmed by ophthalmoscopy, ultrasonography, fluorescein angiography, and ^{32}P test. Biopsy of intraocular tissue seems hazardous and was not carried out. The age distribution is shown in Table 1.

I use concave mirror-like applicators consisting of pure sheet silver 1.0 mm thick which contain $^{106}Ru/^{106}Rh$ in equal distribution. The front window on the concave side is 0.1 mm and allows the electrons to pass practically unhindered. The back is 0.9 mm thick and absorbs nearly 95% of the β-irradiation. These ophthalmic applicators have two ears with the help of which they can be sutured to the sclera. Several types of ruthenium applicators are available (Fig. 1). Physical properties, experimental studies, and surgical procedures were published earlier (LOMMATZSCH 1974).

The tumor-damaging dose – following STALLARD's recommendation – was at least 100 Gy

Table 1. Age distribution of patients suffering from choroidal melanoma

Age (year)	No. of patients
10–20	4
20–30	11
30–40	22
40–50	88
50–60	81
60–70	72
70–80	29
80–90	2
Total	309

P.K. LOMMATZSCH, Prof. Dr. Sc. med., Augenklinik der Universität, Liebigstr. 14, 7101 Leipzig, Germany

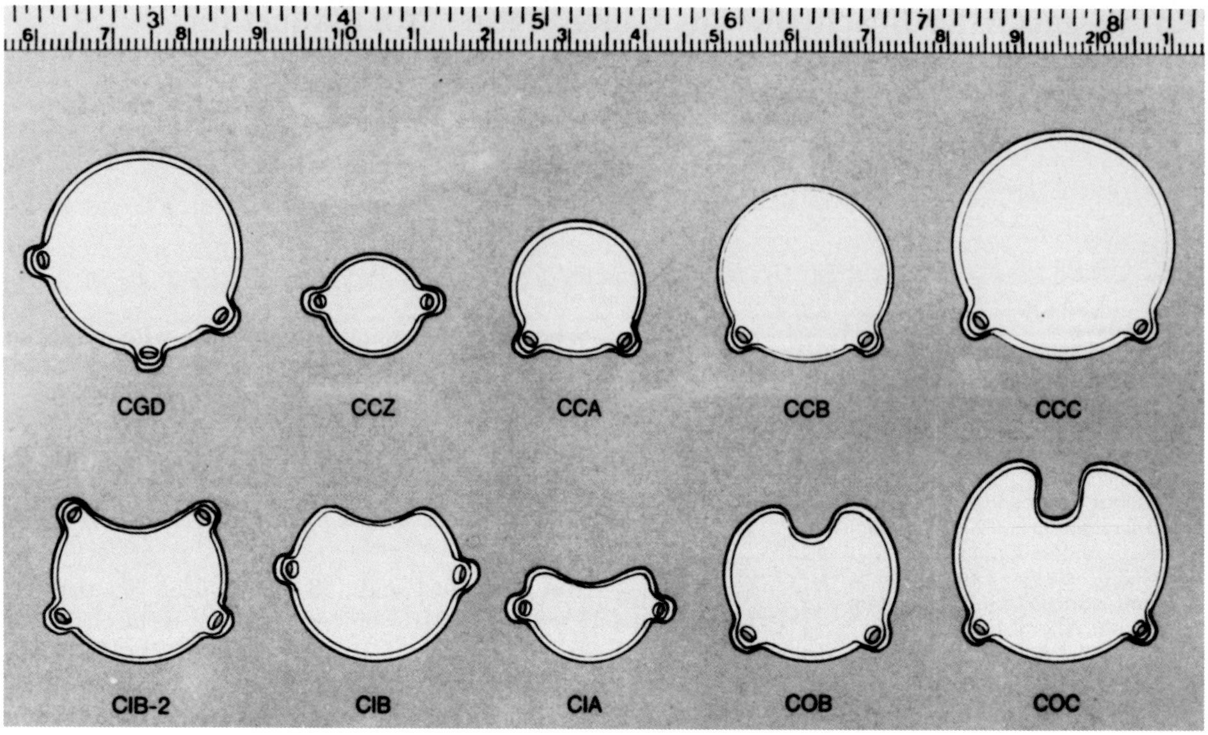

Fig. 1. $^{106}Ru/^{106}Rh$ eye applicators (Isocommerz, Berlin-Buch, Germany)

(10000 rad) at the apex of the tumor. Thus, values above 1000 Gy (100000 rad) may be reached at the tumor base. An irradiation time of 8–14 days seemed to be the optimum exposure time.

After treatment, the patients were examined every month and later, if the tumor began to shrink, every 3 months. Tumor regression was documented by fundus photography, ultrasonography, and fluorescein angiography.

The average follow-up period was 6.7 years (shortest period, 1 year; longest, 21 years); 188 patients were followed up for more than 5 years (average, 9.4 years), and 87 patients were controlled for more than 10 years.

To describe the extent of the primary tumor the pretreatment clinical classification, designated TNM, was used as recommended by the International Union against Cancer (UICC):

- T 1a Tumor not more than 7 mm in its greatest dimension and with an elevation of not more than 2 mm
- T 1b Tumor more than 7 mm but not more than 10 mm in its greatest dimension and with an elevation of more than 2 mm but not more than 3 mm.
- T 2 Tumor more than 10 mm but not more than 15 mm in its greatest dimension and with an elevation of more than 3 mm but not more than 5 mm.
- T 3 Tumor more than 15 mm in its greatest dimension or with an elevation of more than 5 mm.
- T 4 Tumor with extraocular extension.

3 Results

In this study, those patients whose tumors had either changed to a flat scar or had shrunk to a still appreciable grayish or black mass with scarring of the tissue around the tumor, which had remained unchanged for more than 1 year, were considered to have been treated successfully. Moreover, those patients who died later from causes other than metastases and whose choroidal melanoma had shrunk or cicatrized were also included as successfully treated. Only those patients who died from metastases or who had to be enucleated were counted as failures.

Of 309 patients, 216 (69.9%) were treated

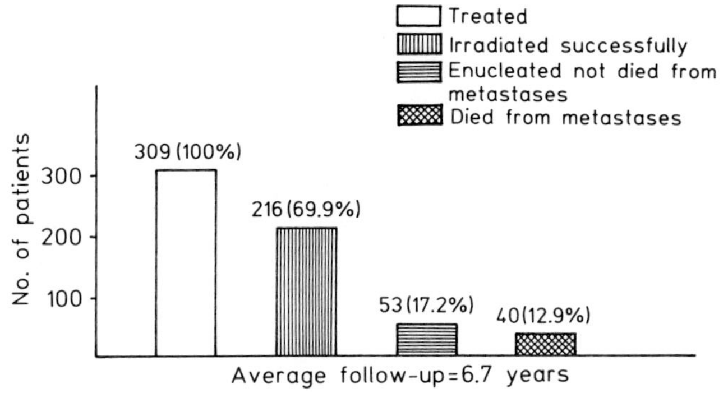

Fig. 2. Results after β-irradiation with $^{106}Ru/^{106}Rh$ plaques

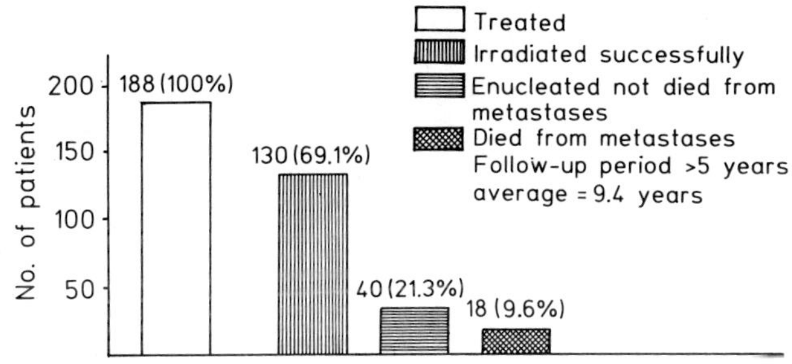

Fig. 3. Long-term results after β-irradiation with $^{106}Ru/^{106}Rh$ plaques of 188 patients

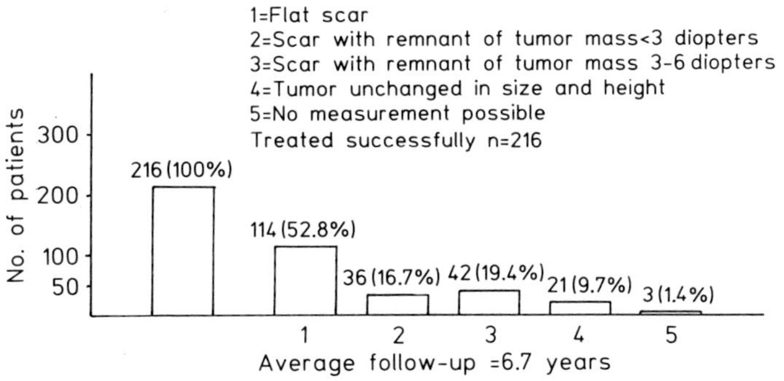

Fig. 4. Tumor regression after β-irradiation 3 diopters = 1 mm tumor thickness

successfully and have been under observation since then, for a mean period of 6.7 years after irradiation (shortest period, 1 year; longest, 21 years). Some 53 (17.2%) patients had to be enucleated because of new tumor growth but did not die from metastases, and 40 (12.9%) died from proven metastases in this period of 20 years (Fig. 2).

A total of 188 patients could be followed up for more than 5 years (101 patients for 5–9 years, 63 patients for 10–14 years, and 24 patients for 15–21 years). Out of the group with long-term follow-up (mean period 9.4 years), 130 (69.1%) patients could be regarded as successfully treated, 40 (21.3%) had to be enucleated and are either alive or died from causes other than metastases, and 18 (9.6%) patients died from proven metastases (Fig. 3).

3.1 Tumor Regression

Of the successfully treated patients, (52.8%) had a flat scar after therapy (Fig. 4). The other eyes showed more or less notable remnants of the former tumor. The fact that tumor regression after this kind of therapy may take many months is important for avoiding any precipitous enucleation. The regression to a flat, pigment-stippled scar depends primarily on tumor size. However, all tumors larger than 3 mm thick left behind an elevated mass, which should be observed carefully

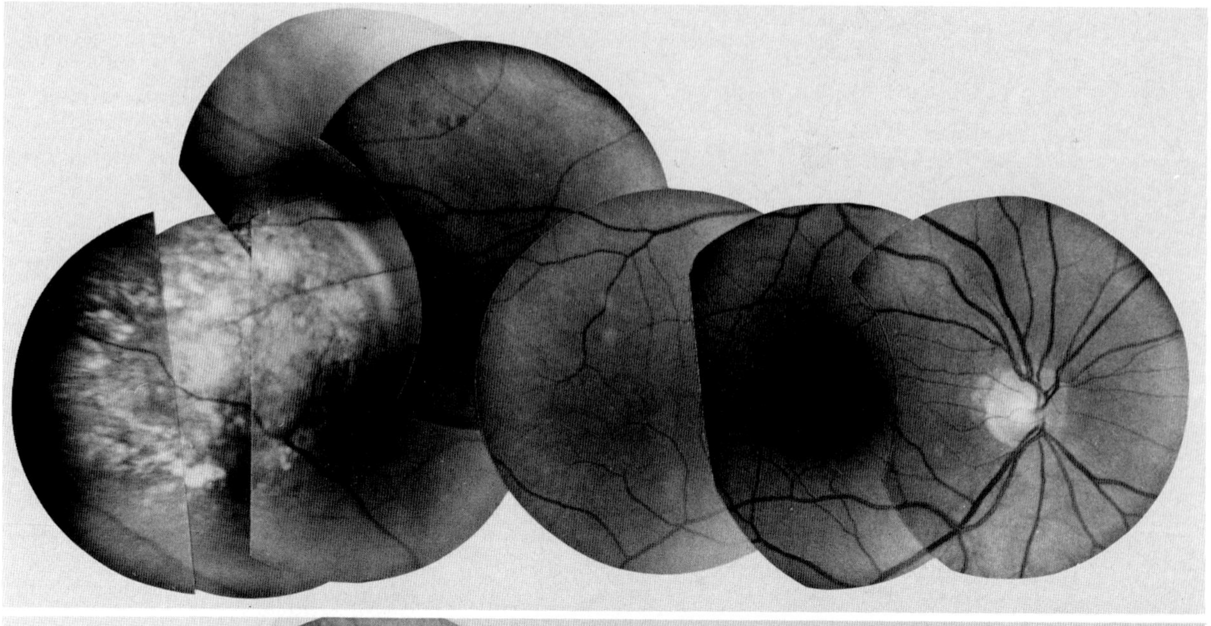

a

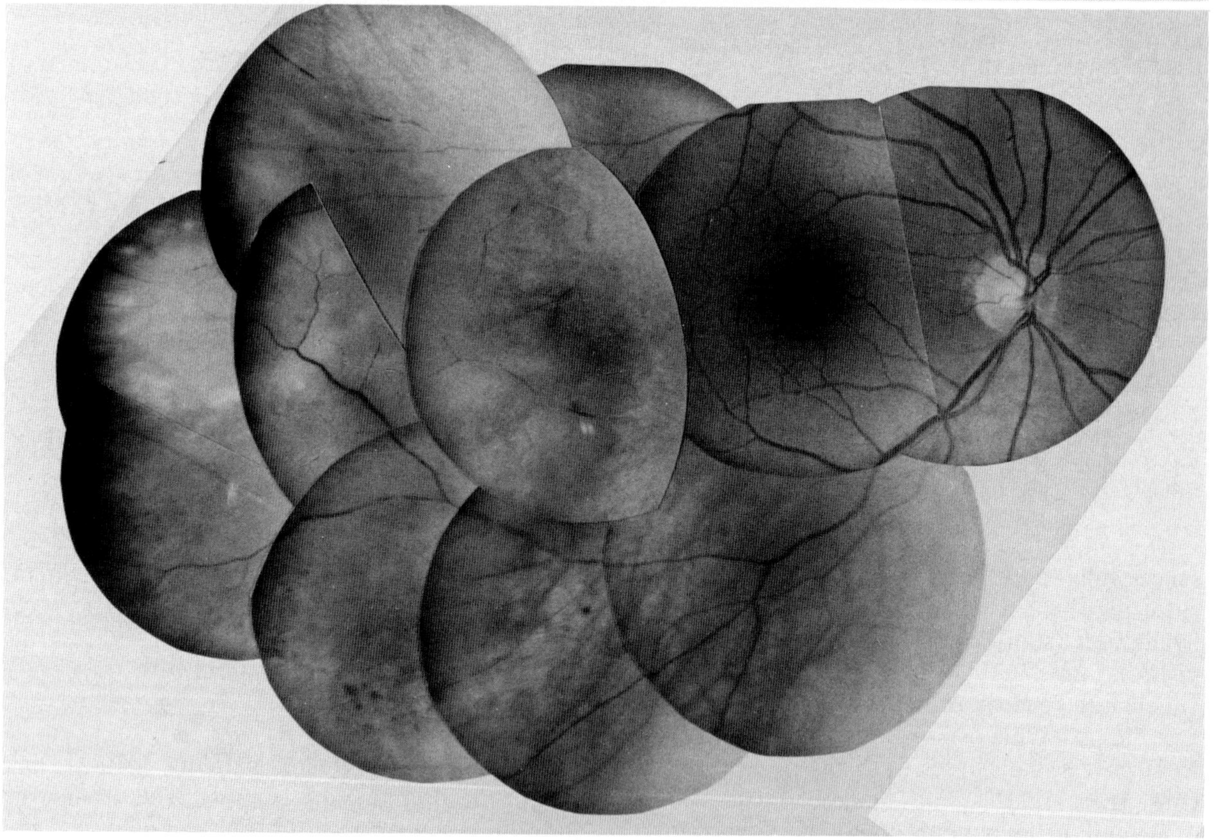

b

Fig. 5. a Choroidal melanoma in right eye of a female patient aged 49 years. Visual acuity = 1.0. **b** 4 weeks after brachytherapy with $^{106}Ru/^{106}Rh$ plaque (820 Gy at the scleral surface). **c** 2 years and 6 months after treatment. **d** 5 years after irradiation. Flat, atrophic, chorioretinal scar with pigmentation in its center. Visual acuity = 0.1 due to macular degeneration

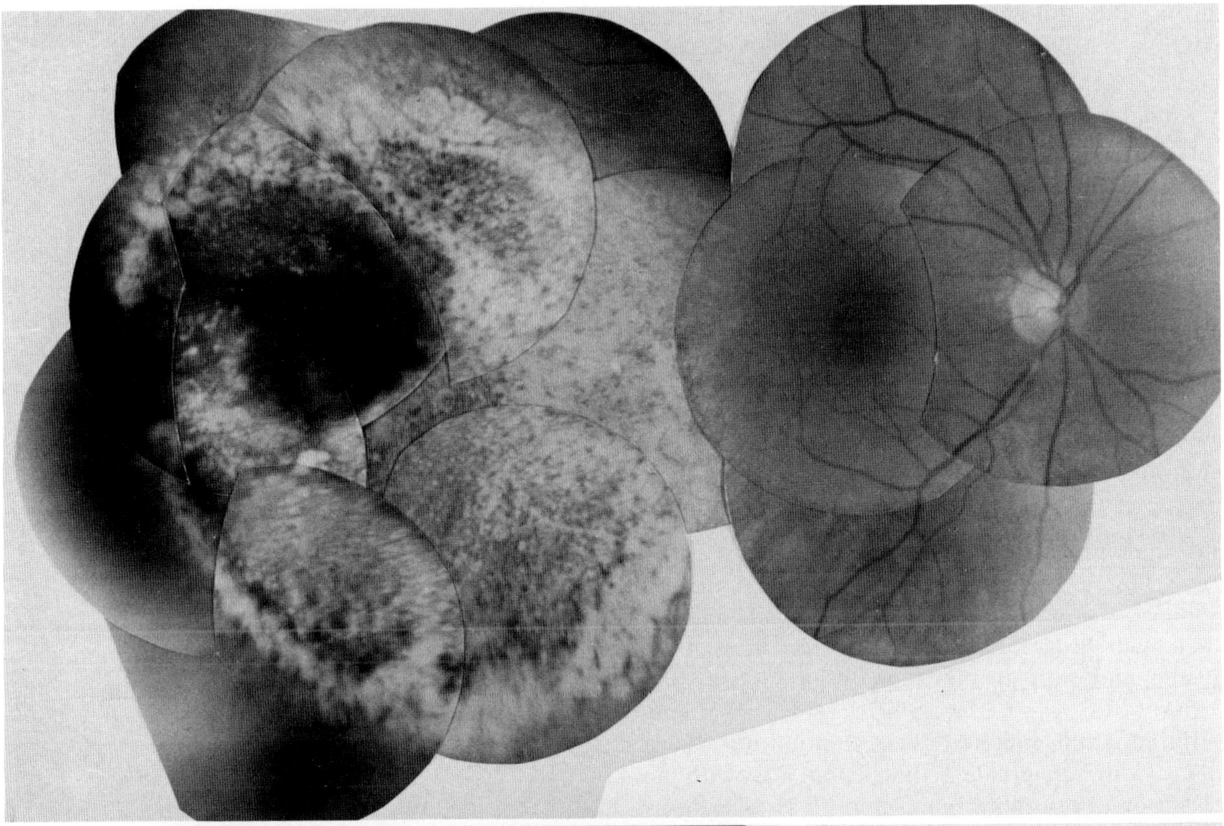

c

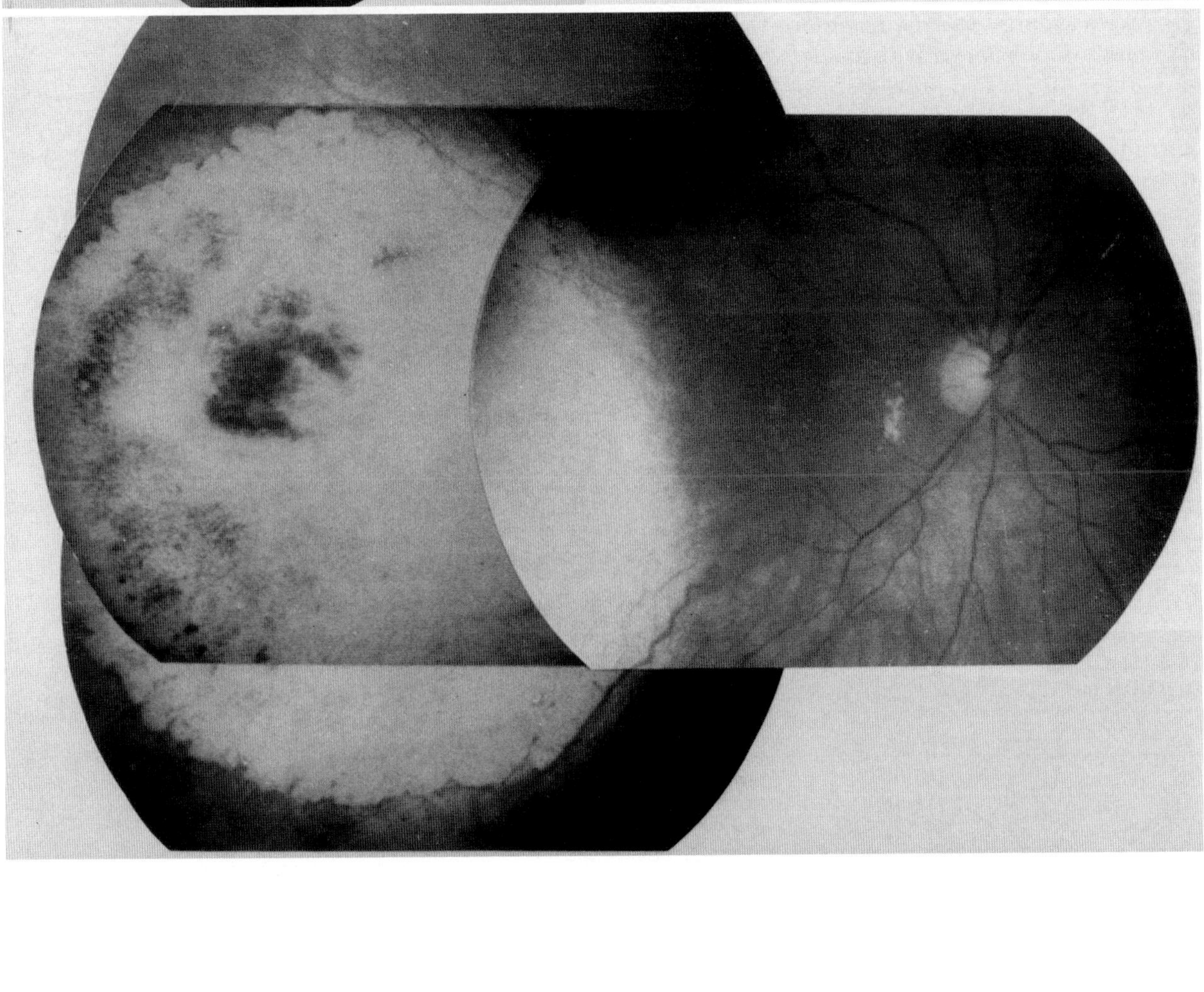

d

during the follow-up period because of some possible reactivation process (Fig. 5).

3.2 Visual Acuity

Preservation of useful visual acuity is a principal aim of β-ray treatment. Forty-nine (22.7%) of all successfully treated patients retained good visual acuity (between 1.5 and 0.5). About half the patients (54.2%) had only a poor central visual acuity of less than 0.2 because the macula had been involved in the cicatricial formation after treatment. Visual field deficiencies corresponding to the retinochoroidal scar could be registered in all our patients.

3.3 Tumor Size

Some 219 (70.9%) patients had small tumors, each with its greatest dimension not more than 10 mm and with an elevation of not more than 3 mm (T1a, T1b). Twenty-six (8.4%) patients suffered from medium-sized tumors not more than 15 mm in greatest dimension and with an elevation of not more than 5 mm (T 2). Sixty-four (20.7%) had large melanomas more than 15 mm in greatest dimension or with an elevation of more than 5 mm (T 3). The prevalence of small tumors (81.5%) is even clearer in the group of 216 successfully treated patients (Table 2). In the group of 40 patients having died from metastases, there was a prevalence of large tumors (42.5%) (Table 3).

Table 2. Distribution of tumor size in 216 successfully treated patients

No. of patients	(%)	Size of tumor
176	(81.5)	Small
13	(6.0)	Medium
27	(12.5)	Large

Table 3. Distribution of tumor size in 40 patients having died from metastases

No. of patients	(%)	Size of tumor
18	(45.0)	Small
5	(12.5)	Medium
17	(42.5)	Large

Table 4. Results of 34 patients treated with additional light coagulation

Status of patients	No. of patients
Treated successfully	21
Living, but enucleated	4
Having died	9
from metastases	4
Total	34

3.4 β-Irradiation and Xenon Coagulation

Light coagulation was additionally performed after β-irradiation in 34 patients to occlude the choroidal vessels around the tumor. First, the central rim of the tumor was surrounded by coagulations. Later, if the tumor was not completely destroyed, the surface of the tumor proper was also coagulated. Especially in eyes with the melanoma close to the optic disc and where the applicator cannot be placed completely around the tumor, I succeeded in partially surrounding the central tumor rim with coagulation effects. The results obtained are shown in Table 4.

3.5 Repeated Radiotherapy

Repeated therapy with radioactive applicators was necessary in 16 patients because the tumor continued growing after the first irradiation. The second treatment was successful in 7 patients. Nevertheless, 4 had to have the eye enucleated, and 5 died from metastases.

Table 5. Radiogenic complications in patients leading to deterioration of visual acuity

Complication	No. of patients
Macular destruction because of scarring around the tumor	83
Atrophy of the optic nerve	23
Macular degeneration	16
Postradiation retinopathy with edema of the optic disc	17
Partial cuneiform cataract without loss of vision	5
Total cataract	7
Vitreous hemorrhage	10
Secondary glaucoma	3
Thrombosis of the central retinal vein	2
Scleral necrosis	1
Exudative reactions such as choroidal and retinal detachment with spontaneous improvement	21

3.6 Radiogenic Side Effects

Radiogenic tissue damage must be expected after each kind of radiotherapy, particularly the serious late effects which may occasionally occur after β-irradiation. Exudative reactions occurring a few days after irradiation, such as chemosis, choroidal detachment, and even transient retinal detachment, should not be regarded as serious complications because of their good prognosis. Against earlier expectations, the primary concern is now the postradiation retinopathy that has frequently appeared after β-irradiation, sometimes even years after initially successful treatment.

Radiogenic complications which led to deterioration of visual acuity are shown in Table 5. Most of the patients (83) with a tumor located 1–2 disc diameters away from the posterior pole showed destruction of the macula because the scarring around the tumor necessarily involved this retinal area.

3.7 Enucleation After β-Irradiation

Enucleation was necessary in 64 patients, mostly within the 1 year of treatment. The reason for removing the eye was new tumor growth in 54 and loss of visual acuity due to radiogenic side effects in 10. Histologic examination of these latter 10 enucleated eyes showed active newly growing melanoma tissue containing some tumor cells in 4, scar tissue without tumor cells in 4, while no histologic examination was available in 2.

Of all 64 enucleated patients, 11 (17.2%) died from metastases within 1–7 years after enucleation, and 9 patients died from other causes.

3.8 Deaths After β-Irradiation

A total of 75 patients of this series died during the follow-up period: 40 from metastases, 27 of diseases other than melanoma (cardiac infarction or apoplectic stroke, 19; other malignant tumors, 6; cirrhosis of liver, 1; suicide, 1), and 8 of unknown causes. The death from metastases had been verified by autopsy in 19, laparoscopy in 4, and by clinical examination in 17. Metastases occurred mainly in the liver, where in some patients an extreme hepatomegaly was produced.

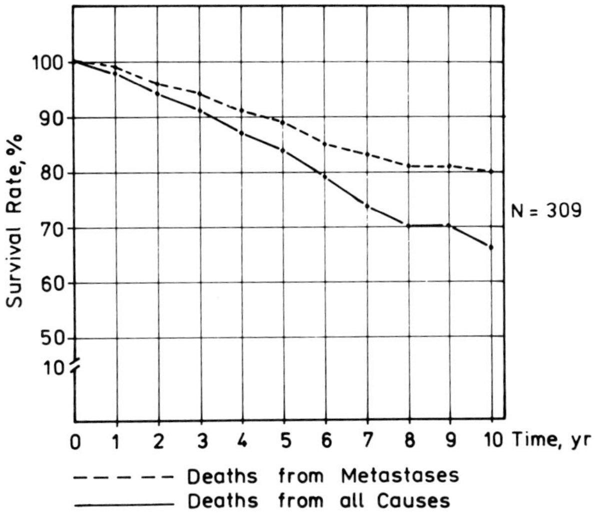

Fig. 6. Survival rate of patients ($n = 309$) with choroidal melanoma treated by $^{106}Ru/^{106}Rh$ β-ray applicator. *Broken line* indicates deaths from metastases; *solid line*, deaths from any cause

3.9 Survival Rate After $^{106}Ru/^{106}Rh$ Therapy

The survival rate was calculated using the life-table method. As shown in Fig. 6, survival rates of 84.3% (deaths from any cause) or 88.7% (deaths from metastases only) after 5 years, and 65.8% (deaths from any cause) or 79.5% (deaths from metastases only) after 10 years could be achieved. In this diagram, the survival rates of patients who died from metastases and those who died from any cause were plotted separately. A comparison with results published in previous studies (Kiehl et al. 1984) shows that survival after local β-irradiation is not worse than after enucleation of eyes containing choroidal melanomas similar in size and extension.

4 Discussion

The traditional treatment of patients suffering from choroidal melanoma was enucleation of the involved eye as soon as diagnosis was made. More recently, the value of enucleation has been questioned, and some authorities have even speculated that this procedure may impair prognosis (Zimmerman et al. 1978; Zimmerman and McLean 1979). This hypothesis has provoked objections (Boniuk 1979; Jackobiec 1979; Manschot and van Peperzeel 1980), and so additional studies will be needed to determine whether or not our ideas on the management of choroidal melanomas are correct.

Although in the past it was believed to be ineffective in the treatment of intraocular melanomas, radiotherapy has recently become more generally accepted, especially the use of radioactive plaques and charged particle irradiation (CHAR et al. 1980; GRAGOUDAS et al. 1980; PACKER et al. 1980). Some investigators, including myself, have shown that patients treated by conservative methods such as photocoagulation or irradiation as alternatives to enucleation have a relatively low incidence of metastases, but convincing, randomized, prospective studies are still lacking (KIEHL et al. 1984).

The choice of therapy depends on certain conditions, and each case should be considered individually. In selecting the most effective therapeutic approach the following factors must be carefully weighed: size of the melanoma, its extent and location, its activity, condition of the other eye, and age, general health, and last but not least psychological status of the patient (SHIELDS 1983).

The results presented in this study lead us to the conclusion that β-irradiation is an effective procedure, especially for the treatment of small and medium-sized choroidal melanomas (T 1, T 2). Additional light coagulation or a second β-irradiation make it possible to treat even larger tumors in two steps.

In previous studies (LOMMATZSCH 1974, 1978, 1979, 1983) 74% (1974), 61.6% (1979), and 64.4% (1983) cure rates were found. These results are in accordance with the therapeutic results after a long-term follow-up presented in this paper, with a healing rate of 69.9%.

The apparent increase in the percentage of successfully treated patients from 64.4% (1964–1980) to 69.9% in this paper may be related to the shorter follow-up of the most recent 104 patients, though the average follow-up of 6.7 years was longer in comparison with 5.4 years in the previous series. The slight differences between the percentages of the successfully treated patients in whom the tumor regressed to a flat scar – 45.5% (1980), 60.5% (1980) of patients observed for more than 5 years, and 52.8% in this paper – are probably caused at random.

These figures suggest that β-irradiation with ^{106}Ru/^{106}Rh plaques will produce a successful therapeutic outcome in about two-thirds of patients, provided the following recommendations are observed. First, because of the physical properties of Ruthenium-106 and Rhodium-106 (at a distance of 6 mm, the dose received is less than 10% of the surface dose), the choroidal melanoma should not exceed 6 mm in height above the scleral surface. Tumor cells at a distance of more than 6 mm from the applicator will not receive the damaging dose of at least 1000 Gy (10000 rad). Secondly, the distance of the posterior margin should be at least 1–2 disc diameters from the nerve head, otherwise radiogenic papillitis with atrophy of the optic disc will prevent useful visual acuity. Thirdly, if the tumor has involved the ciliary body (contrary to former opinion) a special shape of ^{106}Ru/^{106}Rh plaque can be used with some success. Fourthly, there should be no extension of the tumor outside the eye.

Radiogenic side effects and the incidence of complications depend on the dose distribution within the eye. In the case of β-irradiation the severity and extension of postradiation retinopathy is less than after cobalt-60 irradiation, with its higher volume dose (MACFAUL 1977). The high incidence of visually destructive radiation effects after β-irradiation was caused by macular destruction in those patients in whom the tumor grew close to the macular region. If the tumor comes closer than 2 disc diameters to the optic disc, the dose delivered to the nerve head will result in an increasing incidence of optic atrophy after ^{106}Ru/^{106}Rh plaque therapy.

Although it is evidence from these and other statistical data that survival rates and local therapeutic results are mainly dependent on the number of tumor cells within the eye, it may sometimes be necessary to irradiate larger tumors because the patient has refused enucleation or the tumor occurred in the only eye able to see. In some of these patients, I have observed contrary to expectation a surprising shrinkage of the tumors. In questionable situations, therefore, it seems to be justified to try β-irradiation first and to hope for regression before performing enucleation.

Local irradiation with ^{106}Ru/^{106}Rh plaques is not likely to result in a higher incidence of metastases than enucleation, although the only scientific way to answer this question would be to conduct a randomized prospective study. ROTMAN et al. (1977) pointed out that those methods of treatment which left the eye in situ had a greater overall survival rate than enucleation. Perhaps irradiation, in contrast to enucleation, promotes an immune response that may indeed control distal metastasis.

New radiotherapeutic procedures have been published recently that extend the range of local radiotherapy options by iodine-125 seeds (Packer et al. 1980) and proton beam irradiation or heavy helium ions (Gragoudas et al. 1980). Although long-term follow-up examinations are still lacking, the preliminary experience with these new methods is encouraging. The management of choroidal melanomas will probably remain controversial in the near future. In any case β-irradiation with $^{106}Ru/^{106}Rh$ plaques adds to the possibilities of treating choroidal melanoma.

References

Boniuk M (1979) A crisis in the management of patients with choroidal melanoma. Am J Ophthalmol 87: 840–842

Char DH, Castro JR, Quivey J, et al. (1980) Helium ion charged particle therapy for choroidal melanoma. Ophthalmology (Rochester) 87: 565–570

Gragoudas ES, Goitein M, Verhey L, Munzenreider J, Suit HD, Koehler A (1980) Proton beam irradiation: an alternative to enucleation for intraocular melanomas. Ophthalmology (Rochester) 87: 571–581

Jakobiec FA (1979) A moratorium on enucleation for choroidal melanoma. Am J Ophthalmol 87: 842–846

Kiehl H, Kirsch I, Lommatzsch P (1984) Das Überleben nach Behandlung des malignen Melanoms der Aderhaut: Vergleich von konservativer Therapie ($^{106}Ru/^{106}Rh$ Applikator) und Enukleation ohne und mit postoperativer Orbitabestrahlung. 1960 bis 1979. Klin Monatsbl Augenheilkd 184: 2–14

Lommatzsch PK (1974) Treatment of choroidal melanoma with $^{106}Ru/^{106}Rh$ applicators. Surv Ophthalmol 19: 85–100

Lommatzsch PK (1978) Beta-ray treatment of malignant epibulbar melanoma. Graefes Arch Klin Exp Ophthalmol 209: 111–124

Lommatzsch PK (1979) Radiotherapie der intraokularen Tumoren, insbesondere bei Aderhautmelanom. Klin Monatsbl Augenheilkd 174: 948–958

Lommatzsch PK (1983) β-Irradiation of choroidal melanoma with $^{106}Ru/^{106}Rh$ applicators. Arch Ophthalmol 101: 713–717

MacFaul PA (1977) Local radiotherapy in the treatment of malignant melanoma of the choroid. Trans Ophthalmol Soc UK 97: 421–427

Manschot WA, Van Peperzeel HA (1980) Choroidal melanoma. Enucleation or observation? A new approach. Arch Ophthalmol 98: 71–77

Packer S, Rotman M, Fairchild RG, Albert DM, Atkins HL, Chan B (1980) Irradiation of choroidal melanoma with iodine-125 ophthalmic plaque. Arch Ophthalmol 98: 1453–1457

Rotman M, Long RS, Packer S, Moroson H, Galin MA, Chan B (1977) Radiation therapy of choroidal melanoma. Trans Ophthalmol Soc UK 97: 431–435

Shields JA (1983) Diagnosis and management of intraocular tumors. Mosby, St Louis

Stallard HB (1966) Radiotherapy for malignant melanoma of the choroid. Br J Ophthalmol 50: 147–155

Zimmerman LE, McLean IW (1979) Effect of enucleation on uveal melanomas. Am J Ophthalmol 87: 741–760

Zimmerman LE, McLean IW, Foster WD (1978) Does enucleation of the eye containing a malignant melanoma prevent or accelerate the dissemination of tumor cells? Br J Ophthalmol 62: 420–425

3.2 Choroidal Melanoma: Role of Brachytherapy in Management

LUTHER W. BRADY, JERRY A. SHIELDS, JAMES J. AUGSBURGER, JOHN L. DAY, REGINALD WOODLEIGH, ARNOLD M. MARKOE, and ULF L. KARLSSON

CONTENTS

1 Introduction 113
2 Treatment Procedures 117
3 Results 117
4 Conclusions 120
References 121

1 Introduction

The eye is a complex organ made up of the lids, extraocular, and intraocular structures. There is a wide diversity of opinion as to the role for radiation therapy in the management of malignant tumors of the eye. Some authors have stated that all primary intraocular malignant tumors with the exception of retinoblastoma are best treated by surgical techniques (ARNESEN and NORMES 1975; ASHTON and WYBAR 1966; DEL REGIATO and SPJUT 1977; JAM 1964; KELLER 1973; LEDERMAN 1964; STALLARD 1966a; STARR and ZIMMERMAN 1962; WINTER 1963, 1964). They go on to point out that the only possible reason to irradiate the eye in patients with such lesions would be to preserve vision. Further, it has been stated that irradiation to cancericidal levels may produce an iridocyclitis, resulting in an imbalance between aqueous production and absorption ending in glaucoma. Even given circumstances under which definitive radiation therapy may be pursued, it should be limited to lesions that are located in the posterior half of the eyeball.

However, with the advent of contemporary innovative techniques of radiation therapy, particularly the utilization of brachytherapy, reassessment of this position indicates that there are specific circumstances under which malignant intraocular tumors can be treated by radiotherapy.

Even though ophthalmologists have long been familiar with this disease entity, controversy still exists regarding the optimal therapeutic modalities to be used. Classically, enucleation of the affected eye has been the surgical therapy of choice. However, under certain circumstances, alternative surgical management by local resection (FOULDS 1974; LONG et al. 1971; MACFAUL 1977; PEYMAN and APPLE 1974; SAUTTER and NEUMANN 1973; STALLARD 1966b) laser photocoagulation (MEYER-SCHWICKERATH 1957; VOGEL 1972) or cryopexy (LINCO et al. 1967) has been used. Very small lesions felt clinically to be melanomas have been treated by observation alone until the lesion has shown clinical growth (MARKOE et al. 1987).

Enucleation as a modality for treating malignant melanomas has been questioned by ZIMMERMAN (1967) and ZIMMERMAN and MCLEAN (1975) who have suggested that the prognosis following enucleation may be inferior to that for untreated patients. Such a suggestion has led to the development of the so-called no touch technique in enucleation (FRAUNFELDER et al. 1977) although it is not at all clear whether tumor seeding is affected by manipulation of the human globe at the time of removal.

Since the pioneering work of STALLARD (1948, 1959, 1966b), radiation therapy has emerged as an alternative to enucleation. Both external beam radiation therapy (CHAR and CASTRO 1982; CHAR et al. 1982; GRAGOUDAS et al. 1977, 1978; SAUNDERS et al. 1973) as well as brachytherapy techniques (LONG et al. 1971; STALLARD 1966b; BONIUK and GIRARD 1965; BRADY et al. 1984; FITZ PATRICK et al. 1978; LOMMATZSCH 1974, 1983) have been examined and by the modern approach appear to yield grossly similar results in terms of

LUTHER W. BRADY, M.D., JOHN L. DAY, M.D., REGINALD WOODLEIGH, M.S., ARNOLD M. MARKOE, M.D., ULF L. KARLSSON, M.D., Department of Radiation Oncology and Nuclear Medicine, School of Medicine, Hahnemann University, Mail Stop 200; Broad and Vine, Philadelphia, PA 19102-1192, USA

JERRY A. SHIELDS, M.D., JAMES J. AUGSBURGER, M.D., Oncology Service, Wills Eye Hospital, Philadelphia, PA 19107, USA survival, local control, and preservation of vision.

Modern approaches using external beam radiation therapy techniques have emphasized beams of heavy charged particles, taking advantage of the Bragg peak effect (an extremely sharp penumbra of the beam) to treat the tumor effectively to a high, homogeneous dose while sparing to a relative degree the nearby normal structures. Optimization of therapy requires exceptionally accurate tumor localization and demarcation, precise patient immobilization during therapy, verification of the reproducibility on a daily basis, and complex treatment planning. This type of therapy is currently being pursued at only two centers within the USA.

In the USA in 1989, the American Cancer Society anticipates that there will be 1900 new cases of primary malignant tumors of the eye. Of these, 1000 will occur in men and 900, in women. During that same time interval, the data indicate that 300 deaths will result as a consequence of the disease (150 in men and 150 in women). The most common primary malignant intraocular tumor is malignant melanoma of the choroid, comprising about 75% of all lesions. The second most common primary malignant intraocular tumor is retinoblastoma accounting for about 20% of patients. The remaining 5% of primary malignant intraocular tumors includes epithelial tumors of the uvea, connective tissue and other mesenchymal tumors, hematopoietic tumors, and meningiomas of the optic nerve. Metastates to the eye are the most common malignant intraocular tumor, arising most often as secondaries from breast or lung cancer, or as part of lymphomatous and leukemic processes (Table 1).

Table 1. Incidence of malignant intraocular tumors

	Incidence (%)
Primary	
Pigmented tumors	~75
Reinoblastoma and other neuroectodermal tumors of the retina	~20
Epithelial – uvea	<1
Connective tissue and other mesenchymal tumors	<1
Hematopoietic	<1
Meningioma	rare
Secondary (% of patients with metastatic disease)	
Lung	~15
Breast	~58
Lymphoma	
Leukemia	>8

Malignant melanoma of the uveal tract is the most common primary intraocular malignancy. Ophthalmologists have been familiar with these tumors for many years, but despite this, considerable controversy regarding the best approach to diagnosis and treatment remains. Current data have been confusing since most information comes from individual case reports or small series of cases (Arnesen and Normes 1975; Ashton and Wybar 1966; Jay 1964; Keller 1973; Lederman 1964; del Regato and Spjut 1977; Stallard 1966a) However, the composite effort on the part of the Wills Eye Hospital Oncology Service and the Department of Radiation Oncology at Hahnemann University has clearly established the fact that brachytherapy radiation techniques have given rise to results that are similar to those achieved under the most ideal circumstances using enucleation (Brady et al. 1982, 1984, 1988).

New techniques in diagnosis and management of choroidal melanomas offer exciting prospects for a change in the overall treatment approach of the disease process and also improvement in terms of prognosis. In the past, diagnosis has been made by direct ophthalmoscopy alone, and errors resulting in unnecessary enucleation were frequent (Shields and Zimmerman 1973). Recent studies have shown that the adequate use of new techniques such as indirect ophthalmoscopy, fluorescein angiography, ultrasonography, computed tomography, and magnetic resonance imaging can reduce the incidence of diagnostic error so that the diagnosis can be made with better than 97% accuracy (Shields and McDonald 1973).

Melanomas located in the anterior uvea are usually detected earlier than those located posteriorly and may be removed by either iridectomy or iridocyclectomy. Histologic confirmation of the diagnosis is therefore obtained by surgical resection. However, lesions of the posterior uvea are not readily accessible to biopsy, and a clinical diagnosis may occasionally be necessary.

Current techniques have allowed for improvement in the diagnostic accuracy for posterior uveal melanomas. Table 2 and Fig. 1 give the current procedures appropriate to a diagnosis of melanoma of the choroid. If there remains a degree of uncertainty as to the diagnosis, a conservative approach with a period of careful observation photographically documented may be indicated.

Other diagnostic modalities may be potentially useful in the future (Table 3). However, they are not as widely accepted for routine clinical evalua-

Table 2. Current approaches to diagnosis of melanoma

1. History
2. Medical evaluation
3. Examination of the opposite eye
4. Indirect ophthalmoscopy
5. Contact lens
6. Transillumination
7. Fundus photography
8. Fluorescein angiography
9. Visual field examination
10. Ultrasonography
11. Radioactive phosphorus uptake test (^{32}P)
12. Ophthalmologic consultations

Table 3. Diagnostic modalities for possible future use with melanomas

1. Immunologic studies
2. Other radioactive isotopes
3. Biopsy
4. Carcinoembryonic antigen
5. Computed tomography
6. Other diagnostic methods

tion. In summary, many of the diagnostic problems of choroidal melanomas confronting the clinician in the past have been eliminated. This is related to the greater clinical awareness about melanomas and pseudomelanomas as well as to the use of indirect ophthalmoscopy and other ancillary procedures. With further investigation, other diagnostic modalities may be developed for clinical use.

The traditional treatment for malignant melanomas of the posterior uvea has been enucleation of the eye as soon as the diagnosis was established. Studies have shown that patients with small melanomas have an excellent prognosis following enucleation. Recently, investigations have demonstrated that some small melanomas may be dormant for many years and may pose no significant threat to the patient. Others have revealed that enucleation does not appear to improve the prognosis for a patient over 65 years old. The traditional treatment by enucleation has been challenged by a number of authorities, and other approaches have been advocated, particularly for small choroidal melanomas (Zimmerman and McLean 1979; Zimmerman et al. 1978) (Table 4).

It is generally accepted that melanomas of the choroid can be divided into three categories by size. Warren's (1974) definitions have been adopted, namely, a small melanoma being one that is less than 10 mm in diameter and less than 2 mm in elevation as measured by ultrasonography. A medium sized melanoma is larger in either diameter or elevation than a small melanoma but not greater than 15 mm in diameter or 5 mm in elevation. A large melanoma is larger than a medium-sized melanoma in either dimension or elevation (Table 5).

Observation may be indicated when the patient is periodically available. One such example is an asymptomatic tumor not affecting central vision. Often such small melanomas may be quiescent for many years without clinical evidence of enlargement. Other circumstances that the indications for observation include an asymptomatic melanoma in a patient older than 65 years or a slowly growing melanoma in the only remaining eye of an older or chronically ill patient.

Photocoagulation may be carried out when the tumor has been diagnosed accurately and is still small. This method was first used by Meyer-Schwickerath in 1952 (Meyer-Schwickerath 1961). Precise criteria for the selection of these patients must be applied. Further follow-up will be necessary before the true efficacy of this tech-

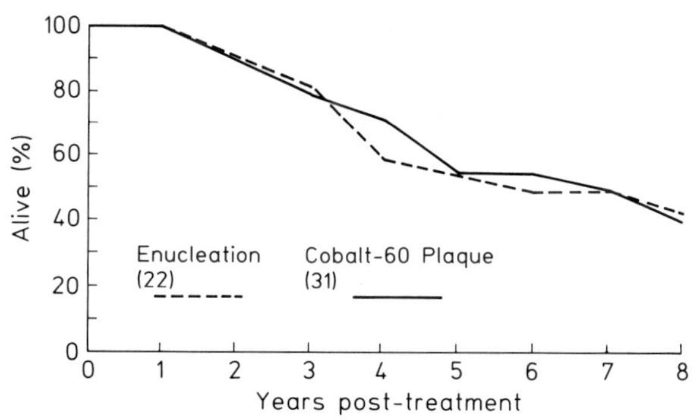

Fig. 1. Comparison of statistically equivalent patients enucleated or treated by radiocobalt eye-plaque therapy (residual patients after exclusion for nonoverlapping parameters and having a hazard ratio >1 for statistically important prognostic factors; these patients are at highest risk of dying from their disease)

Table 4. Current approaches to management of melanoma

1. Observation
2. Photocoagulation
3. Cryotherapy
4. Diathermy
5. Local resection
6. Enucleation
7. Exenteration
8. Radiotherapy

Table 5. Pretreatment and posttreatment tumor dimensions for 100 posterior uveal melanomas managed by cobalt plaque radiotherapy

Tumor size category	*n*	Pretreatment mean tumor dimensions[a]	Posttreatment mean tumor dimensions[a]	% Reduction in tumor volume
Small	7	7.6 × 6.6 × 2.3 ($115.4\,mm^3$)	7.1 × 6.1 × 1.0 ($43.3\,mm^3$)	62.5
Medium	41	9.6 × 8.0 × 4.2 ($322.6\,mm^3$)	9.4 × 7.8 × 2.7 ($198.0\,mm^3$)	38.6
Large	52	11.8 × 10.2 × 7.3 ($878.6\,mm^3$)	10.5 × 9.1 × 4.4 ($420.4\,mm^3$)	52.0

[a] Base dimensions × thickness in millimeters.

nique can be established. It does appear to offer a reasonable alternative to enucleation in selected cases of small melanomas.

Cryosurgical treatment of choroidal melanoma has not been used extensively, and its utilization should be limited to some patients with small melanoma. However, insufficient experimental and clinical data are available to establish its value.

Diathermy has been used to treat selected melanomas of the choroid, but there seems to be little, if any, role for it.

Local resection theoretically represents an ideal approach to the treatment of this disease state where it's possible to remove the tumor and yet salvage the eye. This has been achieved in the case of small iris melanomas for which surgical iridectomy is usually curative. Iridocyclectomy has been of value in small ciliary body melanomas, but surgical complications are more frequent. The eventual role for local resection remains controversial, and further data will be necessary in order to establish its role.

Enucleation remains the preferred technique, but the debate as to the impact of removal of the eye and its ultimate influence on prognosis remains. If vision is significantly decreased secondary to a sizeable tumor, general agreement indicates enucleation. However, surgical management has resulted in 50% 5-year survival rates, and 35% 10-year survival rates. As the methods for conservative management with preservation of the intact eye improve, there will be fewer enucleations for choroidal melanoma. Exenteration is rarely done for malignant melanomas of the choroid, being reserved for those with significant orbital extension.

The development of innovative radiation therapy techniques, particularly using brachytherapy procedures, is emerging with increasing frequency in the management of patients with malignant melanoma of the choroid. These tumors are not especially radiosensitive, and generally external beam radiation therapy techniques are not recommended. However, the use of radioactive plaques allows for the delivery of maximum radiation dosage to the tumor while minimizing the radiation to sensitive ocular structures such as the lens and the retinal vessels.

Moore and Scott (1929) treated a patient's only remaining eye by implantation of gold-radon seeds directly into the tumor. This produced considerable regression of the tumor. Subsequently, Stallard (1966a) used radon seeds and later a cobalt-60 plaque sutured to the sclera as a temporary placement. Favorable results were reported in a series of 100 patients treated by this technique with a 75% 5-year survival. Thereafter, various workers were encouraged by the efficacy of this technique and have pursued it using radon seeds, cobalt-60 plaques, ruthenium 106/rhodium 106 β-applicators, iodine-125 seeds, and iridium-192 seeds.

From the data accrued relative to the various treatment techniques being pursued, factors influencing prognosis can be identified (Table 6). As the tumor increases in size, the outcome becomes graver. Spindle cell lesions have a better prognosis than do mixed cell lesions, and epithelioid cell tumors have the poorest prognosis. Tumors with heavy pigmentation have poor prognoses as do those with increased mitotic activity. Necrosis favorably influences outcome, but reticular content or lymphocytic infiltration does not. If the tumor extends to involve the angle, optic nerve, sclera, extrascleral tissues, or ciliary body, the ultimate likelihood of a successful outcome is diminished.

Table 6. Factors gravely influencing prognosis of melanoma

1. Tumor size (small > medium > large)	Yes
2. Cell type (spindle cell > mixed > epithelial)	Yes
3. Pigmentation (absent > heavy)	Yes
4. Mitotic activity (absent > present)	Yes
5. Necrosis (with > without)	Yes
6. Reticular content	No
7. Lymphocytic infiltration	No
8. Angle infiltration	Yes
9. Optic nerve infiltration	Yes
10. Scleral infiltration	Yes
11. Extrascleral extension	Yes
12. Ciliary body infiltration	Yes
13. Location (anterior > posterior)	Yes

2 Treatment Procedures

In 1976, the Department of Radiation Oncology at the Hahnemann University School of Medicine and the Oncology Service at the Wills Eye Hospital embarked upon a program to evaluate the efficacy of cobalt-60 plaque treatment for melanomas of the choroid. In the interval until November 1987, 2309 patients with posterior uveal melanomas including lesions of the choroid and ciliary body have been seen and treated. Of these, 378 patients presented with iris melanomas, 301 with retinoblastoma, 99 with conjunctival melanoma, and 301 with metastases to the choroid for a total of 3388 malignant tumors of the eye.

Table 7 illustrates the treatment programs that have been assessed using brachytherapy techniques. In the study period, 1060 patients were treated by radioactive plaques, 747 by enucleation, 77 by resection, 11 by exenteration of the orbit, and 48 by photocoagulation.

Table 7. Types of radioactive plaques for treatment of choroidal melanoma

	n
Cobalt 60	604
Iridium 192	201
Iodine 125	118
Ruthenium 106	137
Total	1060

The initial concentration using brachytherapy techniques was with radioactive cobalt-60 episcleral applicators. Previously the statistically adjusted comparison of patients treated by enucleation and patients treated by cobalt-60 eye-plaque brachytherapy had been reported (Fig. 1) (Brady et al. 1988). Statistical adjustments are mandated because the two populations in this nonrandomized, prospective study were disparate. Augsburger et al. (1986) have previously discussed the criteria by which the populations were adjusted in order to achieve statistical similarity. By this analysis, the two treatments appear to be equivalent in terms of survival rate.

More recently, all patients treated through June 1982 have been assessed, thus expanding the data available for analysis with minimum 5-year follow-up figures. The data shown in Table 8, comparing 232 enucleated patients with 178 patients treated by radiocobalt plaque brachytherapy, clearly indicate that the average age at the time of treatment was higher in the enucleation group but not significantly so. However, the other factors were statistically different with a higher risk for a poorer prognosis (Augsburger et al. 1986) in the brachytherapy group of patients. The radiocobalt treatment group had tumors of greater basal diameter which were located more anteriorly, both known to be situations associated with a poorer prognosis (Augsburger et al. 1986).

3 Results

Table 9 gives the raw survival data for these two populations, which have thus far not been subjected to the sophisticated statistical adjustments mentioned above. In this unadjusted population there was a 5% poorer survival rate of the plaque-

Table 8. Comparison of 410 patients with intraocular melanomas treated by different techniques

	n	Basal diameter (mm)	Location[a]	Age (years)
Enucleation	232	11.46 ± 0.21	0.879 ± .053	58.39 ± .88
^{60}Co-plaque therapy	178	12.63 ± 0.30	1.045 ± .066	55.09 ± 1.21

[a] Location: 0 = posterior to equator; 1 = between equator and ora serrata; 2 = anterior to ora serrata.
Figures are mean ± SEM.

Table 9. Unadjusted survival for treated melanoma patients

Time (years)	Enucleation	^{60}Co-plaque treatment
0	1.000	1.000
1	0.987 ± 0.007	0.936 ± 0.019
2	0.947 ± 0.015	0.883 ± 0.025
3	0.888 ± 0.021	0.829 ± 0.029
4	0.822 ± 0.026	0.780 ± 0.032
5	0.776 ± 0.029	0.733 ± 0.035
6	0.757 ± 0.030	0.694 ± 0.037
7	0.732 ± 0.034	0.683 ± 0.038
8	0.702 ± 0.039	0.642 ± 0.042

Figures are fraction surviving ± SEM.

treated patients, relative to enucleation patients in the initial year after therapy, and this difference was maintained at later time intervals. These data will be added to as more of our study group reaches the 5-year follow-up examination, and the crude and fine statistical adjustments of the population remain to be performed.

Figure 2 shows the preservation of visual acuity (defined as vision in the affected eye of 6/60 or better). The percentage of patients with visual acuity in the treated eye better than 6/60 steadily declines with time after brachytherapy at a rate of approximately 10% per year. Therefore, about 50% of brachytherapy patients retain useful vision in the treated eye by 5 years after therapy. This must be contrasted with the complete loss of vision in the affected eye following enucleation (Brady et al. 1988).

With the advent of different techniques relative to brachytherapy plaque preparation, a more sophisticated set of criteria was established for treatment. Table 10 shows the criteria in terms of size used with the various plaques. In general, the choice of plaque was based upon tumor thickness at the time of initial assessment. The patients have been assessed in a prospective trial under the aegis of the Radiation Therapy Oncology Group (RTOG study no. 82-04). Radioiodine (iodine 125) plaques were utilized for tumors of 3 mm or less in thickness, plaques containing iridium 192 for tumors between 3 and 5 mm, ruthenium 106 for those measuring 5–7 mm, and cobalt-60 plaques for tumors greater than 7 mm in thickness and with base diameters less than 12 mm.

The radioactive iodine-125 and iridium-192 plaques were of our own design. The data from the RTOG study 82-04 are shown in Tables 11 and 12 and Figs. 3–6.

Regression patterns following plaque treatment have been described and fall essentially into four categories (Cruera et al. 1984):

Table 10. Criteria for use of various radioactive plaques in choroidal melanoma

Tumor height (mm)	Radioactive plaque type
<3	Iodine 125
3–5	Iridium 192
5–7	Ruthenium 106
>7 and 12 and with base diameter less than 12 mm	Cobalt 60

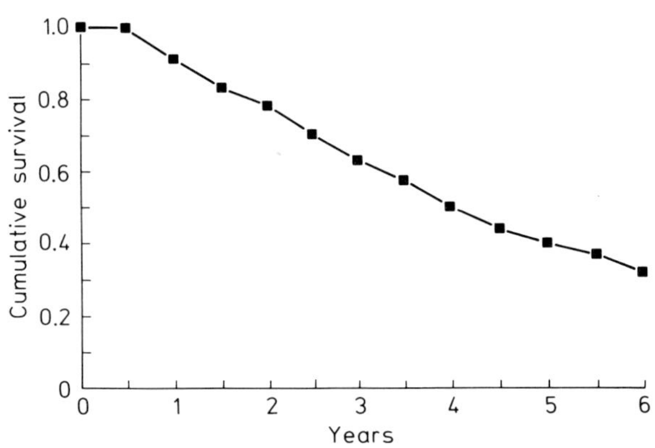

Fig. 2. Life table assessment of visual acuity in 100 patients managed by radiocoblat eye-plaque therapy: *x-axis*, percentage of patients retaining corrected vision in the treated eye of 20/200 (6/60) or better, *y-axis*, years after therapy

Table 11. Crude rate of enucleation during follow-up arranged by tumor size

	≤3 mm (n = 32)	>3–5 mm (n = 82)	>5 mm (n = 68)
Enucleation	5 (16%)	7 (9%)	18 (26%)
No enucleation	27 (84%)	75 (91%)	50 (74%)

Table 12. Crude rate of enucleation during follow-up for patients receiving correct plaque

	Iodine^{125} (n = 21)	Iridium^{192} (n = 59)	Cobalt^{60} (n = 67)
Enucleation	3 (14%)	5 (8%)	17 (25%)
No enucleation	18 (86%)	54 (92%)	50 (75%)

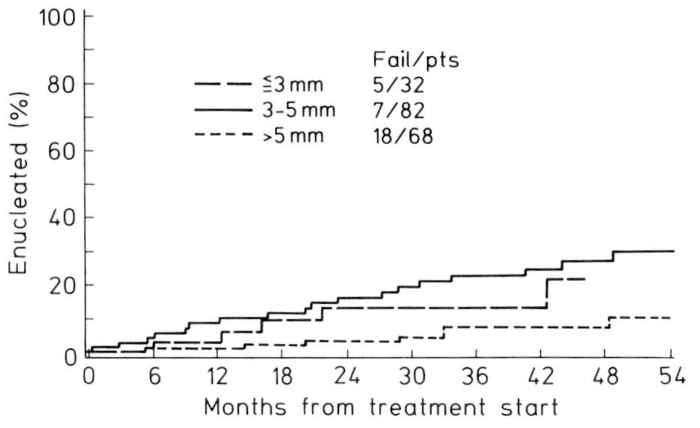

Fig. 3. Time to enucleation (in months) according to size of tumor

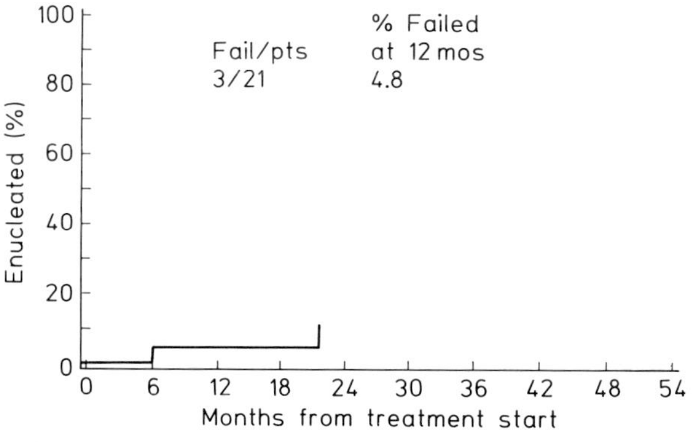

Fig. 4. Time to enucleation (in months) for patients with small (≤3 mm thick) tumors treated by iodine-125 eye-plaque therapy

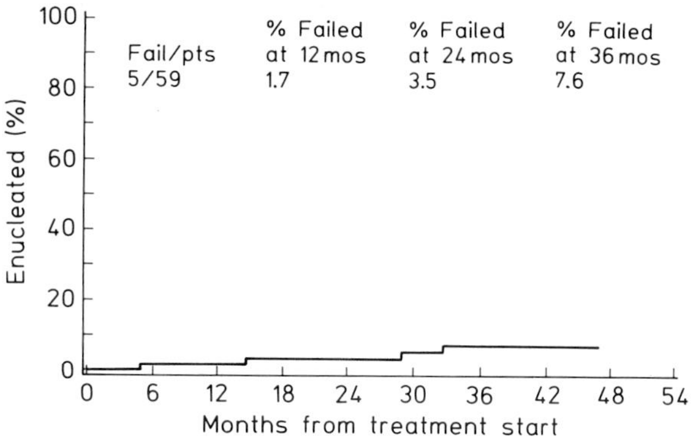

Fig. 5. Time to enucleation (in months) for patients with medium (3–5 mm thick) tumors treated by iridium-192 eye-plaque therapy

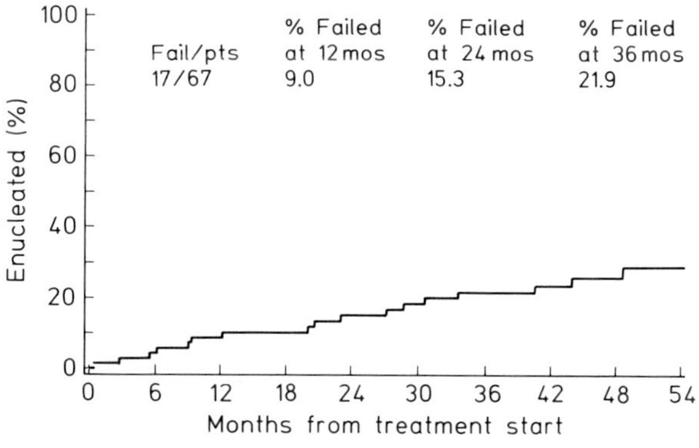

Fig. 6. Time to enucleation (in months) for patients with large (>5 mm thick) tumors treated by cobalt-60 eye-plaque therapy

1. A slow decrease in tumor thickness as determined by serial ultrasonographic measurement
2. Rapid shrinkage with almost total disappearance of the tumor over 1–3 years after treatment
3. Shrinkage with exudative response, especially in eyes with pretreatment bolus retinal detachment
4. Shrinkage with hemorrhagic response

Augsburger et al. (1984) reported that melanomas initially managed by a period of observation, which were documented to grow rapidly prior to enucleation, were more likely to contain epithelioid tumor cells than those which grew slowly. Based on these findings, it is assumed that tumors of equivalent size which demonstrate similar rapid growth rates prior to radioactive plaque application also contain predominantly epithelioid cells. These tumors have been found to display a correspondingly rapid regression pattern following plaque radiotherapy. For this reason, rapid shrinkage of a melanoma following plaque brachytherapy is considered to be a less favorable sign in term of prognosis and the potential for systemic dissemination than slow shrinkage occurring within the tumor.

Ocular complications including radiation vasculopathy of the directly irradiated field, radiation retinopathy and papillopathy, vitreous hemorrhage, cataract, neovascular glaucoma, radiation keratopathy, dry eye, radiation blepharitis, and scleral necrosis do occur. When compared with results following external beam radiation therapy, posterior segment complications seem to occur with a marginally greater frequency following plaque therapy while anterior segment and external ocular complications appear to be greater following external beam radiation therapy. The incidence of complications relative to management using brachytherapy techniques is minimal

Table 13. Complications following cobalt-60 plaque therapy of 100 posterior uveal melanomas

Radiation tumor vasculopathy	18	30
Radiation retinopathy	12	
(with radiation papillopathy)	5	
Intravitreal hemorrhage	11	
Radiation cataract	7	
Punctal occlusion/epiphora	3	
Radiation anterior uveitis	2	
Scleral necrosis/thinning	2	
Persistent diplopia	2	

and do not represent a contraindication to its use (Table 13).

A review of 1019 patients with posterior uveal melanomas treated by plaque radiotherapy (cobalt 60, ruthenium 106, iodine 125, or iridium 192) in the interval between April 1976 and December 1987 revealed that 59 patients (6%) required enucleation of the affected eye. Histopathologically, the mean size of the tumors was 11 mm in basal diameter and 5 mm in thickness. The melanoma was of mixed cell type in 40 patients (68%), spindle B cell in 10 (17%), necrotic in 7 (12%), and epithelioid in 2 (3%) (Shields et al. 1989).

The average number of mitoses per 40 high power fields in these irradiated tumors was 5, but more specifically, in eyes enucleated for tumor regrowth, the average was 6, while in those eyes enucleated for reasons other than tumor regrowth, the average was less than 1. In a matched group of nonirradiated posterior uveal melanomas, the average number of mitoses was 6 per 40 high power fields. Most (64%) of the irradiated uveal melanomas had no identifiable mitoses. In the 10 uveal melanomas classified as spindle B cell type and the 7 classified as necrotic, there was no mitotic activity.

Histopathologic evidence of transscleral extension of the uveal melanoma was found in 10 patients. The intraocular tumor size here averaged 10 × 4 mm, slightly smaller than the average tumor size in the overall series. In four eyes, the extension was purely scleral, in two the vortex vein was involved, in two a small episcleral nodule (<5 mm in size) was found, and in two a large episcleral mass (>5 mm) was noted.

The results suggest that viable tumor cells may exist after plaque radiotherapy, but in most cases, radiotherapy appreciably interferes with the mitotic activity of the melanoma cells. Table 14 gives the incidence of enucleation in various international studies.

4 Conclusions

Conservative radiotherapy using episcleral plaque brachytherapeutic techniques appears to offer an alternative to enucleation in the treatment of appropriately selected patients suffering from primary intraocular choroidal melanoma. Conservative therapy has a finite probability of preservation of useful vision over a protracted period of time following treatment.

Brachytherapy applications offer a significant advantage over enucleation or conventional external beam radiation therapy techniques. With uveal melanomas there is no present indication that plaque brachytherapy is an inferior therapeutic modality to enucleation in terms of patient survival, provided that the selection of patients for conservative therapy is appropriate. However, in terms of retention of useful vision, episcleral plaque brachytherapy appears to provide a definite advantage over enucleation.

Table 14. Enucleation after conservation by radiation therapy for choroidal melanoma

Treatment modality	Enucleated/treated total	Percentage	Center
Radionuclide plaque			
^{60}Co	42/601	7	Wills/Hahnemann
^{106}Ru/^{106}Rh	36/205	17	Berlin
^{106}Ru/^{106}Rh	1/42	2	Munster
^{125}I	7/60	12	Downstate
^{125}I	2/26	8	Mayo Clinic
Charged particles			
Protons	57/1006	5.7	Massachusetts General Hospital (MGH)
Protons	22/270	8	Villigen
Helium ions	29/228	10	Berkeley

References

American Cancer Society (1989) Cancer facts and figures. American Cancer Society, New York.

Arnesen K, Normes M (1975) Malignant melanoms of the choroid as related to coexistent benign nevus. Acta Ophthalmol (Copenh) 53: 139–152

Ashton N, Wybar K (1966) Primary tumors of the iris. Ophthalmologica 151: 97–113

Augsburger JJ, Gonder JR, Amsel J, Shields JA, Donoso LW (1984) Growth rates and doubling times of posterior uveal melanomas. Ophthalmology 91: 1707–1715

Augsburger JJ, Gamel JW, Sardi VF, Greenberg RA, Shields JA, Brady LW (1986) Enucleation vs cobalt plaque radiotherapy for malignant melanomas of the choroid and ciliary body. Arch Ophthalmol 104: 6555–661

Boniuk M, Girard LA (1965) Malignant melanomas of the choroid treated photocoagulation, transscleral diathermy and radon seed implant. Am J Ophthalmol 59: 212–216

Brady LW, Shields JA, Augsburger JJ, Day JL (1982) Malignant intraocular tumors. Cancer 49(3): 578–580

Brady LW, Shields JA, Augsburger JJ, Day JL, Saunders WM, Castro JR, Munzenrider JE, Gragoudas E (1984) Posterior uveal melanomas. In: Phillips TL, Pistenmaa DA (eds) Radiation oncology annual, vol 1. Raven, New York, pp 233–245

Brady LW, Markoe AM, Amendola BE, Karlsson UL, Micaily B, Shields JA, Augsburger JJ (1988) The treatment of primary intraocular malignancy. Int J Radiat Oncol Biol Phys 15: 1355–1361

Char DH, Castro JR (1982) Helium ion therapy for choroidal melanoma. Arch Ophthalmol 100: 935–938

Char DH, Castro JR, Quivey JM (1982) Helium ion charged particle therapy for choroidal melanoma. Ophthalmology 87: 565–570

Cruera AF, Augsburger JJ, Shields JA, Brady LW, Markoe AM, Day JL (1984) Regression of posterior uveal melanomas following cobalt^{60} plaque radiotherapy. Ophthalmology 91: 1716–1719

Del Regato JA, Spjut HJ (1977) Cancer of the eye. In: del Ragato JA, Spjut HJ (eds) Cancer diagnosis, treatment, prognosis, 5th edn. Mosby, St Louis, pp 160–181

Fitzpatrick PJ, Chenergy SAG, Japp B (1978) Treatment of choroidal melanoma with radioisotopes. Presented at the Am Soc Ther Radiol Oncol, Los Angeles

Foulds AS (1974) Local excision of choroidal melanomas. Trans Ophthalmol Soc UK 93: 343–346

Fraunfelder FT, Boozman FW III, Wilson DS (1977) "No touch" technique for intraocular malignant melanomas. Arch Ophthalmol 95: 1616–1620

Gragoudas ES, Goitein M, Koehler AM (1977) Protein irradiation of small choroidal malignant melanomas. Am J Ophthalmol 83: 665–673

Gragoudas ES, Goitein M, Koehler AM (1978) Protein irradiation of choroidal melanomas. Preliminary results. Arch Ophthalmol 96: 1583–1591

Jay B (1964) Current developments in ophthalmology: a follow-up of limbal melanomata. Proc R Soc Med 57: 497–500

Keller AZ (1973) Histology, survivorship and related factors in the epidemiology of eye cancers. Am J Epidemiol 97: 386–393

Lederman M (1964) Discussion of pigmented tumors of the conjunctive. In: Boniuk M (ed) Ocular and adnexal tumors: new and controversial aspects. Mosby, St Louis, pp 24–48

Lincoff H, McLean J, Lang R (1967) The cryosurgical treatment of intraocular tumors. Am J Ophthalmol 63: 389–399

Lommatzsch P (1974) Treatment of choroidal melanomas with ^{106}Rh beta-ray applicators. Surv Ophthalmol 19: 85–100

Lommatzsch P (1983) Beta irradiation with ^{106}Ru/^{106}Rh applicators of choroidal melanomas: sixteen years experience. In: Intraocular tumors. Lommatzsch PK, Blodi FC, (eds) Academic, Berlin, pp 290–30

Long RS, Galin MA, Rotman M (1971) Conservative treatment of intraocular melanomas. Trans Am Acad Ophthalmol Otolaryngol 75: 84–93

MacFaul PA (1977) Local radiotherapy in the treatment of malignant melanoma of the choroid. Trans Ophthalmol Soc UK 97: 421–427

Markoe AM, Brady LW, Grant GD, Shields JA, Augsburger JJ (1987) Radiation therapy of ocular disease. In: Perez CA, Brady LW (eds) Principles and practice of radiation oncology. Lippincott, Philadelphia, pp 453–472

Meyer-Schwickerath G (1957) Further progress in the field of light coagulation. Trans Ophthalmol Soc UK 77: 421–440

Meyer-Schwickerath G (1961) The preservation of vision by treatment of intraocular tumors with light coagulation. Arch Ophthalmol 66: 458–466

Moore RD, Scott RS (1929) Clinical and pathological report of bilateral glioma retinae. Proc R Soc Med (Sect Ophthalmol) 22: 39–50

Peyman GA, Apple DJ (1974) Local excision of a choroidal malignant melanoma. Full thickness eye wall resection. Arch Ophthalmol 92: 216–218

Rotman M, Long RS, Packer S, Moroson H, Galin MA, Chan B (1977) Radiation therapy of choroidal melanoma. Trans Ophthalmol Soc UK 97: 431–435

Saunders WM, Char DH, Quivey JM (1973) Precision high dose radiotherapy: helium ion treatment of uveal melanoma. Int J Radiat Oncol Biol Phys 11: 227–233

Sautter H, Neuman G (1973) Full thicknss scleral resection in iridocyclectomy and choroidectomy for anterior uveal tumors. Ophthalmic Surg 4: 25–31

Shields JA, McDonald PR (1973) Improvement in the diagnosis of posterior uveal melanomas. Trans Am Ophthalmol Soc 71: 194–211

Shields JA, Zimmerman LW (1973) Lesions simulating malignant melanoma of the posterior uvea. Arch Ophthalmol 89: 466–471

Shields CL, Shields JA, Karlsson UL, Markoe AM, Brady LW (1989) Reasons for enucleation following plaque radiotherapy for posterior uveal melanomas: clinical findings. Ophthalmology 96(6): 919–923

Stallard HB (1948) Radiotherapy of malignant intraocular neoplasms. Br J Ophthalmol 32: 618–639

Stallard HB (1959) Malignant melanoma of the choroid treated with radioactive applicators. Trans Ophthalmol Soc UK 79: 373–392

Stallard HB (1966a) The treatment of retinoblastoma. Ophthalmologica 151: 214–230

Stallard HB (1966b) Radiotherapy for malignant melanoma of the choroid. Br J Ophthalmol 50: 147–155

Starr HJ, Zimmerman LE (1962) Extrascleral extension and orbital recurrences of malignant melanomas of the choroid and ciliary body. Int Ophthalmol Clin 2: 369–385

Vogel MH (1972) Treatment of malignant choroidal melanomas with photocoagulation. Evaluation of 10-year followup data. Am J Ophthalmol 74: 1–11

Warren RM (1974) Prognosis of malignant melanomas of the choroid and ciliary body. In: Blodi FC (ed) Current concepts in ophthalmology, vol 4. Mosby, St Louis

Winter FC (1963) Surgical excision of tumors of the ciliary body and iris. Arch Ophthalmol 70: 19–29

Winter FC (1964) Iridocyclectomy for malignant melanomas of the iris and ciliary body. In: Boniuk M (ed) Ocular and adnexal tumors: new and controversial aspects. Mosby, St Louis, pp 341–352

Zimmerman LE (1967) Changing concepts concerning the malignancy of ocular tumors. Arch Ophthalmol 78: 166–173

Zimmerman LE, McLean IW (1975) Changing concepts in the prognosis and management of small malignant melanomas of the choroid. Trans Ophthalmol Soc UK 95: 487–494

Zimmerman LE, McLean IW (1979) An evaluation of enucleation in the management of uveal melanomas. Am J Ophthalmol 87: 741–760

Zimmerman LW, McLean IW, Foster WD (1978) Does enucleation of an eye containing a malignant melanoma prevent or accelerate the dissemination of tumor cells? Br J Ophthalmol 62: 420–425

3.3 Experience with High-Dose β-Irradiation ($^{106}Ru/^{106}Rh$) of Choroidal Melanomas

ROLF-P. MÜLLER, R. PÖTTER, and H. BUSSE

CONTENTS

1 Introduction 123
2 Material and Methods 123
3 Results 125
4 Discussion 126
References 127

1 Introduction

Techniques and physical data of brachycurie-therapy with ruthenium-106 and other radioisotopes for choroidal melanomas have been described (BRADY et al. 1984; BUSSE and MÜLLER 1983; LOMMATZSCH 1974, 1983; MÜLLER et al. 1984, 1985; ROTMAN et al. 1983; STALLARD 1966). In the FRG treatment with ruthenium-106 eye applicators has been favored because of easy handling, low rate of side effects or late complications, and very good clinical results (LOMMATZSCH 1983; LOMMATZSCH et al. 1986; GUTHOFF et al. 1986).

The purpose of the Münster clinical program, which was started in 1981, is to evaluate the effect of *high-dose* β-irradiation (15 000 cGy calculated to the apex of the tumor) in intraocular melanomas, regarding tumor regression, treatment-related side effects, visual acuity, and survival. The high tumor dose, which is significantly higher than that used by other groups (BRADY et al. 1982; HALLERMANN and LOMMATZSCH 1979; LOMMATZSCH 1984; PACKER and ROTMANN 1980; ROTMAN et al. 1983; SHIELDS et al. 1982; STALLARD 1966), was chosen because of the known radioresistance of melanoma cells and the reported rate of about 12% postradiation enucleations, caused by tumor progression after a tumor dose of about 8000 cGy (LOMMATZSCH et al. 1980; LOMMATZSCH 1981, 1983; GUTHOFF et al. 1986).

Concerning the indications for treatment of uveal melanomas with ruthenium-106 plaques, with most patients the recomendations of LOMMATZSCH (1977, 1983) have been followed, with the exception that in patients over 75 years of age, and in patients with a so-called oculus ultimus (last eye) tumors up to 7 or 8 mm thick will be treated with plaque therapy to spare them the psychical trauma of an enucleation. In most of these patients a second course of ruthenium-106 plaque therapy is necessary.

Contraindications for ruthenium-106 plaque therapy are tumors surrounding the optic disc and/or the macula and tumors with a basal diameter exceeding the therapeutic surface of the given plaques.

2 Materials and Methods

From January 1981 to June 1988, a total of 181 patients (106 female, 75 male) with malignant melanoma of the choroid were treated with ruthenium 106 in the Departments of Radiooncology and Ophthalmology of the University of Münster. The patients were between 27 and 86 years of age, with a mean age of 60 years. In 101 patients the right eye was affected, in 80 cases the left eye.

Pretherapeutically in all patients the dimensions of the tumors were determined by ophthalmoscopy (Fig. 1) and A-scan ultrasonography, which was the basis for the thickness calculations of the tumors. In most, additional fluorescein angiography (Fig. 2) and computerized tomography (CT) scans (Fig. 3) were performed, and in the past few years magnetic resonance imaging

ROLF-P. MÜLLER, Prof. Dr., Klinik and Poliklinik für Strahlentherapie der Universität zu Köln, Joseph-Stelzmann-Str. 9, 5000 Köln 41, Germany

R. PÖTTER, Dr., Strahlentherapeutische Klinik der Universität Münster, 4400 Münster, Germany

H. BUSSE, Prof. Dr., Universitäts-Augenklinik, Domagkstr. 15, 4400 Münster, Germany

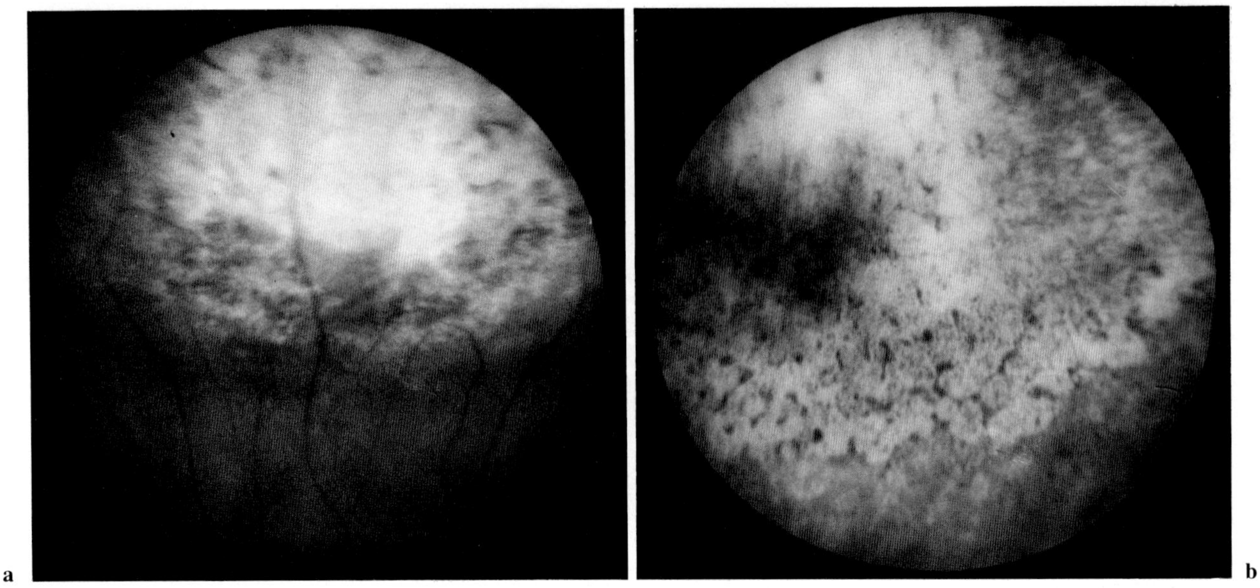

Fig. 1a,b. Ophthalmoscopical findings in a 62-year-old female patient with choroidal melanoma before (a) and 18 months after (b) treatment with ruthenium-106 brachycurietherapy

(MRI) examinations in selected cases. The post-therapeutic shrinkage of the tumors was measured by A-scan ultrasonography and also documented ophthalmoscopically (Fig. 1) and often on CT scans (Fig. 3). In every patient pretherapeutically a general work-up was done to exclude metastatic disease (physical examination, chest X-ray, abdominal ultrasound, bone scan, and laboratory studies).

The surgical procedure usually (92%) can be performed under local anesthesia. Intraoperatively, the exact localization of the tumor is identified by transscleral illumination and the margins

Fig. 2a,b. Same patient as in Fig. 1. Results of fluorescein angiography pretherapeutically (a) and 18 months postirradiation (b)

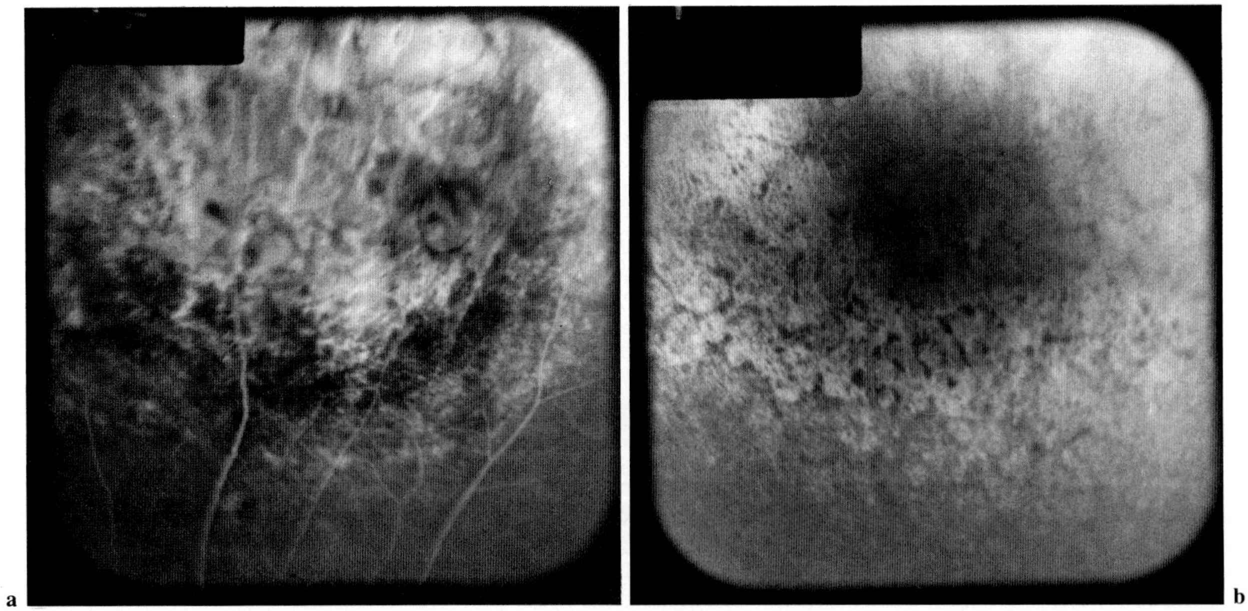

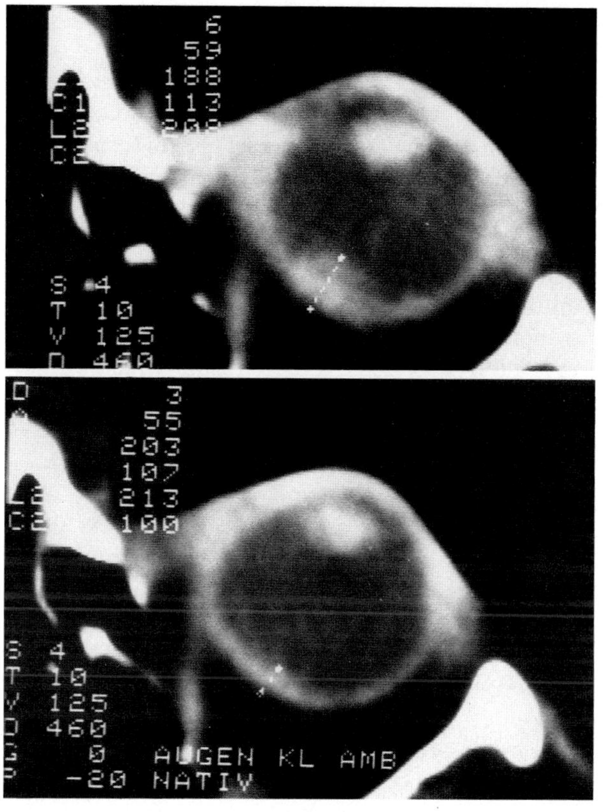

Fig. 3a,b. Computerized tomography findings in a 49-year-old male patient with choroidal melanoma pretherapeutically (**a**; tumor thickness 6 mm) and 14 months after treatment (**b**; tumor thickness 3 mm, including sclera)

of the tumor are marked on the sclera. From the collection of ruthenium-106 plaques one will be selected which covers the tumor with a safety margin of at least 1 mm.

For each patient the actual dose rate of the individual plaque has to be determined as well as the radiation time to apply 15 000 cGy at the *top* of the tumor. This is the standard radiation dose in this ongoing program, but it was decided not to exceed 80 000–90 000 cGy sclera dose in one radiation course. With very large tumors initially 7000–8000 cGy were given to the apex and a second course of irradiation was planned about 6 months after the first.

Depending on the pretreatment measurements of the tumor thicknesss, three groups were formed: Those with *small* tumors (up to 3 mm thick), 63 patients (35%); those with *medium*-sized tumors (3.1–5 mm thick), 58 patients (32%); and those with *large* tumors (>5.1 mm thick), 60 patients (33%). At the present time the number of patients and the observation time in the single groups are too limited to evaluate the prognostic value.

3 Results

In 86% (156/181) the planned radiation dose of 15 000 cGy at the top of the uveal melanoma could be given. Side effects consisted of a circumscribed retinal detachment in 40% (72/181) and a temporary detachment of a quadrant in 8% (15/181), which both receded in a period of 6 weeks to 3 months. In 6% (11/181) a retinal hemorrhage occurred, in 3% (5/181) retrobulbar hemorrhage, and in another 3 patients (2%) a vitreous hemorrhage was seen.

Treatment results refer to 132/181 patients, in whom the observation time was at least 12 months or more. Usually the tumors showed a slow regression and decrease of tumor thickness, which began about 6–8 weeks after irradiation and lasted continuously up to 1 year and longer.

In 83/132 patients (63%) total tumor regression was seen, what is called a "grey mouse", because of permanent pigmentary tissue without vascularisation (Fig. 1b, 2b). In 34/132 patients (26%) a partial regression was assessed, which means a decrease to about 45%–50% of the pretreatment thickness. In 11 patients (8%) there was no measurable change in tumor thickness after the first course of radiation, while 4 patients (3%) showed progression.

In 9/181 patients (4.8%) enucleation had to be performed after brachycurietherapy, in 5 because of the development of a neovascular glaucoma and in 4 because of tumor progression. A total of 13 patients (7.1%) died, 9 due to distant metastases, 1 after myocardial infarction, and 3 because of other nontumor-related diseases.

To preserve vision is one of the aims of brachycurietherapy of intraocular melanomas, but this depends very much on the location of the tumor. In 65% (86/132) visual acuity could be preserved at pretreatment levels, while in 40% (5/132) a posttreatment improvement was found because of resolution of subretinal fluid. In 31% (41/132) visual acuity decreased, (including loss of central vision), mainly because of the central localization of the tumor and the development of radiation vasculopathy and/or retinopathy, which were the

Table 1. Long-term sequelae after ruthenium-106 irradiation of choroidal melanomas (n = 132, observation time > 12 months)

	n
Radiation retinopathy	22 } (24%)
Radiation vasculopathy	9 }
Neovascular glaucoma	5 (4%)
Persistent diplopia	3 (2%)

most frequent posttreatment complications (in 24% of patients) (Table 1).

A second course of ruthenium irradiation was performed, at least 6 months after the first application, in 46 patients, including those 15 who showed no changed after the first course. In 44% (14/32; observation time >12 months), total tumor regression was achieved, and 26% (8/32) were partial responders. No measurable change in thickness occurred in 10/32 (30%).

4 Discussion

For many years enucleation was the unquestioned standard therapy for choroidal melanomas, but since Zimmerman et al. (1978) reported that enucleation in choroidal melanomas might cause an accelerated dissemination of tumor cells, there has been increasing interest in developing conservative, bulb-preserving treatment programs.

Brachycurietherapy in this particular tumor has been performed with cobalt-60 plaques (Brady et al. 1982; Shields et al. 1982; Stallard 1966) and with individually designed iodine-125 (Packer and Rotmann 1980; Rotmann et al. 1983) or iridium-192 plaques (Shields et al. 1982).

In the FRG ruthenium-106 plaques, designed by Lommatzsch in East Berlin, became the most common form of brachycurietherapy for choroidal melanomas. Because of the known relative radioresistance of melanoma cells and the ignorance of the subtype of the individual melanoma histology, in Münster it was decided to evaluate the benefit of high-dose β-irradiation (15000 cGy to the tumor top). We wanted to find out whether the number of progressive tumors could be diminished and the rate of complete remissions raised without resulting in more acute or late morbidity. Regarding this, the rate of acute side effects and long-term sequelae (Table 1) was low and tolerable; in only 5 patients an enucleation because of neovascular glaucoma had to be performed.

In our patients the tumors usually showed a slow and constant decrease in thickness, which lasted up to 1 year or more. This corresponds to the findings of Guthoff et al. (1986) who report a "half-life" (50% regression in thickness) of about 9 months for their ruthenium-treated choroidal melanomas and found a correlation between tumor regression and sclera contact dose.

In 63% of our patients (83/132) total tumor regression was achieved, a result which is better than that reported by groups which did not apply such high radiation doses to the apex of the tumors (8000 vs. 15000 cGy) (Lommatzsch 1983; Hallermann and Lommatzsch 1979; Hallermann and Guthoff 1983; Guthoff et al. 1986). Together with those 34/132 patients (26%) who showed partial tumor regression (which amounted at least to a mean of 45% of the pretreatment thickness), altogether 89% (117/132) of the patients observed 12 months and more responded positively after the first course of ruthenium brachycurietherapy. In our opinion a second course of radiation is opportune in those patients who show only partial remission or no change, but not in those patients with local progression. In 14/32 patients (44%) a total tumor regression was seen after the second course of brachycurietherapy, and in 8/32 (26%) partial tumor regression occurred. Since the clinical diagnosis of choroidal melanoma has not been confirmed histologically, it cannot be excluded that the different pattern of regression may be caused by the various histological subtypes of ocular melanomas.

In 6 of the 10 patients who showed no decrease in thickness after the second irradiation, ultrasonography revealed an increase of the tumor echogenicity. This was estimated to be a reaction after radiation, and because these patients are clinically free of tumor progression and dissemination for at least 18 months, it was decided to observe them closely at 3-month intervals and to perform enucleation only if the tumor shows progression.

To maintain useful vision is one of the aims of bulb-conserving radiation treatment techniques, but this cannot be used as a parameter of successful treatment in every case. The quality of vision after treatment mostly depends on the location of the tumor. In 69% (91/132) of our patients vision could be kept at pretreatment levels (including 5 patients with improvement), a result which corresponds very well with other reports (HALLERMANN and GUTHOFF 1983; GUTHOFF et al. 1986; LOMMATZSCH 1983), despite our high radiation doses.

In conclusion, treatment of choroidal melanoma with high doses of β-irradiation does not lead to increased rates of actual or late side effects but seems to result in better rates of tumor regression. However, any influence on survival is not clearly demonstrable.

References

Brady LW, Shields JA, Augsburger JJ, Day JL (1982) Malignant intraocular tumors. Cancer 49: 578–585

Brady LW, Shields JA, Augsburger JJ, Day JL, Saunders WM, Castro JR, Munzenrider JE, Gradoudas E (1984) Posterior uveal melanomas. In: Phillips TL, Pistenmaa DA (eds) Radiation oncology annual, vol 1. Raven, New York, pp 233–245

Busse H, Müller RP (1983) Techniques and results of 106-Ru/106-Rh radiation of choroidal melanomas. Trans Ophthalmol Soc UK 103: 72–77

Guthoff R, von Domarus D, Steinhorst U, Hallermann D (1986) 10 Jahre Erfahrung mit der Ruthenium-106/Rhodium-106-Behandlung des malignen Melanmos der Aderhaut- Bericht über 264 bestrahl-Tumoren. Klin Monatsbl Augenheilkd 188: 576–583

Hallermann D, Guthoff R (1983) Retrogression of choroidal melanoma after beta-irradiation with Ruthenium-106/Rhodium-106. In: Lommatzsch PK, Blodi FC (eds) Intraocular tumors. Springer, Berlin Heidelberg New York, pp 307–315

Hallermann D, Lommatzsch PK (1979) Langzeitbeobachtungen nach Strahlentherapie des malignen Melanomas der Aderhaut mit dem 106-Ru/106-Rh-Applikator. Ber Dtsch Ophthalmol Ges 76: 177–180

Kiehl H, Kirsch I, Lommatzsch P (1984) Das Überleben nach Behandlung des malignen Melanomas der Aderhaut: Vergleich von konservativer Therapie (106-Ru/106RH Applikator) und Enukleation ohne und mit postoperativer Orbitabestrahlung, 1960–1979. Klin Monatsbl Augenheilkd 184: 2–14

Lommatzsch PK (1974) Treatment of choroidal melanomas with 106-Ru/106-Rh beta-ray applicators. Surv Ophthalmol 19: 85–100

Lommatzsch PK (1977) Die therapeutische Anwendung von ionisierenden Strahlen in der Augenheilkunde. Thieme, Leipzig

Lommatzsch PK (1981) Möglichkeiten und Grenzen bulbuserhaltender Maßnahmen beim malignen Melanom der Uvea. Klin Monatsbl Augenheilkd 179: 393–398

Lommatzsch PK (1983) Beta irradiation with 106-Ru/106-Rh applicators of choroidal melanomas. In: Lommatzsch PK, Blodi FC (eds) Intraocular tumors. Springer, Berlin Heidelberg New York, pp 290–301

Lommatzsch PK, Kiehl H, Kirsch I (1980) Vergleich von operativer und konservativer (106-Ru/106-Rh) Therapie beim Melanom der Aderhaut nach Überlebensraten und-zeiten (berechnet nach der Sterbetafelmethode). Klin Monatsbl Augenheilkd 176: 950–955

Lommatzsch PK, Weise B, Ballin R (1986) Ein Beitrag zur Optimierung der Behandlung des malignen Melanoms der Aderhaut mit β-Applikatoren (106-Ru/106-Rh). Klin Monatsbl Augenheilkd 189: 133–140

Müller RP, Busse H, Fischedick AR (1984) Techniques of 106-Ru/106-Rh radiation of choroidal tumors. J Ocular Ther Surg 3: 130–133

Müller RP, Busse H, Kroll P, Gast E (1985) Le traitement des mélanomes choroidiens par radiothérapie de contact avec le ruthénium 106. J Fr Ophthalmol 8: 639–643

Packer S, Rotmann M (1980) Radiotherapy of choroidal melanomas with iodine-125. Ophthalmology 87: 582–590

Rotmann M, Packer S, Bhutiani I (1983) Management of choroidal melanomas using radioactive ophthalmic applicators. In: Hilaris BS, Batata MA (eds) Brachytherapy, oncology-1983. Memorial Sloan-Kettering Cancer Center, New York, pp 119–124

Shields JA (1982) The management of posterior uveal melanomas. In: Intraocular tumors, diagnosis and management. Mosby, St Louis

Shields JA, Augsburger JJ, Brady LW, Day JL (1982) Cobalt plaque therapy of posterior uveal melanomas. Ophthalmology 89: 1201–1207

Stallard HB (1966) Radiotherapy for malignant melanomas of the choroid. Br J Ophthalmol 50: 147–155

Zimmerman LE, McLean IW, Foster WD (1978) Does enucleation of the eye containing a malignant melanoma prevent as accelerate the dissemination of tumor cells. Br J Ophthalmol 62: 420–25

4 Head and Neck Tumors

4.1 Classification of Oral Cavity and Oropharynx Carcinomas

ROLF SAUER

The classifications of the American Joint Committee on Cancer (AJCC 1983/1988) and of the UICC 1985/1987 indicating the extent of primary tumors and distant metastases are generally similar but differ in specific details for the regional lymph node metastases. UICC and AJCC in their last edition made several major changes in the previous N classifications in order to have a worldwide uniform TNM system.

Primary Tumor

The staging system of the primary tumors did not change between 1982 and 1987/1988.

Lip and Oral Cavity

Tx	Primary tumor cannot be assessed
To	No evidence of primary tumor
Tis	Carcinoma in situ
T1	Tumor 2 cm or less in greatest dimension
T2	Tumor more than 2 cm but not more than 4 cm in greatest dimension
T3	Tumor more than 4 cm in greatest dimension
T4	Tumor invades adjacent structures Lip: e.g. through cortical bone, tongue, skin of neck Oral cavity: through cortical bone, into deep (extrinsic) muscle of tongue, maxillary sinus, skin

Pharynx

Tx	Primary tumor cannot be assessed
To	No evidence of primary tumor
Tis	Carcinoma in situ
T1	Tumor 2 cm or less in greatest dimension
T2	Tumor more than 2 cm but not more than 4 cm in greatest dimension
T3	Tumor more than 4 cm in greatest dimension
T4	Tumor invades adjacent structures, e.g. through cortical bone, soft tissues of the neck, deep (extrinsic) muscle of tongue

N classification

The regional lymph nodes are the cervical nodes. These include:

- Submental nodes
- Submandibular nodes
- Cranial jugular (deep cervical) nodes
- Medial jugular (deep cervical) nodes
- Caudal jugular (deep cervical) nodes
- Dorsal cervical (superficial cervical) nodes along the accessory nerve
- Supraclavicular nodes
- Prelaryngeal and paratracheal nodes
- Retropharyngeal nodes
- Parotid nodes
- Buccal nodes
- Rectroauricular and occipital nodes

ROLF SAUER, Prof. Dr., Strahlentherapeutische Klinik der Universität Erlangen-Nürnberg, Universitätsstr. 27, 8520 Erlangen, Germany

N status	UICC 1985	AJCC 1983	UICC 1987	AJCC 1988
N/pNx	[Regional lymph nodes cannot be assessed]			
N/pNo	[No regional lymph node metastasis]			
N/pN1	Metastasis in movable ipsilateral lymph nodes	[Metastasis in a single ipsilateral lymph node, 3 cm or less in greatest dimension]		
N/pN2	Metastasis in movable contralateral or bilateral lymph nodes	[**N2A** Metastasis in a single ipsilateral lymph node >3 cm but not more than 6 cm]		
		[**N2B** Metastasis in multiple ipsilateral lymph nodes, none more than 6 cm]		
				N2C Metastasis in bilateral or contralateral lymph nodes, none more than 6 cm
N/pN3	Fixed regional lymph node(s)	**N3A** Massive homolateral node(s), at least one >6 cm	[Metastasis in a lymph node >6 cm]	
		N3B Bilateral clinically positive nodes (each side should be staged separately)		
		N3C Contralateral clinically positive node(s) only		

Stage Grouping (UICC)

Stage 0	Tis	No	Mo
Stage I	T1	No	Mo
Stage II	T2	No	Mo
Stage III	T3	No	Mo
	T1–3	N1	Mo
Stage IV	T4	No, N1	Mo
	any T	N2, N3	Mo
	any T	any N	M1

Distant Metastases

Mx	Presence of distant metastasis cannot be assessed
Mo	No evidence of distant metastases
M1	Evidence of distant metastases

4.2 Interstitial Radiotherapy of Oral Cavity and Oropharynx Carcinomas (Paris Technique)

BERNARD PIERQUIN, JEAN JACQUES MAZERON, and LAVAL GRIMARD

CONTENTS

1	Introduction	133
2	Patients and Methods	133
2.1	Lips	133
2.2	Buccal Mucosa	134
2.3	Oral Tongue	136
2.4	Floor of the Mouth	138
2.5	Hard and Soft Palate	140
2.6	Tonsillar Region	142
2.7	Base of Tongue	142
2.8	Neck Metastases	143
2.9	Salvage Brachytherapy for Oral Cavity and Oropharynx	144
	References	144

1 Introduction

Brachytherapy is often indicated in the management of primary or recurrent carcinomas of the lip, oral cavity, and oropharynx. It offers the dual advantage over external beam therapy alone of significant sparing of the salivary glands and other peripheral cervicofacial tissues while delivering a high radiation dose to the tumor. Interstitial irradiation provides better functional and esthetic preservation than surgery in the great majority of situations in which both modalities offer the same cure rate. Brachytherapy also represents a highly effective alternative to radical surgery for patients with locally recurrent disease or who develop new primaries in previously irradiated areas. It is also occasionally indicated for associated neck disease, especially for recurrence after surgery and external irradiation.

When brachytherapy is anticipated, especially when combined with external irradiation, special precautions should be taken to reduce the risk of mandibular and dental complications of irradiation. After evaluation by the dentist of the oral tissues and teeth, any restoration work is done prior to implantation. When teeth in poor condition require extractions, the gingival mucosa should be carefully sutured and allowed to heal completely first. It is often preferable not to extract teeth in borderline condition when adequate time cannot be allowed for mucosal healing prior to the implantation but to repair them later. To preserve the remaining teeth, fastidious oral hygiene, frequent dental prophylaxis, and fluoride applications are obligatory. Dentures can be used 6 months after treatment, provided the gingival mucosa is in good condition and the dentures are adjusted to fit perfectly.

The recommended dosimetric system of implantation is the Paris System (DUTREIX et al. 1982; PIERQUIN et al. 1987). The rules of implantation are relatively simple, and when observed, the subsequent treated volume can be accurately predicted. After implantation, the basal dose rate and the reference isodose (equal to 85% of the basal dose rate) are calculated and displayed in multiple planes by computerized dosimetry (Fig. 5). Treatment times can then be accurately calculated.

2 Patients and Methods

2.1 Lips

The great majority of squamous carcinoma of the lip occur in the lower lip. Irrespective of the extent of the lesion concerned, except when there is associated mandibular involvement, brachytherapy alone is used to administer a dose of 65–70 Gy to the tumor. Additional cervical lymph node dissection and/or neck irradiation is determined by the extensiveness of the primary tumor and/or presence of palpable lymphadenopathy.

BERNARD PIERQUIN, Prof. Dr., JEAN JACQUES MAZERON, Dr., Département de Carcinologie, Service de Radiothérapie, Hôpital Henri Mondor, 51, Avenue de Maréchal de Lattre Tassigny 94010 Créteil, France

LAVAL GRIMARD, Dr., Ottawa Regional Center, Ottawa, Ontario, Canada

For small superficial lesions (≤2 cm in diameter), the hypodermic needle implantation technique is recommended, using local anesthesia with the patient in supine position. For lesions ≤5 mm thick and <1 cm in diameter, a single plant implant consisting of one radioactive line 2 cm long on both sides of the lesion is adequate. For larger tumors, source length and spacing are determined by the Paris System (DUTREIX et al. 1982; PIERQUIN et al. 1987). Geometry is maintained with soft plastic tube spacers. A total dose of 65 Gy at the reference isodose (85% basal dose) is prescribed. For slightly thicker, moderately exophytic lesions, a dose of 70–75 Gy is prescribed in order to irradiate adequately the portion of the tumor extending above the plant of the implant. This is permissible since the normal lip tolerates these doses quite well. When the lesion is highly exophytic, the implant is modified to contain 3 horizontally oriented radioactive lines arranged in a triangular pattern maintained with lucite retaining plates. The added third line is placed in contact with the superior surface of the exophytic portion of the tumor (Fig. 1).

For large infiltrating lesions of the lower lip (>2 cm), the preferred treatment technique varies depending on their morphology. Brachytherapy for these is performed under neuroleptic analgesia or general anesthesia. The rigid guide needle technique provides optimal parallelism while the plastic tube technique is often better tolerated by the patient. The technique chosen is based on clinical judgement. For lesions with surface extensions <4 cm in diameter, either the guide needle or the plastic tube technique is appropriate to establish a triangular or multiplane implant. The entry and exit points of these needles should be at least 0.5 cm beyond the palpable edge of the lesion. For concentrically infiltrating lesions (<4 cm in diameter), the horizontally oriented needles are sequentially implanted in the following locations: the first in the mucosa of the lip near the vermillion, the second just inferior to the junction of the vermillion and the labial skin, and the third just above the labiomental groove. For massive lesions (≥4 cm in diameter), a two or occasionally a three plane implant is required. The results of treatment are summarized in Table 1. Treatment complications are rare. Soft tissue necrosis occurred in 9 of 316 patients (2.8%) at the Institute Gustave Roussy (GERBAULET et al. 1978). There are poor esthetic and functional results in 5% of patients overall (MAZERON and RICHAUD 1984), the majority occurring in T3 and T4 tumors.

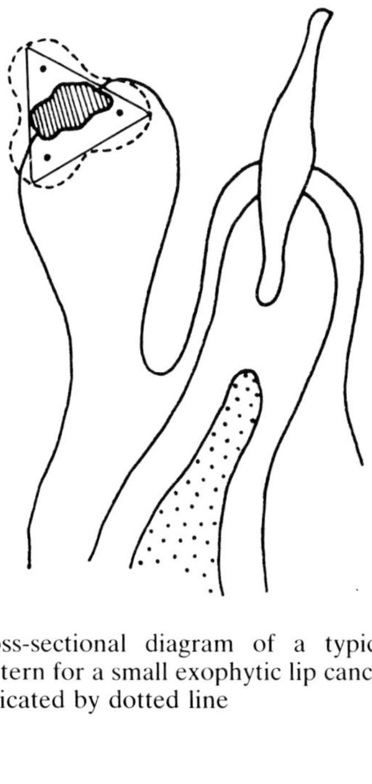

Fig. 1. Cross-sectional diagram of a typical triangular implant pattern for a small exophytic lip cancer. Reference isodose indicated by dotted line

2.2 Buccal Mucosa

Endocurietherapy alone (65 Gy) is adequate for T1 and small T2 (≤3 cm) superficial carcinomas of the central or anterior buccal mucosa. Larger lesions are managed by external irradiation of the primary lesion and ipsilateral neck to a dose of 45–50 Gy supplemented by a boost to the primary of 25 Gy by brachytherapy. Brachytherapy is generally contraindicated if there is deep involvement of the gingivobuccal sulcus, retromolar trigone, or posterior intermaxillary commissure. Neck dissection is added when clinically positive lymph nodes are present.

For lesions ≤1 cm which are noninfiltrative, a double guide gutter without lip is implanted under local anesthesia. This is done through the open mouth directly into the buccal submucosa to encompass the entire lesion. A double hairpin 2.5 cm long is substituted and sutured to the mucosa (see Sect. 2.3). A dose of 65 Gy is administered at the reference isodose (85% basal dose) (MARINELLO et al. 1985). Larger T1 lesions (≤2 cm) with minimal infiltration (<5 cm) require a single plane implant using the plastic tube technique under general anesthesia The guide needles are implanted transcutaneously and nearly parallel to the horizontal portion of the mandible to lie about 3 mm deep to the mucosal surface. Plastic tubes are substituted

Table 1. Results of brachytherapy treatment of oral cavity lesions

Lips: Groupe Européen de Curiethérapie (MAZERON and RICHAUD 1984)

		Local control
Iridium 192 ($n = 1267$)	T1	98%
	T2	97%
	T3	90%
Radium 226 ($n = 603$)	T1	91.7%
	T2	89.2%
	T3	42.0%

Buccal mucosa: Groupe Européen de Curiethérapie (GERBAULET and PERNOT 1985)

		Local Control
Curietherapy only	($n = 226$) (all T)	78%
Curietherapy plus External irradiation	($n = 80$)	51%

Mobile tongue: Créteil, 1970–1985

		T1N0 (70)	T2N0 (85)	T1–2N1–3 (13)
Iridium 192	Primary local control	87%	88%	69%
	Local control after salvage surgery	91%	92%	69%
	5-Year crude survival	51%	42%	1/10
External irradiation (45 Gy) + Iridium 192 (30 Gy)		T1N0 (3)	T2N0 (25)	T1–2N1–3 (9)
	Primary local control	1/3	36%	6/9
	Local control after salvage surgery	1/3	44%	7/9
	5-Year crude survival	0/3	24%	0/9

Floor of mouth: Créteil, 1970–1985

		T1N0 (43)	T2N0 (45)	T2N1–3 (23)
All patients Iridium 192	Primary local control	98%	78%	74%
	Local control after salvage surgery	100%	80%	74%
	5-Year crude survival	69%	40%	26%
Tumors without gingival involvement		T1N0 (43)	T2N0 (37)	T2N1–3 (13)
	Primary local control	98%	86%	85%
	Complications: Grade II–III	T1N0 (42)	T2N0 (21)	T2N1–3 (11)
	Soft tissue necrosis	10%	16%	18%
	Osteoradionecrosis	2%	6%	18%

Tonsillar region: Créteil (MAZERON et al. 1987b)

		Local Control	5-Year survival
External irradiation + iridium 192	T1 (19)	100%	66%
	T2 (26)	96%	

Soft palate: Créteil (MAZERON et al. 1987b)

		Local Control	5-Year survival
External irradiation + iridium 192	T1 (21)	86%	43%
	T2 (9)	78%	

Base of tongue

Necker series, 1974–1981 (HOUSSET et al. 1987)

Local failures:	Cobalt + iridium	Surgery + cobalt	Cobalt only
T1	0/6	2/7	21% (4/19)
T2	26% (6/23)	15% (3/20)	53% (19/35)

Créteil series, 1971–1981 (CROOK et al. 1988)

		Local Control	5-Year survival
External irradiation + iridium 192	T1 (13)	85%	50%
	T2 (35)	71%	

Neck metastases: Institut Gustave Roussy (WIBAULT and CHASSAGNE 1977)
(Local tumor status in 164 patients 3 months following ^{192}Ir implantation for recurrent neck disease)

Complete remission	Partial remission (>50% volume reduction)	<50% vol. reduction
28%	35%	37%

Salvage brachytherapy for oral cavity and oropharynx

		5-Year local control
Créteil (MAZERON et al. 1987a): 70 Oropharynx	Base of tongue	57% (25/44)
	Faucial arch	100% (21/21)

Nancy (LANGLOIS et al. 1988): 26 Mobile tongue + 97 oropharynx 2-Year actuarial local control: 67%

Complications: rate of necrosis	
Tonsils	21%
Base of tongue	28%
Mobile tongue	34%
Soft palate	5%

for the needles using the pushing maneuver and are maintained in position at the skin surface with plastic tube spacers and immobilization buttons. The treated volume should include at least a 1-cm margin of apparently normal mucosa, and the recommended spacing between lines varies from 12 to 15 mm.

External irradiation followed by implantation is recommended for large T2 lesions (≥3 cm) or lesions with deep infiltration of the cheek (>5 mm). Depending on the thickness of the residual tumor, two plane implants may be required. In this situation at least several millimeters must be allowed between the limits of the target volume and the entry and exit points of the needles in the superficial plane. This margin should be even larger for the deep plant to allow for the separation between planes. This can be difficult to obtain posteriorly because of the proximity of the intermaxillary commissure. Thus, it is sometimes necessary to cross the posterior end of the implant with the loop procedure (Pernot technique). In this technique, the guide needles are substituted by nylon filament (Fig. 2a). A Reverdin's needle is then inserted through the exit wound of the lower line and advanced to the exit wound of the upper line (Fig. 2b). This line is pulled back out through the lower hole and a plastic tube is substituted for the nylon lines (Fig. 2c–e). The loop formed should lie at least 0.5 cm anterior to the ascending ramus of the mandible. Finally, since the cheek is thicker posteriorly, to maintain parallelism for two plane implants, the anterior end of the tubes in the superficial plane must often be suspended beyond the skin. In this situation, lucite retaining plates are necessary to maintain correct interplanar separation. The results of treatment are summarized in Table 1.

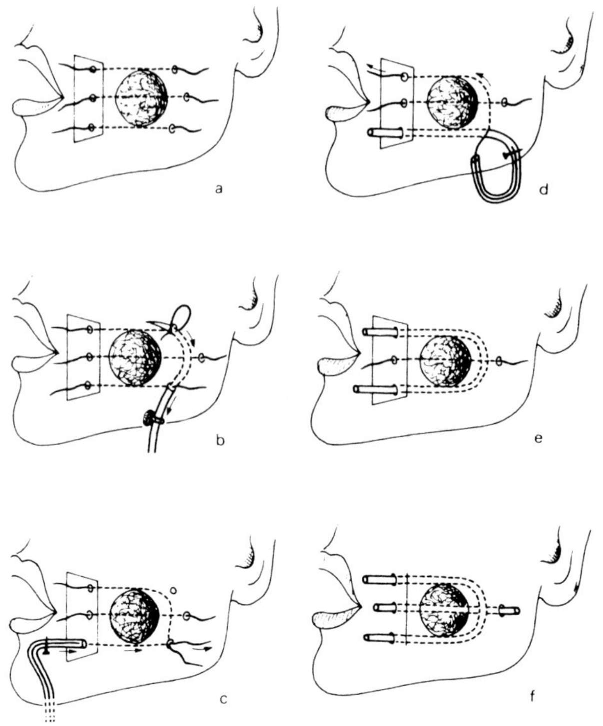

Fig. 2a–f. Plastic tube technique for buccal carcinoma. Formation of a posterior loop for a lesion near the intermaxillary commissure (Pernot technique)

2.3 Oral Tongue

Endocurietherapy alone (65–70 Gy) is recommended for T1NO and T2NO lesions of the mobile tongue (Crook et al. 1988; Pierquin et al. 1987). The combination of external irradiation to the primary and nodal areas (45–55 Gy) followed by interstitial implantation used to deliver a boost (30–35 Gy) to the primary had disappointing results in many European centers and is no longer recommended for stage I and II carcinomas of the mobile tongue (Haie et al. 1983; Mazeron et al. 1982). Small lesions of the tip and immediately adjacent ventral surface of the tongue should be treated by primary excision combined with postoperative endocurietherapy. For clinically negative necks, we recommend close follow-up for T1 lesions. For T2 lesions, elective neck dissection followed by irradiation, excluding the volume implanted, for pathological specimens, or close follow-up with delayed therapeutic neck dissection and irradiation are both valid alternatives (Mazeron et al. 1990; Van Den Brouck et al. 1980).

For lesions on the lateral border of the tongue, the guide gutter technique or the plastic tube technique is recommended. The former is indicated for small lesions (T1 or small T2) in cooperative patients and is done under local anesthesia with the patient sitting in a dental chair. The latter usually requires general anesthesia or at least neuroleptic anesthesia supplemented by local anesthesia. This is done with the patient supine or in a dental chair and allows treatment of larger volumes. A fluoroscopy unit in the operating room greatly facilitates controlling implant geometry and is almost essential for the guide gutter technique.

For the guide gutter technique, atropine 0.5–

0.6 mg is given 20 min prior to the procedure. Because significant bleeding can occur, proper equipment must be available on the instrument tray in the event a hemostatic stitch becomes necessary. Local anesthesia with lidocaine 2% with or without epinephrine (15–20 ml) is administered with a spinal needle angulated at its tip and long enough to reach the posterior third and junctional zone of the tongue. The anesthesia should encompass the entire tumor and a margin of at least 0.5 cm around the lesion. The height of the tumor determines the length of the guide gutters required, which is usually 3–4 cm. The guide is grasped by its lip with long forceps and the posterior one is implanted first, just posterior to the palpable limit of the tumor and perpendicular to the dorsal surface of the tongue. The second guide is implanted just anterior to the lesion, parallel with the first guide, and the third guide is implanted equidistant and parallel to the others. The spacing between sources should not exceed 15 mm nor be less than 10 mm. For small T1 (≤1.5 cm), two hairpins are sufficient. The implantation is then checked on fluoroscopy for parallelism in both orthogonal directions. If the geometry is unsatisfactory, it is corrected by gently extracting and reimplanting the guides. One or more such adjustments are not unusual, even in experienced hands.

When the guides are in optimal position, using a small Reverdin's needle a 20 cm 1/0 silk suture is sewn into the lingual muscle beneath and close to the transverse bar of each guide, starting anteriorly. A small hemostat is clamped on the free ends of each suture. Double hairpin sources from 3.0 to 3.5 cm long are then placed nearby in a shielded container, and mobile lead screens are placed between the patient and the operators. The posterior guide is afterloaded first while using the anterior sutures to pull the tongue forward. The hairpin is picked up in one hand with nonlocking forceps without teeth. The crossbar of the guide is gripped gently with forceps held in the other hand, and the pin tips are brought into the two visible openings in the legs of the guide. The pins should slide easily into the legs of the guide. Once the pin is fully inserted, it is held in position with a small surgical hook while the guide is extracted from the tongue. The hairpin is then secured by tying several knots at the level of the small indentation in the head of the pin. The same procedure is repeated for the middle and anterior guides (Fig. 3). The time of after-loading is recorded, and orthogonal radiographs with known magnification are made for dosimetry (Figs. 4, 5). These films are repeated the following day as some swelling can modify the geometry and dose distribution. The implantation is inspected once daily, and source removal should always be performed in an operating room for adequate control of potential bleeding.

The plastic tube technique is suitable for more extensive infiltrating lesions which are too large for the standard hairpin technique. Transverse loops are formed starting from the submental area. Lateral branches are located just beneath the mucosa of the corresponding area of the lateral border of the tongue and the medial branches as close as possible to the midline, depending on the medial extent of tumor infiltration. Generally, the posterior needles are inserted first for posterior lesions and the anterior needles first for anteriorly located lesions. The plastic tubes are inserted by the pulling method of substitution (Raynal maneuver) after substituting nylon filament for the needles. The separation between the branches of these loops can vary from 12 to 18 mm, and the loops should be equidistant. Spacing greater than 20 mm produces an overdose or hyperdose sleeve (170% basal dose), which is too large for normal tissue tolerance. Finally, a soft rubber tube placed around the plastic tubes at the apex of the loops and sutured to the tongue or alternatively a rubber tube spacer through the plastic tubes at the apex of the loops reduces the incidence of necrosis on the

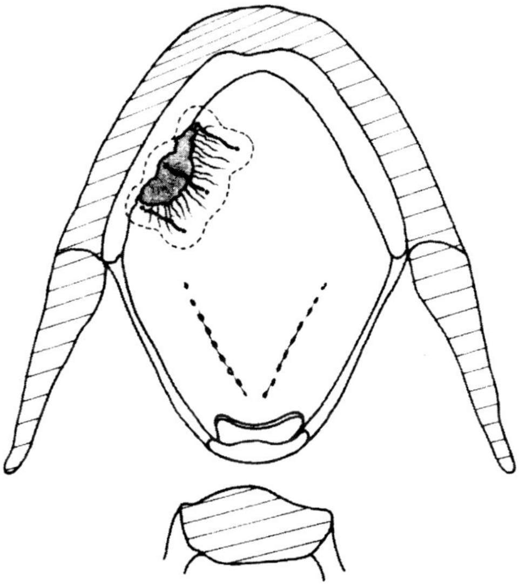

Fig. 3. Schematic representation of hairpin implant, right lateral border of tongue. Reference isodose indicated by *dotted line* around the hairpins

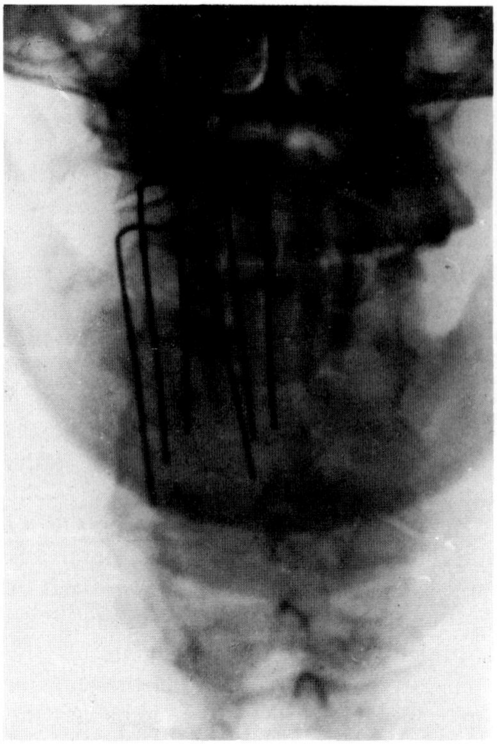
a

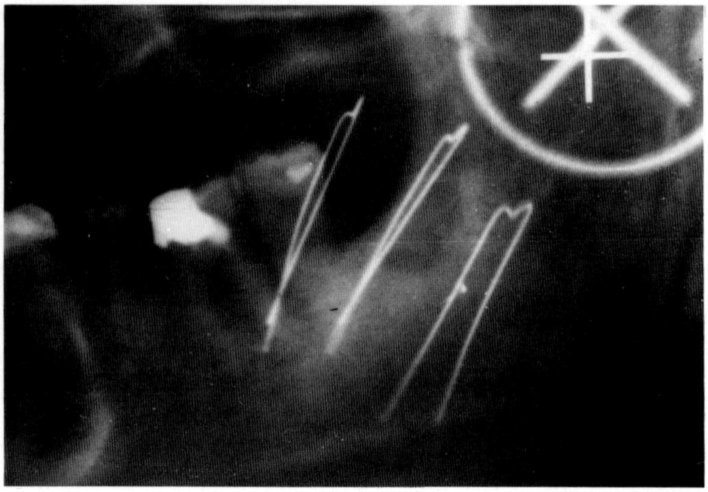
b

Fig. 4a,b. Hairpin implant, lateral border of tongue **a** AP check film, **b** lateral check film

surface of the tongue which, although not dramatic as it always heals, is a source of discomfort for the patient.

For T1 and small T2 superficial tumors of the tip of the tongue, we recommend a surgical excision of the tip and adjacent ventral surface. Once the tongue is healed, 2 or 3 sagittal loops are implanted to pull the tongue remnant down against the floor of the mouth, the anterior branches passing through the floor of the mouth and the posterior ones through both the floor and the tongue. For large T2 and T3 lesions, the methods of implantation are identical to infiltrating anterior pelvilingual tumors.

The results of treatment are shown in Table 1. Complications of interstitial therapy include soft tissue necrosis, which occurred in 16% (23/145) of our patients, and osteoradionecrosis, seen usually in the same patients, 15%: 22/145 (MAZERON et al. 1990). Soft tissue necrosis heals spontaneously within 12 months in the majority of patients. At times a short course of intravenously administered steroids with high-dose parenteral antibiotics is indicated for inflamed and painful ulcerations (necrosis grade II). Sequestrectomy is rarely indicated. The risk of osteoradionecrosis is greater when external irradiation is given, even though the area of implantation is shielded: 25% vs 8% in the Créteil experience (MAZERON et al. 1990).

2.4 Floor of the Mouth

Carcinomas of the floor of the mouth <3 cm in any surface dimension can be managed by endocurietherapy alone. Larger lesions (T2 > 3 cm and small T3) can also be treated by implantation but with a greater risk of postirradiation necrosis of soft tissues and osteoradionecrosis. As in carcinoma of the mobile tongue, the use of external irradiation followed by an interstitial boost has given poor results in Europe. Endocurietherapy is contraindicated in lesions with gingival involvement. Individualized intraoral mandibular shields reduce considerably the risk of osteoradionecrosis but require a dental service with expertise in this area. Because of these limitations, there are relatively fewer indications for brachytherapy of carcinomas of the floor of the mouth than for those of the oral tongue.

For small superficial lesions (≤1.5 cm), the double guide gutter technique (guides 3 cm long with leg separation of 12 mm) with the insertion of

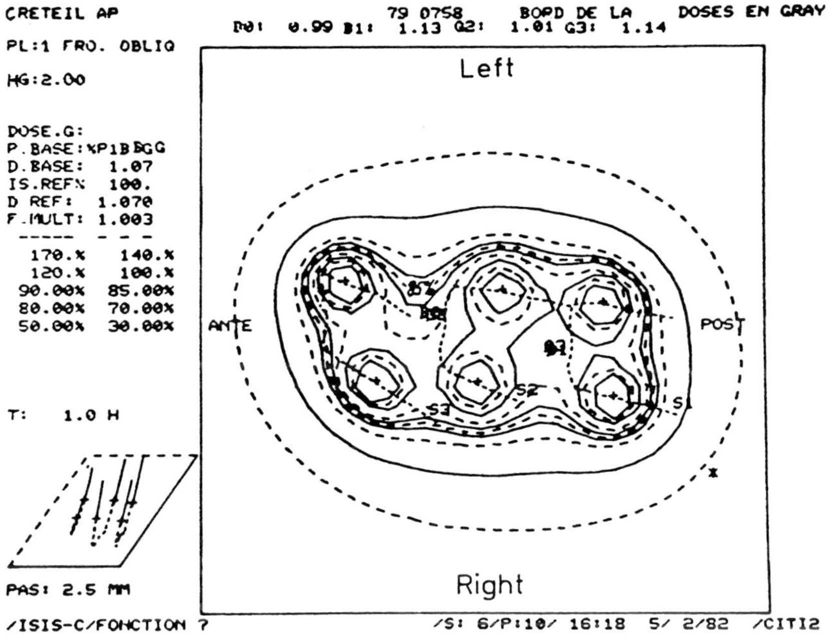

a

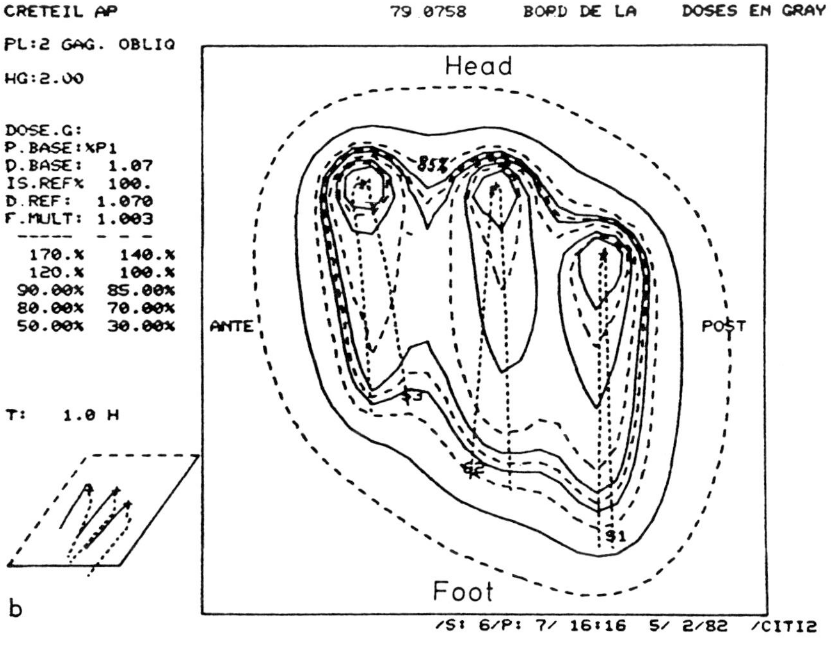

b

Fig. 5a,b. Dosimetry of hairpin implant, lateral border of tongue: **a** calculation in central plane, **b** calculation in sagittal plane

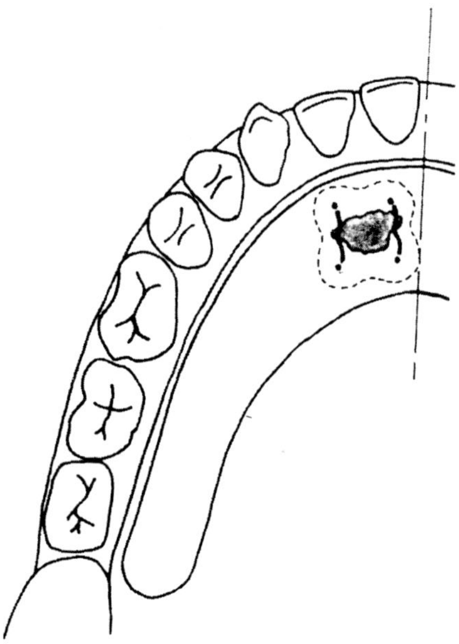

Fig. 6. Stage T1 squamous cell carcinoma, anterior floor of mouth, (9 mm diameter, noninfiltrating). Implantation with two hairpins (2 cm long, leg separation = 9 mm). Reference isodose indicated by *broken line* around sources

two hairpins is recommended. A dose of 65 Gy is delivered at the reference isodose (Fig. 6). For lesions ≤3 cm in diameter, endocurietherapy by the guide gutter technique is appropriate when tumor infiltration does not exceed 0.5 cm. The plastic tube technique is preferred for more infiltrative lesions (>0.5 cm). A dose of 65 or 70 Gy is recommended.

Small lesions of the junction of the floor of the mouth and tongue can be treated using the guide gutter technique used for infiltrating lesions of the lateral border of the tongue. The medial branches of the guide gutters are implanted into the lingual muscles well beyond the medial edge of the tumor, and the lateral branches are implanted into the floor of the mouth at the lateral margin of the tumor. The plastic tube technique is also frequently used. When the pelvilingual lesion is more extensive and infiltrating, the principle of implantation should take into account the trapezoidal shape of the tissues encompassed by the mandibular arch. Five lines are required, made up of an anterior row containing 2 lines and a posterior row or 3 lines. If necessary, a sixth line can be added. The results of endocurietherapy are presented in Table 1. The main complications are soft tissue and bone necroses. Management is conservative in the majority of patients with only 3/85 (3.5%) T1–2 patients without gingival involvement requiring surgery for osteoradionecrosis in Créteil (1970–1985).

2.5 Hard and Soft Palate

For the rare lesions of the hard palate, the guide gutter technique is recommended for small superficial tumors (local anesthesia, sitting position). The concave shape of the hard palate necessitates using curved guides implanted in a sagittal or parasagittal direction. Single guide gutters spaced about 8–15 mm apart are easier to implant than double guide gutters. Two parallel lines are usually adequate to encompass small lesions. The looped heads of the hairpins should be oriented laterally to avoid overdosage at the anterior edge of the target volume. The dose prescribed at the reference isodose is 60 Gy.

Carcinomas of the soft palate are usually moderately advanced and extend beyond the structure of the origin at presentation.

The guide gutter technique of endocurietherapy (local anesthesia, sitting position) is recommended for small T1 lesions of the soft palate. This procedure is done with the patient given nothing by mouth, sitting in a dental chair. Atropine is given 20 min prior to the procedure. The uvula is transected at its base, leaving a smoothly curved inferior surface of the soft palate. The oral cavity is sprayed with 5% lidocaine, then 10–15 ml of lidocaine 2% with epinephrine is injected into the lateral border of the soft palate on the side of the lesion. The palate being adequately ballooned by the local anesthetic, the

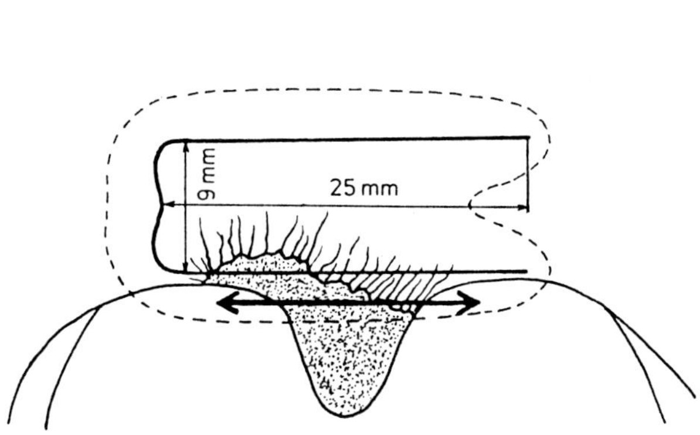

Fig. 7. Carcinoma of soft palate: uvula and right inferior border of palate involved. After transecting the uvula (level of double arrow) endocurietherapy by the guide gutter technique is performed. Single hairpin source 9 × 25 mm

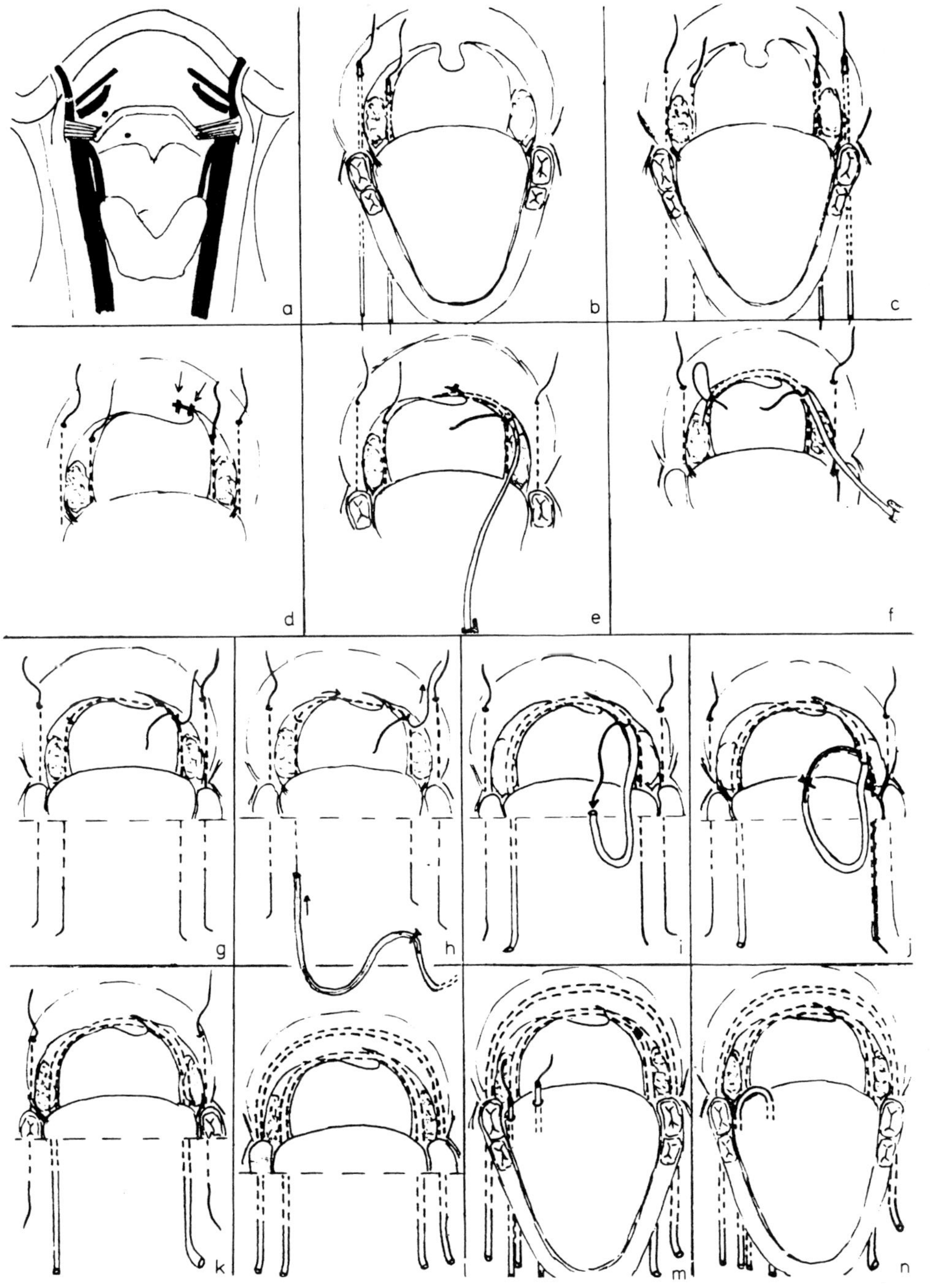

Fig. 8a–n. Plastic tube technique for palatotonsillar implants (see text)

anterior and the posterior surfaces of the soft palate are separated. This allows implantation of the guide gutter in a slightly oblique direction without piercing the posterior surface of the palate. As the lidocaine is absorbed, the branches of the guide gutter automatically assume a transverse position. To afterload the hairpin, firm traction is placed on the crossbar of the guide gutter to bring it into a slightly oblique orientation. One hairpin 3–4 cm long implanted transversely is adequate to encompass a lesion 1 cm in diameter (Fig. 7). The dose prescribed for endocurietherapy alone is 60–70 Gy.

The plastic tube technique (Pernot technique) is indicated for more advanced lesions of the soft palate or palatotonsillar area as a boost (30 Gy) following external irradiation (45–50 Gy) to the primary and neck. This technique is done under general or neuroleptic anesthesia with the patient in a semi recumbent position in a dental chair. The position of vascular structures, hyoid bone, and a projection of the bases of the tonsillar pillars are marked on the skin of the neck (Fig. 8a). Then a 10-cm long guide needle is inserted through the skin just inferior to the hyoid bone at a point corresponding to a projection of the posterior pillar on the skin. The needle is advanced posteriorly about 2.5 cm into the cervical tissues lateral to the pharyngeal wall. A finger in the pharynx monitors the progress of the needle. When the base of the posterior pillar is reached, the needle is redirected upward to transfix the entire length of the posterior pillar, exiting as high as possible, and the nylon filament is threaded through. This needle must be at least 0.5 cm away from the ascending ramus of the mandible. The entry point of the second needle lies anteromedially 1.5–2.0 cm from the first needle and is usually above the hyoid bone. This second needle is introduced about 5 mm into the neck tissues and is then redirected to pass upward and exit through the anterior pillar (Fig. 8b). The same procedure is repeated on the contralateral side (Fig. 8c), and the needle positions are checked by fluoroscopy. If their positions are satisfactory, they are removed, leaving the nylon filaments in place. The next step is to implant the soft palate itself. In order to avoid a geographic miss, the uvula can be sutured to one side (Fig. 8d) or transected at its base at the end of the procedure. A Reverdin's needle is inserted at the exit wound of the nylon filament in the upper part of the posterior pillar. The needle is advanced through the inferior portion of the soft palate across to the contralateral side of the palate. The contralateral monofilament is then gripped in the eye of the needle and pulled back into the palate (Fig. 8e–g). A plastic tube is substituted for the nylon filament using the pulling maneuver (Fig. 8h–k). Finally, to complete the anterior arch, the Reverdin's needle inserted in the anterior portion of the anterior pillar at the exit point of the nylon filament must be kept parallel to the posterior tube already in position in the soft palate (Fig. 8l). When the posterior tongue is involved by direct extension downward along the anterior pillar, a loop is implanted in the posterior portion of the tongue (Fig. 8m,n). Computer dosimetry is derived from orthogonal films. Scans in a sagittal plane passing through the center of the sources in the palate and in a horizontal plane (or slightly oblique) perpendicular to the vertical segment of the sources are required to evaluate the dosimetry of the implant. When the separation between the palatal sources is 1.5–2 cm, the dose distribution is usually satisfactory. Results of treatment are presented in Table 1. Acute complications include hemorrhage and delayed mucosal healing. Chronic complications consisted of the reappearance of unstable epithelialization of the oropharyngeal mucosa (17/105) and bone exposure (6/105) in a series from the Centre Alexis Vautrin in Nancy, France (Pierquin et al. 1987).

2.6 Tonsillar Region

Interstitial irradiation has a limited role in the management of tonsillar carcinomas. Endocurietherapy alone can be considered for very small, noninfiltrating T1 tumors of the tonsillar region and more commonly as a boost technique for more extensive lesions.

For highly selected small lesions (<1 cm), the guide gutter technique is used. A single hairpin, 3 or 4 cm long, is usually adequate. The tonsillar fossa is framed by the anterior branch of the pin which traverses the anterior pillar and the posterior branch within the posterior pillar. For larger lesions confined to the tonsillar region (T1, small T2), the plastic tube technique of implantation is used as a boost to form a single large loop encompassing the tonsillar region. The apex of this sagittal loop lies at a lower level on the soft palate. When there is some soft palate extension, which is the most common situation, a modified Pernot technique of the soft palate is used with the plastic tubes ending in the tonsillar pillars contralateral to the lesion (entry point of the Reverdin's needle used to implant the soft palate). This allows full coverage of the soft palate. Table 1 provides some results of treatment.

2.7 Base of Tongue

Brachytherapy is seldom indicated in the treatment of carcinomas of the base of the tongue. For lesions ≤4 cm confined to the tongue base or vallecular region, endocurietherapy following external irradiation has improved local control in limited series. Endocurietherapy alone has been

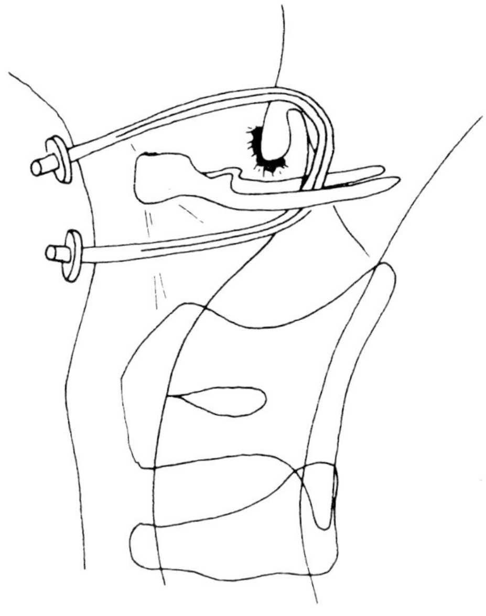

Fig. 9. Plastic tube implantation for carcinoma of the vallecular region The epiglottis is pulled forward against the tongue base during the irradiation

used a salvage therapy or for second primaries in previously irradiated patients.

Under general anesthesia, a projection of the tumor is outlined on the skin of the anterior neck. A guide needle is introduced perpendicular to the skin surface, guiding its penetration into the pharynx with the index finger. Using the pulling or Raynal maneuver of substitution, the plastic tube is introduced to give the inferior branch of the first loop. The superior branch of the loop is then formed, and traction on the monfilament should be just enough to bring the loop apex into contact with the tongue surface. Three or four parallel, sagittally oriented loops are implanted with a spacing of 12–15 mm. The separation between the branches of the loops may exceed 20 mm for large tumors, and a transverse loop with its branches situated inside the target volume established by the other loops and equidistant from them can be added to prevent underdosage centrally. The curved portion of the transverse loop is not loaded as this would create a hot spot on the tongue surface where the loops overlap. Radiographs for dosimetry are not obtained until the day following implantation because of possible post-implantation edema of the tongue. The technique can be modified for carcinoma of the vallecular region. In this situation, the inferior branches of the loops are implanted between the hyoid bone and the thyroid cartilage. After traction, the free portion of the epiglottis is pulled forward and held against the base of the tongue (Fig. 9). Results are shown in Table 1. Soft tissue ulceration occurred in 33% (10/30) of patients treated in Créteil between 1971 and 1981, with 3 patients progressing to osteoradionecrosis (CROOK et al. 1988).

2.8 Neck Metastases

Endocurietherapy is indicated in the following situations for neck metastases: (a) as a boost when surgery is contraindicated; (b) for small, isolated nodal recurrences or subcutaneous tumor nodules when surgery is unlikely to accomplish complete extirpation of the lesions; (c) for massive fixed neck recurrences.

A CT scan of the neck is very useful in planning such implants. The skin should be intact as the presence of ulceration predisposes to delayed carotid exposure. The first technique consists of using straight or curved lines to introduce plastic tubes and is appropriate for superficial as well as relatively deep-seated neck masses which do not extend to the mucosal surface of the pharynx. The thickness of the target volume determines the number of planes (at least two for lesions ≥2 cm). For the deep planes and in the retromandibular area, it is easier to implant the tubes around the deep margins of the lesion with large Reverdin's needles (Fig. 10) rather than straight guide needles. The second technique is recommended for large, deep-seated, retromandibular masses involving the pharyngeal wall. In this situation, horizontal

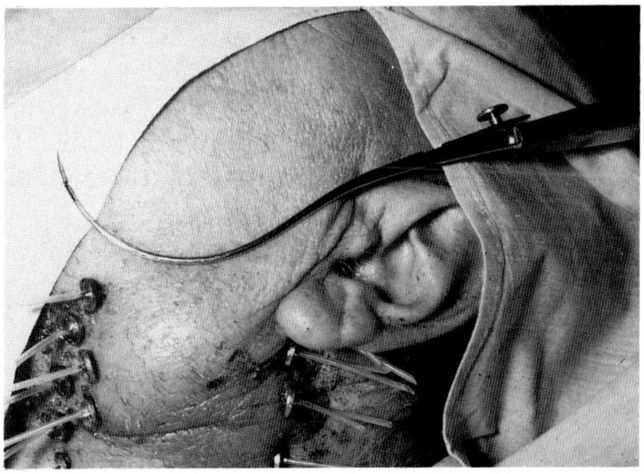

Fig. 10. Treatment of retromandibular cervical lymphadenopathy by two-plane, plastic-tube implant. The large Reverdin's needle was used to implant the deep plane of tubes

loops encompassing the vascular and lymphatic plexus of the superior carotid region are formed at the level of the retromandibular region with posterior branches lying near the cervical spine and anterior branches along the medial surface of the mandibular ramus. The dose prescribed is 60 Gy at the reference isodose, irrespective of previous irradiation. After completion of treatment, the patient is kept NPO, and the tubes are removed in the operating room. The tubes are extracted slowly and gently, one at a time, starting superiorly. If bleeding is observed, firm fingertip compression will usually control the carotid hemorrhage. Results of treatment are shown in Table 1. In the group of patients (164) mentioned in Table 1, the main complication was carotid hemorrhage (18 patients) secondary to late carotid exposure which occurred essentially in those with pre-existing skin ulceration. Excluding this problem related to selection of patients, complications were otherwise limited, with only one perioperative death.

2.9 Salvage Brachytherapy for Oral Cavity and Oropharynx

Patients previously irradiated should be considered for endocurietherapy at the time of a recurrence or a second primary. Brachytherapy controls most tumors and carries little risk of complications for the tonsillar regions and the soft palate. It provides less local control and more complications for patients with tumors of the base of tongue, glossotonsillar sulcus, and mobile tongue, but in some situations it is the only option available for salvage. In general, local control is inversely proportional to the size of the tumor while the incidence of necrosis increases with the diameters of the volumes implanted. Two recent series present typical results achieved (Table 1) (7, 12) (Langlois et al. 1988; Mazeron et al. 1987a).

References

Crook J, Mazeron JJ, Marinello G, Martin M, Raynal M, Calitchi E, Faraldi M, Ganem G, LeBourgeois JP, Pierquin B (1988) Combined external irradiation and interstitial implantation for T1 and T2 epidermoid carcinomas of base of tongue: the Creteil experience (1971–1981). Int J Radiat Oncol Biol Phys 15: 105–114

Dutreix A, Marinello G, Wambersie A (1982) Dosimétrie en curiethérapie. Masson, Paris

Gerbaulet A, Pernot M (1985) Le carcinome épidermoïde de la face interne de la joue. A propos de 748 malades. J Eur Radiother 6(1): 1–4

Gerbaulet A, Chassagne D, Hayem M, Vandenbrouck C (1978) L'épithélioma de la lévre. Une série de 335 cas. J Radiol Electrol 59: 603–610

Haie C, Gerbaulet A, Wibault P, Chassagne D, Marandas P (1983) Résultats de la curiethérapie et de l'association radiothérapie transcutanée-curiethérapie dans 155 cas de cancers de la langue mobile. Actual Carcinol Cervico fac 9: 53–57

Housset M, Baillet F, Dessard-Diana B, Martin D, Miglianico L (1987) A retrospective study of three treatment techniques for T1–T2 base of tongue lesions: surgery plus postoperative radiation, external radiation plus interstitial implantation and external radiation alone. Int J Radiat Oncol Biol Phys 13: 511–516

Langlois D, Hofsetter S, Malissard L, Pernot M, Taghian A (1988) Salvage irradiation of oropharynx and mobile tongue about 192 iridium brachytherapy in Centre Alexis Vautrin. Int J Radiat Oncol Biol Phys 14: 849–853

Marinello G, Wilson JF, Pierquin B, Barret C, Mazeron JJ (1985) The guide gutter or loop technique of interstitial implantation and the Paris system of dosimetry. Radiother Oncol 4: 265–273

Mazeron JJ, Richaud P (1984) Compte rendu de la XVIII° réunion du groupe européen de curiethérapie. Session consacrée aux cancers de la lévre, Padova, Italie, Mai 1981. J Eur Radiother 5: 50–56

Mazeron JJ, Calitchi E, Martin M, Maylin C, LeBourgeois JP, Lobo P, Baillet F, Pierquin B (1982) Analysis of local failures after treatment with curietherapy using iridium 192 for squamous cell carcinomas of the mobile tongue. J Eur Radiother 3(3): 131–138

Mazeron JJ, Langlois D, Glaubiger D, Huart J, Martin M, Raynal M, Calitchi E, Ganem G, Faraldi M, Feuillhade F, Brun B, Marin L, LeBourgeois JP, Baillet F, Pierquin B (1987a) Salvage irradiation of oropharyngeal cancers using iridium 192 wire implants: 5-year results of 70 cases. Int J Radiat Oncol Biol Phys 13: 957–962

Mazeron JJ, Crook J, Mahot P, Martin M, Raynal M, Faraldi M, Juvanon JM, Peynegre R, Pierquin B (1987b) Mise au point sur la radiothérapie exlusive des T1 et T2 de l'arche vélo-amygdalienne. Ann Otolaryng ol Chir Cervicofac 104: 197–203

Mazeron JJ, Crook JM, Benk V, Marinello G, Martin M, Raynal M, Haddad E, Peynegre R, LeBourgeois JP, Pierquin B (1990) Iridium 192 implantation of T1 and T2 carcinomas of the mobile tongue: the Creteil experience. Int J Radiat Oncol Biol Phys (in press)

Pierquin B, Chassagne D, Baillet F, Castro JR (1971) The place of implantation in tongue and floor of mouth cancer. JAMA 215(6): 961–963

Pierquin B, Wilson JF, Chassagne D (1987) Modern brachytherapy. Masson, Paris

Vandenbrouck C, Sancho-Garnier H, Chassagne D, Saravane D, Cachin Y, Micheau C (1980) Elective versus therapeutic radical neck dissection in epidermoid carcinoma of the oral cavity, results of a randomized clinical trial. Cancer 46: 386–390

Wibault P, Chassagne D (1977) L'endocuriethérapie des récidives ganglionnares du cou par iridium 192 (à propos de 164 cas). Arch Ital Otorinolaryngol 5: 389–395

4.3 Interstitial Brachytherapy in Head and Neck Tumors

Basil S. Hilaris, Anca E. Tchelebi, and Chitti R. Moorthy

CONTENTS

1 Introduction 145
2 Planning and Evaluation of Interstitial Brachytherapy 145
3 Principles and Techniques of Brachytherapy 146
4 General Care of Patients Receiving Interstitial Brachytherapy 148
4.1 Before Brachytherapy 148
4.2 During Brachytherapy 148
4.3 After Brachytherapy 148
5 Results of Treatment 148
6 Discussion 150
References 151

1 Introduction

Brachytherapy presently, in spite of the extensive utilization of external megavoltage radiation, plays a significant role in the management of head and neck cancer. The early techniques of interstitial brachytherapy using radium needles were developed in the early 1920s in Paris by Regaud, Lacassagne, and Coutard (see del Regato 1986, 1987). Paterson and parker in 1934 developed the dosage system known as the Manchester, which placed interstitial brachytherapy on a more rational basis. Paterson recognized very early on that the dose that can be safely delivered by brachytherapy varied with the volume implanted and that the smaller the volume, the greater the tolerance. This marked dependency of the tissue tolerance dose on the size of the irradiated volume is the key to understanding the value of interstitial brachytherapy in head and neck cancer. The other advantage of brachytherapy, and we refer to the precise localization of the radiation effect, is more evident in the head and neck because of easy tumor accessibility and visibility. The role of brachytherapy was strengthened by the introduction of afterloading techniques and artificial radionuclides, especially iridium 192, in the mid 1950s by Henschke et al. (1963) in New York. Recent developments in computer technology and the imaging revolution have created the tools to describe the patient's anatomy and to calculate and display dose distributions (Hilaris et al. 1987a). Thus, interstitial brachytherapy continues to be an indispensable adjunct to the management of many head and neck cancers.

Brachytherapy is used as the definitive treatment of early superficial lesions, as long as the size of the tumor is not more than 4 cm. In deeply invasive lesions (T3–4) brachytherapy can be used either as a boost after external irradiation to the primary site, after surgery whenever the margins of resection are involved, in combination with chemotherapy, and after a recurrence either for salvage or for palliation. Patient selection rests on several other factors, such as the general health of the patient, site and extent of the tumor, preference of the patient and his/her family, and familiarity and experience of the physician with this treatment modality. Afterloading techniques are now used to the exclusion of the older radium techniques. Planning brachytherapy treatment utilizing modern imaging techniques, especially computerized tomography (CT), is important in order to define the target volume in relation to the patient's anatomy as well as to display the calculated 3-dimensional dose distribution.

2 Planning and Evaluation of Interstitial Brachytherapy

The objective of treatment planning is to generate an optimal dose distribution, deliver the prescribed dose to the target volume, and minimize the dose to the surrounding normal tissues. The introduction of computers has made it possible to improve planning of interstitial brachytherapy

Basil S. Hilaris, M.D., F.A.C.R. Anca E. Tchelebi, M.D., Chitti R. Moorthy, M.D., New York Medical College, Department of Radiation Medicine, Valhalla, NY 10595, USA by optimizing source arrangement and dose distribution. The steps in treatment planning are as follows.

The tumor volume is determined by clinical examination and increasingly by CT, magnetic resonance imaging (MRI), ultrasound, or at times by endoscopic examination under anesthesia. Whenever possible, the tumor volume should be outlined by small clips to facilitate subsequent identification in localization X-radiography studies.

The target volume includes the tumor volume and a variable margin of normal tissue around it, which is usually 1–2 cm. The target volume is the region within which the delivered dose is equal to or greater than the prescribed dose.

The decision of whether a permanent or temporary implant is indicated and whether one-plane, two-plane, or volume implant is required are based on the location and size of the tumor and its proximity to normal structures. Temporary implants are used in the curative treatment of accessible cancers located in the oral cavity, oropharynx, and neck. Permanent implants are used primarily in the palliative management of head and neck tumors, in the curative management of inaccessible sites such as the nasopharynx, or whenever hospitalization is not possible. Tumor bed implants (temporary or permanent) are valuable in improving local control, if adequate margins cannot be obtained by surgical resection. Planar implants are recommended for target regions less than 1.5 cm thick, two-plane implants for target regions 1.5–2.5 cm thick, and volume implants for targets involving larger tumors.

The determination of the total radioactivity and its distribution within the treatment volume urgently needs standardization. Presently, planning follows rules incorporated either in the classic systems described by Paterson–Parker (Meredith 1967) or Quimby (Glasser et al. 1961) or in modern ones such as the Paris (Dutreix et al. 1982) or New York system (Hilaris et al. 1988a). The New York system aims either at rapid planning for both temporary and permanent interstitial implants, using nomographs and tables, or at "tailoring" the strength and location of the sources by computer-assisted repeated adjustments, using the last-square optimization when a 3-dimensional target contour is available from CT or MRI scans.

The dose prescription is based on prior treatment, dose rate, treatment volume, and the tolerance of adjacent normal tissues. It is generally accepted that for temporary iridium-192 implants, doses in the range of 6000–7000 cGy can be safely delivered over 6–7 days in previously untreated patients. Lower doses are given after prior treatment or in the case of "boost" therapy in conjunction with external radiation. For dose-rate variations in the range of 30–60 cGy/h most radiation oncologists do not make adjustments in the total dose prescribed. A dose of 14 000–18 000 cGy is generally prescribed for permanent iodine-125 implants. Downward total dose adjustments are recommended when a permanent implant is done after external radiation.

The evaluation of a temporary interstitial implant includes the determination of the adequacy of target-dose distribution, determination of a satisfactory normal tissue dose, and identification of the treatment-dose rate used to determine the treatment time. The evaluation of permanent implants aims mainly at determining the adequacy of the target region's minimum peripheral dose and at recording the maximum normal tissue dose. The potential value of volume-dose analysis for evaluating interstitial implants is currently being investigated (Anderson 1986).

3 Principles and Techniques of Brachytherapy

Cancers of the oral cavity have been historically treated with either interstitial brachytherapy alone or by combination with external radiation. The oral cavity refers to the structures contained in the mouth and includes the floor of the mouth, anterior two-thirds of the tongue, buccal mucosa, and hard palate. The selection of a target dose follows the guidelines discussed previously: 6000–7000 cGy over 6–7 days for small lesions and a boost of 2500–3000 cGy over 2–3 days after external radiation. The choice of the interstitial technique varies, however, with the site: Loop technique or its modifications including the guide gutter technique in the floor of the mouth and anterior two-thirds of the tongue (Hilaris et al. 1988b); one- or two-plane temporary implant in the buccal mucosa; customized intraoral applicators using manual or remote afterloading of the radioactive sources in the hard palate. The target volume always includes the visible lesion plus a 1–2 cm margin around it.

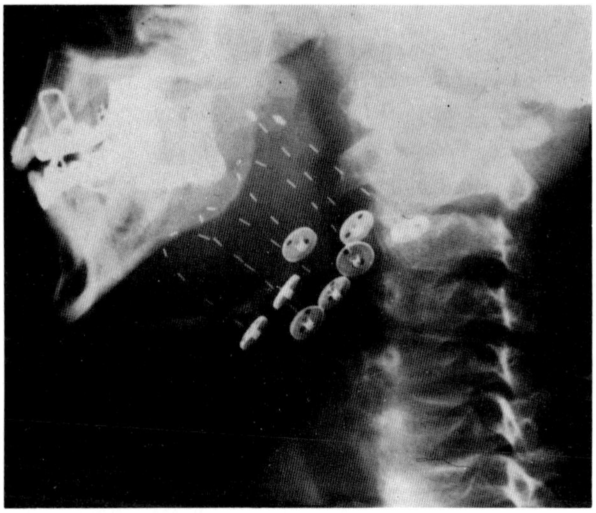

Fig. 1. Loop technique in a patient with carcinoma of the base of the tongue and tonsil

Oropharyngeal tumors are less accessible than tumors of the oral cavity, and therefore, brachytherapy is not as frequently performed in those sites. It is always used in combination with external radiation as a boost therapy in small to moderate size, well-defined, and easily accessible lesions. Structures included in the oropharynx are: the base of the tongue, tonsillar fossa and pillars, soft palate, and lateral pharyngeal wall. The target dose, in general, is in the range of 2500–3000 cGy over 2–3 days. The target volume includes the original extent of the tumor and a 1–2 cm margin around it. The technique varies with the site: the temporary loop technique and its variations in the base of the tongue and tonsillar region (Fig. 1), and the permanent technique and its variations in the soft palate and pharyngeal wall (HILARIS et al. 1988b).

Cancer of the lip may be treated equally well by surgery or radiation. Brachytherapy is applied alone in early lesions and in combination with external radiation in larger lesions. A one- or two-plane temporary implant is recommended. The target volume includes the tumor and a 0.5–1 cm margin beyond it. The target dose is 6000 cGy over 5–7 days for an implant alone and 2500–3000 cGy over 2–3 days as a boost to external radiation.

Cancer of the nasopharynx, nasal cavity, and maxillary sinuses is occasionally treated either by an interstitial or an intracavity approach supplementary to external radiation. More frequently, however, brachytherapy in these locations is used for the treatment of recurrent tumors. Techniques and target doses are tailored to the goal of the treatment, aiming at improved local control with minimal complications.

Neck node metastases are treated by interstitial brachytherapy either for persisting disease after external radiation or surgery, or for recurrent disease after unsuccessful surgery or radiation. A one- or two-plane temporary implant is recommended (Fig. 2). The target volume includes the residual or recurrent disease plus an adequate, frequently individualized tumor margin. The target dose is either 6000 cGy over 5–7 days for an implant alone or 2000–3000 cGy when combined with external radiation.

Permanent iodine-125 implants are used for the treatment of small recurrent lymph nodes that cannot be resected; the recommended target dose is 14000–18000 cGy.

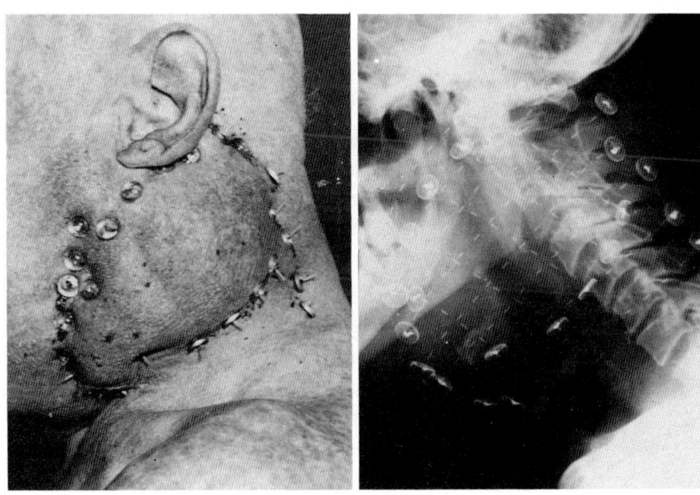

Fig. 2. Two-plane implant in a patient with recurrent carcinoma in the left neck

4 General Care of Patients Receiving Interstitial Brachytherapy

4.1 Before Brachytherapy

In oral and oropharyngeal lesions, to lessen the severity of complications, it is important to initiate proper dental care prior to brachytherapy. A careful and complete dental evaluation should include panoramic radiographs of the mandible. These are important so that the extent of infection, if present, may be adequately recognized and to provide a basis for comparison if subsequent bone changes appear. Teeth with coronal destruction, periodontal disease, or periapical infection, and, in general, all potentially diseased teeth which may have to be removed within a year or so after treatment should be extracted before radiation. All salvageable teeth should be restored and an intensive and supervised dental prophylaxis initiated, which should include fluoride gel applications and daily oral irrigations. Fluoride treatment should be applied early and continued for the lifetime of the patient, especially if external therapy is integrated with brachytherapy. The patient should be discouraged from consuming alcoholic beverages and smoking. Oral infections should be diagnosed and managed promptly. High protein, high vitamin, and high caloric food supplements are used when necessary to maintain optimal nutritional status.

4.2 During Brachytherapy

It is advisable, in the immediate postimplantation period to maintain the patient on intravenous fluids. Tube feeding may begin 12–24 h later. A blenderized tube feeding formula of regular food can be prepared by the dietary department to contain one calorie per cc; 2000 ml are given over a period of 24 h to provide 227 g of carbohydrates, 80 g of protein, and 85 g of fat. The formula can be modified for diabetic patients or patients on a low-sodium diet. The total amount should be divided into three to four separate feedings for the 24-h period. Oral irrigations repeated several times during the day play an important role in preventing infection and maintain good oral hygiene during and immediately after implantation. This consists of directing 1000 ml of a lukewarm alkaline-saline solution through the oral cavity under mild pressure using an irrigating set. During the height of the radiation reactions, patients obtain more relief from oral discomfort through this procedure than from any other means. Care should be taken to cotnact all areas of the mouth and as far back in oropharynx as possible. The patient should be encouraged during the implantation period to walk around the room and spend as much time as possible out of bed. This is particularly important in elderly patients, who otherwise tend to develop pulmonary complications. Adequate sedation and analgesics are recommended during this period so that the patient is as comfortable as possible. To minimize the radiation reaction of the surrounding normal tissues, it is advisable whenever possible to use an intraoral lead shield during the implantation. Such a shield can be easily constructed to fit the individual patient and provide adequate protection of the surrounding tissues because of the relatively low energy of the iridium sources used for temporary interstitial implantation. Thus, a thickness of 2.5 mm of lead will decrease the dose to the normal tissues by almost 50%.

4.3 After Brachytherapy

The local reactions with iridium-192 brachytherapy appear by the end of the 7th and 8th days and reach their peak 10–15 days after the removal of the interstitial implant. Complete healing ordinarily occurs 4–5 weeks later, and the complications are usually acceptable. If soft tissue or bone necrosis develops, they should be treated conservatively. The frequent use of oral irrigations should be continued. Zinc peroxide packs (or similar preparations) over a period of several hours help to prevent the growth of anaerobic bacteria; they also deodorize and may stimulate wound healing. Other conservative measures include gentle debridement of necrotic material and removal of food debris and loose bone fragments. Surgical resection is used as a last resort if more conservative procedures fail.

5 Results of Treatment

Extensive experience with interstitial brachytherapy alone or in combination with external radiation has been accumulated in several European and American centers.

Table 1. Survival in early selected head and neck cancer treated by brachytherapy

Site	Reference	*n*	Stage	Survival
Lip	PIGNEUX et al. (1979)	80	T1–3NO	99% (5-year actuarial)
Tongue, anterior	HAIE et al. (1986)	119	T1–3N0	50% (5-year actuarial)
Floor of mouth	HAIE et al. (1986)	51	T1–3N0	72% (5-year actuarial)
Tonsil	MAZERON et al. (1986)	23	T1–2N0	80% (5-year actuarial)
	PUTHAWALA et al. (1985)	15	T1–2N0	90% (5-year actuarial)
	PERNOT and BAILLET (1978)	17	T1–2N0	53% (3-year crude)
Tongue, base	MAZERON et al. (1988)	49	T1–2	52% (5-year actuarial)
	HOUSSET et al. (1986)	23	T1–2N0–1	52% (5-year actuarial)
	BAILETT (1980)	24	T1–2N0–1	54% (5-year actuarial)
Buccal mucosa	PERNOT and GERBAULET (1978)	211	T1–2N0	64% (3-year actuarial)
Soft palate	PERNOT and BAILLET (1978)	18	T1–2N0	67% (3-year actuarial)
Pharyngeal wall	SON and KACINSKI (1987)	9	T1–2N1	78% (2-year crude)

Table 2. Local control in selected early head and neck tumors treated by brachytherapy

Site	Reference	*n*	Local controls (%)
Lip	MAZERON and RICHAUD (1984)	1870	95
	PIGNEUX et al. (1979)	80	89
Tongue, anterior	HAIE et al. (1986)	119	92
	MAZERON et al. (1982)	103	83
Floor of mouth	BAILLET et al. (1982)	179	86
Tonsil	MAZERON et al. (1986)	23	100
	PUTHAWALA et al. (1985)	18	94
	GOFFINET (1984)	15	100
Tongue, base	PUTHAWALA et al. (1988)	10	100
	BAILLET (1980)	24	71
Uvula	CHUNG et al. (1979)	21	95
	SEALY et al. (1984)	18	89
	MAZERON et al. (1987a)	43	88
Pharyngeal wall	SON and KACINSKI (1987)	9	78
Buccal mucosa	PERNOT and BAILLET (1978)	226	65

Table 1 summarizes the results of treatment in early selected head and neck tumors. Survival is usually calculated by the actuarial method and is specified at periods varying from 2 to 5 years, as indicated in parentheses. In early lesions of the oral cavity (defined as tumors $\leqslant$4 cm) the reported survival from the Institut Gustave Roussy (HAIE et al. 1986) ranges from 50% to 70% at 5 years. In early oropharngeal lesions the survival at 5 years varies widely between 52% and 90% depending on the tumor site (PIGNEUX et al. 1979; HAIE et al. 1986; MAZERON et al. 1986, 1987a, 1988; PUTHAWALA et al. 1985; PERNOT and BAILLET 1987; PERNOT and GERBAUELT 1987; HOUSSET et al. 1986; BAILLET 1980).

Table 2 lists the local control in early selected head and neck tumors according to the anatomical site. Local control in oral cancer ranges between 83% and 95% and in oropharyngeal lesions between 65% and 100% (PIGNEUX et al. 1979; HAIE et al. 1986; MAZERON et al. 1982, 1986, 1987a; PUTHAWALA et al. 1985, 1988; PERNOT and BAILLET 1987; SON and KACINSKI 1987; MAZERON and RICHAUD 1984; BAILLET et al. 1982; BAILLET 1980; CHUNG et al. 1979; SEALY et al. 1984; GOFFINET 1984).

Table 3. Results of brachytherapy in advanced and recurrent head and neck cancers

Site	Reference	*n*	Local controls (%)	Survivals (%)
Advanced				
Tongue, base	VIKRAM et al. (1985)	10	100	80 (median 2 years)
Tonsil	LEBORGNE et al. (1986)	30	66	40 (at 3 years)
Miscellaneous	MAZERON et al. (1987b)	70	69	14 (5 years)
	GOFFINET et al. (1985)	14	79	43 (median 16 months)
	PUTHAWALA et al. (1985)	65	69	60 (median 40 months)
	GOFFINET (1984)	55	60	60 (Mean 33 months)
Recurrent				
Tongue, base	VIKRAM et al. (1985)	10	70	0 (at 2 years)
Miscellaneous	VIKRAM et al. (1983)	125	71	9 (at 2 years)
	GOFFINET et al. (1985)	34	59	29 (median 11 months)

Table 3 lists the results of brachytherapy in advanced and/or recurrent head and neck tumors, treated either by brachytherapy alone or in combination with external radiation. Evaluation of local control of tumor rather than survival in general permits a better appraisal of brachytherapy in the advanced lesions. The local control in previously untreated patients with locally advanced head and neck tumors ranges from 60% to 80% and in tumors recurrent after surgery and/or radiation therapy, from 59% to 71%. The survival is 40%–60% and 0%–15%, respectively. The overall survival at 2 years is in the range of 30%–40%, with some patients surviving longer (PUTHAWALA et al. 1985; GOFFINET 1984; GOFFINET et al. 1985; LEBORGNE et al. 1986; MAZERON et al. 1987b; VIKRAM et al. 1983, 1985).

Table 4 summarizes the acute and late complications observed in the various reports discussed previously. These consist of acute superficial mucosal ulcerations, hemorrhage, infection, late soft tissue necrosis, osteonecrosis and neuropathy, orocutaneous fistula, and carotid rupture. Dental decay, trismus, and xerostomia have not been reported by any of the authors (MAZERON et al. 1986, 1987; PUTHAWALA et al. 1985, 1988; SON and KACINSKI 1987, VIKRAM et al. 1985; GOFFINET et al. 1986; GOFFINET 1984; PARYANI et al. 1985).

Table 4. Acute and late morbidity in head and neck tumors treated by brachytherapy

Complication	Frequency (%)	
	Early tumors[a]	Advanced Tumors[b]
Acute		
Hemorrhage	0–1	1–15
Ulceration	14–35	17
Infection	0–3	0
Chronic		
Soft tissue necrosis	0–12	10–27
Osteonecrosis	0–3	4
Neuropathy	0–14	0
Orocutaneous fistula	0	1–2
Carotid rupture	0	2
Dental decay	0	0

[a] MAZERON et al. (1986); PUTHAWALA et al. (1985, 1988); SON and KACINSKI (1987); GOFFINET (1984)
[b] MAZERON et al. (1987); VIKRAM et al. (1985); GOFFINET et al. (1985); PARYANI et al. (1985).

6 Discussion

The value of interstitial brachytherapy in head and neck cancer remains undetermined because there are no adequately controlled studies to evaluate its merits. Our analysis suggests that control rates ranging from 70% to 100% can be achieved in early head and neck lesions with minimal morbidity. It should be kept in mind, however, that these higher control rates may reflect a better selection of patients included in the various reported series. Lower control rates are observed in early lesions in the oropharynx, especially in tumors of the base of the tongue, because they usually include poorly differentiated tumors and are frequently infiltrating.

The value of brachytherapy in advanced head and neck lesions is more controversial. Brachytherapy is used as a boost following external irradiation in an attempt to increase local control without prohibitively increasing morbidity. Indeed, our analysis shows that the 60%–100% local control achieved by the combination of brachytherapy and external radiation therapy is comparable, if not better, with the 57%–79% obtained by hyperfractionated external radiation (Parsons et al. 1987; Marcial et al. 1987; Wang 1988) and with the 26%–73% obtained by conventional external irradiation (Lo et al. 1987; Wang et al. 1985).

Acute complications observed with interstitial brachytherapy appear during or immediately after completion of the treatment and are self-healing. Late complications appear within the first 2 years after treatment. The majority of the late complications heal with conservative management, although some may require surgical intervention. Our review demonstrated that chronic complications with interstitial brachytherapy arise with a similar or at times lower frequency than complications observed with high-dose, external megavoltage radiation: e.g., soft tissue necrosis 10%–27% with interstitial brachytherapy versus 10%–22% with external irradiation (Larson 1983; Gardner et al. 1987); osteonecrosis 4% and 19%–40%, respectively (Lo et al. 1987; Larson 1983); xerostomia 0% and 56%, respectively (Lo et al. 1987); and trismus 0 and 24%, respectively (Lo et al. 1987).

References

Anderson LL (1986) A "natural volume-dose histogram for brachytherapy. Med Phys 13: 898–903

Baillet F (1980) Tumeurs de la base de langue. C R Soc Fr Radiol 1: 140–141

Baillet P, et al. (1982) Résultats d'une étude multicentre concernant 966 épitheliomas due plancer de la bouche. J Eur Radiother 3: 147–152

Chung CK, et al (1979) Squamous cell carcinoma of the soft palate and uvula. Int J Radiat Oncol Biol Phys 5: 845–850

Del Regato JA (1986) History and heritage; Antoine Lacassagne. Int J Radiat Oncol Biol Phys 12: 2165–2173

Del Regato JA (1987) History and heritage; Henri Coutard. Int J Radiat Oncol Biol Phys 13: 433–443

Dutreix A, et al. (1982) Dosimétrie en curiethérapie. Masson, Paris, pp 109–138

Gardner KE, et al. (1987) Time-dose relationships for local tumor control and complications following irradiation of squamous cell carcinoma of the base of tonque. Int J Radiat Oncol Biol Phys 13: 507–510

Glasser O, et al. (1961) Physical foundations of radiology, 3rd edn. Harper and Row, New York

Goffinet DR (1984) Iridium 192 removable oropharyngeal interstitial implants of the tonsillo-palatine and base tongue regions. In: Hilaris BS, et al. (eds) Brachytherapy oncology update. Memorial Sloan-Kettering Cancer Center, New York, pp 69–80

Goffinet DR, et al. (1985) 125 I vicryl suture implants as a surgical adjuvant in cancer of the head and neck. Int J Radiat Oncol Biol Phys 11: 399–402

Haie C, et al. (1986) Results of brachytherapy in the management of oral cavity cancer (Abstr). Endocuriether Hypertehermia Oncol 2: 60

Henschke UK, et al. (1963) After-loading in interstitial and intracavitary radiation therapy. AJR 90: 386–395

Hilaris BS, et al. (1987a) New approaches to brachytherapy. In: DeVita VA, et al. (eds) Important advances in oncology 1987, vol 12/2. Lippincott, Philadelphia, pp 237–261

Hilaris BS, et al. (1987b) Brachytherapy treatment planning. Front Radiat Ther Oncol 21: 94–106

Hilaris BS, et al. (1988a) Interstitial Brachytherapy Planning and Evaluation In: Hilaris BS, et al. (eds) An atlas of brachytherapy. Macmillan, New York, pp 70–95

Hilaris BS, et al. (1988b) Brachytherapy Techniques In: Hilaris BS, et al. (eds) An atlas of brachytherapy. Macmillan, New York, pp 46–69

Housset M, et al. (1986) A retrospective study of three treatment techniques for T1–T2 base of tongue lesions: surgery plus postoperative radiation, external radiation plus interstitial implantation and external radiation alone. Int J Radiat Oncol Biol Phys 13: 511–516

Larson DL (1983) Major complications of radiotherapy in cancer of the oral cavity and oropharynx. A 10 year retrospective study. Am J Surg Oncol 146: 531–536

Leborgne JH, et al. (1986) The place of brachytherapy in the treatment of carcinoma of the tonsil with lingual extension. Int J Radiat Oncol Biol Phys 12: 1787–1792

Lo K, et al. (1987) Results of irradiation in the squamous cell carcinomas of the anterior faucial pillar retromolar trigone. Int J Radiat Oncol Biol Phys 13: 969–974

Marcial VA, et al. (1987) Hyperfractioned photon radiation therapy in the treatment of advanced squamous cell carcinoma of the oral cavity, pharynx, larynx, and sinuses, using radiation therapy as the only planned modality (preliminary report by the Radiation Therapy Oncology Group RTOG). Int J Radiat Oncol Biol Phys 13: 41–47

Mazeron JJ, Richaud P (1984) Compte rendu de la XVIII réunion du Groupe European de Curiethérapie. Session consacrée aux cancers de la lévère. J Eur Radiother 5: 50–56

Mazeron JJ, et al. (1982) Analysis of local failures after treatment with curietherapy using iridium 192 for squamous cell carcinoma of the mobile tongue. J Eur Radiother 3: 131–138

Mazeron JJ, et al. (1986) Interstitial radiation therapy for squamous cell carcinoma of the tonsillar region. The Creteil experience (1971–1981). Int J Radiat Oncol Biol Phys 12: 895–900

Mazeron JJ, et al. (1987a) Definitive radiation treatment for early stage carcinoma of the soft palate and uvula: the indications for iridium 192 implantation. Int J Radiat Oncol Biol Phys 13: 1829–1837

Mazeron JJ, et al. (1987b) Salvage irradiation of oropharyngeal cancers using iridium 192 wire implants: 5-year results of 70 cases. Int J Radiat Oncol Biol Phys 13: 957–962

Mazeron JJM, et al. (1988) Limited external irradiation and iridium-192 implantation in the treatment of squamous cell carcinoma of the oropharynx (Abstr). Hospital Henri Mondor, Creteil, France. 2nd International Conference on Head and Neck Cancer, July 31–Aug 5, Boston

Meredith WJ (ed) 1967 Radiation dosage, the Manchester system, 2nd ed. Livingstone, Edinburgh

Parsons JT, et al. (1988) Hyperfractionation for head and neck cancer. Int J Radiat Oncol Biol Phys 14: 649–658

Paryani SB, et al. (1985) Suture implants in the management of advanced tumors in the neck attached to the carotid artery. J Clin Oncol 3: 809–812

Paterson R, Parker HM (1934) A dosage system for gamma ray therapy. Br J Radiol 7: 592–612

Pernot M, Baillet F (1987) Hard and Soft Palate In: Pierquin B, et al. (eds) Modern brachytherapy. Masson, Paris, 137

Pernot M, Gerbaulet A (1987) Buccal mucosa: In: Pierquin B, et al. (eds) Modern brachytherapy. Masson, Paris, pp 101–106

Pigneux J, et al. (1979) The place of interstitial therapy using iridium-192 in the management of carcinoma of the lip. Cancer 43: 1073–1077

Puthawala AA, et al. (1985) Limited external irradiation and interstitial iridium-192 implant in the treatment of squamous cell carcinoma of the tonsillar region. Int J Radiat Oncol Biol Phys 11: 1595–1602

Puthawala AA, et al. (1988) Limited external beam and interstitial iridium 192 irradiation in the treatment of carcinoma of the base of the tongue: a ten year experience. Int J Radiat Oncol Biol Phys 14: 839–848

Sealy R, et al. (1984) The treatment of cancer of the uvula and soft palate with interstitial radioactive wire implants. Int J Radiat Oncol Biol Phys 10: 1951–1955

Son YH, Kacinski BM (1987) Therapeutic concepts of brachytherapy/megavoltage in sequence for pharyngeal wall cancers. Cancer 59: 1268–1273

Vikram B, et al. (1983) Permanent iodine-125 implants in head and neck cancer. Cancer 51: 1310–1314

Vikram, B, et al. (1985) A non-looping afterloading technique for base of tongue implants: results in the first 20 patients. Int J Radiat Oncol Biol Phys 11: 1853–1855

Wang CC (1988) Local control of oropharyngeal carcinoma after two accelerated hyperfractionation radiation therapy schemes. Int J Radiat Oncol Biol Phys 14: 1143–1146

Wang CC, et al. (1985) Twice-a-day radiation therapy for cancer of the head and neck. Cancer 55: 2100–2104

4.4 Manual Afterloading in Brachytherapy (Endocurietherapy with ^{192}Ir Wires and Needles)

PETER C. VERAGUTH

CONTENTS

1 Introduction 153
2 Nylon or Teflon Tubes 153
2.1 Loading of Radioactive ^{192}Ir Wires 153
3 Direct Introduction of Charged Teflon Tubes 154
3.1 Reverdin's Needle 155
3.2 Big Operation Needle with a Thread 155
3.3 Insertion of ^{192}Ir Wires in Sewn Silk Thread 155
4 Stainless Guide Gutters for Single or Double Wires 155
5 Subcutaneous Injection Needles as Support for the Radioactive Sources 156
6 Radioprotective Precautions When Loading the ^{192}Ir Wires Manually 156
7 Advantages and Disadvantages of Manual Loading 157
References 157

1 Introduction

Although an automatic device for the afterloading implant of radioactive sources in brachytherapy (interstitial curietherapy) seems to be ideal, there are often situations in which manual loading is practised, e.g. when using curved guide needles or tubes (base of the tongue) or practising an implants while the tumour is surgically exposed (= peroperative implant – bladder carcinoma, unresectable salvary gland's tumour etc.). In my opinion, manual afterloading seems justified as well in smaller radiotherapy centers which do not have an automatic afterloading apparatus at their disposal. Thus, I shall describe the different ways in which radioactive sources may be introduced manually into the tissues.

2 Nylon or Teflon Tubes

After the introduction of straight or curved steal needles into the tumour bed, plastic tubes of 1.2-mm diameter are pulled through by means of rigid or flexible mandrels. On both sides, the ends of the tube are fixed by a lead stopper over a small plastic ball. This is the usual arrangement for an implant of the breast, lymph node or soft-tissue sarcoma. The Teflon tube may be introduced into the tissue with one end remaining open. At the point of insertion, the tube is then fixed by means of a leucoplast or a plastic plate. Such an arrangement is suitable for implants of the perineum or of the parametrium (Fig. 1). In both situations, the entire length of the tube and of its inner part lying in the tissue must be covered by a mandrel.

2.1 Loading of Radioactive ^{192}Ir Wires

Usually, the radioactive wire (0.15 mm thick and from 3 to over 12 cm long) is embedded in a second thin Teflon tube (outer diameter 0.55 mm). The length of this inner tube is chosen to extend sufficiently far out both ends (10–15 cm); these contain a blue mandrel and will be slightly fused at two points in order to maintain the radioactive wire in the correct position. The length of the ^{192}Ir wire is chosen in accordance with the arrangement of the tubes, e.g. 4–6 cm for a breast implant.

These inner Teflon tubes are prepared in advance and thereafter threaded in the tube already introduced into the tissue by the operator standing behind a 3-cm lead shield using long forceps. The loaded inner tube is then pushed forward with little movements (Fig. 2a). As soon as one end appears on the opposite side, the inner tube is quickly pulled through, until the end of the ^{192}Ir wire has joined the outer end of the bigger guide tube (Fig. 2b). If it is planned for the wire to be located in the middle of the two points of incision, the inner protruding tube is cut at a distance C, calculated by the following formula:

$$C = \frac{\mathrm{A} - \mathrm{L} - \mathrm{li} + \mathrm{re}}{2}$$

PETER C. VERGUTH, Prof. Dr med., Universitätsklinik für Strahlentherapie, Inselspital, 3010 Bern, Switzerland

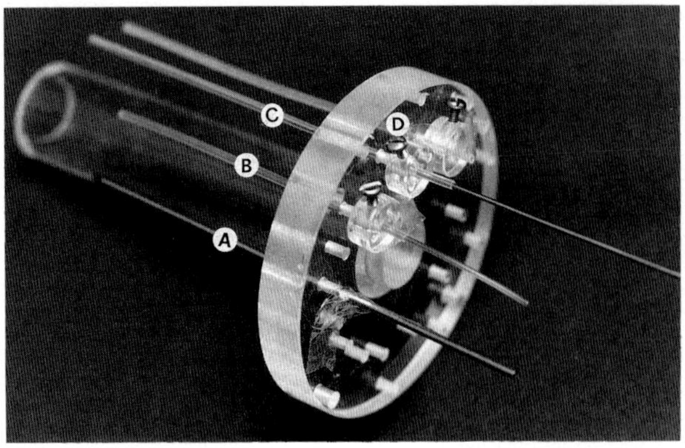

Fig. 1. Plastic template for an anal carcinoma implant. *A*, Insertion of a metallic needle; *B*, plastic tube with a plastic mandrel inserted; *C*, similar tube with a metallic mandrel; *D*, fixed empty plastic tube. Note the small plastic disc with the screw to be fixed on the template

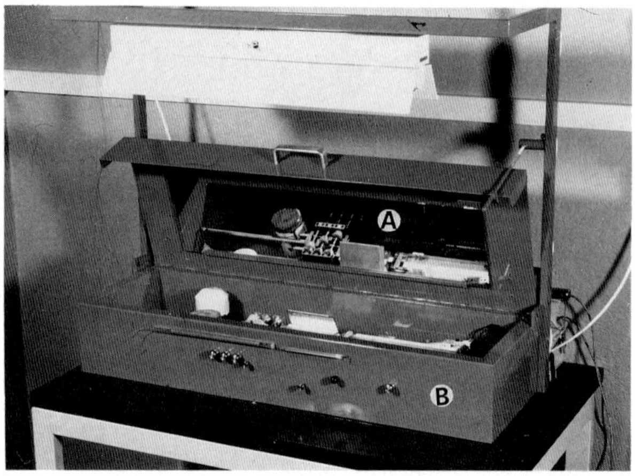

Fig. 3. Automatic encapsulator for ^{192}Ir wires. The operator observes the manipulation in a mirror (*A*) while being protected by a leaden wall (*B*)

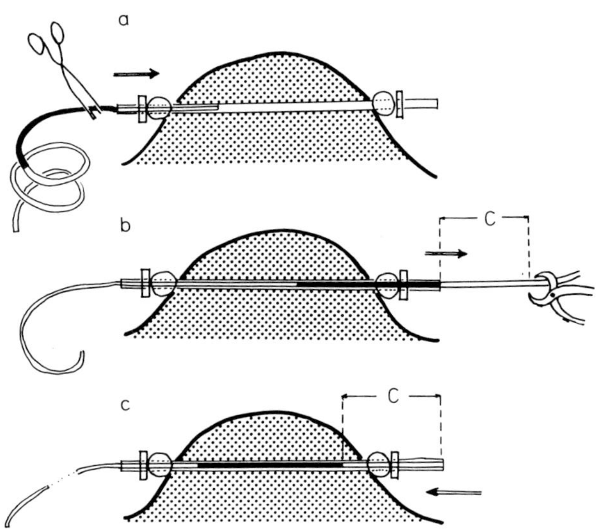

Fig. 2a–c. Shcematic representation of loading ^{192}Ir wire in Teflon tube. For explantation, see text

A = total length of the tube (1.2-mm diameter)
L = length of the ^{192}Ir wire
li = protruding end on the left side of the Teflon tube
re = protruding end on the right side of the Teflon tube

This distance determined in advance, *C*, is measured with a caliper square, and the inner tube together with the mandrel is cut by a sharp pinch. It is withdrawn until the two ends are at the same level (Fig. 2c). In this way, the position of the wire is correct, and the protruding end is melted with a small flame or by a heating forceps. At the opposite end, the inner tube is then cut at a distance of 1–2 cm from the introduced thicker tube. The whole procedure should not last more than 1–3 min.

The introduction of the ^{192}Ir wire into the inner tube is greatly facilitated by a so-called wire encapsulator. Such a machine is available commercially or may be constructed by the center's own mechanics (Fig. 3). With this machine the preparation of a wire with the appropriate length and position is possible without any exposure of the technician to radiation and is finished within 10–15 min.

If guiding Teflon tubes with an open end in the tissue are used, e.g in the perineum, the inner end of the charged tube is prepared in advance to the right length and then welded over the iridium wire. This charged tube is pushed forward to the inner end of the guide tube until a certain resistance is felt by the operator. There is a similar situation is outer tubes which are more sharply bent or angulated, e.g in the base of the tongue or in the mesopharynx wall. In this situation, the fine inner tube cannot be pushed through the entire length of the guide tube because of the severe curvature. The guide tube, especially when severely bent, may crack when the an esthetist extubates the patient.

3 Direct Introduction of Charged Teflon Tubes

Instead of introducing a guide tube into the tissue first, one may use the charged thin tubes with a diameter of 0.85 mm directly. To do this, some technical aids are needed which are described below.

3.1 Reverdin's Needle

For implants of superficial tumors (lymph nodes, angle of the eye) it is often helpful to use sharply bent needles. In these instances, a Reverdin's needle-holder is very convenient and much better than a curved steel needle. When the point of the needle comes out of the skin, the loaded small Teflon tube is inserted in the eyelet hole and withdrawn through the needle-trace, and the ^{192}Ir wire of 2–4 cm length is inserted at the end of the small Teflon tube and fused there; it remains in a lead container near the operating field until the very moment the Reverdin's needle is withdrawn (Fig. 4).

3.2 Big Operation Needle with a Thread

The thin Teflon tube may be fixed on an ordinary bent operating needle of 6–8 cm length if instead of the eyelet it is prepared by a tiny thread with an outer diameter of 6 mm, similar to the usual Redon needle. The threaded end is inserted into the tube (with a similar diameter) which is pulled through the implanted tissue. The further preparation of the 30–40 cm long fine Teflon tube is very similar to the technique described for the Reverdin's needle above.

Raynal (1977) has proposed a procedure in which, instead of the needle, a nylon thread is introduced in the inner part of the tube, for at least 10 cm. This thread is then pinched with forceps within the guide tube, and the nylon thread bound up with an ordinary surgical needle is drawn through the tissues and brings the loaded end of the long Teflon tube to the right place.

3.3 Insertion of ^{192}Ir Wires in Sewn Silk Thread

Black sewn silk threads of standard execution and of caliber 0–4 may be widened in their inner part (with a small mandrel) so that they may be charged with ^{192}Ir wires 3–6 cm long. These preparations must be done by a skilful person behind a leaden glass. When the radioactive wire has been introduced, the thread is immersed in a solution of 30% perchlorvenyl. Thus, the ^{192}Ir wire is kept fixed in the thread. In this way, a thread of at least 50 cm is stored in a lead container, similar to the one described above for a thin plastic tube and sterilized and prepared for the operating theater (Sahatchiev and Mouchmov 1971).

This manner of applying iridium sources seems suitable for implants in regions with difficult access, e.g. in the fornix vaginae.

4 Stainless Guide Gutters for Single or Double Wires

This method, originally developed by Henschke (1955) and later by Pierquin et al. (1964) in the

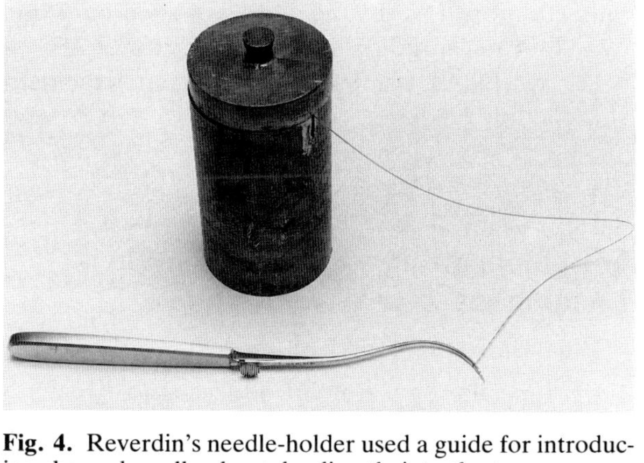

Fig. 4. Reverdin's needle-holder used a guide for introducing charged small nylon tube directly into the tumour area. Prior to insert the charged part of the tube, the ^{192}Ir wire is kept in the lead container near the operation field.

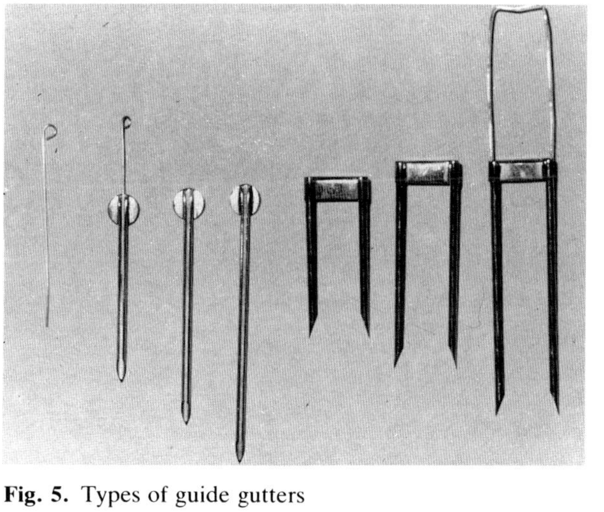

Fig. 5. Types of guide gutters

1950s and 1960s, consists of introducing an inactive, hollow steel needle into the tumour. After verifying the correct position of the needles (fluoroscopic control), the active ^{192}Ir wires of 0.5-mm thickness are introduced into the gutter, and the guide gutter is withdrawn (Fig. 5).

The gutters commercially available are prepared for a single wire or for wires in the form of a hairpin 3–6 cm long. The distance between the two branches of the hairpin is standardized and brought to 12 mm. The ^{192}Ir wires are delivered long before they are introduced for this safety reasons, is performed while the wire is submerged in a sterile cup of water. In this way the part which is cut will not be lost.

This method of implantation is very useful for the mobile part of the tongue. The guide gutters, usually straight, may be slightly curved, e.g. for use in the anus or cheek in order to follow better the anatomical form.

The manual loading of these guide gutters is done during the operation itself under sterile conditions and lasts per wire not more than 1–3 min. Before the wires are introduced, two threads per wire are prepared around the gutters; by this means the radioactive wire is tightly fixed immediately after the withdrawal of the gutters. Beside this, a thick silk suture is fixed on the iridium wire (before its introduction) in order to facilitate its withdrawal at the end of the treatment. When the wire is introduced or withdrawn the operator stands behind a leaden wall which is 3 cm thick and slightly bent towards the patient (Fig. 6).

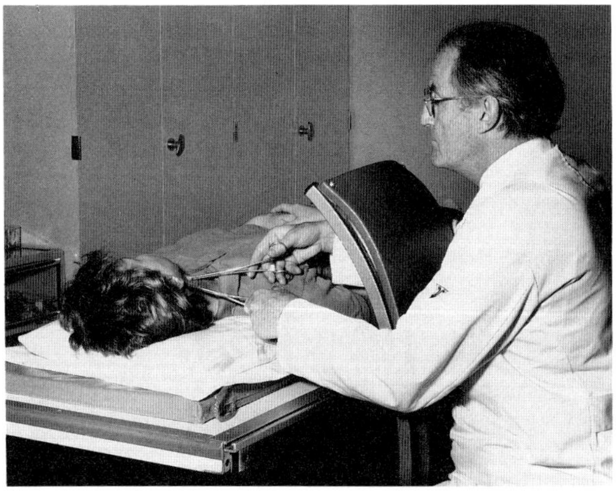

Fig. 6. Leaden wall used during insertion of radioactive wires to protect the operator

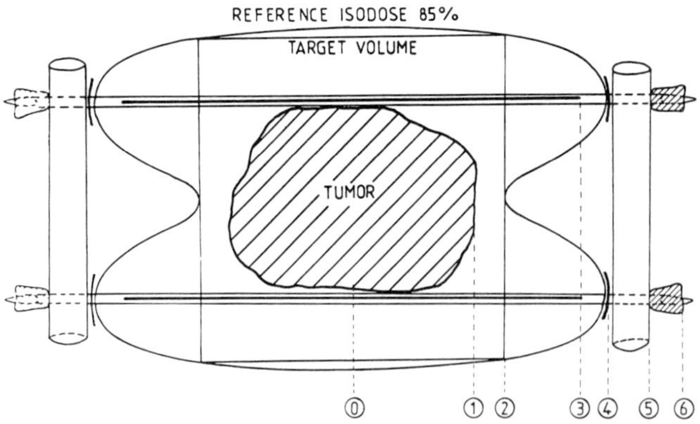

Fig. 7. Schematic drawing of an implant with hypodermic needles. *0*, Apparent tumour; *1*, limit of apparent tumour; *2*, limit of target volume; *3*, ends of radioactive wires; *4*, skin exit of the hypodermic needles; *5*, plastic tube spacer; *6*, lead immobilization button crimped on end of each needle. (From Pierquin et al. 1987)

5 Subcutaneous Injection Needles as Support for the Radioactive Sources

Pierquin has mentioned another possibility of implanting long subcutaneous injection needles instead of Teflon tubes. They are arranged in a similar manner within the tumour but are fixed on both sides across the plastic template with holes or using thick plastic tubes (of 3–4 cm diameter) perforated across at predetermined distances (Fig. 7). Into these needles the bare iridium wires are introduced. The two ends are closed by a lead stopper. This method is recommended especially for skin tumours and in the penis.

Instead of the injection needles, one can use ordinary steel needles. It must be considered that the radiation dose of the ^{192}Ir wire is diminished by 3% (measured transmission factor 9.256 ÷ 9.455 = 0.969) compared with an arrangement with Teflon tubes.

6 Radioprotective Precautions When Loading the ^{192}Ir Wires Manually

Each time the operator introduces the ^{192}Ir wires manually, he or she is exposed to a radiation dose, but this is very small if the appropriate precautions are taken. The following should be observed:

1. Mobile leaden walls, 3 cm thick and 40 cm wide, are put between the patient and the operator.
2. All manipulations of the radioactive sources are conducted with long (25–30 cm) forceps. Their end is covered with a small plastic cup in order not to damage the tiny Teflon tube or the wires themselves.
3. The threading the introduction of the radioactive sources must be done under good conditions, in bright light. This maneuver is done frequently in the patients's room. It is necessary to take more than 2 min for one wire. For a more extensive implant with 8–10 wires, e.g. in the case of the base of the tongue, the filling up of the tubes rarely lasts more than 15 min.

Repeated measurements of the received doses on the hands and body of the operator yielded values between 10 and 25 mrem (100–250 μSv).

These radiation doses seem to be acceptable for a radiotherapist who accomplishes an implant no more frequently than 2–4 times a week. The risk may be more considerable for an operator who carries out 6–10 similar treatments per day. He or she then receives approximately a dose of 250 mrem = 2.5 mSv per day or 5 rem (0.05 Sv) a month or 60 rem (0.6 Sv) over a whole year.

The average dose received by a technician during the preparation of the ^{192}Ir charged tubes is practically zero if an encapsulator, as described above, is available, or it reaches approximately the same values as given in Table 1, when under exceptional circumstances the fine Teflon tubes must be charged by hand behind a leaden glass. The introduction of the wire into a sewn silk thread (see Sect. 3.3) exposes the hands to a higher dose.

Table 1. Measured radiation dose to operator during endocurietherapy

	Average dose/ min	Average dose/ application
Hands/fingers	50–120 μSv (5–12 mrem)	100–250 μSv (10–25 mrem)
Body (behind leaden protection)	<10 μSv (<1 mrem)	<20 μSv (<2 mrem)

7 Advantages and Disadvantages of Manual Loading

The main disadvantage consists of the radiation exposure to the therapist. This is even smaller when the wires are withdrawn and put back into the container at the end of the implantion and amounts to approximately half or one-third of the doses above mentioned if this manoeuvre is carried out rapidly.

The main advantage is the simplicity of application under direct visual control of the correct position of the radioactive sources. In the majority of situations, it is also easy to adjust the wires after radiologic verification if they are not placed in an ideal position.

Especially for smaller centres where curietherapy is carried out only once to three times a week, it is a further advantage, if the costs of an automatic loading machine and its maintenance can be saved. There are also situations (blunt ending tubes or tubes arranged as sharply bent loops) when the automatic afterloading machine encounters some difficulties. For these reasons, I believe that there are several situations in which manual loading cannot be avoided.

References

Henschke UK (1955) Artifical radioisotopes in nylon ribbons for implantation in neoplasms. Proc Internat Conf Genf, Vol X, 48

Pierquin B, Chassagne D, Perez R (1964) Précis de curiethérapie. Masson, Paris

Pierquin B, Wilson JF, Chassagne D (1987) Modern brachytherapy. Masson, Paris

Raynal M (1977) Endocuriethérapie des cancers cutanés par mise en place directe des gaines plastiques porte-fils d'iridium 192. J Radiol Electrol 58: 713–714

Sahatchiev A, Mouchmov A (1971) L'emploi des fils de soie radioactifs dans la technique de brachyradiothérapie par tubes plastiques en boucles. Ann Radiol (Paris) 14: 491–495

4.5 Temporary and Permanent Brachytherapy in Advanced Head and Neck Cancer – The Erlangen Experience

RAINER FIETKAU, MANFRED WEIDENBECHER, WOLFGANG SPITZER, and ROLF SAUER

CONTENTS

1 Introduction 159
2 Patients 159
3 Therapeutic Concept 160
4 Results 162
4.1 Permanent Implantation of Iodine 125 162
4.2 Temporary Iridium 192 Implantation 163
4.3 Temporary Iridium 192 Implantation in the Treatment of Recurrent Diseaase 163
5 Discussion 164
5.1 Permanent Implantation of Iodine 125 164
5.2 Temporary Iridium 192 Implantation 164
5.3 Soft Tissue Necrosis 165
6 Conclusion 166
References 166

1 Introduction

The local control of advanced tumours of the head and neck continues to be unsatisfactory. Despite all the advances made in the fields of surgery, radiation treatment and chemotherapy, and combinations of these classic therapeutic modalities, a regional recurrence rate of up to 60% must still be expected (VIKRAM et al. 1984; VAN DEN BOGAERT et al. 1985; HORIOT et al. 1988; MARCIAL et al. 1988).

With the aid of interstitial brachytherapy, it is now possible to apply within a short space of time a high local dose to a well-defined area while sparing neightbouring healthy tissue and thus to approach the first therapeutic aim of local tumour control. Under this premise in the period between 1981 and 1985, we carried out initial permanent implantation of iodine 125 in combination with external beam radiotherapy. Since 1986 we have been employing temporary implantation using iridium 192 within the framework of a multimodal therapy concept comprising surgery, chemotherapy and radiation therapy.

For the present study, we have evaluated these two groups of patients, giving particular consideration to local tumour control after a minimum follow-up period of 19 months.

RAINER FIETKAU, Dr. med., ROLF SAUER, Prof. Dr. med., Strahlentherapeutische Klinik der Universität Erlangen-Nürnberg, Universitätsstr. 27, 8520 Erlangen, Germany

WOLFGANG SPITZER, Priv.-Doz. Dr. med. dent., Klinik und Poliklinik für Kieferchirurgie, Glückstr. 11, 8520 Erlangen, Germany

MANFRED WEIDENBECHER, Prof. Dr. med., Klinik und Poliklinik für Hals-Nasen-Ohrenkranke, Waldstr. 1, 8520 Erlangen, Germany

2 Patients

Between 1.1.1981 and 1.3.1988, a total of 109 patients with tumours of the head and neck region were given interstitial brachytherapy. Twenty-two patients received primary treatment consisting of permanent implantation of iodine 125, 87 patients (62 with primary and 25 with recurrent tumours) were treated with a temporary iridium 192 implant (tumour localisations, see Table 1). For the most part, the tumours were locally advanced lesions (Table 2, staging in accordance with AMERICAN JOINT COMMITTEE ON CANCER 1983). After treatment, the patients were followed-up on a regular basis. The evaluation date was 1.10.1987 for the iodine 125 treated patients and 1.11.1989 for the

Table 1. Localisation of the primary tumours or recurrent tumours treated with permanent iodine 125 or temporary iridium 192 implants

	Iodine 125	Iridium 192	
		Primaries	Recurrences
Oral cavity	10	32	8
Oropgarynx	12	26	8
Larynx	0	2	0
Nasopharynx	0	1	0
Lip	0	0	2
Lymph node metastases	0	1	7
Total	22	62	25

Table 2. Stage distribution (AMERICAN JOINT COMMITTEE ON CANCER 1983) of the tumours in patients treated with interstitial brachytherapy employing iodine 125 or iridium 192 in the period between 1981 and 1987

	Iodine 125	Iridium 192	
		Primaries	Recurrences
Stage I	0	3	2
Stage II	0	11	3
Stage III	14	21	5
Stage IV	8	27	9
Unknown	0	0	6
Total	22	62	25

iridium 192 treated patients (median follow-up period 20 and 27 months, respectively).

3 Therapeutic Concept

Since interstitial therapy was considered to be a boost applied to the region of the primary tumour, external beam radiotherapy was also employed. Using conventional techniques, a mid-plane dose to the primary tumour of 40– (usually) 50 Gy was applied using cobalt 60 or 6 MV photons. Depending upon the patient's lymph node status, the cervical and supraclavicular lymphatic area was irradiated with 50–60 Gy. Wherever possible, a neck dissection was carried out (functional, radical or limited radical) after radiation therapy.

The *iodine 125 implantation* was carried out 10–14 days after external beam irradiation with doses of up to 40 Gy. For implantation, the primary tumour region was exposed via a lateral pharyngotomy or by splitting the tongue. The iodine 125 seeds were applied in the afterloading mode with the aid of a Mick applicator (HILARIS 1975). The

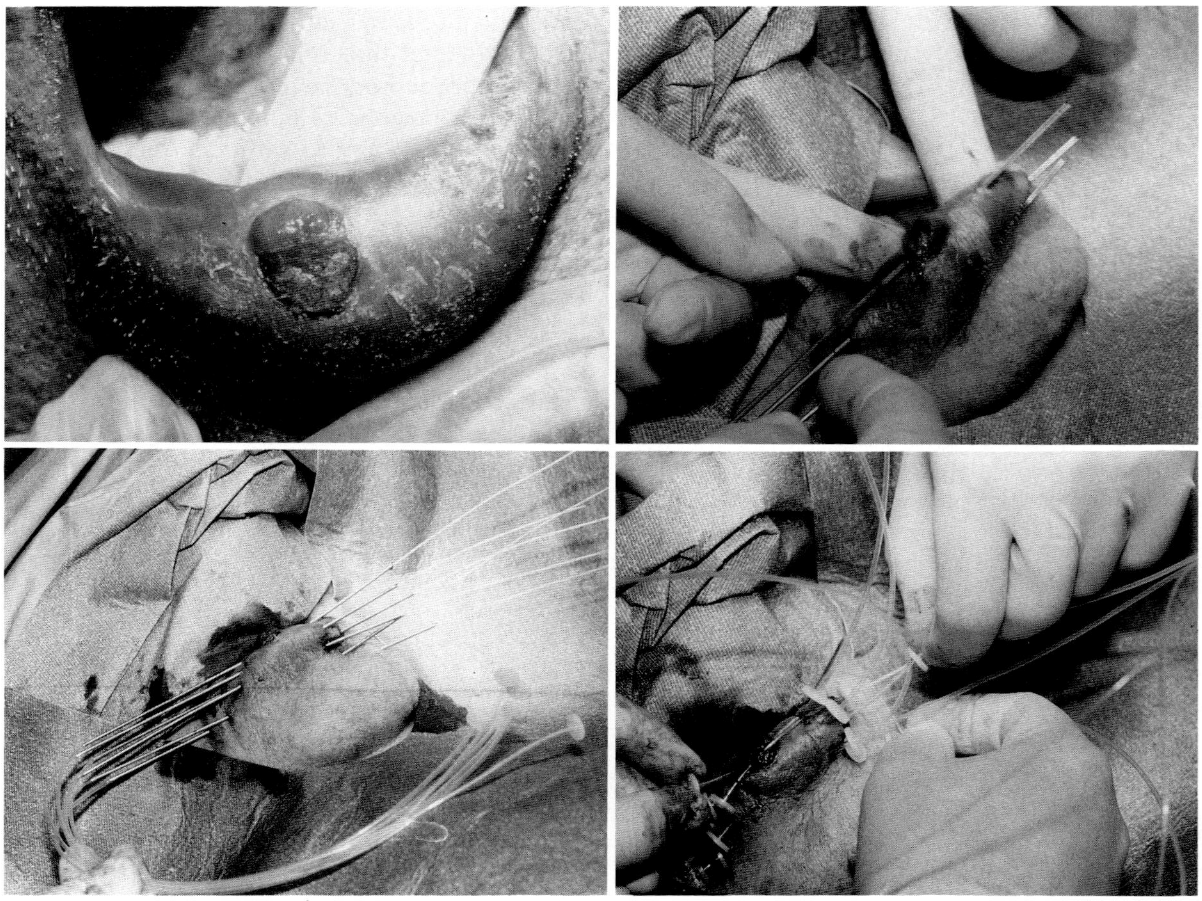

Fig. 1. a Single-needle technique employed in the treatment of a recurrent carcinoma of the lower lip: tumour status prior to radiotherapy. **b** Empty hollow steel needles are inserted into the appropriate tissue volume such that the two ends of the needles project from the skin. **c** Plastic afterloading tubes with plastic buttons on the sealed ends are threaded through the needles. **d** The needles are withdrawn, leaving in place the plastic tubes that are secured with the aid of small plastic buttons

a b c d

median tumour surface dose applied was 175 Gy (160–300 Gy) over 1 year. Clinical dosimetry was employed intra-operatively and prospectively with the aid of a so-called Anderson normogram from which, using the estimated volume of the tumour and the mean seed activity, the recommended overall activity, the required number of seeds, their spacing along the needles and the distances between the needles can be obtained. Post-operatively, orthogonal radiographs were obtained with reproducible geometry and transverse sections in a computerized tomography (CT) unit for retrospective computer-assisted dosimetry and to verify the correct localisation of the seeds.

The *temporary implantation of iridium 192* was usually done prior to external beam irradiation, and a dose of 20–30 Gy was given. For the interstitial implantation, we employ the plastic tube technique described by SYED and FEDER (1977). The single-needle technique is used for implantation in tumours of the region of the floor of the mouth, tongue, base of the tongue, wall of the pharynx, lips and in the case of cervical lymph node metastasis. In principle, under general anesthesia, 5–20 hollow steel needles are introduced into the clinically defined target volume. Plastic afterloading tubes provided with plastic buttons on the closed ends are then introduced through the needles. These latter are withdrawn, leaving the plastic tubes in place. To anchor the plastic tubes, plastic buttons are fitted over the tubes and pushed up to the skin (Figs. 1, 2). Afternatively, in the paired needle technique a pair of hollow needles is employed. Here, a distinction is made between the loop technique in the region of the mandible, and occasionally of the base of the tongue, and the arch technique used to treat lesions located in the region of the soft and hard palates. In principle, a couple of hollow needles are inserted into the target volume from each side of the neck. The two ends of each plastic tube are threaded through the intra-oral openings of the needles and passed to the outside. The needles are then removed, and

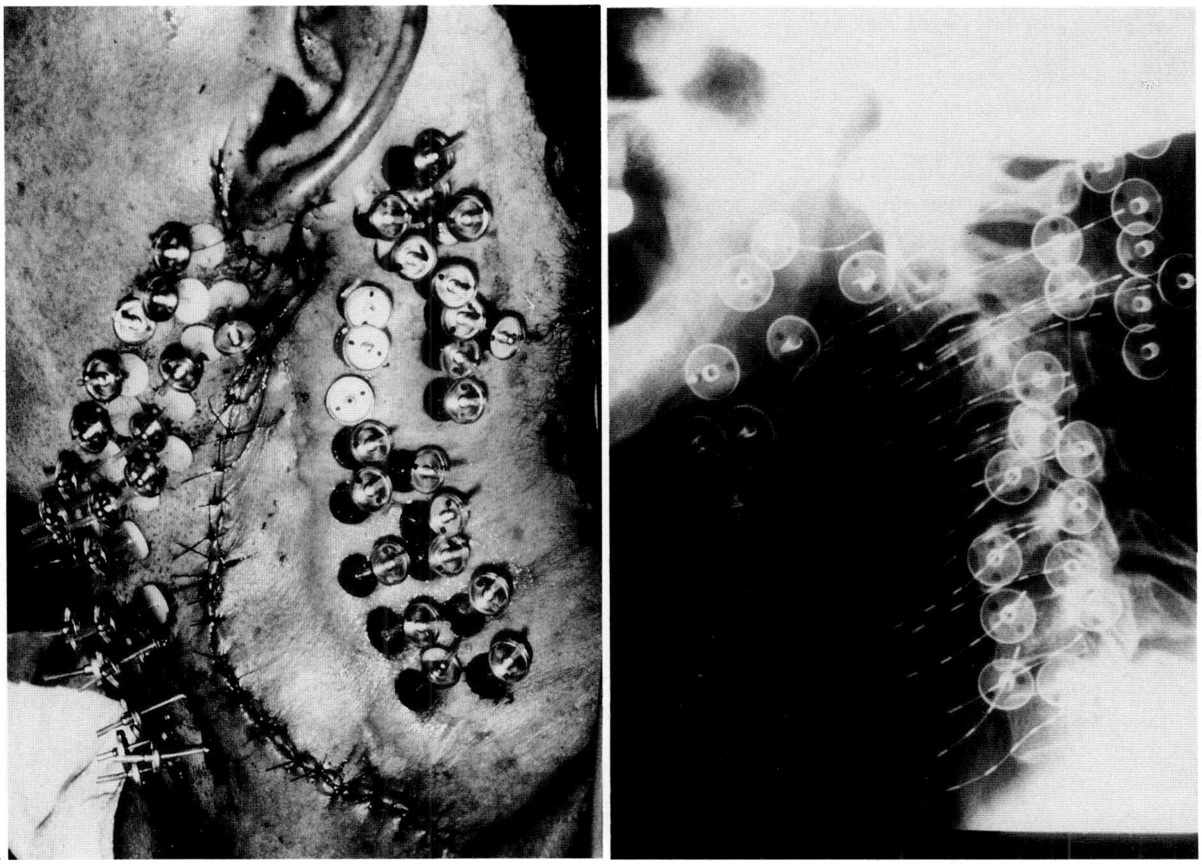

a b

Fig. 2. a Implant in position in macroscopic residual tumour of the left side of the neck. A dissection of the neck was first performed to remove lymph node metastases; the remaining tumourous tissue was implanted during surgery. **b** Localisation radiograph of copper dummy seed ribbons in LSD projection

Fig. 3. Lateral localisation radiograph of the implant in a T3 carcinoma of the tonsils that has spread to the soft palate. The pillars of the fauces have been implanted using the arch technique (*arrow*)

the plastic tube, which remains in situ, forms an inverted U in the lower jaw, or over the soft and hard palates (Fig. 3). For fixation purposes, plastic buttons are fitted over the ends of the plastic tube and pushed up to the skin.

When this procedure has been completed, inactive copper dummy ribbons are introduced into the applicators for computer-assisted dosimetry, and orthogonal radiographs are obtained in the therapy simulator under conditions of defined geometry. Loading with radioactive iridium seed ribbons is effected either manually or automatically. When the applicators are explanted, it must be ensured that the facilities and staff are available for immediate anesthesia and surgery, so that any resulting haemorrhage can be treated and aspiration prevented without delay.

Chemotherapy was given in 25 patients. In 6 patients simultaneous radio-chemotherapy comprising two courses of 5-fluorouracil and cisplatin was carried out. Nineteen patients were sequentially treated with at least two courses of cisplatin and bleomycin prior to the external radiotherapy.

Table 3. Results of analysis of recurrent disease in 22 patients with tumours of the head and neck treated with permanent implants of iodine 125 (n = 22)

Alive, disease-free	4 (19%)
Dead	18 (81%)
local recurrences	11 (50%)
regional recurrences	0
distant metastases	0
intercurrent disease	7 (31%)

4 Results

4.1 Permanent Implantation of Iodine 125

Four of the 22 patients treated (20%) are still alive and tumour-free (Table 3). Seven patients died tumour-free of the intercurrent diseases that are commonly observed in patients with tumours of the head and neck (cirrhosis of the liver and sequelae of alcohol abuse). Eleven patients (50%) developed local recurrent disease, to which they subsequently succumbed. Neither regional nor distant metastases were observed. The 2-year survival and the actuarial 5-year survival rates (according to Cuttler and Ederer 1958) were 35% and 20%, respectively (Fig. 4).

Acute side reactions were restricted to a painful fibrinous inflammation which reached a peak some 6–8 weeks after completion of the treatment. Two patients with osteoradionecrosis (13%) and one with glossoplegia were observed.

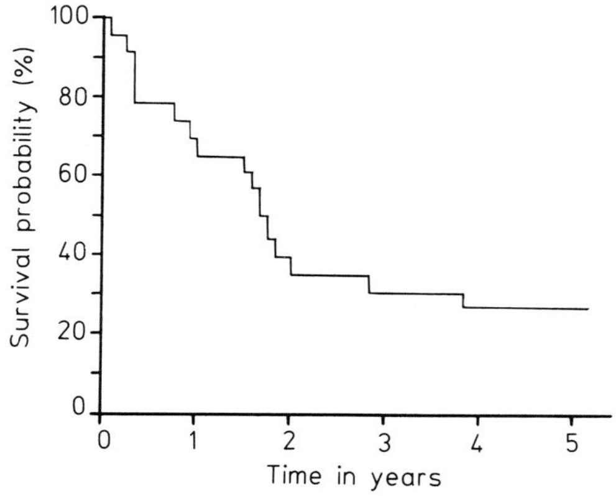

Fig. 4. Survival curve calculated in accordance with Cuttler and Ederer (1958) for 22 patients treated with a combination of permanent iodine 125 implantation and external irradiation

4.2 Temporary Iridium 192 Implantation

Interstitial iridium 192 therapy was employed within the framwork of an aggressive, multimodal treatment concept as a boost to the region of the primary tumour. Depending upon the individual situation, chemotherapy, surgery of the primary tumour or neck dissection was first applied, followed by interstitial iridium 192 implantation and then external radiation treatment with a dose of up to 50–60 Gy. With this concept, complete remission was achieved in 60 of 62 patients (96%) and partial remission in 2. With a minimum follow-up period of 19 months (median 27 months) after completion of the interstitial therapy, 32 of 62 patients (52%) are still alive and tumour-free (Table 4). Twenty-seven (43%) have died, 19 of their tumour, 1 each of the treatment-related causes, intercurrent disease and a second malignancy. In 5 patients who were lost to follow-up, the cause of death is unknown. The 2-year survival rate as computed in accordance with CUTTLER and EDERER (1958) (Fig. 5) is 61% for the overall group. Marked differences are observed between stages I/II (92.5%) and III and IV (59% and 49%, respectively).

Further analysis of recurrent disease (Table 4) revealed that in the 2 patients who experienced partial remission, tumour progression again occurred. Following a complete initial remission, local recurrent disease was observed in 7 of 60 patients. In 5, the recurrent tumour was located at the margin of the implant and in 2, within the region of the erstwhile implant. The local control rate after 2 years was thus 85%. Regional recurrent tumours occurred in 5, and distant metastases in a further 9 patients.

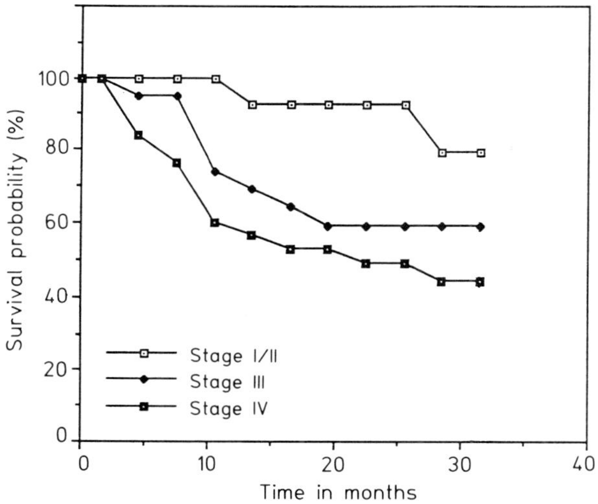

Fig. 5. Survival curves calculated in accordance with CUTTLER and EDERER (1958) for patients treated with temporary iridium 192 implantation within the framework of primary tumour therapy. Fourteen patients had stage I/II tumours, 21 stage III, and 27 stage IV lesions

Acute side effects were limited to a fibrinous mucositis that persisted for a number of weeks but which responded well to antiphlogistic treatment. In 31 patients (50%), post-treatment soft tissue necrosis occurred; in the meantime, this has cleared up in response to aggressive conservative treatment. In 6 patients (10%), osteoradionecrosis made a partial resection of the mandible necessary. In 1 female patient in complete remission, an acute haemorrhage from the carotid artery occurred 14 months after completion of treatment.

Table 4. Results and analysis of recurrent disease in 77 patients treated with temporary iridium-192 implants within the framework of primary treatment (n = 62) or recurrent lesion treatment (n = 25)

	Primaries	Recurrences
Alive, tumour-free	32 (52%)	4 (16%)
Alive, with tumor	3 (5%)	1 (3%)
local recurrences	3	–
distant metastases	–	1
Dead	27 (43%)	20 (81%)
local recurrences	6 (10%)	8 (32%)[a]
regional recurrences	5 (8%)[b]	5 (20%)[a]
distant metastases	9 (14%)[b]	1 (3%)
therapy-related	1 (2%)	0
secondary primary	1 (2%)	3 (12%)
intercurrent disease	1 (2%)	1 (3%)
unknown	5 (8%)	3 (12%)

[a] One patient with local and regional recurrence.
[b] One patient with regional recurrence and distant metastases.

4.3 Temporary Iridium 192 Implantation in the Treatment of Recurrent Disease

In 20 of 25 patients (80%) with local or regional recurrent lesions or second tumours, complete remission was again achieved with radiation therapy (interstitial radiotherapy alone with 30–40 Gy; interstitial radiotherapy in combination with external irradiation applying 30–50 Gy). With a minium follow-up period of 19 months after termination of treatment, 4 patients are still alive and free of tumour (Table 4). Of the 6 patients in whom no prior radiation treatment had been carried out, 2 (33%) are still alive and tumor-free. In contrast, only 2 of the 19 patients (8%) who received repeat radiotherapy have survived over the longer term, with a follow-up period of 26 months each. Twenty patients (80%) have since died, including 13 due to their tumour disease, 3 from a second or third tumour and 1 of intercurrent disease. In 3 patients, the exact cause of death is unknown. A further analysis of recurrent disease revealed 8 local recurrent or residual tumours following treatment, 5 regional recurrent lesions and distant metastases in a single case. One male patient is still alive 39 months after treatment of recurrent disease with pulmonary and mediastinal metastases. Thus, the local control rate (over the long term) is 14 of 25 (56%) after 2 years. The 2-year survival calculated in accordance with Cuttler and Ederer (1958) (Fig. 6) is 26%, the median survival rate is 9 months (1.3–35+ months). By way of late sequelae, soft tissue necrosis developed in 5 patients (20%) and osteoradionecrosis in 2 (8%).

5 Discussion

The aim of interstitial radiation treatment is to reduce the local recurrence rate in the case of advanced tumours. For this reason, in the following discussion the major criterion for success is the local control rate achieved.

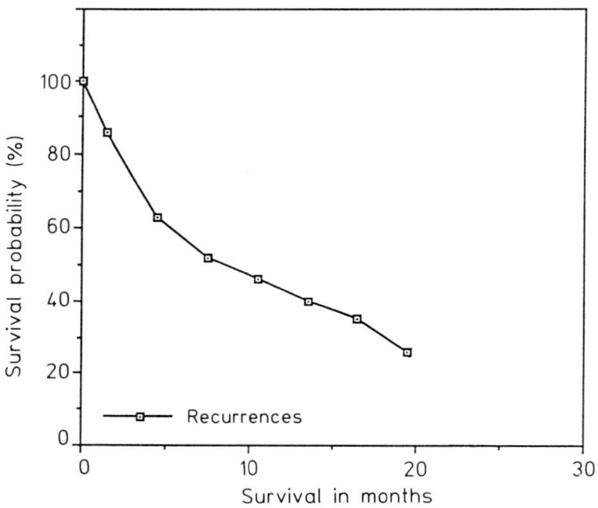

Fig. 6. Survival curve calculated in accordance with Cuttled and Ederer (1958) for 25 patients who received temporary iridium 192 implantation within the framework of recurrent disease treatment

5.1 Permanent Implantation of Iodine 125

In comparison with the literature, we observed a high local recurrence rate of 50%. Kim and Hilaris (1975) palliatively treated lymph node metastases in the neck by implanting iodine 125 (minimal peripheral tumour surface dose 160 Gy). By this means they were able to achieve local control in 38 of 49 patients (78%), who survived for more than 4 months. Martinez et al. (1983) achieved local control in 4 of 18 patients with tumours of the head and neck by using a combination of external radiation therapy and interstitial irradiation with iodine 125. Goffinet et al. (1985b) reported a local control rate of 79% (11/14) after combined interstitial and external radiotherapy. In the case of treatment of recurrent disease done with a curative intent, a local control rate of 59% (20/34) was achieved. Son et al. (1989) implanted, intra-operatively, iodine 125 (dose 75–250 Gy) in 25 patients with advanced tumours of the head and neck (18 recurrences, 6 stage IV, 1 stage III). The local 3-year control rate was 78%, the 3-year survival rate 48%. Liberman et al. (1989) treated some 42 patients with recurrent tumours using iodine 125 implantation (mean dose 82.63 Gy) in addition to surgical resection. The determinate 5-year survival rate was 29%; 18 patients (43%) died of their tumour disease.

We consider the following factors to be the reasons for our poorer local control rates:

1. Due to the shielding of the implantation centre after applying (externally) 40–50 Gy, an underdosage at the margin of the implant can occur.
2. In our series, only inoperable patients were selected for implantation treatment. This means that patient selection differed among the various institutions, so that the data reported cannot be compared.
3. Furthermore, it has not yet been clarified whether in view of the low-dose rate of iodine 125 (5–10 cGy/h) an adequate dose per cell cycle in the rapidly proliferating head and neck tumours can actually be accomplished. There is thus a danger that destroyed cells might be replaced by repopulation through tumour cells (Hall 1985).

5.2 Temporary Iridium 192 Implantation

Since 1986, we have stopped implanting iodine 125 in the head and neck region. This means that the intra-operative radiation exposure of surgical staff is completely obviated. With the aid of suitable afterloading devices, exposure of the care-providing staff can also be reduced. Moreover, the flexible plastic applicators that are employed for temporary iridium-192 implantation are more easily implanted in the head and neck region, and pharyngotomy or tongue splitting can be avoided. Furthermore, the temporary implantation we practise makes it possible to carry out dosimetry before the applicators are loaded with radioactive material. In this way, monitoring or correction of the loading of the seed ribbon is possible.

The primary tumour was controlled in 85% of patients by the interstitial approach. Moreover, it must be taken into account that 49 of 62 patients (80%) had advanced tumours (stages III and IV).

While we employ surgery, interstitial radiotherapy and external radiation therapy plus concomitant chemotherapy as a multimodal therapeutic concept, the literature has so far reported only the results obtained with a combination of interstitial and external irrdiation. PUTHAWALA et al. (1981) reported local control rates of 69% and 46% in the case of T1–2 and T3–4 cancers of the tongue, respectively. Subsequently, the same group (PUTHAWALA et al. 1985a and 1988) achieved a local control rate of 83% (58/70 patients) in carcinoma of the base of the tongue and of 84% (57/80 patients) in carcinoma of the tonsils. In both reports, the percentage of patients with stage III/IV disease was 81%–83%. GOFFINET et al. (1985a) reported complete remission in 12 of 14 patients (84%), 11 of whom had stages III/IV disease. In a comparative group also containing 14 patients, who were submitted to surgery with subsequent external radiation treatment, only 9 patients remained free of tumour. VIKRAM et al. (1985) reported local tumour control in each of their 10 patients (1 × T1, 2 × T2, 7 × T3). For small tumours in the stages T1–2, HOUSSET et al. (1987) reporting on cancer of the base of the tongue, MAZERON et al. (1987b) reporting on cancer of the soft palate and uvula and ESCHE et al. (1988), who treated carcinoma of the uvula, achieved local control in 80%–90% of the patients using interstitial radiotherapy, either alone or in combination with external irradiation. WENDT et al. (1989) and CROOK et al. (1989), employing combined interstitial radiotherapy and external irradiation to treat carcinoma of the tongue, achieved control rates of 78%–86% in stage T1 tumours and 62%–89% in stage T2a–b lesions. In the case of tumours of the floor of the mouth, GRIMARD et al. (1989) reported local control in 71% (T2) to 96% (T1), using interstitial radiotherapy alone.

With the second (interstitial) radiotherapy of recurrent lesions we were, in the majority of patients, able to achieve only palliative results. Although 14 of 19 patients (73%) again achieved complete clinical remission, in the long-term only 2 patients survived. Although over the long-term a local control rate of 42% (8/19 patients) was reached, in the majority of patients this was of no importance for their survival since 7 died of regional recurrent disease, distant metastases, a second or third tumour or intercurrent disease.

Other authors have reported more favourable results. The only explantation we can offer for this is that they possibly treated patients with more favourable disease states. In Erlangen, patients with regional recurrent tumours are submitted initially to radical surgery, and only inoperable patients are sent for radiotherapy. MAZERON et al. (1987a) and LANGLOIS et al. (1988) reported local control rates for recurrent and second tumours that ranged from 69% to 72% after 2 and 59% to 69% after 5 years using interstitial therapy following prior external irradiation. The reported survival rates were 36%–48% after 2 and 14%–28% after 5 years. PUTHAWALA et al. (1985b) achieved a 75% local control rate for carcinomas of the tonsils treated with interstitial radiation therapy. Two years after treatment, 42% of their patients were still free of recurrent disease. FONTANESI et al. (1989) achieved local control in 21 of 23 patients (90%) re-irradiated for recurrent tumours or second primaries. Here, however, only 6 patients were still alive 12–23 months after implantation. Most of their patients died of distant metastases.

5.3 Soft Tissue Necrosis

In our series we observed a high rate of adverse reactions; 50% soft tissue necrosis and 10% osteoradionecrosis. These figures are higher than those reported by others and may be explained by the following points:

1. Owing to the fact that the tumours concerned were very advanced, we were obliged to implant the radioactive sources into large tissue volumes which, in turn, meant high total activities. In their analysis of radiation necrosis

observed following implantation of carcinoma of the tongue, CROOK et al. (1989) also found a correlation with tumour stage. The incidence of necrosis observed was 17% in stage T1, 29% in stage T2a and 49% in stage T2b tumours.
2. For reasons that have to do with transport and storage of the radioactive source, we initially selected a high specific activity of the individual seeds, namely 1–2 mCi. Later, however, we limited the specific activity to a maximum of 1.0 mCi.
3. In the initial phase, our radioactive sources were implanted too densely. We have since gone over to using a spacing, wherever possible, of 1–1.5 cm between the individual seed ribbons. CROOK et al. (1989) also reported an increase in necrosis when the spacing between the seed ribbons was too large. These authors, too, recommend a spacing of between 12 and 14 mm.
4. Taking into account the dose of 50 Gy delivered with external beam irradiation, the initially selected reference dose of 30–36 Gy was too high. The multimodal therapeutic concept (surgery, chemotherapy and irrdiation) leads to an increase in scarring and thus to a reduction in local perfusion. This, in turn, probably leads to a reduction in the regeneration capacity of the tissue as compared with untreated normal tissue. For this reason, we have now limited the interstitial boost dose to 20–25 Gy.
5. Biopsy-taking and other massively traumatic interventions (e.g. extraction of teeth) promote collapse of the mucous membranes, which are already burdened to the limit, with the result that soft tissue ulcers and necrosis in the sense of a combination injury may develop. Such interventions are thus permissible only if they are absolutely necessary.
6. The neglect of oral hygiene and dental care is equally as important for the development of soft tissue necrosis and osteoradionecrosis as the dose-time factors of the radiation treatment applied (OLCH et al. 1988). For this reason, in the vast majority of patients we now demand extraction of all the teeth in the lower jaw. Single teeth left in place for subsequent anchorage of a prosthesis are of little use to the patient if they promote soft tissue necrosis through chronic traumatisation of the tongue, mucous membranes and gums.
7. CROOK et al. (1989) emphasize the importance of the dose rate for the incidence of radionecrosis. With increasing dose rate, they observed an increase in the incidence of necrosis (<0.5 Gy/h, 9% necroses; >0.5 Gy/h, 34% necroses).

6 Conclusions

In the case of head and neck tumours, temporary implantation of iridium 192 is to be preferred over permanent implantation of iodine 125. In general, we employ this technique as a boost subsequent to external radiation treatment, using shielding of the region of the implant after delivery of an external dose of 50 Gy.

Interstitial boosting with iridium 192 is indicated in the following cases:

1. R1 or R2 resections (R_1 = Microscopic residual tumour; R_2 = Macroscopic residual tumour), or resection with only a small margin of clearance, of the primary tumour or lymph nodes in the neck (avoidence of mutilating resections).
2. In the case of definitive radiotherapy of head and neck cancers, for local boosting, irrespective of subsequent or simultaneous chemotherapy or hyperthermia.
3. This technique can be used alone, that is without external irradiation, only in the individual patient with small, highly differentiated tumours with a low risk of lymphatic spread.
4. Within the framework of palliative treatment to improve the tumour response and reduce external radiation dose after prior high-dose radiation treatment (reduction of morbidity).

References

American Joint Committee on Cancer (1983) Manual of staging of cancer, 2nd edn. Lippincott, Philadelphia, p 33
Crook J, Mazeron JJ, Marinello G, Walop W, Pierquin B (1989) Prognostic factors of local outcome for T1, T2 carcinomas of oral tongue treated by iridium-192 implantation – the Creteil experience. Int J Radiat Oncol Biol Phys (Suppl 1)17: 170
Cutter SJ, Ederer F (1958) Maximum utilization of the life table method in analyzing survival J Chronic Dis 8: 699
Esche BA, Haie CM, Gerbaulet AP, Eschwege F, Richard JM, Chassagne D (1988) Interstitial and external radiotherapy in carcinoma of the soft palate and uvula. Int J Radiat Oncol Biol Phys 15: 619–625
Fontanesi J, Hetzler D, Ross J (1989) Effect of dose rate on local control and complications in the reirradiation of head and neck tumors with interstitial iridium-192. Int J Radiat Oncol Biol Phys 17: 365–369

Goffinet DR, Martinez A, Fee WE (1985b) 125-I vicryl suture implants as a surgical adjuvant in cancer of the head and neck. Int J Radiat Oncol Biol Phys 11: 399–402

Goffinet DR, Fee WE Jr, Wells J, Austin-Seymour M, Clarke D, Mariscial JM, Goode RL (1985a) 192-Iridium pharngoepiglottic fold interstitial implants. The key to successful treatment of base tongue carcinoma by radiation treatment. Cancer 55: 941–948

Gimard L, Mazeron JJ, Martin M, Raynal M, Marinello G, Pierquin B (1989) Iridium-192 curietherapy alone for T1 and T2 carcinomas of the floor of the mouth. Int J Radiat Oncol Biol Phys (Suppl 1)17: 132

Hall EJ (1985) The biological basis of endocurietherapy. The Henschke Memorial Lecture 1984. Endocuriether Hyperthermia Oncol 1: 141–152

Hilaris BS (1975) Handbook of interstitial brachytherapy. Publishing Science Group Inc., Acticn, MA

Horiot JC, Le Fur R, Nguyen TN, Schraub S, Chenal C, de Pauw M, van Glabbeke M (1988) Two fractions per day versus single fraction per day in the radiotherapy of oropharynx carcinoma: results of an EORTC randomized trial (Alostr 438). 7th Annual Meeting of the European Society for Therapeutic Radiology and Oncology, The Hague, The Netherlands, 4–8 Sept.

Housset M, Baillet E, Dessard-Diana B, Martin D, Miglianico L (1987) A retrospective study of three treatment techniques for T1–T2 base of tongue lesions: surgery plus postoperative radiation. external radiation plus interstitial implantation and external radiation alone. Int J Radiat Oncol Biol Phys 13: 511–516

Kim JH, Hilaris BS (1975) Iodine-125 source in interstitial tumor therapy. Clinical and biological considerations. AJ 123: 163–169

Langlois D, Hoffstetter S, Malissard L, Pernot M, Taghian A (1988) Salvage irradiation of oropharynx and mobile tongue about 192-iridium brachytherapy in Centre Alexis Vautrin. Int J Radiat Oncol Biol Phys 14: 849–853

Liberman F, Park R, Lee DJ, Goldsmith M (1989) Resection plus iodine-125 seed implantation for recurrent head and neck carcinomas. Int J Radiat Oncol Phys (Suppl 1)17: 227

Marcial VA, Pajak TF, Kramer S, Davis LW, Stetz J, Laramore GE, Jacobs JR, Al-Sarraf M. Brady LW (1988) Radiation Therapy Oncology Group (RTOG) studies in head and neck cancer. Semin Onco 15: 39–60

Martinez A, Goffinet DR, Fee W, Goode R, Palos B, Cox R, Pooler D (1983) 125-Iodine suture implants as an adjuvant to surgery and external beam radiotherapy in the management of locally advanced head and neck cancer. Cancer 51: 973–979

Mazeron JJ, Langlois D, Glaybiger D, Huart H, Martin M, Raynal M, Calitchi E, Ganem G, Fraldi M, Feuilhade F, Brun B, Marin L, Le Bourgeois JP, Baillet F, Pierquin B (1987a) Salvage irradiation of oropharyngeal cancers using iridium 192 wire implants: 5-year results of 70 cases. Int J Radiat Oncol Biol Phys 13: 957–962

Mazeron JJ, Marinello G, Crook J, Marin L, Mahot P, Raynal M, Calitchi E, Peynegre R, Ganem G, Feraldi M, Huart J, Le Bourgeois JP, Pierquin B (1987b) Definitive radiation treatment for early stage carcinoma of the soft palate and uvula: the indications for iridium 192 implantation. Int J Radiat Oncol Biol Phys 13: 1829–1837

Olch AF, Beume F, Schwartz HC, Kayan AR (1988) Proposition that oral hygiene is as important as dose-time factors inthe prevention of osteoradionecrosis in the murclible. Endocunether Hyperther Oncol 4: 11–16

Puthawala AA, Syed AMN, Neblett D, McNamara C (1981) The role of afterloading iridium (Ir-192) implant in the management of carcinoma of the tongue. Int J Radiat Oncol Biol Phys 7: 407–412

Puthawala AA, Syed AMN, Eads DL, Neblett D, Gillin L, Gates TC (1985a) Limited external irradiation and interstitial 192-iridium implant in the treatment of squamous cell carcinoma of the tonsillar region. Int J Radiat Oncol Biol Phys 11: 1595–1602

Puthawala AA, Syed AMN, Gates TC (1985b) Iridium-192 implants in the treatment of tonsillar region malignancies. Arch Otolaryngal 111: 812–815

Puthawala AA, Syed AMN, Eads DL, Gillin L, Gates TC (1988) Limited external beam and interstitial 192-iridium irradiation in the treatment of carcinoma of the base of the tongue: a ten year experience. Int J Radiat Oncol Biol Phys 14: 839–848

Son YH, Ariyan S, Sasaki CL, Goodwin WJ, Kacinski BM, August D, Ponn RB (1989) Intraoperative iodine-125 brachytherapy in the management of advanced or recurrent head and neck and thoracic – abdominal tumors. Endocuriether Hyperthermia Oncol 5: 9–19

Syed AMN, Feder BH (1977) Technique of afterloading interstitial implants. Radiol Clin (Basal) 46: 458–475

Thiel HJ, Müller R, Weidenbecher M, Sauer R (1987) Interstitielle Brachycurietherapie von HNO – Tumoren. In: Sauer R, Schwab W (eds) Kombinationstherapie der Oropharynx – und Hypopharynxkarzinome. Urban and Schwarzenberg Mūnich pp 69–89

Van den Bogaert W, van der Schueren E, van Togelen C, Horiot JC, Chaplain G, Arangeli G, Gonzalez D, Svoboda V (1985) Late results of multiple fractions per day (MFD) with Misonidazole in advanced cancer of the head and neck. A pilot study of the EORTC radiotherapy group. Radiother Oncol 3: 139–144

Vikram B, Strong EW, Shah JP, et al. (1984). Failure at the primary site following multimodality treatment in advanced head and neck cancer. Head Neck Surg 6: 720–723

Vikram B, Strong E, Shah J, Spiro R, Gerold F, Sessions R, Hilaris B (1985) A non-loping afterloading technique for base of tongue implants: results in the first 20 patients. Int J Radiat Oncol Biol Phys 11: 1853–1855

Wendt CG, Peters LJ, Delclos L, Ang KK, Morrison WH, Maor MH, Byers RM, Oswald BS (1989) Primary radiotherapy in the treatment of stage I and II oral tongue cancers: importance of the proportion of therapy delivered with interstitial therapy. Int J Radiat Oncol Biol Phys (Suppl 1)17: 130

4.6 ^{198}Au Implantation of Carcinoma of the Mobile Tongue

JOACHIM SLANINA, KLAUS KUPHAL, and MICHAEL WANNENMACHER

CONTENTS

1 Introduction . . . 169
1.1 Anatomy . . . 169
2 Natural History . . . 170
2.1 Primary Tumor . . . 170
2.2 Regional Lymphatic Spread . . . 170
3 Epidemiology and Risk Factors . . . 170
4 Diagnostic Work-up and Staging System . . . 171
4.1 Methods of Diagnosis . . . 171
4.2 Staging System . . . 171
5 Pathological Classification . . . 171
6 Prognostic Factors . . . 171
7 General Management . . . 171
7.1 Primary Tumor . . . 171
7.2 Regional Lymph Nodes . . . 172
7.3 Radiotherapy . . . 172
8 Special Indications for Interstitial ^{198}Au Radiotherapy . . . 172
9 Technique of ^{198}Au Interstitial Radiotherapy . . . 172
9.1 Physical Properties of ^{198}Au Seeds . . . 172
9.2 Spatial Seed Arrangement and Dose Distribution . . . 172
9.3 Radiation Protection . . . 174
10 Results . . . 174
11 Complications . . . 175
References . . . 175

1 Introduction

Radioactive gold seeds for permanent interstitial radiation therapy were introduced in 1951 by COLMERY. His technical application was rapidly perfected by SINCLAIR in 1952. Since then, the method has been applied in the USA on various tumors, especially by HENSCHKE (1958; HENSCHKE et al. 1953), and in France by PIERQUIN (1964, PIERQUIN et al. 1962). Today, interstitial radiotherapy with direct implantation of ^{198}Au seeds is rarely used. It has been replaced either by interstitial afterloading therapy or interstitial direct implantation of ^{125}I seeds, because the hard γ-rays of ^{198}Au seeds give less protection to the treating personnel. Nevertheless, ^{198}Au interstitial implantation offers without doubt a method of optimal local cure by intratumoral irradiation. Furthermore, it has – compared with afterloading therapy – the advantage of a continuous effective dosage, which accumulates to 90% within 8 days. This means that a tumor cell with a life cycle of between 8 and 24 h will be irradiated at least once in its sensitive phase of development, resulting in a better selective influence of the irradiation on the tumor cells, while cells of the connective tissue are spared (SCHUMACHER 1974). The encyclopedical character of the present volume therefore justifies the dedication of one chapter to ^{198}Au-seed implantation, as in the future this method might have a renaissance.

JOACHIM SLANINA, PROF. DR., KLAUS KUPHAL, Dr. rer. nat., Albert-Ludwigs-Universität, Radiologische Klinik, Abteilung Strahlentherapie, Hugstetter Str. 55, 7800 Freiburg, Germany

MICHAEL WANNENMACHER, Prof Dr. Dr., Radiologische Universitätsklinik, Abteilung Klinische Radiologie (Schwerpunkt Strahlentherapie), Im Neuenheimer Feld 400, 6900 Heidelberg, Germany

1.1 Anatomy

The tongue is a muscular organ located in the oral cavity (anterior two-thirds) and in the oropharynx (posterior third, i.e., base of the tongue). Its borderline is the V-shaped sulcus terminalis of the circumvallate papillae. Arterially it is supplied by the two lingual arteries, which are branches of the external carotid arteries. Ligature of only one of them is allowed without running the risk of tongue necrosis. Figure 1 shows the 3-dimensional muscle system of the tongue, its innervation, and arterial supply.

Figure 2 describes the lymphatic capillary flow of the tongue (FEIND 1972). Numerous collecting vessels lead to the submental, submandibular, and jugular lymph nodes.

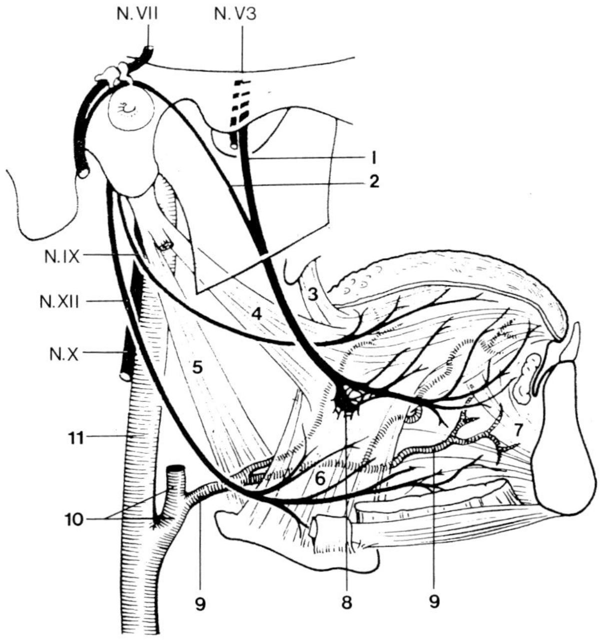

Fig. 1. Topography of the tongue, its innervation, and the lingual arteries. *1* N. lingualis, *2* Chorda tympani, *3* M. palatoglossus, *4* M. styloglossus, *5* M. stylopharyngeus, *6* M. hyoglossus, *7* M. genioglossus, *8* Ganglion submandibulare, *9* A. lingualis *10* A. carotis externa, *11* A. carotis interna, *N.V3* N. mandibularis, *N.VII* N. facialis, *N.IX* N. glossopharyngeus, *N.X* N. vagus, *N.XII* N. hyoglossus (TÖNDURY 1981)

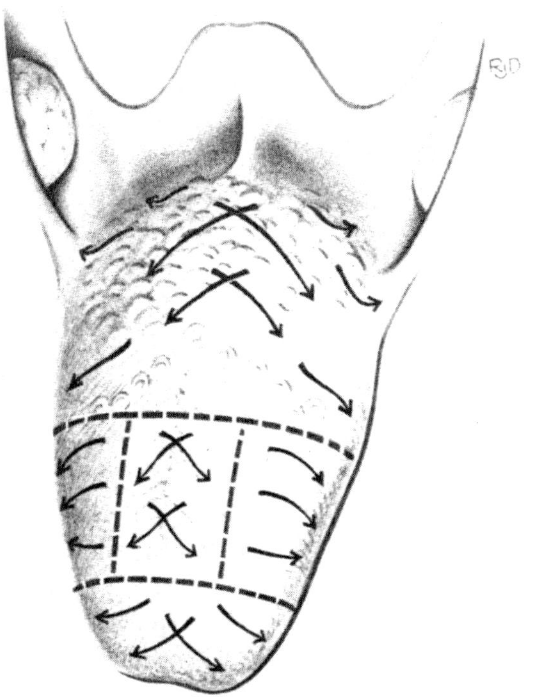

Fig. 2. Lymphatic capillary flow on the dorsum of the tongue (FEIND 1972)

2 Natural History

2.1 Primary Tumor

Most frequently, the primary tumor is located at the margin of the medium third of the tongue, followed by the base of the tongue and anterior third of the lower surface of the tip, and then spreading to the floor of the mouth. Symptoms are irritation and pain. Finally, extensive infiltration of the muscle may cause fixation which results in difficulty in speaking and eating.

2.2 Regional Lymphatic Spread

Approximately 35% of patients with tumors of the mobile tongue have lymph node metastases primarily, and another 30% with no lymph node affection at the time of diagnosis will later metastasize, the incidence being proportional to the primary tumor stage (LINDBERG 1972). In most patients the lymph node metastases will be found ipsilateral, in the area of the subdigastric and submandibular lymph nodes (Table 1 and Fig. 3). Some 5%–10% have primary metastases in the contralateral lymph nodes, and the percentage may increase in the case of surgical irritations or disturbances of the lymphatic flow or because of tumor growth (BOHNDORF 1987).

3 Epidemiology and Risk Factors

With reference to all human tumors the incidence of carcinoma of the tongue is reported between 1% and 5% in the FRG and 7%–8% in the USA

Table 1. Nodal distribution on admission (LINDBERG 1972; FEIND 1972)

Location of positive nodes	Ipsilateral		Contralateral	
	L[a]	F[b]	L[a]	F[b]
Upper jugular	53%	37%	5%	7%
Midjugular	9%	23%	2.5%	5%
Submandibular triangle	20%	14%	2.5%	1%
Lower jugular	2.5%	6%	0.6%	3%
Submental triangle	3%	0.5%	1%	0.5%
Posterior traingle	0.6%	2%	0%	0%

[a] Whole tongue: 62/125 patients with positive nodes = 49%; total number of positive nodes = 228
[b] Oral tongue: 105/302 patients with positive nodes = 35%, total number of positive nodes = 164

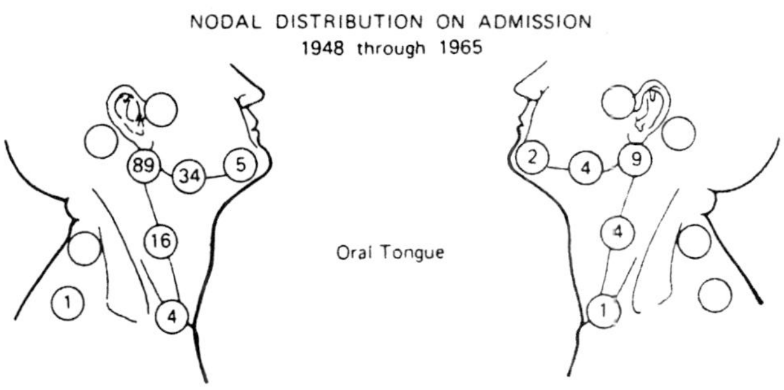

N_0	N_1	N_{2A}	N_{2B}	N_{3A}	N_{3B}	N_1-N_3 / Total
197	40	9	32	8	16	105 / 302 ≈ 35%

Fig. 3. Carcinoma of the anterior two-thirds of the tongue: incidence of metastatic lymph nodes (LINDBERG 1972)

(DREWS 1977). In 1951, the mortality rate in the USA was 0.8% (SCHRÖDER 1968). Among the head and neck tumors, carcinoma of the tongue ranges between 20% and 50%. Male patients predominate. The age group between 40 and 60 years is mainly affected; however, peak incidence is in the 6th decade. Predispositioning factors are mechanical and chemical irritations. Leucoplakias often represent the precancerous stage.

4 Diagnostic Work-up and Staging System

4.1 Methods of Diagnosis

Diagnosis must be confirmed histologically. The size of the primary tumor can be determined by inspection, palpation, and computerized tomography (CT). When planning ^{198}Au-seed implantation of tumors in the middle and/or the base of the tongue, palpation under anaesthesia is necessary. Only this allows an exact tumor localisation under the same conditions which the operator will find during implantation. The diagnosis of lymph node involvement is helped by sonography and CT in addition to palpation.

4.2 Staging System

Staging should follow the recommendations of the TNM classification of the UICC (1987) (Table 2). This system has widely been adopted by the AMERICAN JOINT COMMITTEE FOR CANCER STAGING AND END RESULTS REPORTING (1977).

Table. 2. TNM classification of the tongue (summary): anatomical sites and subsites with ICD-O-Key (UICC 1987)

Oral cavity
Tongue
(a) Dorsal surface and lateral borders anterior to vallate papillae (anterior two-thirds) (141.1, 2)
(b) Inferior surface (141.3)

Oropharynx
Anterior wall (glosso-epiglottic area)
(a) Tongue posterior to the vallate papillae (base of tongue or posterior third) (141.0)
$T_1 \leq 2$ cm
$T_2 > 2-4$ cm
$T_3 > 4$ cm
T_4 adjacent structures
N_1 Ipsilateral single ≤ 3 cm
N_2 Ipsilateral single between 3 and 6 cm
Ipsilateral multiple $\leq$ 6cm
Bilateral, contralateral $\leq$6 cm
N_3 >6 cm

5 Pathological Classification

Some 95% of the lesions are tumors located in the middle and anterior part of the tongue (ectodermal origin) and are classified as squamous cell carcinoms. Most tumors of the base of the tongue (entodermal origin) are either undifferentiated or transitional cell carcinomas (NOLTENIUS 1981).

6 Prognostic Factors

Prognosis deteriorates with growth of primary tumor, lymph node metastases, involvement of the base of the tongue, progressive Karnowsky index, and the refusal of the patient to cooperate. Male patients carry a poorer prognosis.

7 General Management

7.1 Primary Tumor

A treatment with curative intention normally consists in a combination of surgery and radiotherapy. Surgical intervention is limited by the size of the tumor and/or extent of lymph node metastases. Small primary tumors of the mobile tongue (T1–2) can be removed in toto with a safety margin. Surgical intervention in this case should be followed by postoperative local regional radiotherapy. Primary tumors with larger dimensions or located in the base of the tongue cannot totally be removed due to functional and technical reasons. They are treated with primary radiotherapy, if possible in combination with chemotherapy, e.g., cisplatin activated by the Radiation Therapy Oncology Group (RTOG) (AL-SARRAF et al. 1987; CRISSMAN et al. 1987).

7.2 Regional Lymph Nodes

A combination of radiotherapy of the ipsilateral as well as the contralateral cervical and supraclavicular lymph node areas must be recommended – with the exception of previous radical neck dissection – since deposits in these areas are very likely. In the case of primary involvement of the lymph nodes, a functional – not radical – neck dissection should be followed by radiotherapy. Radical neck dissection with ligation of the vena cava and skeletisation of the external carotid arteries may cause tissue hypoxia and bleeding, both unfavourable conditions for efficient postoperative radiotherapy.

7.3 Radiotherapy

When using percutaneous irradiation with photons and fast electrons, a tumor dose of 60 Gy should be administered. Occassionally an additional boost up to 70 Gy in a limited area can be given. As an alternative to surgical treatment of the primary tumor – with the aim of better local and functional conservation – interstitial radiotherapy methods have been developed. These methods are: interstitial afterloading procedures, ^{125}I-seed implantation, application of radium needles, iridium wire, and radon or ^{198}Au seeds. Their range of indications can be compared with that for surgery, being furthermore very appropriate for small volume boost irradiation of the tumor and/or tumor layer.

8 Special Indications for Interstitial ^{198}Au Radiotherapy

Favourable indications for the ^{198}Au implantation are primary tumors with exact knowledge of their dimensions located in the anterior two-thirds of the tongue and, under certain premises, also small tumors of the anterior part of the base of the tongue. ^{198}Au therapy can also be applied in exceptional cases of tumor recurrence.

9 Technique of ^{198}Au Interstitial Radiotherapy

9.1 Physical Properties of ^{198}Au Seeds

^{198}Au seeds for interstitial radiotherapy are small cylinders 2.5 mm in length and with a diameter of 0.8 mm. The radioactive gold is mantelled by a thin wall of inactive platinum in order to protect the environment from contamination. ^{198}Au radiation mainly emits 411 keV photons while the β-radiation is absorbed in the platinum wall. In the neighborhood of a seed, the attenuation of the radiation by the tissue is nearly compensated for by scattering radiation. Thus, the dose distribution around a source may be expressed simply by the inverse quadratic law. As a consequence, the dose decrease with distance from the source is not as rapid as with lower energy radiation, e.g., with ^{125}I, therefore the positioning of the ^{198}Au seeds need not be very accurate.

^{198}Au seeds have an initial activity of 370 MBq (10 mCi) at the time of implantation. As the half-life of 2.7 days is rather short, preparation and implantation dates have to be coordinated very well. Another consequence of the short half-life is that half the dose planned at any point in the tissue will be reached within 2.7 days and 97% of the dose within 2 weeks.

9.2 Spatial Seed Arrangement and Dose Distribution

The target volume is determined by measuring length, width, and thickness of the palpable tumor enlarged by 0.5–1 cm in each direction as a safety margin. In order to have a quick method of calculation of the total initial activity, i.e., the number N of seeds needed, the fomula $N = 0.2\ F + 5$ developed by BUSCH and WOLFART (1968) in Freiburg can be used, where F (in cm^2) refers to the surface of the target volume. The formula allows a good approximation for a minimum total dosage of 60 Gy at the surface of the target volume if the seeds are arranged nearly equidistantly. The proportionality between the activity needed and the surface of the implanted volume is well known (PARKER 1938).

The implantation of the seeds is carried out with an applicator described by HODT et al. (1952). In order to deposit several seeds per single insertion of the hollow needle, a spatial lattice is planned consisting of one or several cubes of about 1.5-cm edge length, depending on the size of the target volume. The seeds are implanted in the corners and centers of the cubes. The basic model of a cube with eight seeds in the corners and one seed in the center has proven to be optimal with regard to an acceptable dose distribution, as computer calculations have revealed (RENNER et al. 1979).

Figure 4 shows as an example of a 14-seed model with two additive cubes, each having a 1.5-cm edge length resulting in a spatial lattice of $3.0 \times 1.5\,\mathrm{cm}^3$ for a target volume block of $5.0 \times 2.5 \times 2.5\,\mathrm{cm}^3$. The initial total activity is 5200 MBq (140 mCi) with 14 seeds. When the tumor volume is irregular, this principal lattice arrangement can be supplemented by individual additional seeds, and thus cold spots are avoided.

The isodose distribution of the 14-seed model was calculated in the brachytherapy program of a treatment planning system (Philips Company, Eindhoven, Netherlands) as described by DUTOU et al. (1979) in two plane sections through the middle of the block and along the edges. The 60 Gy isodose covers the outer dimensions of the target volume in all areas (Fig. 5).

This dose distribution has to be checked by the radiotherapist with a dose-rate meter in order to correct underdosage in some points with additional seeds, if necessary.

In practice, an ideal seed arrangement as described above is neither possible nor necessary because of the deformability of the tongue. It has been shown that approximately equidistant distribution of the seeds in the target volume is sufficient.

SCHUMACHER (1974) proposed to arrange the ^{198}Au seeds only at the surface of the target volume because of enlarged necrosis in the center of the target volume when too many seeds were

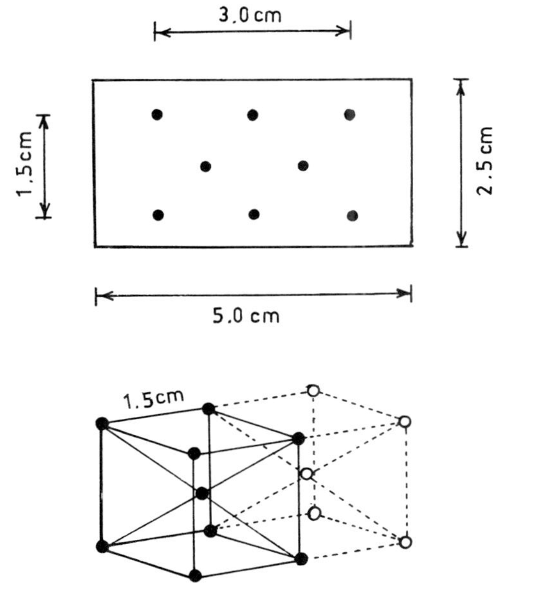

Fig. 4. Fourteen ^{198}Au-seed model: target volume two cubes. Minimal tumor dose 60 Gy (SLANINA et al. 1982)

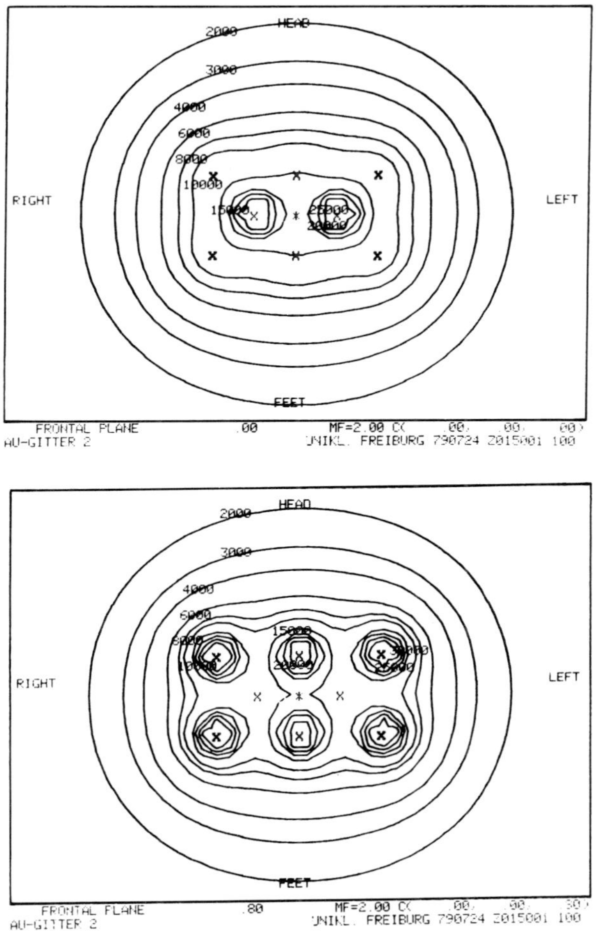

Fig. 5. Isodose distribution of the 14-^{198}Au-seed model in the mid-plane (*above*) and parallel plane through the edge points (*below*) (SLANINA et al. 1982)

concentrated there. However, this experience was made with larger target volumes in different kinds of organs; in the small target volume treated by the Freiburg method this complication was not observed.

9.3 Radiation Protection

Due to the high dose rate emitted by the patient, special measures for radiation protection are necessary. The applicator has to be filled in a radium packing facility accessible only by the hands. The implantation of the seeds can be performed in about 5 min. The treating personnel (radiotherapist, surgical assistant, anesthetist, physicist, and surgical nurse) are protected by lead shieldings (5 cm thick) placed at both sides of the patient's head.

Typical values of radiation exposure measurements performed with thermoluminescence dosimeters (TLD) are shown in Table 3 for an average application of 14 ^{198}Au seeds.

After implantation, the patient has to stay in a special room with sufficient radiation protection until the total activity has declined to 185 MBq (5 mCi). With a half-life of 2.7 days and 14 seeds, this takes 2 weeks. During this time, nursing is reduced as far as possible, and visiting is allowed only in special cases for a limited time. Nurses who have taken care of the patient are examined for exposure level during the 2 weeks after implantation. This was repeated with seven other patients over the course of 2 years. The measured dose values were averaged per nurse and per patient, resulting in a dose of 15 mR (Table 3).

Table 3. Radiation exposure during an interstitial implantation with ^{198}Au seeds (doses are given in mR)

	Head	Chest	Right finger	Left finger	Gonades	Foot
Radiotherapist	21	26	38	159	10	4
Medical physicist (prepares applicator)	17	11	60	nd	10	19
Surgical assistant			170	nd	nd	nd
Anesthesist	nd	17	nd	nd	nd	nd
Surgical nurse	nd	15	nd	nd	13	nd
Nurses	nd	15[a] (5–35)	nd	nd	nd	nd

[a] Average value per nurse and per patient measured during the care of 8 patients with a medium number of 14 (11–18) seeds who were cared for by 4–6 nurses (average 5) per patient for about 2 weeks.
nd, Measurements were not done.

10 Results

There are very few published reports on comparable series of individual therapy of carcinoma of the tongue. Therefore, these results can only be compared approximately. Table 4 shows 5-year survival rates of patients treated with ^{198}Au seeds or other interstitial therapies; Table 5 gives the local control rates.

Table 4. 5-year survival rates of carcinoma of the mobile tongue (staging T1–3N0M0) treated by interstitial radiotherapy

Reference		*n*	Therapy: Tongue	Therapy: Neck	T1	T2	T3
Gilbert et al.	(1975)	11	Rn ± E	E	73%		
Decroix and Ghossein	(1981)	287	Ra	S	59%	45%	
Slanina et al.	(1982)	30	Au ± E	O, E, S E + S	75%	44%	
Vermund et al.	(1982)	73	Ra ± S	O	60%	50%	
Horiuchi et al.	(1982)	33	Au, Rn, Ra (± E) Ra	O	76%		
Vermund et al. (1984)	(1984)	96	Ra + S (± E)	(E)	58%	39%	

Au, ^{198}Au seeds; Rn, radon seeds; Ra, radium needles; E, external megavolt irradiation; S, surgery; O, no treatment

Table 5. Local control rates of carcinoma of the mobile tongue (staging T1–T3) treated by interstitial radiotherapy

Reference	n	Therapy	Follow-up (months)	T_1	T_2	T_3
Pierquin et al. (1970)	29	Ir	36	97%		
Chu and Fletcher	79	Ra + E		100%	89%	64%
(1973)	26	E	28–268		56%	41%
	113	Ra		94%	84%	70%
Slanina et al. (1982)	25	Au ± E	8–198 (68)		96%	
Horiuchi et al. (1982)	166	Au, Rn, Ra (± E)	24	85%	57%	27%
Vermund et al. (1982)	118	Ra + S	5-year ac.s	90%	55%	35%
Vermund et al. (1984)	96	Ra + S	5-year ac.s	89%	55%	
Jingu et al. (1986)	136	Ra (± E)	5-year ac.s	100%	86%	100%
Feroldi et al. (1986)	55	Ir	5-year ac.s	88%	65%	

ac.s, actuarial survival; Au, ^{198}Au seeds; Rn, radon seeds; Ir, ^{192}Ir afterloading; Ra, radium needles; E, external megavolt irradiation; S, surgery

11 Complications

As short-term complications, approximately in all patients local mucositis and in most patients circumscribed ulceration or necrosis in the tumor area are seen, which will heal within some weeks or months. Disturbances of sensory functions, such as hyperesthesia or impairment of the sense of taste were observed in about one-third of patients (Slanina et al. 1982) with the probability of this persisting for months or even years. A severe (long-term) complication is osteonecrosis, which was reported after treatment with ^{198}Au seeds or radium needles in about 5% of patient by Arnal (1968).

References

Al-Sarraf M, Pajak TF, Marcial VA, Mowry P, Cooper JS, Stetz J, Ensley JF, Velez-Garcia E (1987) Concurrent radiotherapy and chemotherapy with cisplatin in inoperable squamous cell carcinoma of the head and neck. Cancer 59: 259–265

American Joint Committee for Cancer Staging and End Results (1977) Reporting: Manual for staging of cancer. AJC, Chicago

Arnal ML (1968) Spezielle Probleme bei der Strahlentherapie der Mundhöhlentumoren am Beispiel des Zungenkarzinoms. Fortschr Kiefer Gesichtschir 13: 163–176

Bohndorf W (1987) Tumoren des Larynx und Hypopharynx. Spezielle Metastasenprobleme am Hals. In: Scherer E (ed) Strahlentherapie – Radiologische Onkologie. Springer, Berlin Heidelberg New York, pp 521–542

Busch M, Wolfart J (1968) Zur Dosierung implantierter Gammastrahler. Strahlentherapie 136: 437–447

Chu A, Fletcher GH (1973) Incidence and causes of failures to control by irradiation the primary lesions in squamous cell carcinomas of the anterior two-thirds of the tongue and floor of mouth. Amer J Roentgenol 117: 502–508

Colmery BH (1951) Thesis, Ohio State University

Crissman JD, Pajak TF, Zarbo RJ, Marcial VA, Al-Sarraf M (1987) Improved response and survival to combined cisplatin and radiation in non-keratinizing squamous cell carcinomas of the head and neck. Cancer 59: 1391–1397

Decroix Y, Ghossein NA (1981) Experience of the Curie Institute in treatment of cancer of the mobile tongue. I. Treatment policies and result. Cancer 47: 496–502

Drews B (1977) Zur Strahlentherapie des Zungenkarzinoms – Behandlungsergebnisse, prognostische Faktoren und lokale Spätfolgen. Dissertation. Albert-Ludwigs-University, Freiburg

Dutou L, Lacroze M, Ginstet C (1979) A brachytherapy program used with the Philips Treatment Planning System. Medicamundi 24: 93–97

Feind CR (1972) The head and neck. In: Haagensen CD (ed) The lymphatics in cancer. Saunders, Philadelphia, pp 59–230

Feroldi P, Frata P, Baroncelli G, Belletti S (1986) Statistical evaluation of a total dose – dose rate relationship in oral tongue cancer 192 Ir implants. Strahlenther Onkol 162: 561–564

Gilberg EH, Goffinet DR, Bagshaw MA (1975) Carcinoma of the oral tongue and floor of mouth: fifteen years experience with linear accelerator therapy. Cancer 35: 1517–1524

Henschke UK (1958) Interestitial implantation in the treatment of primary bronchogenic carcinoma. AJR 79: 981–987

Henschke UK, James AG, Myers WG (1953) Radiogold seeds for cancer therapy. Nucleonics 11: 46–48

Hodt JJ, Sinclair WK, Smithers DW (1952) A gun for interstitial implantation of radioactive gold grains. Br J Radiol 25: 419–421

Horiuchi J, Okuyama t, Shibuya H, Takeda M (1982) Results of brachytherapy for cancer of the tongue with special emphasis on local prognosis. Int J Radiot Oncol Biol Phys 8(5): 829–835

Jingu K, Hayabuchi N, Miyoshi M, Wada S, Matsui M, Masuda K, Matsuura S (1986) Interstitial radiotherapy for lingual cancer. Local control rate and complications. Fukuoka Acta Med 77(7): 357–365

Lindberg R (1972) Distribution of cervical lymph node metastases from squamous cell carinoma of the upper

respiratory and digestive tracts. Cancer 29: 1446–1449
Noltenius H (1981) Systematik der Onkologie. Klassifizierung – Morphologie – Klinik, vol 1 Urban and Schwarzenberg, Mūnich
Parker HM (1938) A dosage system for interstitial radium therapy. Br J Radiol 11: 313–340
Pierquin B (1964) Précis de curiethérapie. Masson, Paris
Pierquin B, Cuccia CA, de Plaen P, Kelling R, Müller JH, Benussi E (1962) Curiéthérapie interrstitielle par radioisotopes. Am Radiol 5: 395–406
Pierquin B, Chassagne D, Cachin Y, Baillet F, Fournelle le Bus F (1970) Carcinomes épidermoides de la langue mobile et du plancher buccal. Etude de 245 cas traités à l'Institut Gustave-Roussy. Acta Radiol [Ther] (Stockh) 9: 465–480
Renner WD, O'Connor TP, Paulauskas NM (1979) Computer assistance in planning radiotherapy seed implant. Int J Radiat Oncol Biol Phys 5: 427–432
Schröder F (1968) Die Chirurgie des Karzinoms der Zunge und des Mundbodens. Fortschr Kiefer Gesichtschir 13: 125–130
Schumacher W (1974) Neue Möglichkeiten der Radionuklidtherapie ausgedehnter Tumoren von Thoraxwand, Lunge, Vulva, Blase, Rektum und anderen Lokalisationen. In: Pabst HW (ed) Nuklearmedizin – Ergebnisse in Technik, Klinik und Therapie. Schattauer, Stuttgart pp 608–616
Sinclair WK (1952) Artificial radioactive sources for interstitial therapy. Br J Radiol 25: 417–419
Slanina J, Wannenmacher M, Kuphal K, Knüfermann H, Beck C, Schilli W (1982) Interstitial radiotherapy with ^{198}Au seeds in the primary management of carcinoma of the oral tongue: results in Freiburg/Breisgau from January 1964 to July 1980. Int J Rad Oncol Biol Phys 8: 1683–1689
Töndury G (1981) Angewandte und topographische Anatomie, 5th edn Thieme, Stuttgart
UICC (1987) TNM classification of malignant tumors, 4th edn Springer, Berlin Heidelberg New York
Vermund H, Brennhovd IO, Kaalhus O, Poppe E (1982) Preoperative radiation therapy in squamous cell carcinoma of the anterior two-thirds of the tongue at the norwegian radium hospital. Int J Radiat Oncol Biol Phys 8: 1263–1269
Vermund H, Brennhovd I, Kaalhus O, Poppe E (1984) Incidence and control of occult neck node metastases from squamous cell carcinoms of the anterior two-thirds of the tongue Int J Radiat Oncol Biol Phys 10: 2025–2036

5 Breast Cancer

5.1 Brachycurietherapy in Breast Cancer

François Baillet

CONTENTS

1 Introduction 179
2 Materials and Methods 179
2.1 Target Volume and Forecast Dosimetry 180
2.2 Dose and Definitive Dosimetry 181
3 Indications and results 182
3.1 Conservative Management of Breast Cancer Using Lumpectomy, Telebeam Irradiation and Interstitial Brachytherapy Boosts 182
3.2 Conservative Treatment by Exclusive Irradiation Using Telebeam Techniques and Interstitial Brachytherapy 183
3.3 Salvage Brachytherapy 184
References 185

1 Introduction

Interstitial (radium) implants were used experimentally to treat "inoperable" breast cancers in the early 1920s (Keynes 1929). The routine use of interstitial brachytherapy in conservative breast cancer management became possible with the development of iridium-192 after-loading methods during the 1950s and 1960s (Pierquin et al. 1987b). While presently endocurietherapy methods are used less frequently than telebeam techniques to irradiate cancerous breast lesions, it is our opinion that interstitial brachytherapy can be particularly useful and effective in the conservative management of breast cancer when used in association with telebeam techniques, surgery, and systemic treatment.

2 Materials and Methods

Iridium-192 sources have proved to be the most practical for temporary interstitial implants. Continuous, think, flexible wires are available and may be used with afterloading techniques, causing little mechanical damage to surrounding tissues. The wires may be cut to any desired length. I use this radionuclide exclusively with a mean linear activity of 1 $mRh^{-1}m^2cm^{-1}$ (range 0.8–1.3)

Both plastic tubes and stainless steel guide needles are available as inert guide systems for breast implants. The use of guide needles makes it possible to obtain a more perfect parallelism between lines. Plastic tubes cause less tissue damage and follow anatomic curves better. Since I use a method of dose calculation which does not require that the sources be placed absolutely parallel to each other, I prefer to use a plastic tube guide system.

The spacing between radioactive lines may be maintained by plexiglass retaining plates or soft perforated plastic tubes. I prefer the less rigid immobilization of the plastic tube spacers. In addition, if the implant consists of two or more planes, spacers are used only between lines of the same plane. Interplanar separation is not maintained by spacers between planes. This causes less damage to breast tissue, particularly at the plastic tubes' entry and exit points. I use silicone plates to hold plastic tubes for plesiocurietherapy of superficial recurrent disease. This device maintains a plane of radioactive lines 0.5 cm from the skin surface.

Prior to implantation of the guide system the initial primary tumor volume is defined, and the radioactive source lengths necessary to cover this volume are drawn on the skin. The sources are not inserted until after the implant procedure was been completed, the patient returned to her room, dosimetry performed, and the duration of application calculated. This makes it possible to control the time of source removal and to insure greater radiation protection of the recovery room personal.

The implantation of the guide system is performed under general anesthesia. Local anes-

Francois Baillet, Prof. Dr., Centre des Tumeurs, Hôpital Salpêtrière, 47–83 Boulevard de l'Hôpital, 75634 Paris, France thesia is used only if general anesthesia is contraindicated. MAYLIN et al. (1980) developed a technique of perioperative implantation in which plastic tubes are inserted in the tumor bed immediately following tumor excision.

The plastic tubes are generally inserted horizontally to decrease the risk of displacement of the guide system and of the radioactive sources within the guides. The entry and exit points are placed 1 cm from the extremities of the radioactive sources; thus, the length of the plastic tube inserted into the breast should be 2 cm longer than the radioactive source lengths. To facilitate surveillance the plastic guide tubes are cut 4 cm beyond the entry and exit points. Nylon wires are inserted into each end of the plastic tubes to block the radioactive sources in place. The length of each nylon wire is equal to one-half the length of the radioactive sources subtracted from one-half the total length of the plastic tubes.

2.1 Target Volume and Forecast Dosimetry

The volume to be treated (target volume) includes the tumor plus a 1-cm safety margin in all directions. If previous treatment has modified the tumor's size and shape, determination of the treatment volume is based on the initial tumor dimensions. It is preferable that the radiotherapist examine the patient before surgery or systemic treatment to determine the initial tumor size and location. When this is not possible and the exact tumor size and location before surgery and/or chemotherapy is uncertain, one systematically adds 2 cm to the tumor dimensions noted in the patient's records.

In practice, the patient work-up before beginning conservative treatment for breast cancer includes a frontal view photograph of the chest with the patient in a sitting position (this facilitates later evaluation of cosmetic results); and determination of the tumor location and size with the patient in treatment position, that is, lying supine. The tumor is outlined with a felt pen and a frontal photograph of the chest is taken from above. In addition, the tumor dimensions and situation in relation to the nipple are noted on a diagram. The initial mammograms are reviewed. Exceptionally, an atypical extension of the tumor is noted on the mammograms, making it necessary to enlarge the tumor volume.

The active lengths and number of implant planes used are determined by the dosimetric system adopted. The Paris system (DUTREIX et al. 1982) is often used in France. Two plane implants are routinely required, and the spacing between the sources and the planes must be chosen so that the sources are equidistant and arranged as the apices of equilateral triangles. An elementary basal dose rate may be determined for each equilateral triangle at the intersection of perpendicular bisector lines projected from the sides of the triangles (formed by the intersection of the line sources with the central plane). The arithmetic mean of those elementary basal dose rates is termed the basal dose rate. The dose along the isodose surface which has been chosen as 85% the basal dose rate is termed the reference isodose. I use a modified Paris system requiring less rigid adherence to certain basic principles. This new system was presented in 1979 at the Journées Nationales de Radiologie in Paris. It is well adapted to the protocols developed in this treatment center for the conservative management of breast cancer (BAILLET et al. 1988; MAYLIN et al. 1988). It does not require that the triangular arrangement of the sources be absolutely equilateral. An optimal isodose is sought for by referring to the isodose surface which is 85% the minimum dose rate inside (MDI) the radioactive application (Fig. 1). The isodose chosen as the reference isodose must cover the target volume as defined above and must not result in hyperdose sleeves around the wires of greater than 1 cm. A hyperdose sleeve refers to the tissue immediately surrounding the radioactive line which is included in the isodose surface corresponding to twice the value of the reference isodose. When this modified system is used and an equilateral triangular source arrangement is obtained at implanta-

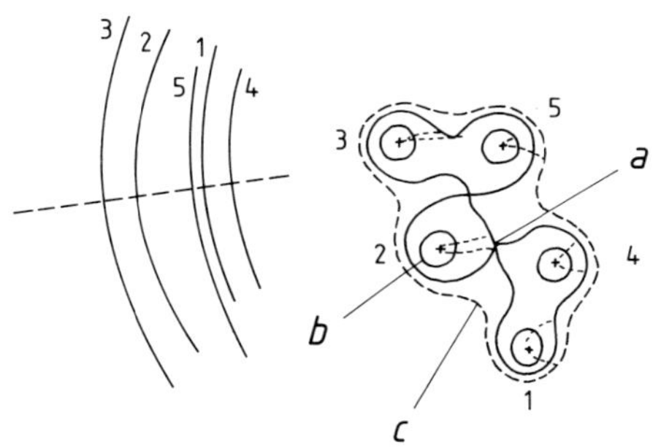

Fig. 1. Dosimetry in the reference plan. *1–5*, radioactive wires; *a*, minimal dose inside the application = first continuous isodose; *b*, hyperdose sleeve; *c*, reference isodose

tion, MDI corresponds to the lowest basal dose (BD) of the Paris system.

The radioactive length chosen is calculated from the maximal transverse diameter of the tumor. Generally, all the sources placed in the same plane have the same length. The active length of the deep plane is determined first and is equal to the maximum tumor diameter multiplied by 1.4 plus 1 or 2 cm for safety. When the tumor measures more than 5 cm, the active length is equal to the maximum tumor diameter plus 2 cm plus 1 or 2 cm for security. For example, if the tumor measures 4 cm, the active length will be $(4 \times 1.4) + 1.5 = 8$ cm, and if the tumor measures 6 cm the active length will be $6 + 2 + 1.5 = 9.5$ cm. The active length of the superficial plane is 1 cm shorter than the active length of the deep plane. If a large tumor is to be treated in a small breast the active length may be calculated for each separate source according to the size of the tumor at the level of each source.

The number of sources in the deep plane is determined by the spacing and the necessity of placing an active line exterior to the cephalad and caudad limits of the tumor volume in a routine horizontal implant. Since the active lines in the superficial plane are placed approximately halfway between the lines of the deep plane, the superficial plane contains one less line than the deep plane.

The spacing between lines of the same plane in a routine implant setup is 2 cm. The average spacing between planes is 1.7 cm (1.3–2 cm). It is occasionally possible to use a single plane implant if the breast is small and the breat tissue flattens against the chest wall when the patient is supine. Three plane implants are rarely necessary and are used only for tumors larger than 7 cm in very large breasts.

The use of this system of forecast dosimetry defines large treated volumes which in my opinion increases the efficacy of treatment. For example, for a 6-cm tumor we would use 5 lines with a length of 9.5 cm in the deep plane and 4 lines with a length of 8.5 cm in the superficial plane. This adds up to a total active length of iridium-192 wire of 81.5 cm.

2.2 Dose and Definitive Dosimetry

The doses prescribed in early attempts to use interstitial brachytherapy to boost tumor dose after telebeam therapy gave effective local tumor control but caused unnecessarily extensive damage to healthy tissues (Pierquin 1975, 1984). I have found that when the breast as received 45–50 Gy by telebeam techniques the dose delivered by interstitial brachytherapy should not exceed 20 Gy if primary lumpectomy has been performed, and 25 Gy if the tumor is in place and being treated exclusively by irradiation.

The radiation damage to heathy tissues varies with the dose and volume of tissue treated. This treated volume may be represented by the total length of radioactive wire used in the implant setup. We studied the radiation damage in 44 patients with the tumor in place treated exclusively by irradiation and interstitial brachytherapy boosts. Telebeam techniques were used to deliver 45 Gy in five fractions of 1.8 Gy per week. An additional 35, 30, or 25 Gy were given using iridium-192 implants. All patients who received 35 Gy by implantation showed evident radiation sequelae no matter what the total length of radioactive wire implanted. Patients who received 30-Gy and 25-Gy implants consistently showed no trace of radiation damage when the total lengths of implanted wire used were less than 30 cm and 65 cm, respectively. For total radioactive lengths exceeding these lengths and dosages, good results without radiation damage were inconstant.

I later used 20-Gy implants in our protocols of exclusive irradiation for breast cancer. Patients receiving this boost dose consistently showed no trace of radiation damage for total active lengths of up to 100 cm.

I use 20-Gy implant boosts after lumpectomy and particularly for perioperative implants for tumors of 3–5 cm. Since the total active length used for these setups is always less than 100 cm, these patients showed no major radiation sequelae. The only minor radiation sequel occasionally noted is a zone of telangiectasis outlining the course of one or more of the radioactive lines if the lines were placed too close to skin so that the skin was included in the hyperdose sleeve.

In my experience, the local tumor control does not vary with dosage when interstitial brachytherapy boosts of 25–35 Gy are used. I am presently studying local tumor control when 20-Gy implants are used to boost the dose to primary tumor sites with no remaining palpable tumor after telebeam therapy.

Actually, since I use treated volumes which are large compared with the measured tumor size, I worry more about radiation damage to the normal tissue than local tumor control. The use of 25-Gy implant boosts for large tumors (T3, T4 >5 cm) could cause inacceptable tissue damage. For this reason, if the initial telebeam therapy leaves no palpable tumor I use 20-Gy implants. However, if palpable tumor remains, I employ a newly developed technique of split-course brachytherapy with modification of the source pattern for the second half of the treatment (BAILLET 1985). Thus a total of 25–30 Gy by two interstitial implants of 12.5–15 Gy 1 month apart is delivered. The position of the radioactive lines for the second implant is shifted in relation to the position of the lines for the first implant. Since there is partial repair of healthy tissue between implants and the hyperdose sleeves around the lines are smaller, it is hoped that split-course brachytherapy will decrease radiation damage to normal tissue while maintaining effective local tumor control. I feel that it is possible to use exclusive irradiation to treat very large tumors (>10 cm) effectively without undue normal tissue damage even when palpable tumor remains before interstitial brachytherapy. The long-term results of this approach are presently under study.

Definitive dosimetry is performed by computer from two orthogonal X-radiographs using lead wires to simulate the length and position of the active lines. For these routine implants, one is rarely obliged to choose a reference isodose surface other than 85% of the MDI. If the Paris system of dosimetry is used with rigid steel guides and plexiglass retaining plates, the reference isodose will always be 85% of the (mean) BD. In order to deliver the same irradiation dose as with the modified system, it is necessary to increase the dose calculated using the Paris system by 5%. For example, 25 Gy calculated by the modified system is equivalent to 26.25 Gy calculated using a strict Paris system.

3 Indications and Results

3.1 Conservative Management of Breast Cancer Using Lumpectomy, Telebeam Irradiation and Interstitial Brachytherapy Boosts

This form of conservative management is classically indicated for tumors of up to 3 cm (PUJOL 1984) as are other protocols based on lumpectomy.

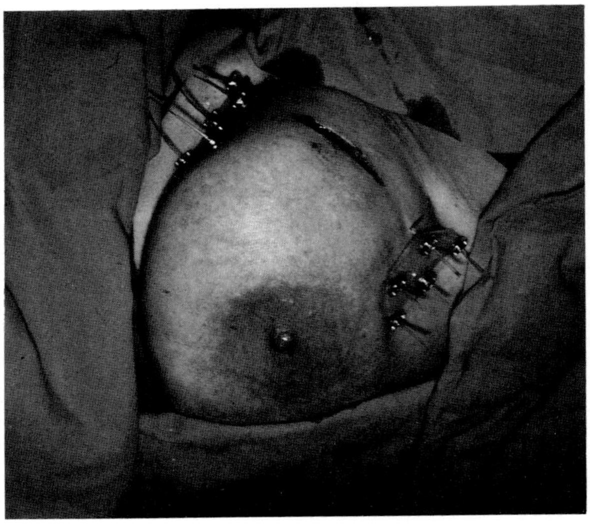

Fig. 2. Lumpectomy with intraoperative brachycurietherapy

When compared with the exclusive use of telebeam techniques, brachytherapy boosts are infrequently included in treatment protocols. Iridum-192 techniques were used for many years in Lyon. Despite a low 5% local recurrence rate, this protocol was abandoned because of poor cosmetic results in 30% of the 408 patients receiving 50 Gy by telecobalt and 20 Gy by implant (MONTBARBON et al. 1987). The local tumor control was greater than when telecobalt techniques were used alone. As discussed above, cosmetic results depend on the total dose and the position of the radioactive lines in relation to the skin. PIERQUIN (1984) followed 109 patients for a minimum of 5 years after they had received 45 Gy by telecobalt and 25 Gy by iridium-192 implant. He noted very good cosmetic results (no sequelae or only subtle changes) in 90% of T1 patients and in 65% of T2 patients. I have followed 104 patients for a minimum of 5 years after 45-Gy telecobalt irradiation and 20-Gy implant boosts and have observed practically no radiation sequelae other than a few exceptional cases of telangiectasis.

Perioperative brachytherapy was originally developed to permit the use of lumpectomy for tumors of 3–5 cm. This method has now been applied to smaller tumors (MAYLIN et al. 1988) (Fig. 2). Among the 455 T1 and T2 patients (30% T1, 70% T2) treated by this method and followed from 3 to 8 years, 8% have developed locoregional recurrences and 85% show very good cosmetic results (no sequelae or only subtle changes). Local recurrences were noted in 5% and 9% of T1 and T2 patients, respectively. The local tumor control and cosmetic results in T2 patients with tumors of greater than 3 cm are practically the same as in T2 patients with tumors of 3 cm or less. In sum, lumpectomy with perioperative implant is a good conservative method of treatment for breast cancer up to 5 cm. It is worth nothing that the survival rates in all these series of patients treated by lumpectomy, telebeam irradiation, and brachytherapy were normal. Effectively, survival rate is not dependent on the choice of locoregional treatment techniques (BLUMING 1982; FISHER et al. 1983).

Lumpectomy may also be performed after telebeam irradiation. If after delivering 45 Gy to a tumor larger than 3 cm the residual tumor is not greater than 3 cm, it is possible to carry out a lumpectomy followed by interstitial brachytherapy. This method is recommended by PIERQUIN (1984).

Finally, lumpectomy may be used to treat local recurrences after initial conservative therapy when the recurrence shows well-defined limit and is situated outside the boosted tissue volume. I have performed 8 such salvage lumpectomies for recurrences of 4 cm or less followed by perioperative interstitial implantation to deliver a dose of 30 Gy. Only one patient has developed a new local recurrence.

3.2 Conservative Treatment by Exclusive Irradiation Using Telebeam Techniques and Interstitial Brachytherapy

Exclusive irradiation is much less often used in the conservative management of breast cancer than irradiation in combination with lumpectomy (Fig. 3). This is particularly true for protocols using iridium implant boosts, which have a reputation for causing undue normal tissue damage. PIERQUIN (1984) found that exclusive irradiation with 45 Gy delivered by a telecobalt beam and 37 Gy given by iridium implant gave poorer cosmetic results than lumpectomy followed by 45-Gy telecobalt irradiation and 20-Gy iridium implants. Only 40% of patients treated for small T3 tumors showed good comestic results at 10 years. However, local tumor control was good, so that after 13 years' follow-up 90% and 85% of T2 and T3 patients, respectively, still had their breast. To improve these results, the author recommends lumpectomy followed by telecobalt irradiation to 45 Gy and an iridium implant to 20 Gy for tumors of less than 3 cm. Lumpectomy may also be used for larger tumors after initial telecobalt irradiation to 45 Gy if the palpable residual tumor does not exceed 3 cm. When there is no palpable tumor after 45-Gy cobalt therapy, the patient is treated exclusively by irradiation with a 25-Gy iridium implant. When the residual tumor is too large for lumpectomy, either mastectomy is performed or, if the patient refuses mastectomy, the treatment is again exclusive irradiation, and the dose delivered by iridium implant is increased to 37 Gy.

The local tumor control after iridium-192 interstitial therapy is quite good. Among the patients with small T3 tumors treated by PIERQUIN, only 15% subsequently underwent salvage mastectomy, while 40% of patients treated by exclusive irradiation at the Curie Institute, where the boost dosage is given by cobalt therapy, subsequently required secondary surgery (VILCOQ et al. 1984, 1988). A recent, randomized study carried out with 250 patients treated at the Curie Institute compared boost dosage by cobalt therapy and iridium implant. The results showed that iridium

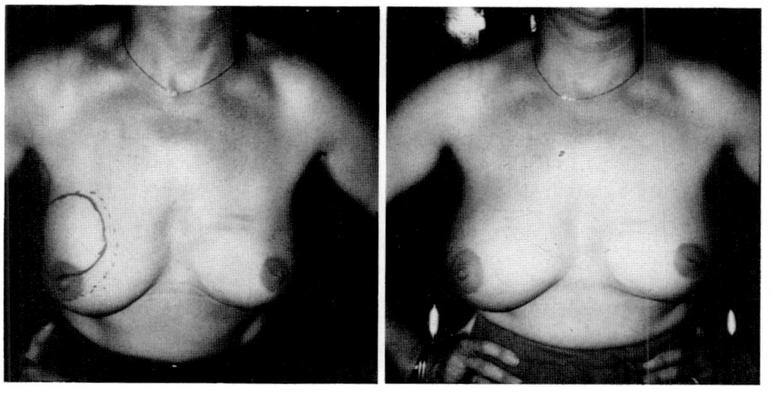

Fig. 3. Conservative treatment by radiotherapy without surgery. Tumor before chemotherapy (12.5 cm) (*left*) and after chemotherapy (10.5 cm) and 8 years after radiotherapy (*right*) (45-Gy equivalent cobalt 60 + 25-Gy brachycurietherapy with iridium 192)

implant gives significantly better local tumor control than cobalt therapy (VILCOQ, personal communication).

In order to improve the results obtained with iridium implants, I analyzed the dose-effect relationship. As a result, I stopped using doses of 35 Gy, which cause too much normal tissue damage, and modified the dose according to the presence or absence of residual tumor (30 Gy and 25 Gy, respectively). Primary chemotherapy was then started with the double objective of controlling the metastatic risk as much as possible and facilitating subsequent irradiation therapy by reducing tumor volume. The chemotherapy regimen was elaborated by JACQUILLAT et al. (1988). Between 1980 and 1983 I used this new protocol to treat 143 patients (7 T1, 53 T2, 63 T3, and 20 T4). Among the patients of this series, 35% had palpable nodes and 31% showed signes of an acutely evolving primary tumor. The mean tumor diameter was 6 cm. Primary chemotherapy was given over a period of 6 weeks. If patients had more advanced tumors (>7 cm, N2, acute evolution) Adriamycin was included in the polychemotherapy regimen. After chemotherapy, the mean tumor diameter decreased to 3 cm (from mean 6 cm before treatment). The primary tumor showed clinically complete regression in 18% of patients. Telebeam cobalt irradiation was then given to a dose of 45 Gy by classic fractionation or by an equivalent dose using a hypofractionation regimen (BAILLET et al. 1987). Iridium implant was used to deliver a boost dose of 30 Gy if there was residual tumor and of 25 Gy if the primary tumor had completely regressed (68% of patients). After irradiation all patients showed clinically complete regression of the primary tumor. Over a mean follow-up of 5 years (range 3–7 years), 11.5% of the patients developed locoregional recurrences. The locoregional recurrences rates according to tumor size were 3.5%, 9%, and 15% for T1 + T2 ≤3 cm, T>3 cm, and T3 + T4, respectively. In 50% of patients the recurrences developed outside the tissue volume which had received the iridium implant. Six of the 16 patients with locoregional recurrences were treated using conservative techniques (2 simple axillary dissections, 4 lumpectomies with perioperative interstitial brachytherapy). Among the remaining 10 patients, 3 underwent mastectomy and 7 developed distant metastases which were treated by chemotherapy exclusively. A total of 98% of patients conserved their breast. Very good cosmetic results (no sequelae or only subtle changes) were noted in 89%. The detailed results are shown in Table 1. The use of interstitial brachytherapy in this protocol is highly effective. However, for patients with T3 tumors >7 cm the rate of very good cosmetic results falls to 72%. Therefore, since 1984 I have decreased the dose delivered by iridium implant to 25 Gy if there is residual tumor and 20 Gy if the tumor has completely regressed.

3.3 Salvage Brachytherapy

When local recurrence develops after conservative management, it may be possible to perform a salvage lumpectomy in association with a perioperative iridium implant, if the recurrence is small and located at a distance from the tissue volume which had received the boost dose.

Iridium implant may also be used alone if the recurrent tumor is small and shows well-defined limits. A multi-institution study presented by the

Table 1. Results of conservative treatment using primary chemotherapy, then telebeam cobalt irradiation, followed by iridium implant

	T1 + T2 ≤ 3 cm	T2 > 3 cm	T1 + T2	T3 + T4
Number of patients	27	33	60	82
Palpable nodes	19% (5)	18% (6)	18.5%	47.5% (39)
Acute evolution	15% (4)	6% (2)		47.5% (39)
Mean tumor size (cm)	2.5	4	3.5	8
5-Year survival (actuarial)	91%	85%	88%	70%
Recurrences T or N	3.5% (1)	9% (3)	6.5%	15% (12)
Breast conservation	100%	97% (32)	98.5%	97.5% (80)
Very good cosmetic result (no sequelae or only subtle changes)	93% (25)	91% (30)	92%	79.5% (65)

( ) = number of patients.

Groupe Européen de Curiethérapie at its 1978 meeting in Paris, showed that this technique was effective in 50% of patients. On the other hand, the cosmetic results were poor. The use of split-course brachytherapy with shifting of the source pattern could reduce normal tissue damage.

Local recurrences after mastectomy may be treated by brachytherapy if only one or a few are present. If the patient has not received postoperative irradiation, it is preferable to irradiate the chest wall to 45 Gy using wide-field telebeam techniques, then to boost the dosage to the recurrent lesion by brachytherapy. If the recurrences have developed on a previously irradiated chest wall, it is possible to try split-course brachytherapy with a source shift in order to decrease normal tissue damage while maintaining treatment effectiveness. It is also possible to try an association of brachytherapy and hyperthermia or an iridium implant with very weakly active wires to decrease the dose rate.

References

Baillet F (1985) Une nouvelle méthode de curiethérapie plus efficace et mieux tolérée: la curiethérapie en 2 temps avec changement de position des sources et radiosensibilisant (in French). Bull Acad Natl Med (Paris) 169: 231–238

Baillet F, Dessard-Diana B, Diana C, Housset M, Maylin C, Thomas F (1987) Therapeutic and cost-effectiveness of a special type of hypofractionated radiotherapy. In: Joint Commission on Accreditation of Hospital (ed) Proceedings of an international symposium on quality assurance in health care. Shanahan, Chicago, p 112

Baillet F, Alapetite C, Boisserie G, et al. (1988) Intérêt d'un protocole original de traitement conservateur pour le cancer du sein associant chimiothérapie première et une modalité particuliére d'irradiation. Bilan des 142 premiers cas. In: Société Francophone de Cancerologie. Le traitement conservateur du cancer due sein. Cedem, Paris, pp 159–178

Bluming AZ (1982) Treatment of primary brest cancer without mastectomy. Review of the literature. Am J Med 72: 820–828

Dutreix A, Marinello G, Wambersie A (1982) Dosimétrie en curiethérapie. Masson, Paris

Fisher B, Bauer M, Margolese R et al. (1983) Five year results of randomized clinical trial comparing total mastectomy and segmental mastectomy with or without radiation in the treatment of breast cancer. N Engl J Med 312: 665–673

Jacquillat C, Baillet F, Weil D, et al. (1988) Results of a conservative treatment combining induction (neo-adjuvant) and consolidation chemotherapy, hormonotherapy and external and interstitial irradiation in 98 patients with locally advanced breast cancer (IIIA–IIIB). Cancer 61: 1977–1982

Keynes G (1929) The treatment of primary carcinoma of the breast with radium. Acta Radiol (Stockh) 10: 393–402

Maylin C, Baillet F, Clot P, Mignot L (1980) Intérêt de l'association tumorectomie-curiethérapie per-opératoire dans le traitement conservateur du cancer du sein. J Eur Radiother 1: 139–141

Maylin C, Socie G, Baillet F, et al. (1988) Expérience Saint-Louis/Necker. 455 cas traités par l'association tumorectomie-curage-curiethérapie per-opératoire-cobalt (oct. 1979 à déc. 1984). In: Société Francophone de Cancerologie. Le traitement conservateur du cancer du sein. Cedem, Paris, pp 99–112

Montbarbon X, Gérard JP, de Laroche G, et al. (1987) Le traitement conservateur du cancer du sein à Lyon. In: Société Francophone de Cancerologie. Le traitement conservateur du cancer du sein. Cedem, Paris, pp 159–178

Pierquin B (1975) Les techniques d'irradiation exclusive des cancers du sein. J Radiol Electrol 56: 443–449

Pierquin B (1984) La radiothérapie exclusive des cancers du sein. In Société Française de Senologie et de Pathologie Mammaire. Les traitements conservateurs du sein. Montpellier: Sauramps Medical pp 77–79

Pierquin B, Raynal M, Otmezguine T, et al. (1987a) Le traitement conservateur des cancers du sein. Résultats à 13 ans. In: Société Française de senologie et de Pathologie Mammaire. Le traitement conservateur du cancer du sein. Cedem, Paris, pp 67–71

Pierquin B, Wilson JF, Chassagne D (1987b) Modern brachytherapy. Masson New York

Pujol H (1984) Conclusions. In: Société Francophone de Cancetologie Les traitements conservateurs du cancer du sein. Sauramps Montpellier, pp 157–159

Vilcoq JR, Fourquet A, Julien D, Gautier C, Calle R, Ghossein NA (1984) Prognostic significance of clinical nodal involvement in patients treated by radical radiotherapy for a locally advanced breast cancer. Am J Clin Oncol 7: 625–628

Vilcoq JR, Fourquet A, Campana F, Julien D, Schlienger P (1988) Cancer du sein: traitement conservateur à l'Institut Curie. In: Le traitement conservateur du cancer du sein. Cedem, Paris, pp 149–158

5.2 Interstitial Boost Irradiation: Indications, Technique, Complications

BERNARD PIERQUIN, FRANK WILSON, and DANIEL CHASSAGNE

CONTENTS

1 Clinical Review 187
2 Indications 187
2.1 All T1 and T2 Tumors 3 cm or Less in Diameter . . 187
2.2 Primary Tumors Larger Than 3 cm and Selected T3 Lesions 187
3 Recommended Technique 187
4 Dosimetry 189
5 Results 189
References 190

1 Clinical Review

The female breast is highly accessible to interstitial brachytherapy despite its large surface area and considerable volume. Because of the necessity to irradiate relatively large tissue volumes, after-loading techniques and particularly important in order to minimize the radiation hazard associated with brachytherapy (HARRIS and HELLMAN 1983; PIERQUIN et al. 1987). This chapter summarizes current indications for and results of interstitial irradiation in the management of small to moderate size breast carcinomas. In these patients endotherapy is used to deliver a boost dose to a limited volume of the breast tissue following external irradiation of the entire breast and associated regional lymphatic areas.

Primary breast carcinomas are usually situated in the midst of very soft mobile tissues, where strict parallelism or radioactive lines is best achieved using a rigid implant system. Our preference is for a guide needle technique in which needles are fixed in position in both ends of the implant by lucite retaining plates (PIERQUIN et al. 1987).

BERNARD PIERQUIN, Prof. Dr. med., FRANK WILSON, Dr., DANIEL CHASSAGNE, Dr., Professor and Chairman, Département de Carcinologie, Service de Radiothérapie, Hôpital Henri Mondor, 51, Avenue de Maréchal de Lattre de Tassigny, 94010 Créteil, France

2 Indications

The role for endocurietherapy in conservative breast cancer management is essentially to deliver a boost dose of irradiation following external irradiation, preceded or not by surgical excision of the primary tumor. The current treatment policies at Henri Mondor Hospital are outlined.

2.1 All T1 and T2 Tumors 3 cm or Less in Diameter

Primary tumorectomy is followed by 45 Gy of external irradiation (calculated at the rib cage) and a boost of 15–25 Gy by ^{192}Ir implant depending on the adequacy of surgical margins. If entirely clear of tumor, the dose from the implant is 15 Gy; if margins were positive or questionable, then 25 Gy is given.

2.2 Primary Tumors Large Than 3 cm and Selected T3 Lesions

Initial teleradiotherapy to 45 Gy is followed by iridium implant to deliver 25 Gy to the tumor bed if the lesion regresses completely. If the tumor regresses adequately enough to become easily excisable, then tumorectomy is performed, followed by an implant to give 20 Gy. In the event of minimal or no regression, either modified mastectomy (Patey's) or an implant to deliver 35–40 Gy is performed.

3 Recommended Technique

The implantation procedure is performed under general anesthesia with the patient in supine position and the ipsilateral arm slightly extended.

Prior to implantation, the original boundaries of the primary lesion are drawn on the skin of the breast with indelible ink by referring to initial clinical descriptions and mammograms. An appropriate target volume is then designated, centered on the lesion, and including a generous area of surrounding breast tissue, generally constituting 25%–30% of the total glandular substance. Subareolar tissue adjacent to the implant, which in our experience is a common site of microscopic extension of breast carcinoma, is always included. Resultant target volumes are therefore oriented radially around the nipple.

Two plane implants are routinely required to adequately irradiate the tissues to be boosted. Single plane implants are occasionally adequate, if the breast is small and the implant is performed in the upper inner quadrant where the breast tissue thins out towards the midline. Final dimensions of the target volume are determined in correspondence with the number and spacing of the needles to be utilized in the implant.

In the Paris system spacing between individual needles in the same plane may vary at 15, 18, or 20 mm. Interplanar separation to produce the required triangulation between adjacent sources is 13, 16, and 18 mm, respectively, for these spacings. Lucite retaining plates perforated in these patterns are stocked in sterile packets in the operating room.

For primary tumors which initially measured less than 4 cm in diameter, a typical target volume measures approximately 6 cm on each side and is 2 cm thick. This volume can be obtained with seven needles divided into two planes: four in the deep plane and three in the superficial plane. In 2 plane implants, it is unusual for fewer than five needles (three deep, two superficial) to be required. A 2-plane implantation is performed, routinely beginning with the deep plane, as follows:

1. Guide needles should be at least twice as long as the target volume. For example, for a target volume 6 or 7 cm long, 15-cm needles are used. Needles at the two ends of a plane are implanted first and are preferentially implanted in a medial to lateral direction. They should pierce the skin about 1 cm beyond the designated target volume on both sides.
2. If the breast is large and the tumor bed is deply situated, needles are advanced perpendicular to the skin until their points reach a sufficient depth in the glandular substance to provide adequate irradiation at depth prior to turning them horizontally to exit through the skin.
3. The superficial plane of needles is implanted last, alternating with those in the deep plane to obtain the necessary triangulation between adjacent sources. To the extent possible, the superficial plane is kept sufficiently deep to the skin to avoid including it in the high-dose sleeve around the needles and consequent late radiation effects.
4. Once all needles are implanted, the portions of the needles protruding through the skin at the ends of the implanted volume are evened up until about 3 cm of needle is visible on both sides.
5. Then the needles are crushed flat with a surgical clamp along several millimeters of their length at their exit through the skin. The length of this flattened segment may vary depending on the conditions of the implant.
6. Needles are retracted until the flattened segment rests just beneath the skin surface. This maneuver blocks the needle in order to prevent the radioactive sources, to be afterloaded later, from advancing too far into the needle, thus causing intersection of the hyperdose sleeve with the skin around the exit point of the needle.
7. Next, perforated lucite plates are applied over both ends of the needles to obtain strict parallelism (Fig. 1). The implantation technique for single-plane implants is identical to that for 2-plane implants except that plastic tube spacers, which are less traumatic to the skin, are substituted for the lucite plates.
8. Starting with the end needles of the deep plane, fixation buttons are slid over the pointed end of the needles to contact with the template and are clamped onto the needles to immobilize them and to further occlude their lumen. The pointed portion of the needles is then snipped off at the button with cutting pliers.
9. Lead segments are slid over the blunt open ends of the needles until in contact with the template. Prior to clamping these onto the needle shafts a solid steel obturator is inserted into the needle so that when the lead segments are compressed onto the needles an adequate

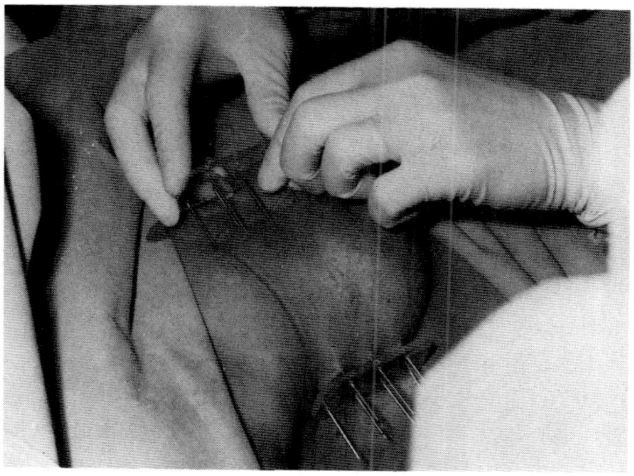
1

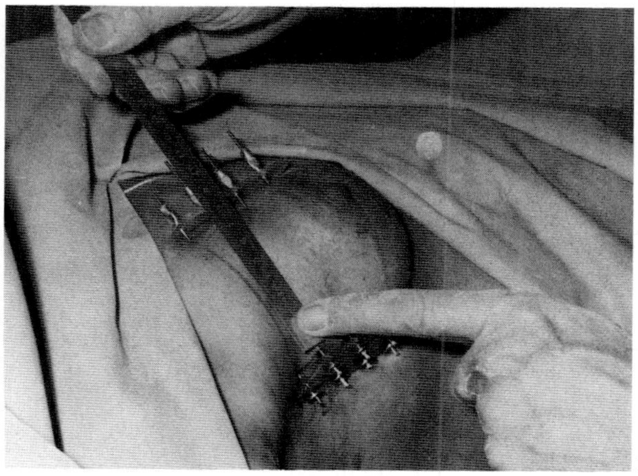
2

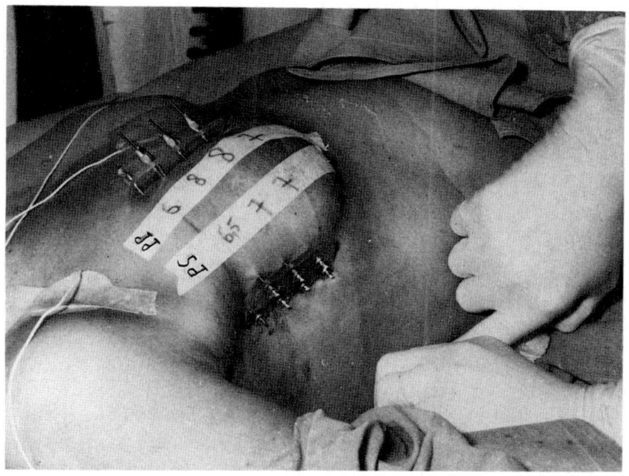
3

Fig. 1. Lucite templates applied over both ends of the needles

Fig. 2. Complete 2-plane implant; confirmation of the source length required by direct measurement

Fig. 3. Complete implant: tape strips on the breast indicate the correct source lengths to be afterloaded. *PP*, deep plane; *PS*, superficial plane

lumen is preserved. When steps 8 and 9 are completed, the implant system is stabilized, and all remaining needles can be immobilized in the same manner.

10. Source lengths are determined by inserting the obturator into each needle until it abuts the occluded segment. The portion that was inserted is measured, and the length of the extracutaneous portion of the needle is subtracted from this measurement to obtain the source length required. This is also checked by direct measurement of the distance between points a few millimeters inside the entry and the exit of the needle through the skin (Fig. 2). Source lengths are recorded on a diagram entered into the permanent record and are designated on tape strips applied to the breast to be double-checked at the time of afterloading (Fig. 3).

4 Dosimetry

Computer generated dosimetry is performed based on orthogonal X-radiographs and a sagittal tomogram. Since implants of this type are optimally in accordance with the Paris system, only the central plane of the implant is obtained. Tomography permits display of the relationship of the target volume to the chest wall and overlying skin (Pierquin et al. 1987).

5 Results

During the past 27 years, we have treated more than 1500 patients with breast cancer with definitive irradiation employing interstitial implanta-

Table 1. Absolute disease-free survival by stage at 5, 7, and 10 years following conservative management of breast cancer

Stage	5 years	7 years	10 years
T1	84%	76%	65%
T2	75%	71%	64%
T3	65%	51%	45%

tion. This series included 26% T1, 57% T2, and 17% T3 lesions. Follow-up at 5, 7, and 10 years is available for 408, 268, and 126 patients, respectively (Table 1) (PIERQUIN and MARIN 1986). Local or regional recurrences to data have been identified in a total of 8% of T1, 9% of T2, and 15% of selected T3 patients followed for at least 10 years.

Breast preservation was achieved in a very high proportion of the surviving patients. Moreover, a very good to excellent aesthetic result was noted following treatment of 87% T1, 54% T2, and 40% T3 lesions. The aesthetic result was at least acceptable to the patient in most other instances.

Until 1981, our treatment policy called for initial excisional biopsy of most T1 lesions (80%) and some T2 lesions (20%) followed by radical irradiation. Telecobalt therapy was used to administer 45 Gy to the mammary gland and associated regional lymphatic areas followed by boosts of 15 Gy to the internal mammary area and of 24 Gy to the low axillary region with 10–13 MeV electron beams. The boost dose to the breast by interstitial implant has gradually been reduced to 15 Gy since 1979.

Most T2 and T3 lesion in this early series were treated with irradiation alone, in which case a larger dose of 37 Gy was administered by implant. Since 1981, in an effort to improve aesthetic results, tumorectomy combined with limited axillary dissection was performed in selected T2 and T3 patients either before or after external irradiation. It should be stressed, however, that even without axillary dissection only three isolated axillary lymph node recurrences have been observed in the entire group of 1500 patients.

At this time conservative management appears to be appropriate therapy for patients with T1–T3 primary breast cancer with an associated axillary lymph node status of N0, N1a, or N1b nodes less than 2 cm in diameter.

At a recent meeting of the European Group of Curietherapists, the experience from 35 European centers with 11300, patients with breast cancer treated conservatively was presented (PIERQUIN et al. 1988). These included 35% T1, 50% T2, and 15% T3 lesions. End results similar to those indicated in the Henri Mondor experience appear to have essentially been reproduced in the group experience. Treatment most often consisted of external irradiation to the entire breast and regional nodes to 50 Gy followed by a boost by iridium implant. Boost doses were of the order of 30 Gy (range 25–35 Gy) if the tumor was left *in situ* or approximately 20 Gy (range 15–25 Gy) if excision had been performed. Local or regional recurrences were observed by the end of 7 years in 5% of patients who underwent tumorectomy and irradiation to a combined dose at the center of the target volume of the implant of 75 Gy and in 10% of patients treated just with irradiation to a total dose of 90 Gy. Aesthetic results at 5 years were excellent in 80%–85% of all T1 patients and in 75% of T2 patients who had had tumorectomy. In contrast, very good aesthetic results were obtained in only 55% of T2 patients and in 45% of T3 lesions irradiated primarily, without prior tumorectomy.

References

Harris J, Hellman S. (1983) The results of primary radiation therapy for early breast cancer at the Joint Center for Radiation Therapy. In: Harris J, Hellman S, Silen W (eds) Conservative management of breast cancer. Philadelphia, Lippincott

Pierquin B, Marin L (1986) The past and future of conservative treatment of breast cancer. Am J Clin Oncol 9: 476–480

Pierquin B, Wilson JF, Chassagne D (1987) Modern brachytherapy. Masson, New York

Pierquin B, Huart J, Raynal M, Otmezguine Y, Mazeron JJ, Calitchi E, Le Bourgeois JP, Marinello G (1988) Radiation therapy as primary treatment of cancer of the breast". Br J Radiol [Suppl] 22: 67–76

5.3 Interstitial Boost Irradiation Technique and Dosage

KARSTEN ROTTE and KURT BAIER

CONTENTS

1 Introduction 191
2 Technique 191
3 Basic Dosimetry (Quality Assurance Program) 193
4 Source Preparation 194
5 Clinical Dosimetry 195
6 Dose-Rate Correction 197
References 198

1 Introduction

In many institutions interstitial brachytherapy was abandoned for several years due to the radiation hazard associated with it. There is now a renascence of this mode of treatment with the introduction of remote controlled afterloading machines and the development of miniaturized sources.

2 Technique

Today the vast majority of radiotherapists (JACOBS et al. 1984, LINDNER 1988; MEERTENS and BARTELINK 1985; ROTTE and LÖFFLER 1986; SCHULZ et al. 1984) are convinced that in interstitial brachytherapy rigid implant systems are superior to systems using bendable plastic tubes as far as dose distribution and homogeneity of the dose administered are concerned. There is also no doubt that only remote controlled afterloading therapy can adequately protect the staff involved from the radiation hazard. Although the basic homogenous irradiation of the breast is carried out quite uniformly in technique and in dosage, there is no unanimous opinion among clinicians which dose rate should be perferred in brachytherapy.

KARSTEN ROTTE, Prof. Dr. med., KURT BAIER, Dipl.-Phys., Strahlenabteilung der Universitäts-Frauenklinik Würzburg, Josef-Schneider-Str. 4, 8700 Würzburg, Germany

Radiobiological calculations, using the linear-quadratic model for example, show that late effects will increase by more than 60%, if a boost treatment of a single high dose-trate fraction of 15 Gy is used instead of a low dose-rate treatment of 20 Gy with a dose-rate of 50 cGy/h or irradiation with fast electrons of 20 Gy in 10 daily fractions of 2 Gy. Yet clinical results published so far do not bear this out (HAMMER and SEEWALD 1988; JACOBS 1989; LINDNER 1988). To be on the safe side, we use a low dose-rate regime with iridium-192 wires for interstitial brachytherapy. The method will be outlined in the following.

In accordance with the Paris system (DUTREIX et al. 1982; PIERQUIN et al. 1978) 1.5 mm rigid guide needles are implanted in parallel order with the help of perspex templates (Fig. 1). The templates are perforated in such a way that the needles are kept in a triangular geometry for a homogeneous dose distribution and are symmetrical on both sides. According to the size of the target volume we insert the needles within a range from 14 to 20 mm apart from each other. In most patients the implantation of two layers of needles is sufficient. Before the position of each needle is finally adjusted in order to provide a constant distance between the proximal end of the needles and the skin, stainless steel clips are clamped onto the needles. These are needed to mark the contour of the skin on the radiographs taken later on. Subsequently, each needle is fixed in its place with the help of immobilisation buttons. Finally, by using a cutting plier the distal ends of the unloaded needles are trimmed in such a way that they are a few centimeters longer than the radioactive implant (Fig. 2).

After the procedures outlined above are done, a set of isocentric stereoradiographs (FITZGERALD and MAUDERH 1975; LÖFFLER and SAUER 1988; STORCHI and VAN KLEFFENS 1979) of the needles inserted is taken (Fig. 3). These radiographs are used in the therapy planning system to reconstruct the position of the needles within the patient, their

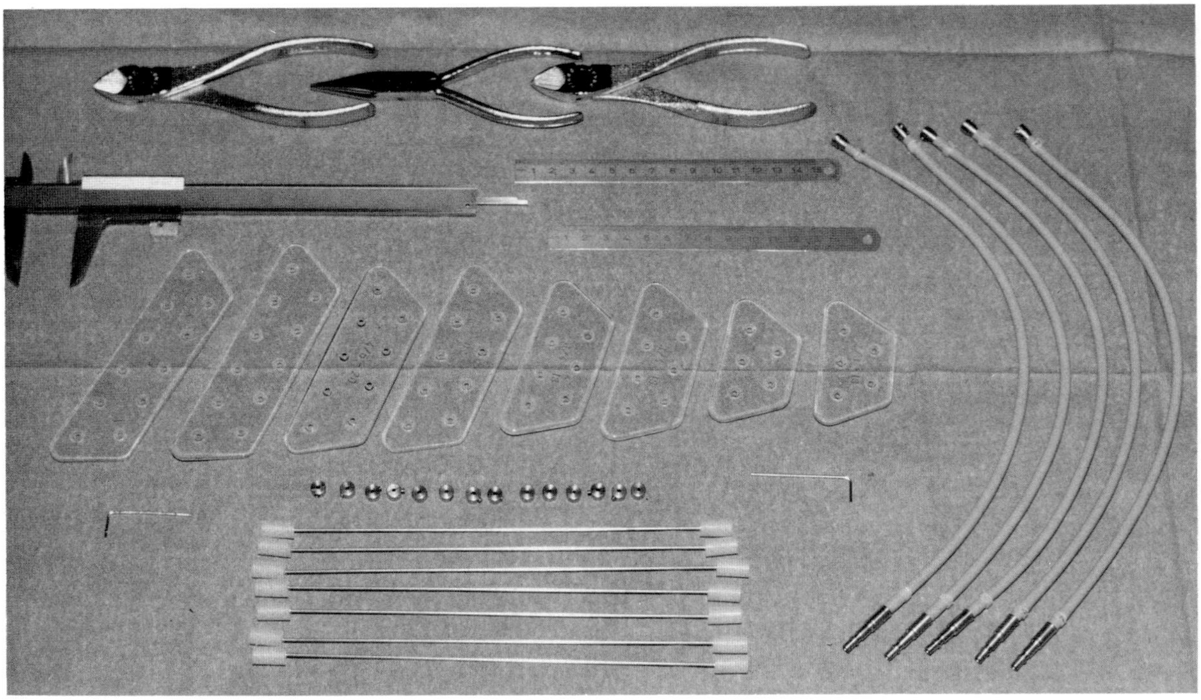

Fig. 1. Instruments needed for interstitial brachytherapy, from *top* to *bottom*: cutting pliers for trimming of the needles, measuring devices to determine the length of each needle segment, templates to assure a geometrically defined application of the needles, needle applicators and buttons tightened on needles to assure the immobilization of applicators in situ. *Right side*: special tubes to connect the needles to the source distributor

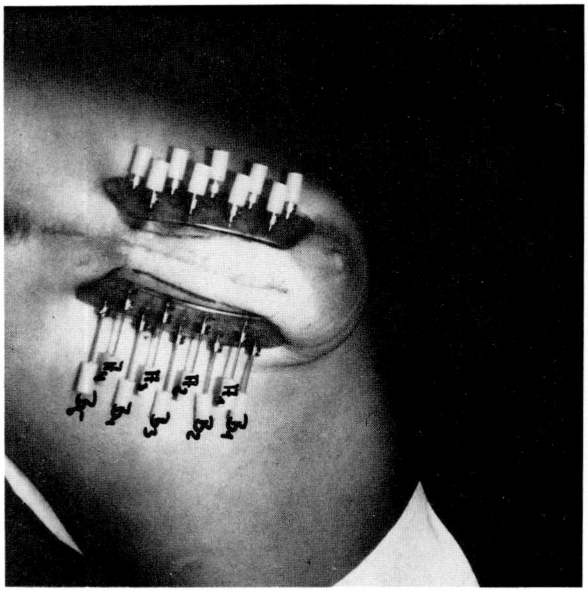

Fig. 2. Needle applicators in situ. Patient readied for stereo X-radiographs for therapy planning

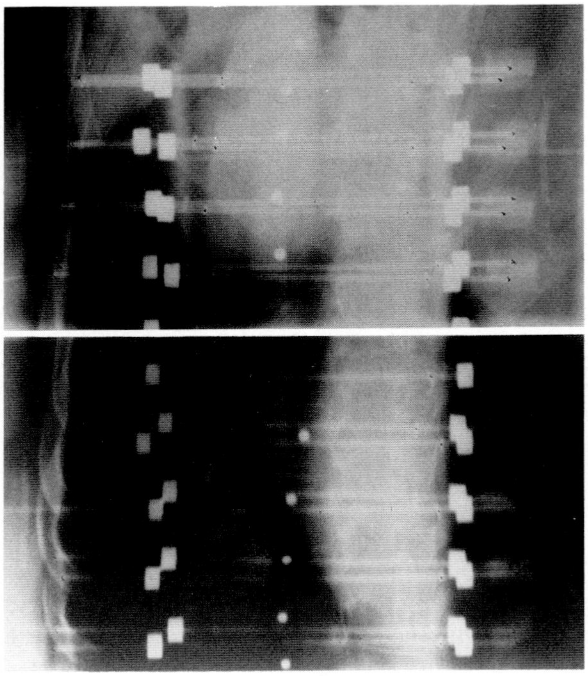

Fig. 3. Isocentric stereoradiographs for reconstruction of a 2-plane breast implant with nine needles. Stainless steel clips are clamped over the entry and exit of each needle. Lead markers show the breast contour near the central plane

entry points into the skin, and with additional lead markers placed on the breast, the contour of this organ. Thereafter the exact length of each iridium wire is computed. Some 5 mm of healthy tissue should be left unirradiated at the entry and exit point of each needle to spare the skin.

3 Basic Dosimetry (Quality Assurance Program)

For treatment with sealed sources a quality assurance program is needed which in its details is quite different from an external beam program. For brachytherapy, the quality assurance program centers mainly around source calibration.

According to ICRU report 38 (1985) γ-ray brachytherapy sources should be specified in terms of the air kerma rate free space at 1 m from the center and perpendicular to the long axis of the source. The long distance measurement geometry minimizes the dependence of the calibration upon the construction of the source and detector. Both source and detector can be treated as points, and oblique transmission of the γ-rays through the capsule and applicator is negligible.

Such measurements are quite difficult at low dose rates. Therefore, the use of a modified well-type ionisation chamber (Fig. 4) for calibration and a ring-type one (Fig. 5) for homogeneity control of wires was inaugurated by Löffler et al. (1986). These ionization chambers are calibrated with a standard ^{192}Ir wire from the Laboratoire de Mètrologie des Rayonnements Ionisants (LMRI, Bureau National de Mètrologie, France).

At first, the source strength deviation of the whole batch should be checked. In a second initial test the linear activity has to be measured using a ring-shaped ionisation chamber. The length of the collimator opening is 10 mm, thus determining the spatial resolution. This allows us to check the linear activity distribution of the whole ^{192}Ir wire. The reference air kerma rate is the basic figure for treatment planning. This value is measured for

a 

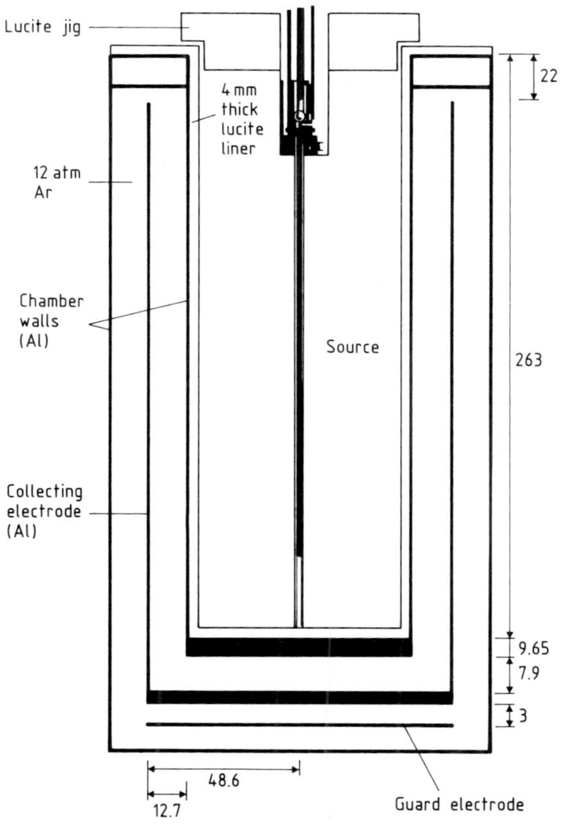

b

Fig. 4a,b. Well-type ionization chamber. **a** View of the chamber; **b** diagram of the chamber for initial dosimetry of iridium source assemblies

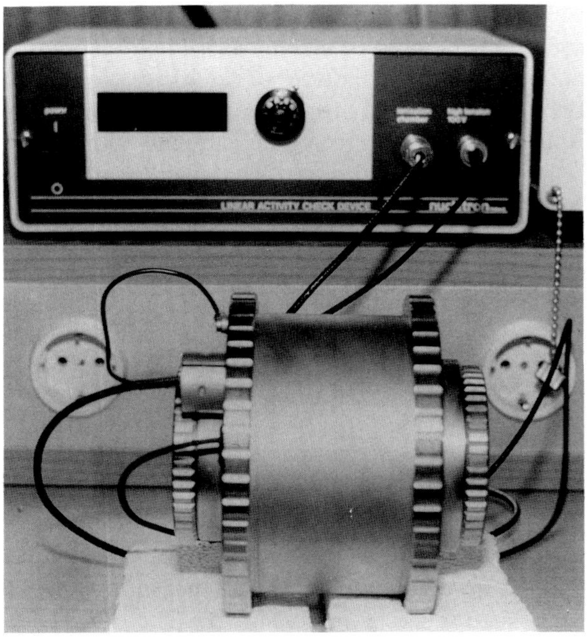

a

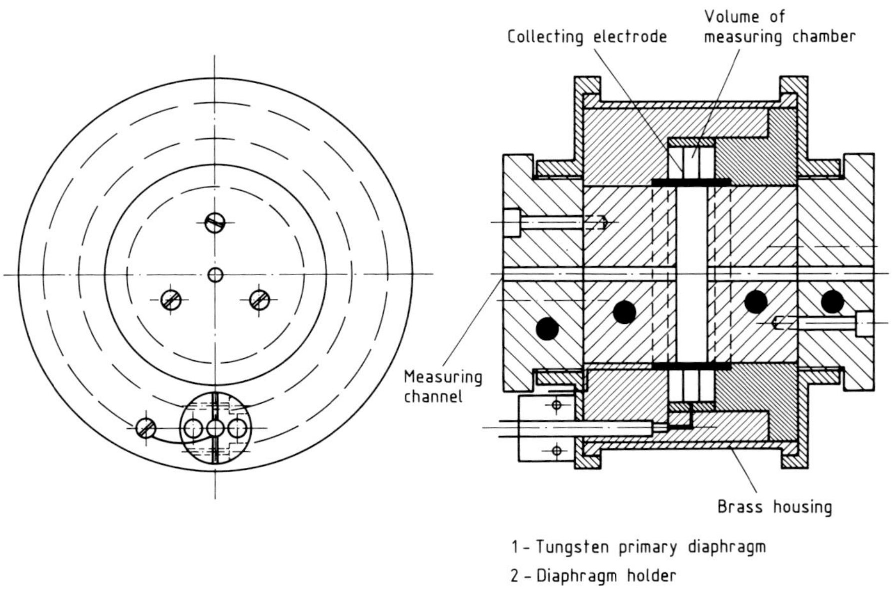

b

Fig. 5a,b. Ring-type linear activity meter. **a** View of the cylindrical ionisation chamber; **b** diagram of the chamber

all prepared sources in the well-type ionisation chamber.

Many users of brachytherapy sources rely on calibration done by manufacturers. However, investigations by Löffler et al. (1986) have shown that the reliance on such source specifications leads very often to systematic errors. Initial hospital dosimetry of each new batch to be used to establish the accuracy and reliability of the manufacturer's source specifications therefore is very important as part of quality assurance.

4 Source Preparation

In order to reduce the number of sources to be prepared wire cuts of active lengths ranging from 20 to 100 mm are assembled in numbers according to the frequency of their need. The sources are cut from 1250 mm iridium wire coils in a special preparation station (Fig. 6) without exposure to the personnel and sealed in a thin plastic tube which is crimped on an inactive guide part.

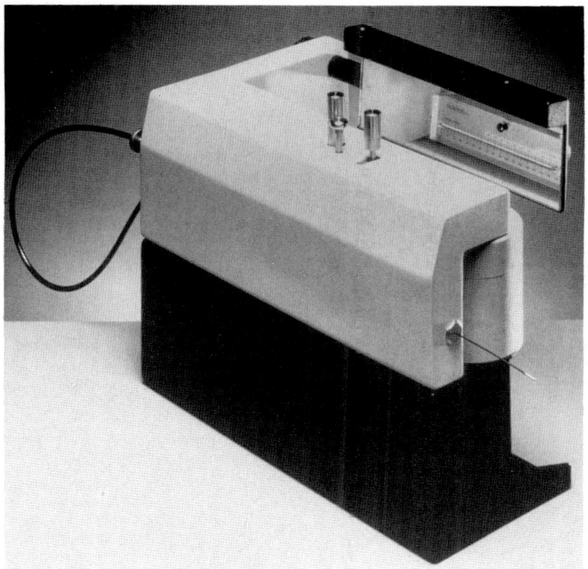

Fig. 6. Exposure-free station for preparation of iridium sources

5 Clinical Dosimetry

In high dose-rate regimes with short treatment times, the clinical dosimetry has to be carried out quickly. Therefore, it is very seldom based on isocentric stereo X-radiographs but mostly is more standardized. For this, different software programs have been developed (BAIER 1988; JACOBS et al. 1984; KILIAN et al. 1990; KRIEGER 1986, 1988). In a low dose-rate regime the time needed to calculate dose distribution, treatment time, etc. is not so critical. Therefore, we prefer a planning system based on individual X-radiographs. In computerized dosimetry based on a set of isocentric stereoradiographs taken of the needles inserted, the iridium core and the platin capsule of the sources are handled together like an unencapsulated line source (VAN DER LAARSE 1985).

In accordance with the Paris system of dosimetry the reference dose is assessed from the mean dose rates in the points midway between the triangular position of the wires in a plane perpendicular to the parallel running needles (Fig. 7). The reference dose rate is calculated as 85% of the mean basic dose. The calculation of the treatment time is based on this dose rate. The dose distribution midway between the layers of the needles (Fig. 8) is used for the determination of the reference volume, which ranges from 30 to 120 ml. It has to be assessed jointly by the surgeon and radiotherapist before radiotherapy commences. A lateral view of the dose distribution of the iridium implants is used to check the parallelism of the needles (Fig. 9).

The position of the active wires relative to the specified skin points on the radiograph allows accurate control of the needle position. All source assemblies should be prepared with the same inactive length relative to the positioning point of afterloading device, i.e., the distance of the proximal sides of the needles to the skin has to be the same for all implants. Therefore, all needles are marked at the same distance on their proximal side. This is important if the sources are to be reused. Before starting treatment a check film

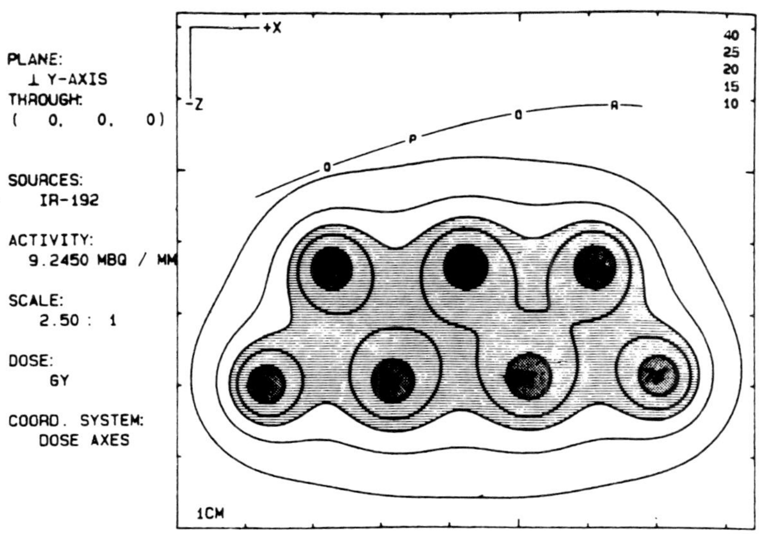

Fig. 7. Dose distribution in a view perpendicular to the needle axes. Reference volume *shaded*

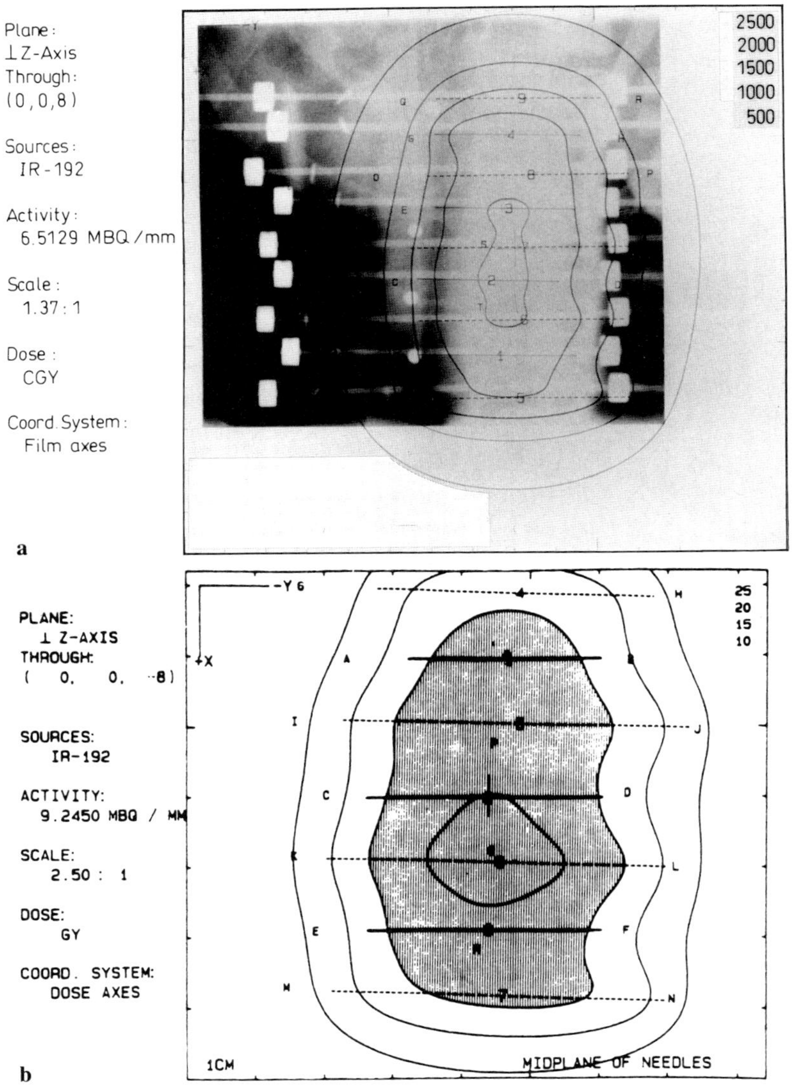

Fig. 8 a Radiograph with superimposed dose distribution midway between the two needle planes. **b** Dose distribution of iridium implants midway between the two needle planes. Reference volume *shaded*

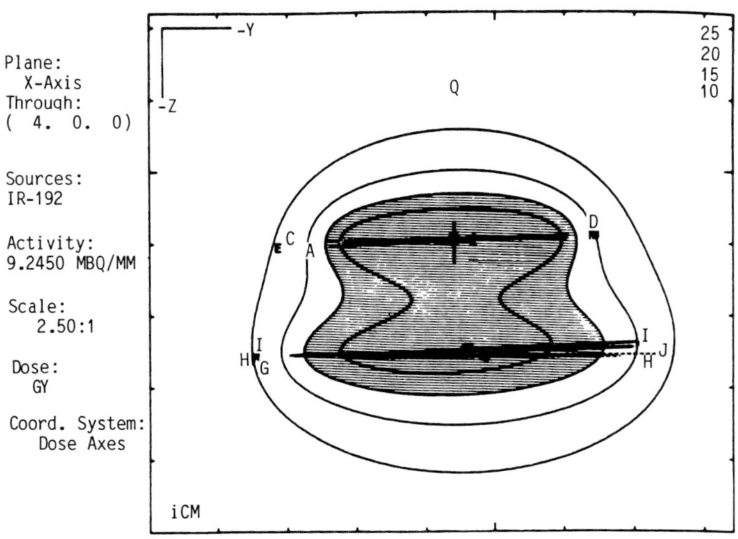

Fig. 9. Lateral view of dose distribution of iridium implants to check parallelism of the needles. Reference volume *shaded*

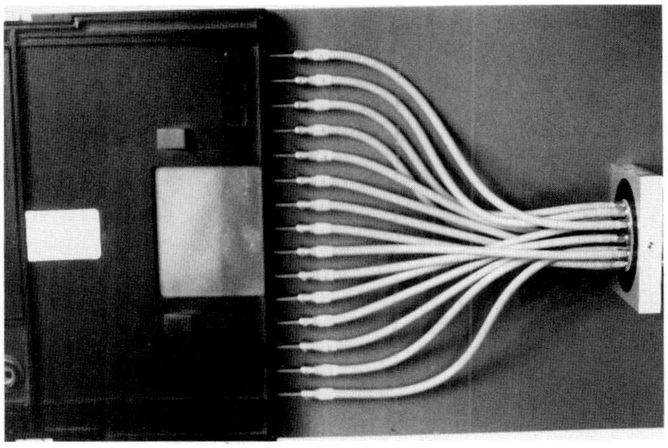

a

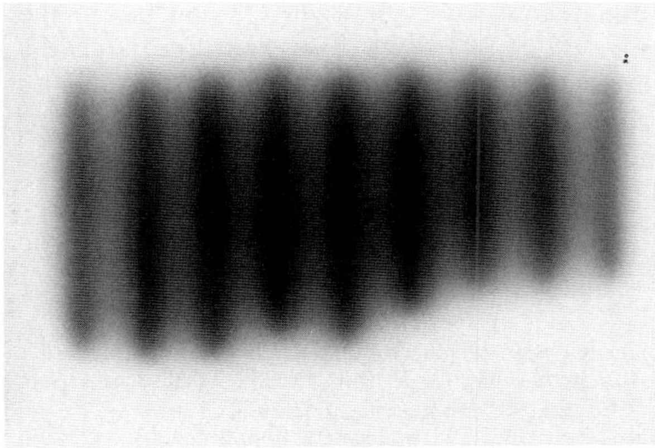

b

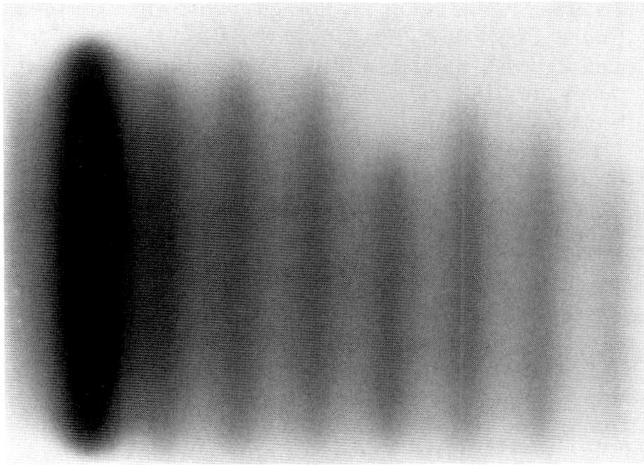

c

Fig. 10a–c. Autoradiographic check film of the loaded sources for confirmation of length, position, and equal activity of the sources. **a** Film-holder for autoradiographic check of the loaded sources; **b** correct source assembly; **c** incorrect source assembly: one wire with higher activity than the rest of the set

of the sources should be taken to confirm their calculated lengths, positions, and radioactivity (Fig. 10).

6 Dose-Rate Correction

For economical reasons it is advisable to reuse the iridium source assemblies for about 100 days. Thereby the linear activity of the sources in an application later on is considerably lower than in an earlier application. This, together with the variations of the reference volume, the number of needles, the source length, and the needle separation distance for each treatment, will result in different reference dose rates ranging between 25 and 100 cGy/h. This marked dose-rate variation can be compensated with the help of biological treatment planning, using the linear quadratic model. On the basis of this simple formula Kellerer and Rossi (1972) introduced a time effect correction factor. They assumed an exponential function of sublethal lesions with a half-life of about 2 h. Barendsen (1981) suggested that the 2 parameters alpha and beta of the LQ-model should be chosen in a way, that the ratio of both will give 5 Gy, to describe connective tissue as the critical one. Normalizing the extrapolated tolerance dose to 50 cGy/h, i.e, 20 Gy in 40 h, the correction factor for the treatment time or pre-

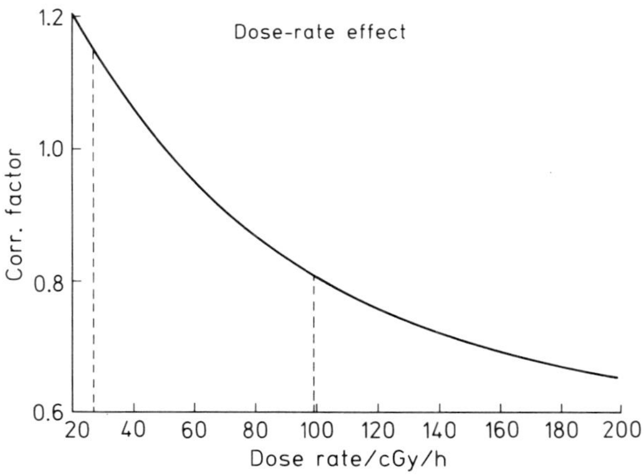

Fig. 11. Nomogram for evaluation of isoeffects of a set of prepared sources within the time span of their use, i.e., 100 days

scribed dose varies between 1.2 and 0.8 in the range of 25–100 cGy/h. For clinical purposes the corrected treatment time of the prescribed dose can easily be found in a nomogram (Fig. 11) for the dose rate in question.

Rerfences

Baier K (1988) Möglichkeiten der schnellen Faltung in der Brachytherapie. In: Nusslin F (ed) Medizinische Physik 1988. DGMP, Tübingen, pp 435–440

Barendsen GW (1981) Dose fractionation, dose-rate and iso-effect relationships for normal tissue response. Int J Radiat Oncol Biol Phys 8: 1981–1997

Cohen L (1980) Biological models and computed isoeffect tables for continuous low dose-rate and intermittent fractionated radiation therapy. Br J Radiol [Special Rep] 17: 138–145

Dale RG (1985) The application of the linear-quadratic dose-effect equation to fractionated and protracted radiotherapy. Br J Radiol 58: 515–528

Dutreix A, Marinello G, Wambersie A (1982) Dosimétrie en curiethérapie. Masson, Paris

Ellis F (1980) Low to high dose-rate by the TDF system, Br J Radiol [Special Rep] 17: 146–156

Fitzgerald LT, Mauderli D (1975) Analysis of errors in three-dimensional reconstruction of radium implants from stereo radiographs. Radiology 115: 455

Fowler JF (1989) The linear-quadratic formula and progress in fractionated radiotherapy – a review. Br J Radiol 62: 679–694

Hammer H, Seewald HD (1988) Use of HDR afterloading method in the treatment of breast cancer. In: Vahrson H, Rauthe G (eds) High dose rate afterloading in the treatment of cancer of the uterus, breast and rectum. Urban and Schwarzenberg, Münich p 271

International Commission on Radiation Units and Measurements (1985) Dose and volume specification for reporting intracavitary therapy in gynecology. ICRU, Bethesda (Report no 38)

Jacobs H (1989) Experience with interstitial HDR-afterloading brachytherapy in breast conserving management of mammary carcinoma. In: Rotte K, Kiffer J (eds) Changes in brachytherapy. Wachholz, Nürnberg p 139

Jacobs H, Teusch P, Schleppi V, Schmieder A, Moncke G (1984) Interstitielle Brachytherapie – Erste Erfahrungen mit 192-Iridium in der Kurzzeittherapie. Strahlentherapie 160: 8

Kellerer AM, Rossi HH (1972) The theory of dual radiation action. Curr Top Radiat Res Q 8: 85

Kilian H, Baier K, Löffler E, Süßenbach K, Dörner K (1990) A comparision of different planning algorithms used in interstitial radiotherapy of 192-Ir wires. In: Bartelink H (ed) Brachytherapy 1988. Proceedings of the 5th International Selectron Users Meeting 1988. Nucletron, Leersum (in press)

Krieger H (1986) Ein schnelles Planungsprogramm zur interstitiellen und intrakavitären 192-Ir-Afterloading-Therapie. Strahlenther Onkol 162: 179

Krieger H (1988) A fast planning program for interstitial and gynaecological use. In: Vahrson H, Rauthe G (eds) High dose rate afterloading in the treatment of cancer of the uterus, breast and rectum. Urban and Schwarzenberg, Munich p 72

Lindner H (1988) HDR interstitial afterloading-therapy of mammary carinoma: methods of localisation, planning and implantation. In: Vahrson H, Rauthe G (eds) High dose rate afterloading in the treatment of cancer of the uterus, breast and rectum. Urban and Schwarzenberg, Münich p 266

Liversage WE (1969) A general formula for equating protacted and acute regimes of radiation. Br J Radiol 42: 432

Löffler E, Sauer O (1988) 3-D Rekonstruktion von Brust-Implantaten zur Applikationskontrolle und Bestrahlungsplanung mittels isozentrischer stereoskopischer Röntgenaufnahmen (ISR) unter Berücksichtigung einer ferngesteuerten interstitiellen Nachlademethode. Strahlentherapie 164: 48

Löffler E, Sauer O, Rotte K (1986) Interstitielle Low-Intensity Afterloading-Therapie mit ^{192}Ir-Drähten. In: von Klitzing L (ed) Medizinische Physik 1986. DGMP, Lübeck, pp 433–441

Meertens H, Bartelink H (1985) First experience with the MicroSelectron in breast conserving therapy implants. In: Mould RF (ed) Brachytherapy 1984. Proceedings of the 3rd International Selectron Users Meeting 1984. Nucletron, Leersum, p 271

Pierquin B, Dutreix A, Paine CH, Chassagne D, Marinello G, Ash G (1978) The Paris system in interstitial radiation therapy. Acta Radiol [Oncol] (Stockh) 17: 33

Rotte K, Löffler E (1986) Strahlentherapeutische Möglichkeiten bei der Behandlung des primär operablen Mammakarzinoms. Gynakol Prax 10: 695–707

Schulz U, Busch M, Bormann U (1984) Interstitial high dose-rate brachytherapie: principle, practice and first clinical experiences with a new remote-controlled afterloading system using Ir-192. Int J Radiat Oncol Biol Phys 10: 915–920

Storchi PRM, van Kleffens HJ (1979) Evaluation of Cartensian coordinates and radiation doses in points determined with stereo X-ray techniques. Comput Prog Biomed 9: 141–148

Thames HD (1985) An "incomplete-repair" model for survival after fractionated and continous irradiations. Int J Radiat Biol 47: 319–339

Van der Laarse R (1985) Treatment planning of interstitial radiotherapy (MPS) with the Selectron treatment planning system. In: Mould RF (eds) Brachytherapy 1984. Proceedings of the 3rd International Selectron Users Meeting 1984. Nucletron, Leersum, p 286

5.4 Iridium-192 Afterloading for Boosting the Tumor Bed

J.H. BORGER and H. BARTELINK

CONTENTS

1 Introduction 199
2 Theoretical Considerations 199
3 Clinical Experience 201
4 Conclusions and Radiotherapeutic Implications . . . 202
References 203

1 Introduction

Interstitial radiotherapy for breast cancer as an alternative to radical mastectomy was first recommended by Janeway in 1917; in 1939 Keynes (ROBINSON 1986), who was also using radium needles, stated that interstitial radiotherapy for breast cancer was as good as radical mastectomy with less mutilation and less edema of the arm (ROBINSON 1986). With the introduction of cobalt machines and linear accelerators, interstitial therapy was confined to the delivery of boost doses, especially in France, where a new dosimetry system was developed by PIERQUIN and DUTREIX (1967) using iridium wires in an afterloading technique (Paris system). Many thousands of patients have had breast conservation therapy (BCT) since then, and the results of prospectively randomized trials (FISHER et al. 1985; VERONESI et al. 1983; SARRAZIN et al. 1983; VAN DONGEN et al. 1987) as well as the data from retrospective studies (AMALRIC et al. 1983; CALLE et al. 1983; PIERQUIN 1983; BARTELINK et al. 1988; LOPER et al. 1987; VAN LIMBERGEN et al. 1987; CLARK et al. 1987) clearly confirm the conclusions of the pioneers in the field: For early breast cancer up to 5 cm in diameter BCT gives results equal to radical mastectomy with respect to survival as well as local control. Local control rates in T1 and T2 breast cancers treated by BCT vary from 91% to 98% at 5 years', 87% to 92% at 10 years' (FISHER et al. 1985; VERONESI et al. 1983; SARRAZIN et al. 1983; VAN DONGEN et al. 1987; AMALRIC et al. 1983; CALLE et al. 1983; PIERQUIN 1983; BARTELINK et al. 1988; LOPER et al. 1987; VAN LIMBERGEN et al. 1987; CLARK et al. 1987), and 80% to 85% at 15 and 20 years' follow-up (HARRIS et al. 1984; KURTZ et al. 1987).

As far as local control is concerned, many prognostic factors have been identified which predict higher recurrence rates: younger age (CALLE et al. 1983; BARTELINK et al. 1988; CLARK et al. 1987; NOBLER and VENIT 1985; RECHT et al. 1988), larger tumor size (LOPER et al. 1987; LEUNG et al. 1986), incomplete tumor excision (VAN LIMBERGEN et al. 1987; NOBLER and VENIT 1985; RECHT et al. 1986), high tumor grade (HARRIS, personal communication; CLARKE et al. 1985), lymphatic invasion (LOPER et al. 1987; HARRIS, personal communication), and extensive ductal carcinoma in situ (BARTELINK et al. 1988; CONNOLLY et al. 1983). Until now, none of these factors has become a clear contraindication for BCT; however, local control rates are lower, the causes of which remain to be clarified. From a radiotherapeutical point of view, the main factors to analyze in this context are target volume, dose level, fraction size, and dose rate. Their involvement with tumor-related factors as well as their effect on normal tissues still remain under investigation.

2 Theoretical Considerations

Breast cancer is a multifocal disease with a high incidence of additional tumor foci directly around the primary but also with some tumor foci as far as 10 cm away (HOLLAND et al. 1985) (Fig. 1). In a zone 2 cm around the primary tumor, foci of invasive and noninvasive carcinomas are found in more than 40% of patients (zone A). In the next zone (2–4 cm around the primary), this prob-

J.H. BORGER, M.D., H. BARTELINK, M.D., Ph.D., Chairman, Netherlands Cancer Institute, Antoni van Leewenhoekziekenhuis, Plesmanlaan 12, 1066 CX Amsterdam, The Netherlands

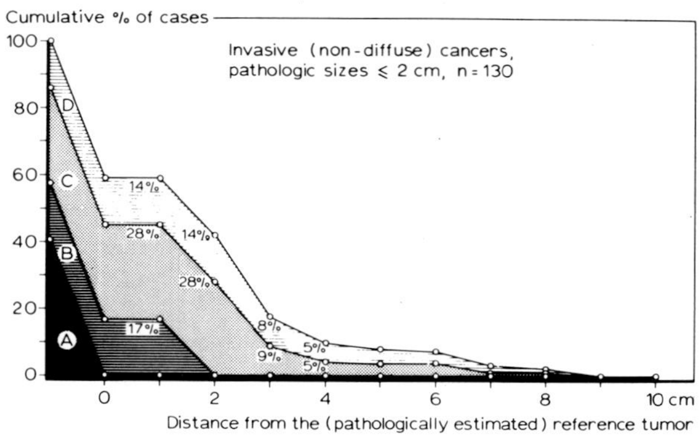

Fig. 1. Additional tumor foci of brèast cancer as a function of the distance from the primary tumor edge; in situ (*C*) and infiltrative lesions (*D*). (Courtesy of HOLLAND et al. 1985)

ability rapidly diminishes to about 10% (zone B). If tumorectomy is performed, the excision margin will be in zone A, and depending on the size of the margins, tumor will be left behind in as much as 40% of patients. If a larger excision is performed (quadrantectomy), the excision margin will be in zone B or even outside this zone and leave behind tumor in approximately 10% of patients. Although it is hard to quantify the volume of these tumor remnants, it seems reasonable to qualify it as microscopic disease. Only if the excision is incomplete and the primary tumor is transected can one expect macroscopic tumor to remain in situ. In patients with invasive carcinoma mixed with extensive ductal carcinoma in situ (DCIs), the same histological pattern is observed in a more outspoken way (HOLLAND, personal communication). In clinical practice a radical excision is clearly much harder to obtain than in pure invasive tumors (BARTELINK et al. 1988).

From these facts one may derive the target volume to be treated in BCT, and it is clear that this very much depends on the extent of the tumorectomy. However, even if a quadrantectomy is performed, tumor is left behind in 10% of patients (5% invasive/5% in situ), and a no further treatment policy should result in at least 5% recurrences (the impact of radiotherapy on DCIS and the biologic behavior of the latter are still unknown). For reasons of cosmetic outcome, however, a more limited tumor excision is advised by many oncologists, which means that the probability of residual tumor remaining near the tumor bed may rise to about 40% (28% in situ, 14% invasive). No further treatment would result in at least 14 + 5% = 19% recurrences. From the NSABP 06 study (FISHER et al. 1985) we know that this figure actually amounts to about 28%, indicating a considerable contribution of in situ carinomas to local breast recurrence rates.

The radioresponsiveness of human breast cancer has been studied by many investigators (FLETCHER 1972; TIMOTHY et al. 1979; ARRIAGADA et al. 1985; DENHAM 1986; BATINI et al. 1978; GHOSSEIN et al. 1978). Experience in exclusive radiotherapeutic treatment of breast tumors shows that high tumor doses are required (70–80 Gy) to obtain reasonable control rates (about 75%) (ARRIAGADA et al. 1985; BATINI et al. 1978; GHOSSEIN et al. 1978). As far as treatment of microscopic tumor extension is concerned, FLETCHER (1972) and TIMOTHY et al. (1979) estimated 50 Gy in 5 weeks to be sufficient for a control rate of over 90% after mastectomy. DENHAM (1986) after extensive revision of the literature including BCT concludes that 50 Gy in 5 weeks given to the breast after complete tumorectomy results in a 75% control rate. He is referring to BCT studies and finds a somewhat lower control rate compared with postmastectomy studies, which might be explained by differences in tumor volume. Arriagada found a

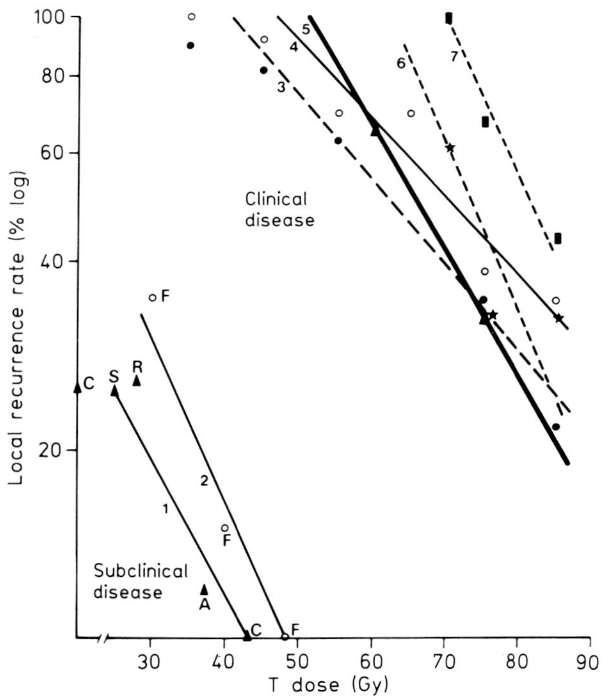

Fig. 2. Dose-response relationship in human breast cancer as a function of tumor volume. (Courtesy of ARRIAGADA et al. 1985)

dose-response relationship in patients treated primarily by radiotherapy for breast cancer (Fig. 2). He predicts a 50% reduction in local recurrence rate for each dose escalation of 15 Gy (ARRIAGADA et al. 1985). Clearly, macroscopic tumor load requires 70 Gy or more for a 90% control rate (TIMOTHY et al. 1979; GHOSSEIN et al. 1978) (Fig. 2). If one applies this knowledge to the abovementioned biologic data, BCT after quadrantectomy aims at microscopic tumor control in the remaining breast; with 50 Gy whole breast irradiation one would expect at least 0.5%–1.3% recurrences (75%–90% of 5%), without considering the contributions of the noninvasive tumor foci. If a more limited tumor excision is done, this figure amounts to 2%–6% again without taking the in situ carcinomas into account and assuming microscopic tumor load. Quantifying the contribution of noninvasive tumor foci to local recurrence rates and deciding whether or not it is justified to assume that tumor remmants in all cases are microscopic clearly remain important questions to be answered in order to decide on dose levels and treatment volumes in BCT. Of course, this is all theoretical, and the role of boost irradiation in BCT after complete tumorexcision remains a matter of debate and awaits clarification, which may be very hard to obtain through a prospectively randomized study. Incomplete tumorectomies clearly justify boost irradiation theoretically to a total dose of about 70 Gy.

3 Clinical Experience

It should be emphasized that the prospective studies in BCT still have a limited follow-up period. From the retrospective studies with longer follow-up periods (HARRIS et al. 1984; KURTZ et al. 1987), a very striking time course of local breast recurrences is seen. In contrast to mastectomized patients, the local recurrence rates in women treated by BCT show a continuous rise even after 10 and 15 years. In the Milan study (VERONESI et al. 1983) local breast recurrences were analyzed according to their localisation in relation to the primary tumor site, revealing that true local recurrences in the tumor bed and recurrences elsewhere in the breast gave an equal contribution to the total breast recurrence rate (4%). It was argued that the latter were not true failures of treatment but de novo developing cancers comparable with the occurrence of contralateral breast cancers. However, considering the low incidence of local failures in this study together with the already mentioned multifocality of the disease and the 50 Gy given to the whole breast, one may speak of a treatment failure which is totally in concert with what may be expected from theoretically considerations. A considerable contribution of in situ cancers left behind after quadrantectomy, however, has to be taken into account with regard to this local recurrence rate. The fact that local recurrence rates in clinical practice in most studies are somewhat higher than would be expected from theoretical considerations concerning infiltrating tumor foci only can also be very easily explained by considering a contribution of in situ carcinoma foci. The radioresponsiveness and biologic behavior of in situ breast cancer are still unknown, but from this, one may derive at least a comparable contribution to local recurrence after radiation therapy as with infiltrating tumors. This would also imply a certain radiosensitivity of the carcinoma in situ foci.

The Boston group recently analyzed the time course of local recurrence following lumpectomy with a 1-cm margin and whole breast irradiation with a boost to the tumor bed (RECHT et al. 1990). The local recurrences were analyzed according to their localisation in relation to the primary tumor bed and boosted area. True tumor-bed recurrences and marginal misses were defined, as were recurrences elsewhere in the breast. True recurrences and marginal misses not only made up the majority of the total local recurrence rate but also showed a different time course in comparison with recurrences elsewhere in the breast. The former occurred in the first 7 years of follow-up whereas the latter occurred only after 4 years, showing a rising pattern of incidence rate which levels off at 8 years. The incidence of recurrences elsewhere in the breast was 2% at 10 years and was equal for purely infiltrating tumors and those containing extensive DCIS. This figure is very much in concert with the Milan data. However, the incidence of the true local recurrences and marginal misses differed very much between purely infiltrating and tumors containing extensive DCIS (8% vs. 33% at 10 years). These results clearly show that the area directly around the primary tumor is at the highest risk for local recurrence and that the tumors containing extensive DCIS follow the same pattern with higher recurrence rates in or near the boosted area. We feel that this can be explained in terms of tumor load and dose levels.

The Amsterdam experience (BARTELINK et al. 1988) confirms the role of the DCIS component in local breast recurrence rates; however, the recurrence rates are lower than in the Boston experience (9% vs. 37%). Both groups use iridium implants for boost irradiating, but the implant volumes were larger and dose levels higher (25 Gy) in Amsterdam, which might explain the difference.

The role of age as a prognostic indicator for local breast recurrence following BCT can only partially be explained by the association of younger age and occurrence of extensive DCIS (RECHT et al. 1988). Tumors in younger patients may represent more aggressive and less radioresponsive tumors, as probably is the case in the high grade breast cancers, which also show a higher recurrence rate than well-differentiated tumors (HARRIS, personal communication; CLARKE et al. 1985). Larger tumor burdens require higher radiation doses, and incomplete tumor excision represents a higher tumor load than radical tumor excision. Several studies have confirmed this simple fact (VAN LIMBERGEN et al. 1987; NOBLER and VENIT 1985; RECHT et al. 1986). However, microscopic evaluation of margin involvement has been studied by only a few investigators (FISHER et al. 1985; BARTELINK et al. 1988; CLARKE et al. 1985). The NSABP 06 study (FISHER et al. 1985) used the microscopic margin involvement as an exclusion criterion, and CLARKE et al. (1985) mention microscopic evaluation of the resection margins without detailing the method. In the latter study, no correlation between microscopic margin involvement and local breast recurrence was found. BARTELINK et al. (1988) mention only a minor contribution of microscopic margin involvement to local failures rate. From these studies the role of microscopic margin involvement in local treatment failure cannot be defined. The evaluation of microscopic margin involvement is subject to interobserver bias, as noted by FISHER et al. (1986), and from the studies mentioned one used it as an exclusion criterion, leaving no room for conclusions (FISHER et al. 1985), while the other two did not find a significant role of this in breast recurrence at dose levels in the tumor bed of 60 (CLARKE et al. 1985) and 75 Gy (BARTELINK et al. 1988). Most studies on BCT do not mention microscopic evaluation of the resection margins, but some mention tumor size (LOPER et al. 1987; LEUNG et al. 1986) as a prognosticator, which might be explained by the fact that in larger tumors the probability of microscopic margin involvement is higher than in smaller tumors, as suggested by MCCORMICK et al. (1987).

Lymphatic invasion is mentioned in some studies as a prognosticator for local breast failures after BCT (KOPER et al. 1987; HARRIS, personal communication). It has to be clarified whether this feature is an indication of higher tumor load in the remaining breast or related to other factors such as grade. If there is a higher tumor load in the remaining breast, greater irradiation doses would be needed and probably preclude aesthetic, breast-conserving therapy.

In some studies the tumor-bed dose is mentioned as a prognostic factor for local breast failure (VAN LIMBERGEN et al. 1987; CLARK et al. 1987; NOBLER and VENIT 1985; HARRIS, personal communication; HELLMAN et al. 1980; PIERQUIN et al. 1986) with a significant reduction of local failure at dose levels varying from 1700–2000 ret. (Rad equivalent therapy according to the Ellis formula); however, most studies lack a multivariate analysis for all prognostic factors concerned. CLARKE et al. (1985) gives a multivariate regression analysis and mentions only two independent prognostic factors: dose level in the tumor bed (1840 ret.) and high grade.

4 Conclusions and Radiotherapeutic Implications

From the numerous studies mentioned above, no firm conclusions can be drawn with regard to dose levels needed for optimal tumor control in relation to most prognostic factors, especially concerning the tumor bed (whether to boost or not). A prospectively randomized study with careful stratification for the prognosticators already known clearly seems necessary. However, the feasibility of such a study should be critically analyzed since very large numbers of patients are needed to make firm conclusions possible. Especially after complete tumor excision for purely infiltrating carcinomas, the role of additional boost irradiation is very hard to define and probably will yield only a minor contribution to local control and survival, which could prove negligible if a cost-benefit analysis is done. For macroscopically incomplete tumor excisions and/or tumors containing extensive DCIS, boost irradiations seem to be necessary, but again the dose levels are not clearly defined. The Amsterdam experience shows a very good local control with 25 Gy boosts in incomplete tumor excision and an acceptable control in tumors containing extensive DCIS, which could mean an overtreatment in the former and a slight undertreatment in the latter. If one assumes macroscopic tumor left behind after incomplete tumorectomy and takes the already mentioned dose-response figures into consideration, a dose of a least 70 Gy in these patients would seem appropriate, which would mean a 20 Gy boost after 50 Gy whole breast irradiation. At these dose levels, brachytherapy theoretically seems to have the advantage of lower volumes of normal tissue irradiated compared with both electron and photon beams. Whether or not cosmetic results indeed are better with iridium implants than with external beam treatment still remains to be clarified as does the optimal dose rate for boost irradiation with iridium implants. As shown by MEERTENS and BARTELINK (1984) reference dose rate and implanted volume are clearly correlated, and larger implants implicate higher dose rates because of the greater number of needles, longer source lengths, and larger spacings. In their study, no influence of dose rate (25–75 cGy/h) on the occurrence of breast fibrosis was found if they corrected for implant volumes. MCRAE et al. (1987) studying dose-volume relations and complications in interstitial implants for breast carcinoma pointed out that higher implant volumes bear a greater risk of complications because of the more extensive tissue volumes inside the very high dose rate isodoses directly around the sources. Thus, larger implant volumes should be modified in order to keep the overdosage volumes as small as possible, which implies that for instance spacing of the sources should be modified (decreased), with careful consideration of the resulting reference dose rate. In some patients the usual 2-plant implant will have to be replaced by a 3-plane implant. The importance of the role of brachytherapy in the treatment of early breast cancer would be increased if one were to consider an even more conservative approach by treatment of the primary tumor by tumorectomy followed by an implant of the tumor bed without external beam treatment of the whole breast. From theoretical considerations one could expect a local control rate of 86%–93.6%, which would still be acceptable compared with the reported series.

In conclusion, many questions remain to be answered concerning dose levels, dose rates, and treatment volumes in BCT. In situ carcinoma of the breast seems to be radioresponsive; however, the evidence is only circumstantial. Further clinical research is needed and should be carried out with special attention to all prognostic factors involved and careful microscopic evaluation of tumor margins. In our view, interstitial boost irradiation has advantages over external beam treatment if a higher boost dose is used (20 Gy or higher).

References

Amalric R, Santamaria F, Robert F, Seyle J, Altshuller C, Pietra JC, Amalric F, Kurtz JM, Spitalier JM, Brandare H, Ayme Y, Pollet JF, Bressac C, Fondarai J (1983) Conservation therapy of operable breast cancer – results at five, ten and fifteen years in 2216 consecutive cases. In: Harris JR, Hellman S, Silen W (eds) Conservative management of breast cancer. New surgical and radiotherapeutic techniques. Lippincott, Philadeliphia, pp 15–23

Arriagada R, Mouriesse H, Sarrazin D, Clark RM, DeBoer G (1985) Radiotherapy alone in breast cancer. Analysis of tumor parameters, tumor dose and local control: the experience of the Gustave Roussy institute and the Princess Margaret Hospital. Int J Radiat Oncol Biol Phys 11: 1751–1757

Bartelink H, Borger JH, von Dongen JA, Peterse JL (1988) The impact of tumor size and histology on local control after breast conserving therapy. Radiat Oncol 11: 297–305

Batini JP, Picco C, Martin M, Calle R (1978) Relationship between time-dose and local control of operable breast cancer treated by tumorectomy and radiotherapy or by radical radiotherapy alone. Cancer 42: 2059–2065

Calle R, Vilcoq JR, Pilleron JP, Schlienger P, Durand JC (1983) Conservative treatment of operable breast carcinoma by irradiation with or without limited surgery – ten year results. In: Harris JR, Hellman S, Silen W (eds) Conservative management of breast cancer. New surgical and radiotherapeutic techniques. Lippincott Philadelphia, pp 3–9

Clark RM, Wilkinson RH, Miceli PN, MacDonald WD (1987) Breast cancer. Experiences with conservation therapy. Am J Clin Oncol 10(6): 461–468

Clarke DH, Le MG, Sarrazin D, Lacombe MJ, Fontaine F, Travagli JP, May-Levin F, Conesso G, Arriagada R (1985) Analysis of local-regional relapses in patients with early breast cancers treated by excision and radiotherapy: experience of the Institut Gustave-Roussay. Int J Radiat Oncol Biol Phys 11: 137–145

Connolly JL, Schnitt SJ, Harris JR, Hellman S, Cohen RB (1983) Pathologic correlates of local tumor control following primary radiation therapy in patients with early breast cancer. In: Harris JR, Hellman S, Silen W (eds) Conservative management of breast cancer. New Surgical and radiotherapeutic techniques: Lippincoff, Philadelphia, pp 123–135

Denham JW (1986) The radiation dose-response relationship for control of primary breast cancer. Radiother Oncol 7: 107–123

Fisher B, Bauer M, Margolese R, Poissin R, Pilch Y, Redmund C, Fisher E, Wolmark N, Deutsch M, Montague E, Suffer E, Weekerham L, Luner H, Glass A, Shibata M, Deckes P, Ketcham A, Oischi R, Russel I (1985) Five year results of a randomized trial comparing total mastectomy and segmental mastectomy with or without radiation in the treatment of breast cancer. N Engl J Med 312: 665

Fisher ER, Sass R, Fisher B, Gregorio R, Brown R, Wickerham L (1986) Pathologic findings from the national surgical adjuvant breast project (protocol 6) II, relation of local breast recurrence to multicentricity. Cancer 37: 1717–172

Fletcher GH (1972) Local results of irradiation in the primary management of localized breat cancer. Cancer 29: 545–551

Gerard JP, Montbaron JF, Chassard JL, Romestaing R, Ardiet JM, Delaroche G, Talon B, Papillon J (1985) Conservative treatment of early carcinoma of the breast: significance of axillary dissection and iridium implant. Radiother Oncol 3: 17–22

Ghossein NA, Stacey P, Alpert S, Ager PJ, Krishnaswany V (1978) Local control of breast cancer with tumorectomy plus radiotherapy or radiotherapy alone. Radiology 121: 455–459

Harris JR, Recht A, Amalric R, et al. (1984) Time course and prognosis of local recurrence following primary radiation therapy for early breast cancer. J Clin Oncol 2: 37

Hellman S, Harris JR, Levene MB (1980) Radiation therapy for early carcinoma of the breast without mastectomy. Cancer 46: 988–994

Holland R, Veling SHJ, Matrunac M, Hendriks JHCL (1985) Histologic multifocality of Tis, T1–2 breast carcinomas. Implications for clinical trials on breast conserving therapy. Cancer 56: 979–991

Koper P, van Putten W, The SK, Treurniet Donker AP, Riechgelt BA, Helle PA, Meerwald GH, Seldenrath JJ, Subandono AJ, Wijnmaalen AJ, van Geel AW, Wiggers T, Meischke-de Jone ML (1987) Treatment results of lumpectomy and radiotherapy for early breast cancer, a survey of 966 patients. 4th EORTC Breast Cancer Working Conference, London

Kurtz JM, Amalric R, Delouch G, Pierquin B, Roth J, Spitalier JM (1987) The second ten years: long time risks of breast conservation in early breast cancer. Int J Radiat Oncol Biol Phys 13: 1327–1332

Leung S, Otmetzguine Y, Calitchi E, Mazeron JJ, Le Bourgeois JP, Pierquin B (1986) Locoregional recurrences following radical external beam irradiation and interstitial implantation for operable breast cancer – a twenty three year experience. Radiat Oncol 5: 1–10

McCormick B, Finne D, Petrek J, Osborne M, Cox L, Shank B, Hellman S, Yahalom J, Rosen PP (1987) Limited resection for breast cancer: a study of specimen margins before radiotherapy. Int J Radiat Oncol Biol Phys 13: 1667–1671

McRae D, Rodgers J, Dritchillo A (1987) Dose-volume and complications in interstitial implants for breast carcinoma. Int J Radiat Oncol Biol Phys 13: 525–529

Meertens H, Bartelink H (1984) First experience with the microselectron in breast conserving therapy implants. Proceedings of the 3rd International Selectron Users Meeting 1984, Innsbruck

Nobler MP, Venit L (1985) Prognostic factors in patients undergoing curative irradiation for breast cancer. Int J Radiat Oncol Biol Phys 11: 1323–1331

Pierquin B (1983) Conservation treatment for carcinoma of the breast: experience of creteuil – ten year results. In: Harris JR, Hellman S, Silen W (eds) Conservative management of breast cancer. New surgical and radiotherapeutic techniques. Lippincott, Philadelphia, pp 11–15

Pierquin B, Dutreix A (1967) Towards a new system in curietherapy (endocurie therapy and plesiocurie therapy with non radioactive preparation). Br J Radiol 40: 184–186

Pierquin B, Mazeron JJ, Glaubiger D (1986) Conservative treatment of breast cancer in Europe: report of the Group Europeen de Curietherapie: Radiat Oncol 6: 187–198

Recht A, Connolly JL, Schnitt SJ, Cody B, Love S, Osteen RT, Patterson WB, Shirley R, Silen W, Come S, Henderson G, Silver B, Harris JR (1986) Conservative surgery and radiation therapy for early breast cancer: results, controversies and unresolved problems. Semin Oncol 13: 435–449

Recht A, Connolly JL, Schnitt SJ, Silver B, Rose MS, Love S, Harris JR (1988) The effect of young age on tumor recurrence in the treated breast after conservative surgery and radiotherapy. Int J Radiat Oncol Biol Phys 14: 3–10

Recht A, Silen W, Schnitt SJ, Connolly JL, Gelman RS, Rose MA, Silver B, Harris JR (1990) Time course of local recurrence following conservative surgery and radiotherapy for early stage breast cancer. Int J Radiat Oncol Biol Phys (in press)

Robinson JO (1986) Treatment of breast cancer through the ages. Am J Surg 151: 317–332

Sarrazin D, Le M, Fontaine MT, Arriagada R (1983) Conservative treatment versus mastectomy in T1 or small T2 breast cancer – a randomized clinical or al. In: Harris JR, Hellman S, Silen W (eds) Conservative management of breast cancer. New surgical and radiotherapeutic techniques. Lippincott, Philadelphia, pp 101–111

Timothy AR, Overgaard S, Overgaard M, Wang C (1979) Treatment of early carcinoma of the breast. Lancet 11: 25

Van Dongen JA, Bartelink H, Aaronson N, Fentiman I, Hayward JC, Lernt T, Millis R, Olthuis G, Peterse JL, Rotmensz N, van der Schueren E, Silvester R, The SK, Winter J, van Zijl K (for the EORTC Breast Cancer Group) (1987) Randomized clinical trial to assess the value of breast conserving therapy (BCT) in stage I and stage II breast cancer; EORTC trial 10801 (Abstr.) Proceedings of the 4th European Conference in Clinical Oncology and Cancer Nursing, Madrid

Van Limbergen E, van der Bogaert, van der Schueren E, Rijnders A (1987) Tumor excision and radiotherapy as primary treatment of breast cancer, ananlysis of patients and treatment parameters and local control. Radiother Oncol 8: 1–9

Veronesi K, Delvechio H, Greco M, Luni A, Muscolini G, Rosponi A, Saccozzi R, Zucali R (1983) Results of quadrantectomy axillary dissection and radiotherapy (Quart) in T1N0 patients. In: Harris JR, Hellman S, Silen W (eds) Conservative management of breast cancer. New surgical and radiotherapeutic techniques. Lippincott, Philadelphia, pp 91–101

6 Anal Canal Cancer

6.1 Carcinomas of the Anal Canal: An Introduction

Rolf Sauer and Jürgen Dunst

When talking about anal carcinoma, it is necessary to distinguish between anal *canal* cancer (70%–80%) and anal *margin* cancer (20%–30%): The anal canal extends 3–5 cm from the rectum to the perianal skin. Its upper limit is the anorectal ring and the boundary between the rectal mucosa and the transitional epithelium. The lower limit is the junction to the perianal (hairy) skin, as defined by the UICC.

According to the WHO classification, 90%–95% of the anal canal cancers are squamous cell carcinomas and so-called basaloid carcinomas (synonyms: transitional cell carcinoma, cloacogenic carcinoma, basosquamous carcinoma). The frequency of basaloid cancers varies from 10% to 40% because it is often difficult to distinguish this entity clearly from the squamous cell type. Differentiation between these types is mostly subjective, and mixed histologic types often occur. On the other hand, there is no evidence that the histologic subtype is a prognostic factor, and therefore the histologic subtype (squamous versus basaloid) has no impact on the therapeutic decision.

The lymphatic spread of anal canal tumors goes via two routes. The first one is upwards via the lymphatic vessels along the superior hemorrhoidal artery, causing involvement of the pararectal nodes. This type of spread is mainly seen in tumors of the upper part of the anal canal. The overall incidence of pararectal node involvement in surgical series is 20%–30% and depends on tumor size and grade of malignancy. The lymphatic drainage of the lower part of the anal canal and of the anal margin goes along the second route: Via the lymph vessels along the inferior hemorrhoidal artery to the inguinal nodes. The incidence of inguinal node involvement in anal canal cancer seems to be lower than pararectal node involvement.

Table 1. TNM staging system for anal canal tumors (4th Edn. 1987)

T1:	Tumor 2 cm or less in greatest dimension.
T2:	Tumor greater than 2 cm but not more than 5 cm in greatest dimension.
T3:	Tumor more than 5 cm in greatest dimension.
T4:	Tumor of any size with invasion of adjacent organs (e.g. bladder, vagina, urethra). Sphincter muscle involvement alone is not classified as T4.
N0:	No regional lymph nodes.
N1:	Metastasis in pararectal lymph node(s).
N2:	Metastasis in unilateral internal iliac and/or inguinal lymph node(s).
N3:	Metastasis in perirectal and inguinal nodes and/or metastasis in bilateral iliac and/or inguinal lymph nodes.

Note: Sphincter muscle involvement may be present in T1–3 tumors. Regional nodes are the pararectal, internal iliac, and inguinal nodes. The pT and pN classification corresponds to the clinical T and N classification.

In the actual TNM staging system, the T stage is defined by the clinically measureable tumor diameter (Table 1) and no longer by the infiltration of the sphincter muscle and rectum. The new T classification is easy to use even without radical surgery and is better applicable in organ-sparing treatment procedures. Regional nodes of the anal canal are the pararectal, internal iliac, and inguinal nodes.

The anal margin is the perianal skin. The most common tumors of this region are squamous cell cancers or basal cell carcinomas (not basaloid cancers!). The regional lymph nodes are the inguinal nodes.

Anal cancers account for about 1% of all bowel malignancies. Women are more often affected than men (female: male ratio = 1.5–3:1). Anal margin cancers are rare with a predominance in men. An increased incidence is seen in homosexual men, and an association with chronic dis-

Rolf Sauer, Prof. Dr., Jürgen Dunst, Dr. Strahlentherapeutische Klinik der Universität Erlangen-Nürnberg Universitätsstr. 27, 8520 Erlangen, Germany eases of the anus like hemorrhoids, fistulae, or condylomata acuminata has also been reported.

For practical clinical use, the differentiation between anal canal and anal margin is mandatory. Anal margin tumors are often accessible to primary surgery without mutilation whereas anal canal tumors require a colostomy in the case of definitive surgical treatment. The following articles deal with conservative treatment strategies in anal canal cancer.

References

Daling JR, Weiss NS, Hislop TG, Maden C, Coates RJ, Sherman KJ, Ashley RL, Beagrie M, Ryan JA, Corey L (1987) Sexual practices, sexually transmitted diseases, and the incidence of anal cancer. N Engl J Med 317: 973–977

Hager T, Hermanek P (1986) Maligne Tumoren der Analregion. In: Gall FP, Hermanek P, Tonak J (eds) Chirurgische Onkologie. Histologie – und stadiengerechte Therapie maligner Tumoren. Springer, Berlin Heidelberg New York, pp 581–589

Hermanek P, Sobin LH (eds) TNM Classification of malignant tumours, 4. fully revised edition, International Union Against Cancer Geneva, Springer, Berlin Heidelberg New York

Mitchell E (1988) Carcinoma of the anal region. Sem Oncol 15: 146–153

6.2 Role of Combined Radiochemotherapy: The Lyon Experience

J. PAPILLON, J.F. MONTBARBON, J.P. GERARD, J.L. CHASSARD, J.M. ARDIET, and P. TOURAINE-ROMESTAING

CONTENTS

1 Introduction . . . 209
2 Technique and Patients . . . 209
3 Results . . . 210
3.1 Toxicity . . . 210
3.2 Combined External Beam and Interstitial Irradiation . . . 211
3.3 Radiochemotherapy . . . 211
4 Discussion . . . 212
References . . . 213

1 Introduction

The management of carcinoma of the anal canal has long been a subject of controversy between surgeons and radiation oncologists. During the past few years the view of most enterologists and surgeons has changed dramatically. Radical surgery as the initial treatment has been abandoned or questioned in most institutions (BOMAN et al. 1984; PARKS 1981; QUAN et al. 1978), and local excision is reserved for clinically benign lesions to be followed by irradiation if malignant tissue is found in the operative specimens.

Progress made in the field of radiotherapy has improved its effectiveness, and there is now a strong tendency in favor of it, preferably associated with concomitant chemotherapy. The present problem is the choice of the most effective strategy capable of obtaining the highest rates of control and anal preservation. Several series of patients treated by external beam irradiation alone have been published (SALMON et al. 1984; ESCHWEGE et al. 1985). The protocol commonly used aims at delivering a dose of 50 Gy in 5–6 weeks to the whole pelvic cavity by two opposed parallel fields or box technique, plus a boost of up to 60 Gy to anal and/or inguinal areas. This method of regularly fractionated irradiation gives a rather high rate of control for small, slightly infiltrating tumors, but the results for tumors larger than 4 cm and those that are deeply infiltrating are not satisfactory.

Carcinoma of the anal canal is a very unique disease which should not be treated by irradiation according to the same guidelines as those applied to most pelvic tumors, such as uterine cervix carcinoma. In the absence of inguinal metastases, the lymphatic drainage areas are situated in the posterior part of the pelvic cavity, and the use of an anterior field is not justified. Moreover, irradiation of the anal area by two opposed fields with the patient in the prone position often gives rise to severe acute reactions with moist desquamation of the perineal skin, which can be easily avoided by using better adapted fields.

2 Technique and Patients

In the selection of the most effective technique of irradiation it is necessary to take into consideration the main particular features of anal canal carcinoma such as (1) perfect accessibility, which means easy assessment of the local spread and easy interstitial curietherapy; (2) frequency of pelvic metastatic lymph nodes, which must be systematically searched for by palpation and CT scan even at an early stage makes irradiation of the posterior lymphatic drainage areas compulsory in all cases; (3) high radiosensibility; (4) proneness of anal and perineal structures to radionecrosis; (5) one of the most peculiar and overlooked characteristics is the slow rate of regression after irradiation, especially with large, infiltrating tumors.

J. PAPILLON, Dr., Professeur à la Faculté, Radiologiste des Hòpitaux, 12, Quai Général-Sarrail, 69006 Lyon, France

J.F. MONTBARBON, Dr., J.L. CHASSARD, and Dr., J.M. ARDIET, Dr., Centre Léon Bérard, Départment de Radiotherapie, 28, rue Laënnec, 69008 Lyon, France

J.P. GERARD, Dr. and P. TOURAINE-ROMESTAING, Dr., Département de Radiotherapie, Hôpital Lyon Sud, 69310 Pierre Bénite, France

Optimal shrinkage is achieved 2 months after completion of irradiation. After such a delay, large tumors may have shrunk such that palpation reveals no residual lesion.

The protocol of irradiation used at the Centre Léon Bérard since 1974 takes into account all these characteristics. In this study are excluded the 43 patients treated between 1971 and July 1974. At that time external beam was only directed to the anoperineal area. Eight instances of pelvic failure prompted us to irradiate the posterior part of the pelvis in all cases. A total of 278 patients have been seen and treated between July 1974 and March 1985 and followed for more than 3 years. Unresectable or disseminated tumors were seen in 32 (11.5%) and treated palliatively. Resectable tumors not suitable for conservation because of complete anal stenosis or involvement of the vaginal mucosa were treated by radiotherapy and delayed abdominoperineal resection. The rate of control in this group of 25 patients is 64%.

In 221 patients with tumor suitable for conservation, a split-course treatment combining external beam and interstitial irradiation was applied. The first stage of this protocol is based on external beam irradiation, which delivers a minimum depth dose of 35 Gy in 15 fractions within 19 days through perineal and sacral fields (Fig. 1).

Fig. 1. Split-course treatment combining external beam and interstitial irradiation by single-plane, iridium-192 implant 2 months later. Cobalt 60 or photons aim to give a minimum tumor dose of 3500 rad in 15 fractions over 19 days through perineal and sacral fields

The technique has been previously described (Papillon 1982; Papillon and Montbarbon 1987). The second stage is the iridium-192 single-plane implant, which takes place after a 2-month rest. At that time, in most patients, there is no clinical evidence of tumor or only a small area of induration representing the point of origin of the lesion. Interstitial curietherapy aims to give a booster dose of 10–20 Gy to the initial site of the tumor.

3 Results

3.1 Toxicity

Patient tolerance of external beam irradiation is good with a short period of mild proctitis and irritation of the anal area without moist desquamation. After iridium-192 implant there is no reaction at all. As to the long-term side effects, it must be emphasized that there is never any perianal fibrosis. The perineal skin and soft tissues remain supple, without any change in consistency or patient discomfort in the sitting position. This demonstrates the perfect tolerance or irradiation through the perineal portal. Four patients (1.8%) developed severe anal radionecrosis requiring abdominoperineal resection or colostomy. Two patients developed severe rectal bleeding due to telangiectasia of the rectal mucosa, related to faulty irradiation technique or a high dose to pelvic metastatic masses. Both these patients are alive and well after more than 5 years, one having

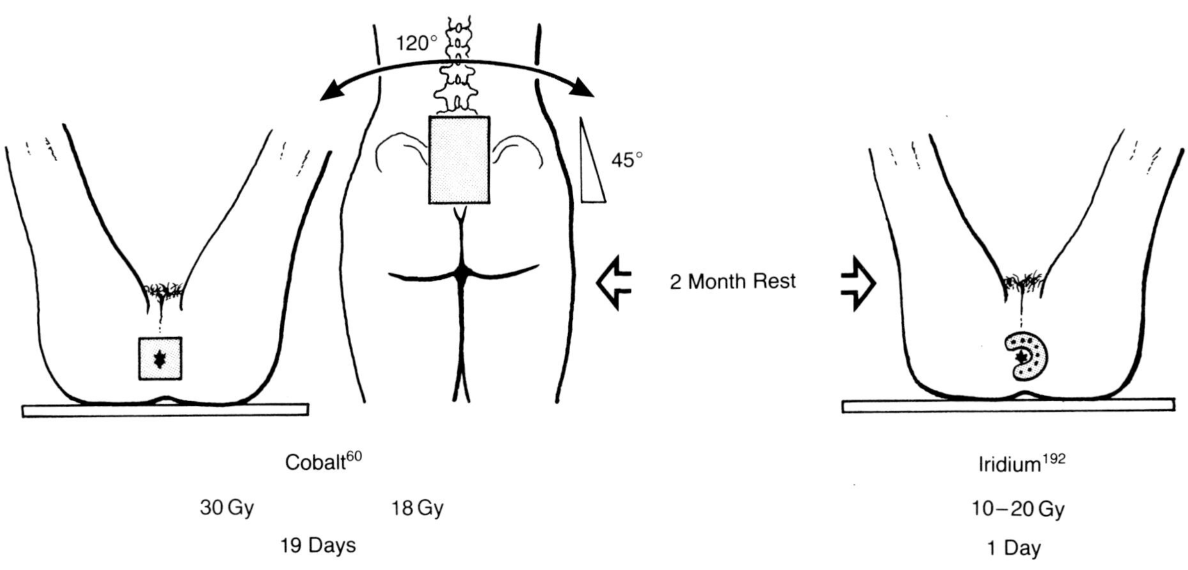

Table 1. Results of combined treatment by cobalt 60 plus iridium 192 ± 5FU, MTC for resectable epidermoid carcinoma of the anal canal (series of the Centre Léon Bérard, Lyon, 1974–1985)

	n	Alive and well	Alive with anal preservation	Death of cancer	Death of intercurrent disease	Postoperative death
At 3 years	221	153 (69.2%)	140 (63.3%)	37[a] (16.7%)	28 (12.6%)	3 (1.3%)
At 5 years	179	118 (65.9%)	110 (61.4%)	33[a] (18.4%)	25 (13.9%)	3 (1.7%)

[a]Among the deaths of cancer are included patients alive at 3 and 5 years with incurable disease.
5FU, 5-fluorouracil; MTC, mitomycin C.

Table 2. Results of combined treatment as a function of tumor size for resectable epidermoid carcinoma of the anal canal (series of the Centre Léon Bérard, Lyon, 1974–1983)

Tumor size	*n*	Alive and well	Alive with anal preservation	Death of cancer	Death of intercurrent disease	Postoperative death
≤4 cm	66	53 (80.3%)	50 (75.7%)	5 (7.5%)	7 (10.6%)	1 (1.5%)
>4 cm	113	65 (57.5%)	60 (53%)	28 (24.7%)	18 (15.9%)	2 (1.7%)

undergone a colostomy, the other an abdominoperineal resection.

The irradiation technique may be adapted according to each individual patient. In those with early superficial lesions the dose delivered by interstitial curietherapy is only 15 Gy. When a pelvic metastatic lymph node is discovered at palpation, a booster dose up to 45 Gy in 3 weeks is given through a direct fixed sacral field and/or by iridium-192 implant.

3.2 Combined External Beam and Interstitial Irradiation

In Tables 1 and 2 give the overall results and the incidence of tumor size on the prognosis, respectively.

In this protocol the use of perineal field combined with iridium-192 implant has proven to be efficient in the control of the primary tumor. The low rate of pelvic failures (5%) demonstrates the efficacy of this external beam irradiation technique. The results at 3 and 5 years are similar. This suggests that patients alive and well at 3 years have the best chance of being definitively cured. The rate of death of cancer at 5 years is 18.4%. In the interpretation of this rate one must take into consideration that 63% of patients had infiltrating tumors larger than 4 cm. In this subgroup the rate of death of cancer is 24.7%. Among patients treated conservatively and clinically disease-free at 5 years, 92% had retained a normal anal function (94% with small tumors and 91% with deeply infiltrating cancers larger than 4 cm).

A comparison of the results obtained by external beam irradiation using conventional fractionation with those achieved by combined external beam and interstitial irradiation in a split-course therapy shows a clear superiority of the protocol used at the Centre Léon Bérard, especially for sphincter preservation in patients with deeply infiltrating tumors larger than 4 cm in size.

3.3 Radiochemotherapy

One of the major changes which occurred in the history of the management of anal canal cancer dates from 1974 with the first report by NIGRO et al. about an original protocol of preoperative radiochemotherapy using concomitantly radiotherapy with 5-fluorouracil (5FU) and mitomycin C (MTC). This experience conducted by a surgeon in an attempt to improve the results of abdominoperineal resection showed that a moderate dose of radiation associated with chemotherapy was able to control some tumors which would have not been controlled by irradiation or chemotherapy alone. The NIGRO (1984) report had a decisive impact on the treatment strategy applied to carcinoma of the anal canal. Many investigators confirmed the original data, and combined radiochemotherapy was from that time on used in an attempt to minimize surgery and potentially to avoid abdominoperineal resection.

The regimens used by the several authors differ markedly in many regards, such as the dosage of radiation, the number of courses, the dosage of chemotherapy, as well as the role and type of surgery. There are big differences in treatment tolerance. CUMMINGS et al. (1984) reported a series of 55 patients treated by whole pelvic irradiation at a dose of 50 Gy associated with two courses of chemotherapy. Fifteen patients (27%) developed severe toxicity, seven requiring colostomy, whereas many authors do not mention any severe complications.

Presently it is admitted that the efficacy of radiochemotherapy is due to an additive effect of both modalities or to a potentiation of radiotherapy by chemotherapy. In most studies the first course of chemotherapy was performed during the first days of radiotherapy. MICHAELSON et al. (1983) at the Memorial Sloan Kettering Cancer Center used chemotherapy prior to irradiation. The results obtained by them demonstrate that induction chemotherapy followed by irradiation is less effective than concomitant radiochemotherapy as initiated by NIGRO et al. (1974).

The current problem is to identify the most effective radiochemotherapy plan able to give the highest rate of sphincter preservation.

Radiotherapy by itself must be capable of controlling a large proportion of tumors. In this regard the split-course regimen of a short course of external beam irradiation directed to the primary tumor by direct perineal field and to the posterior pelvic drainage areas followed 2 months later by an iridium-192 implant has demonstrated its efficacy. The chemotherapy program should be defined in order to prevent acute or late toxicity without altering the timing of the radiotherapy protocol. At the Centre Léon Bérard the dosage of chemotherapy was intentionally reduced to allow a good tolerance. Between 1977 and 1984, 5FU was used at a dose of 600 mg/m^2 per day from day 1 to day 4, and MTC was used at a dose of 12 mg/m^2 on day 1. This combined treatment was applied to 89 robust, middle-aged patients with T3 tumor larger than 4 cm. Retrospectively, the results of radiochemotherapy were compared with a group of 78 patients with matched tumors treated with the same protocol of irradiation whithout chemotherapy because of their age or general condition (Table 3). The study of the rates of local failure shows a statistically significant difference. These results prompted us to use systematically concomitant radiochemotherapy since 1985 in all patients with anal canal carcinoma, regardless of stage of the disease and age of the patients. In case of elderly, frail patients in poor condition, the dosage of 5FU is reduced to 500 mg/m^2 for 4 days and MTC is excluded. In robust patients 5FU is applied at a dose of 750 mg/m^2 and MTC at 12 mg/m^2.

Table 3. Radiotherapy alone vs radiotherapy plus concomitant chemotherapy in epidermoid carcinoima T3 of the anal canal (series of the Centre Léon Bérard, Lyon)

	n	Rate of local failures
Radiotherapy alone (^{60}Co + ^{192}Ir)	78	30%
Radiotherapy (^{60}Co + ^{192}Ir) plus 5-fluorouracil and mitomycin C	89	10% ($P = 0.02$)

4 Discussion

In this strategy surgery has no place in the initial treatment except in 2 cases: (a) When the involvement of the entire circumference of the sphincter results in obstructive symptoms, a permanent colostomy must be carried out before irradiation; (b) with synchronous inguinal metastases a superficial groin dissection is justified beforehand. Lymphangiography is advisable to investigate the status of retrocrural and external iliac lymph nodes and to guide the extent of the surgical dissection. Irradiation of inguinal areas combining photons (30 Gy) and electrons (15 Gy) should be started 2 or 3 weeks later without interfering with the usual protocol of treatment of the primary tumor.

In the absence of clinically involved inguinal nodes, elective groin dissection is not recommended because of the small number of positive nodes found and because of the side effects of this operation. On the other hand, prophylactic irradiation is advisable in case of T_3 tumor or tumor extension to the anocutaneous junction or when the patient is unable to be followed up regularly.

Subsequent abdominoperineal resection is necessary if lack of control of the primary tumor, local recurrences, or severe radiation-induced complications arise.

During the past 5 years some authors have published enthusiastic reports with control rates ranging from 71% to 100%. Usually the number of patients is small and the follow-up short. No mention is made of unresectable cancers or of the presence of pelvic metastic lymph nodes. This suggests that some selection of patients, even if not mentioned, has been applied in these statistics.

In the interpretation of the results of the Lyon experience it must be taken into account that most of our records date from the previous decade. At that time computerized tomography scanning was not readily available in our institution. This investigation method has substantially improved the detection of pelvic metastases and clinical staging.

To conclude, if one accepts that split-course association of external beam and interstitial irradiation is more efficient than external beam radiotherapy alone and that chemotherapy potentiates the irradiation effect, it becomes logical to associate a sophisticated protocol of irradiation with concomitant chemotherapy. In our opinion it is the best way to give each individual patient both the highest chance of cure and good quality of life after treatment. Such a strategy implies great care in patient and tumor selection, in irradiation technical details, and in follow-up.

References

Boman BM, Moertel CG, O'Connell MJ, Scott M, Weiland LH, Beart RW, Gunderson LL, Spencer RJ (1984) Carcinoma of the anal canal. A clinical and pathologic study of 188 cases. Cancer 54: 114–125

Cummings B, Keane T, Thomas G, Harwood A, Rider W (1984) Results and toxicity of treatment of anal canal carcinoma by radiation therapy or radiation therapy and chemotherapy. Cancer 54: 2062–2068

Eschwege F, Lasser P, Chavy A, Wibault P, Kac J, Rougier P, Bognel C (1985) Squamous-cell carcinoma of the anal canal: treatment by external beam irradiation. Radiother Oncol 3: 145–150

Michaelson RA, Magill GB, Quan SHQ, Leaming RH, Nikrui M, Stearns MW (1983) Preoperative chemotherapy and radiation therapy in the management of anal epidermoid carcinoma. Cancer 51: 390–395

Nigro MD (1984) An evaluation of combined therapy for squamous-cell cancer of the anal canal. Dis Colon Rectum 27: 763–766

Nigro MD, Vaitkevicius VK, Considine BJ (1974) Combined therapy for cancer of the anal canal. A preliminary report. Dis Colon Rectum 17: 354

Papillon J (1982) Rectal and anal cancers. Conservative treatment by irradiation. An alternative to radical surgery. Springer, Berlin Heidelberg New York

Papillon J, Montbarbon JF (1987) Epidermoid carcinoma of the anal canal. A series of 276 cases. Dis Colon Rectum 30(5): 324–333

Parks A (1981) Squamous carcinoma of the anal canal. Ann Gastroenterol Hepatol (Paris) 17: 103–107

Quan SHQ, Gordon B, Magill MD, Leaming RM, Majdu SI (1978) Multidisciplinary preoperative approach to the management of epidermoid carcinoma of the anus and anorectum. Dis Colon Rectum 21: 89–91

Salmon RJ, Fenton J, Asselain B, Mathieu G, Girodet J, Durand JC, Decroix Y, Pilleron JP, Rousseau J (1984) Treatment of epidermoid anal canal cancer. Am J Surg 147: 43–48

6.3 Combination Therapy of Anal Canal Cancer: A Report on External Irradiation with or Without Chemotherapy Followed by Interstitial Iridium 192

GUDRUN PIPARD

CONTENTS

1 Introduction 215
2 Patient Selection 215
3 Treatment Methods 216
4 Results 217
5 Conclusions 218
References 218

1 Introduction

This report is based on 10 years of personal experience. The purpose of the analysis is to document the value of conservative radiotherapy with the aim of curing cancer of the anal canal. The anal canal was defined according to the UICC recommendations, comprising a region distal from the anorectal ring to the anal verge. This divides the anal canal in a suprapectinate and an infrapectinate part. Carcinoma of the perianal skin (anal orifice) was excluded from this presentation. The potential role of rescue abdominoperineal resection will be documented and the favorable general outcome in patients presenting with synchronous inguinal metastatic lymph nodes analyzed.

2 Patient Selection

A total of 95 patients suffering from carcinoma of the anal canal were seen in consultation at our institution between October 1976 and September 1987. Treatment modalities and respective numbers of treated patients are shown in Table 1. Of these, 89 were accepted for primary radiotherapy (RT) with or without chemotherapy (CT) with intent to cure: 86 patients were treated by external irradiation with or without chemotherapy followed by an interstitial iridium implant, 2 patients by interstitial iridium alone, and –1 patient by external irradiation combined with chemotherapy. This accounts for 91% of the patients seen.

Table 1. Treatment modalities used from October 1976 to September 1987 with respect to numbers of patients and period of follow-up (n = 95)

	Treatment modality	Patients at follow-up	
		>24 months	≤24 months
Squamous or basaloid carcinoma	RT + Ir without CT	39	1
	RT + CT + Ir	29	15
	Ir alone	1	1
	Preoperative RT + CT followed by APR	3	
	RT alone + CT	1	
	Palliative RT	2	1
Adenocarcinoma	RT + CT + Ir	1	1

RT, external irradiation; CT, chemotherapy; Ir, interstitial iridium; APR, abdominoperineal resection.

Six patients were not treated conservatively: Three were scheduled for preoperative pelvic and perineal irradiation followed by planned abdominoperineal resection (APR), indicated by pelvic lymph node metastases, and the three others received palliative irradiation, indicated on behalf of their poor medical or psychological status.

Only the group of 86 patients treated by RT with or without CT followed by interstitial iridium was considered large enough for meaningful analysis. The elapsed time ran from start of treatment to the last follow-up in September 1987. Some 68 patients have at least 24 months of

GUDRUN PIPARD, MD, Division of Rdiotherapy, University Hospital of Geneva, 21, rue Alcide Jentzer, 1211 Geneva 4, Switzerland. Address for correspondence: 23, rue St. Martin, 74160 Julien en Genevois, France follow-up, forming the basis of this analysis. The longest time to relapse seen in the Geneva series was 18 months after radiotherapy. Thus, it is hoped that the results reported here will provide reliable information concerning the value of primary curative radiotherapy in anal canal cancer.

Carcinoma were classified according to standard criteria into well or poorly differentiated squamous, basaloid, or transitional cell forms. All pathologic subtypes of anal canal cancers were accepted for primary radical radiation therapy.

The TNM/UICC classification in use during the reported treatment period was not accepted by us as a valuable guide for prognosis and end-result evaluation in our primary nonsurgical treatment procedure. Tumor stage classification was done according to the Centre Leon Berard's classification, recommended by PAPILLON and MONTBARBON (1987)

T1 – Tumors less than 2 cm in maximum diameter
T2 – Tumors from 2 to 4 cm
T3 – Tumors >4 cm
T4a – Tumors ulcerating the vaginal mucosa
T4b – Tumors fixed to adjacent pelvic and perineal structures

Encroachment of a small tumor (e.g., 2–3 cm) on the perianal skin or the lower rectum did not classify it as T4.

The age of the patients ranged from 35 to 92 years, with a mean age of 65 years; 26 patients were older than 70 years. There were 52 female and 16 male patients. Some 16 patients had tumors up to 4 cm in maximum diameter, while 52 patients had tumors >4 cm. Six had synchronous inguinal metastatic nodes, one bilateral and five unilateral ones. Fine needle puncture or excisional biopsy proof of nodal involvement was obtained. Since 1980 most of our patients undergo computerized tomography (CT) examination in order to disclose eventually enlarged perirectal or hypogastric nodes at initial work-up. Only three patients with enlarged perirectal nodes and two with iliac nodes were seen. There was no pretherapeutic biopsy proof of the perirectal and iliac nodes suspected to be metastatic. None of the 68 patients had distant metastases at the time of presentation.

3 Treatment Methods

The radiotherapy modalities used are summarized in Table 2. Small tumors up to 4 cm in diameter

Table 2. Methods of external irradiation with or without chemotherapy followed by interstitial iridium

Method	Tumor size	Period	*n*
Cobalt plus iridium	≤4 cm	Oct. 76–Oct. 85	16
	>4 cm	Oct. 76–June 80	23
Photons (10 MV) plus chemotherapy plus iridium	>4 cm	July 80–Oct. 85	29

(*n* = 16 patients) were treated by radiotherapy alone without chemotherapy according to Papillon's technique. A direct perineal field, usually 8 by 8 cm, in the lithotomy position on the cobalt machine was established. Then 30 Gy in 10 fractions were given over 2.5 weeks, supplemented by 18 Gy to a transsacral field directed to the mesorectal nodes. After a planned rest period of 6 weeks, interstitial iridium was implanted for an additional dose of 10–15 Gy to the primary tumor bed.

Big tumors >4 cm were treated up to June 1980 by the same technique as for small tumors without any chemotherapy (*n* = 23). Since July 1980 cancers >4 cm in maximum diameter received concomitant radiochemotherapy followed by interstitial iridium (*n* = 29 patients). In the concomitant radiochemotherapy series the external irradiation was given by parallel opposed pelvic and perineal fields on a linear accelerator, 10 MeV, the patient lying in the prone position. The upper limit of the fields was at the level of S1–2, the lower limit 2 cm beyond the most distal part of the tumor. Laterally the fields comprised the pelvic rim. No blocking of the femoral heads was established in order to give elective irradiation to the medial inguinal nodes, even when clinically not involved. Absorbed dose was calculated in the mid-pelvis at axis; dose per fraction was 2 Gy and total dose, 30 Gy. 5-fluorouracil was given in a 24-h drop infusion during the first 5 days of radiotherapy (800–1000 mg/m^2 daily) and mitomycin C was given in a unique bolus injection on day 1 (0.4 mg/kg, maximum dose 20 mg). No second cycle of chemotherapy was given.

The method used for the interstitial iridium implant was always the same and was done with a standard active needle length of 7 cm. Only the total number of needles implanted (usually 5–6) varied according to the initial tumor extent. In general, a single range of transperineal needles was implanted in a submucosal position. Volume implants were unusual and only done in the very rare clinical condition of insufficient tumor regression 6 weeks after external irradiation in patients formally refusing radical surgery. Then, 15 Gy were given to the reference isodose (Paris system calculation) over 20–25 h.

Treatment of synchronous inguinal nodes consisted in excisional biopsy of enlarged nodes without any radical inguinofemoral lymph node dissection and subsequent postoperative irradiation of 60 Gy to the involved inguinal fold and 45 Gy to the ipsilateral intrapelvic iliac draining route. This irradiation to the involved inguinal area was done at the same time as the external irradiation for the primary anal canal carcinoma.

Metachronous inguinal nodes were managed individually depending on the existence or lack of previous inguinal irradiation. Only in the radiochemotherapy group for T > 4 cm (n = 29) was the medial part of the inguinal fold prophylactically irradiated. In patients treated according to Papillon's technique N0 inguinal patients did not receive elective inguinal radiation therapy.

Table 3. Treatment results of primary conservative irradiation with or without chemotherapy

	Tumor ≤4 cm	Tumor >4 cm	
	RT + Ir (n = 16)	RT + Ir (n = 23)	RT + CT + Ir (n = 29)
Alive, no evidence of disease	62.5% (10/16)[a]	65% (15/23)	75.8% (22/29)
Dead of intercurrent disease without cancer	18.8% (3/16)	8.7% (2/23)	10.3% (3/29)
Dead of cancer or intercurrent disease with persistent local and/or metastatic cancer	18.8% (3/16)[a]	26.1% (6/23)	13.8% (4/29)
Local control and sphincter preserved	75% (12/16)[a]	69% (16/23)	79.3% (23/29)
Final local control after APR	87.5% (14/16)[a]	78.3% (18/23)	89.6% (26/29)
Severe necrosis requiring surgery	12.5% (2/16)	4.3% (1/23)	0%

[a] One patient lost to follow-up after 36 months was reported dead of disease and supposed to have local recurrence. APR abdominoperineal resection; RT, external irradiation; CT, chemotherapy; Ir, interstitial iridium.

4 Results

Treatment results are shown in Table 3, tumors being subdivided according to the above-described irradiation modalities with or without concomitant chemotherapy. It is surprising to see that the differences in local control of advanced tumors >4 cm in maximum diameter was only slightly improved in the concomitant radiochemotherapy arm. This is a nonrandomized series with small patient numbers, and formal conclusions are not possible. Local control with normal sphincter preservation was possible in 75% of patients (n = 16) presenting with T up to 4 cm treated by radiationtherapy alone without any chemotherapy. After rescue surgery the level of local control rose in this group to 87.5%. In patients presenting with T > 4 cm local control was 69% without (n = 23) concomitant chemotherapy and 79.3% with (n = 29). Final local control after radical rescue surgery was 78.3% and 89.6%, respectively.

Few patients died of local recurrence despite rescue surgery: 14 of 68 patients had a component of local recurrence. Ten were fit for radical rescue surgery, of whom five were saved, and five died of anal canal cancer. Two patients refused surgery for psychological reasons, and two others were in too poor a medical condition for surgery.

A total of 18.8% of patients with small tumors and 25.6% patients tumors >4 cm died of cancer or intercurrent disease with persistant local and/or metastatic cancer. Thus, 8 of 68 patients (12%) died of intercurrent disease without cancer.

Amongst the six bearers of synchronous metastatic inguinal nodes, two died of metastases to the liver and lung, while locoregionally controlled. Some 66% are alive without evidence of disease.

Amongst the four patients with metachronous inguinal nodes (two without and two with recurrent local cancer), two are alive and well, while the other two died of disease. Primary radiation therapy combined or not with chemotherapy did not jeopardize survival chances when compared with survival figures for primary radical surgery. Furthermore, this series comprises very old and

Table 4. Complications at least 6 months after the end of radiation therapy

Severe radiation necrosis requiring radical surgery, no cancer on pathological specimen (2/68)	2.9%
Necrosis of the perianal skin managed by a graft, no amputation (1/68)	1.5%
Moderate stenosis of the anal canal (3/68)	4.4%
Telangiectasia, occasional bleeding, no anemia requiring transfusions (14/68)	20.6%
Partial incontinence (1/68)[a]	1.5%

[a] This patient had wide local excision which positive margins before postoperative irradiation.

poor-risk patients not fit for radical surgery. There was no death related to treatment.

Acute complications were essentially seen in the concomitant radiochemotherapy series. Dermatitis, diarrhea, mucositis, and hematological toxoicity were tolerable and mostly managed on an outpatient basis. In three patients with a prior history of pelvic surgery a small-bowel occlusion required interruption of the external irradiation and laparotomy. They had local control despite a slightly reduced amount of external irradiation. The dose of the subsequent interstitial treatment was increased in order to compensate. The interstitial iridium-192 implant was well tolerated. It can be carried out under local anesthesia. Table 4 shows the late complication rate determined at least 6 months after the end of radiation therapy.

5 Conclusions

Anal canal carcinoma is no longer treated in Geneva by primary radical surgery, based on the favorable results of function-preserving radiotherapy. As many as 91% of all patients could benefit from primary conservative irradiation. The use of interstitial iridium as a boost to the primary tumor preserves excellent healing conditions in case of the necessity of rescue surgery. Surgery is limited to our institution to the role of assessing the diagnosis by a simple biopsy. Attempts to debulk the tumor by partial excision are strongly discouraged. This would definitely compromise the functional end result of radiotherapy. Radical surgery is in our view indicated for severe radiation necrosis and recurrence of the primary tumor. The presence of synchronous

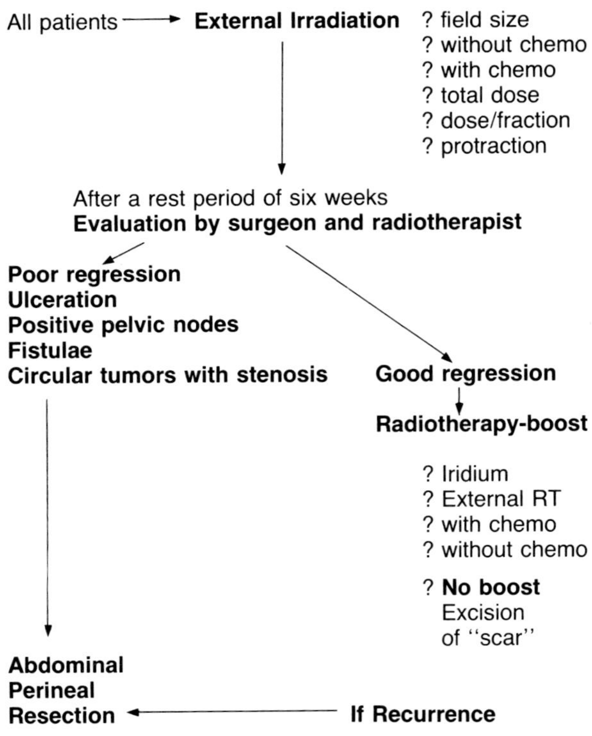

Fig. 1. Treatment proposal for cancer of the anal canal. The question marks show the numerous issues left unresolved when trying to optimize radiation therapy of anal canal carcinoma. Randomized trials with international collaboration are required

positive inguinal nodes does not prevent conservative management of anal canal cancer. Preoperative radiotherapy, 30–40 Gy, combined with chemotherapy should be discussed in patients with rectovaginal fistulae, circumferential tumors with severe stenosis of the anal canal, and when pelvic nodes have been diagnosed on pretherapeutic CT scan. Radical surgery should also be considered in the case of poor tumor regression of persistent large ulceration 6 weeks after the external irradiation with or without chemotherapy. Figure 1 reviews these indications. In the right margin the numerous question left unresolved when trying to optimize primary radical radiation therapy for anal canal carcinoma are shown. It is not feasible to suggest a sole and unique treatment strategy for all stages of these cancers.

References

Boman BM, Moertel CG, O'Connell MJ, Scott M, Weiland LH, Beart RW (1984) Carcinoma of the anal canal. A clinical and pathologic study of 188 cases. Cancer 54: 114–125

Cantril ST, Green JP, Schall GL, Schaupp WC (1983) Primary radiation therapy in the treatment of anal carcinoma. Int J Radiat Oncol Biol Phys 9: 1271–1280
Cummings B, Kean TJ, Thomas G, Harwood A, Rider W (1984) Results and toxicity of the treatment of anal canal carcinoma by radiationtherapy or radiationtherapy and chemotherapy. Cancer 54: 2062–2068
Greenall MJ, Quan SHQ, DeCosse JJ (1985) Epidermoid cancer of the anus. Br J Surg [Suppl] 72: 97–103
Greenall MJ, Magill GB, Quan SHQ, De Cosse JJ (1986) Recurrent epidermoid cancer of the anus. Cancer 57: 1437–1441
John MJ, Flam M, Lovalvo L, Mowry PA (1987) Feasibility of non surgical definitive management of anal canal carcinoma. Int J Radiat Oncol Biol Phys 13: 299–303
Nigro MD (1984) An evaluation of combined therapy for squamous cell carcinoma of the anal canal. Dis Colon Rectum 27: 763–766
Papillon J, Montbarbon MD (1987) Epidermoid carcinoma of the anal canal. Dis Colon Rectum 30: 324–334
Pipard G (1990) Malignant anal tumors. In: Marti MC, Givel JC (eds) Surgery of ano-reclal diseases. Springer, Berlin Heidelberg New York, pp 163–184
Salmon RJ, Fenton J, Asselin B, Mathieu G, Girodet J, Durand JC (1984) Treatment of epidermoid anal canal cancer. Am J Surg 147: 43–48
Schank B (1985) Treatment of anal canal carcinoma. Cancer 55: 2156–2162

6.4 Radiochemotherapy of Anal Canal Cancer With and Without Interstitial Implants

JÜRGEN DUNST, GERHARD GRABENBAUER, NORBERT WOLF, FRANZ PAUL GALL, and ROLF SAUER

CONTENTS

1 Introduction . . . 221
2 Patients and Treatment Protocol . . . 221
2.1 Patients . . . 221
2.2 Radiochemotherapy . . . 221
2.3 Interstitial Irradiation . . . 222
2.4 Reevaluation . . . 222
3 Treatment Results . . . 222
3.1 Response . . . 222
3.2 Interstitial Boost and Local Control . . . 222
3.3 Complications . . . 223
3.4 Salvage Surgery . . . 223
3.5 Survival and Sphincter Function . . . 224
4 Discussion . . . 224
4.1 Value of Interstitial Implants . . . 224
4.2 Value of Rebiopsies After Treatment . . . 225
References . . . 226

1 Introduction

Radiochemotherapy (external irradiation plus simultaneous chemotherapy) for anal canal cancer was introduced by NIGRO et al. (1974) as a preoperative regimen prior to radical surgery and has been further developed to a primary treatment with the aim of function preservation. In Erlangen, we started a radiochemotherapy-based treatment protocol for anal canal carcinoma in 1985. The following article summarizes our preliminary results.

ROLF SAUER, Prof. Dr. med., JÜRGEN DUNST, Dr. med., GERHARD GRABENBAUER, Dr. med., Strahlentherapeutische Klinik, Universitätsstr. 27, 8520 Erlangen, Germany

NORBERT WOLF, PD Dr. med., FRANZ PAUL GALL, Prof. Dr. med., Chirurgische Klinik, Maximiliansplatz, 8520 Erlangen, Germany

2 Patients and Treatment Protocol

The anal canal was defined in accordance with the UICC definition (HERMANEK and SOBIN 1987) as extending from the transitional zone to the junction of the perianal hairy skin.

2.1 Patients

From June 1985 to July 1989, 36 patients with anal canal carcinoma confined to the pelvis (primary tumor or local recurrence after local excision) were treated. Two of them received no chemotherapy (one patient with HIV infection, one patient refused). Four additional patients are not included in this analysis for the following reasons: progressive systemic disease during therapy (two patients), synchronous metastatic breast cancer (one patient), intercurrent death during therapy (one patient).

Of the 30 patients evaluated, 26 received radiochemotherapy as primary treatment, and 4 were treated for pelvic recurrences after local excision. A total of 27 patients had macroscopic tumor at the start of treatment, while 3 had undergone a visibly complete but histologically incomplete local excision prior to radiochemotherapy. There were 25 women and 5 men with a mean age of 59 years (range 26–73 years). Five tumors were classified as basaloid, the others as squamous cell type. Stage distribution (HERMANEK and SOBIN 1987) was 2 T1, 14 T2, 12 T3, and 2 T4 tumors.

2.2 Radiochemotherapy

Radiochemotherapy consisted of external irradiation and simultaneous chemotherapy in the 1st and 5th treatment weeks. Patients were irradiated with a 6-MV or 10-MV photon beam after simulator treatment planning. Five fractions per week with single doses of 1.80–2.00 Gy were given by opposing AP/PA portals or in the form of a box technique. The upper border of the treatment fields was in the L5/S1 interspace or in the middle of the sacrum. The lateral borders included the inner part of the groins. The total dose was 45–50 Gy over 5 weeks (five patients received only 30 Gy).

In the 1st and 5th treatment weeks, simultaneous chemotherapy was given according to NIGRO et al. (1974, 1983; SISCHY 1985). 5-Fluorouracil (5-FU) was administered as a continuous infusion over 24 h on days 1–4 and 29–32 in a daily dosage of 1000 mg/m^2. Mitomycin C (10 mg/m^2) was given as an i.v. push on days 1 and 29.

2.3 Interstitial Irradiation

During the first 2 years of the protocol, only 2 of 16 patients received an interstitial implant as boost treatment. During the last 2 years, implants have been used more extensively, and 7 of 14 patients had an interstitial boost. The implant was carried out after the external irradiation was completed and a rest period of about 6 weeks. We used the template technique (SYED et al. 1973; THIEL 1989) with iridium-192 seeds and wires in all patients. The implant volume covered about one-third to one-half of the anal circumference, i.e., the residual palpable mass with a safety margin of 1–2 cm (Fig. 1). Single-plane as well as double-plane implants were used. The dose rate on the reference isodose covering the target volume was about 30–45 cGy/h, delivering a total dose of 12–15 Gy to the target volume in about 30–50 h.

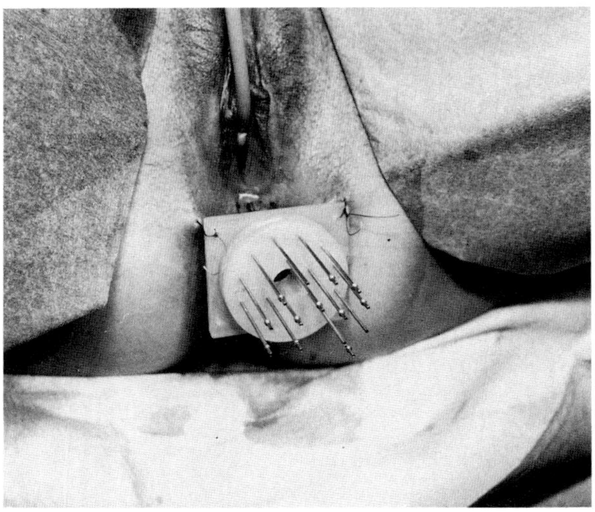

Fig. 1. Interstitial boost in a 45-year-old female patient with anal canal carcinoma involving the posterior vaginal wall. The patient had a histologically complete remission after 45 Gy and simultaneous chemotherapy and received an interstitial boost with 15 Gy. She is in complete remission 42 months after treatment with preserved sphincter function and intermittent signs of mild proctitis

2.4 Reevaluation

All patients were reevaluated 6 weeks after radiochemotherapy and then at 3-month intervals. Some 23 patients also had biopsies from the primary tumor region for histological verification of the response.

3 Treatment Results

3.1 Response

All patients responded to the initial radiochemotherapy with an at least 50% reduction in measurable tumor size. Of 30 patients 20 had a clinically complete response (66%). In patients with macroscopic tumor, 17 of 28 (63%) achieved a complete clinical response.

In 25 patients histological verification of the response was obtained (23 patients with control biopsies after treatment, 1 patient at abdominoperineal resection, 1 by autopsy). There was no evidence of tumor in 80% (20/25). In patients with a clinically complete remission, 89% (16/18) had a negative control biopsy as compared with 57% (4/7) of the partial responders. The overall complete response rate was 77% (23/30).

3.2 Interstitial Boost and Local Control

Sixteen patients were complete responders after radiochemotherapy and received no further treatment. All of them are locally controlled, one has died of distant metastases. Six complete responders were treated with an interstitial boost. Four of them are in complete remission, one is alive with a second local recurrence despite salvage surgery, and one has died of systemic disease.

Three patients received an implant after a positive biopsy. One of them is free of tumor (biopsy-proven) 6 months after the implant. The other two patients were not controlled (one persistent

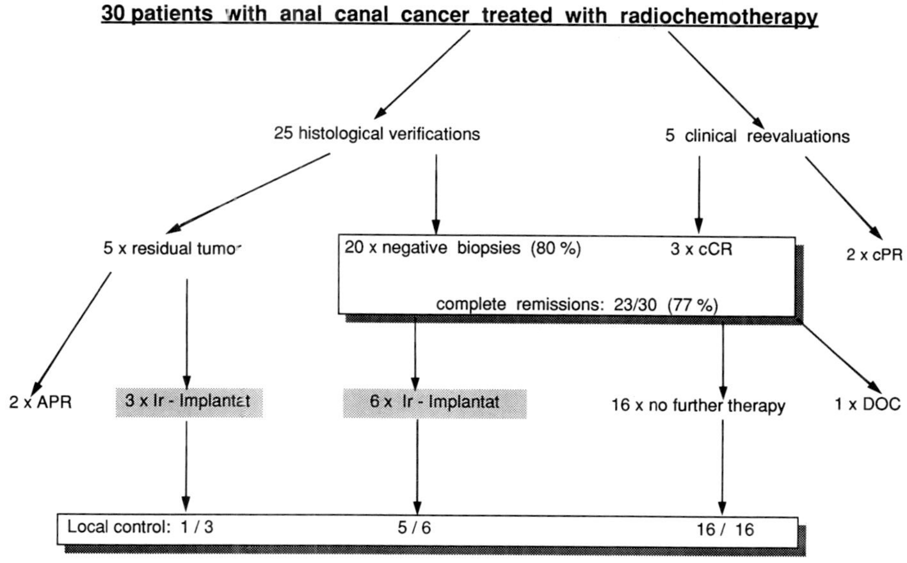

Fig. 2. Treatment outcome in 30 patients with anal canal carcinoma after radiochemotherapy (University of Erlangen, June 1985 through July 1989). *cCR*, clinically complete remission; *cPR*, clinically partial remission; *APR*, abdominoperineal resection; *DOC*, dead of treatment-related complications. Local control means definitive local control after 18 months median follow-up

tumor, one local recurrence after 6 months). Both were treated with salvage abdominoperineal resection and are alive without evidence of disease. In one of these two failures, there was a long rest period of 3 months between external and interstitial irradiation. The results concerning local control and the use of an implant are shown in Fig. 2.

Two patients with histological evidence of persistent tumor were treated with immediate salvage abdominoperineal resection and are alive. Two patients were considered as partial responders. One refused further treatment and died 8 months later with progressive local disease. One patient with a residual mass on CT scans after treatment for a pelvic recurrence did not receive any further therapy because of advanced age and is alive 12 months later without any signs of progression.

3.3 Complications

One lethal complication occurred (1/30 = 3%) due to a sepsis after implantation of a permanent intravenous catheter system. Mild to moderate bone marrow suppression occurred frequently, but no severe toxicity (WHO grade IV). Two patients required a treatment break because of a severe perineal skin reaction. Two radiation-related chronic complications were noted in patients with preserved anal sphincter: one mild proctitis (after an interstitial implant) and one moderate sphincter stenosis (no interstitial boost). The incidence of chronic complications in disease-free survivors was 14% (1/7) after interstitial boost treatment and 6% (1/18) if no boost treatment had been given.

3.4 Salvage Surgery

Five abdominoperineal resections were performed, all without severe complications. Three were necessary because of persistent tumor after radiochemotherapy without interstitial implant (two patients) or with interstitial implant (one patient). One was performed because of a recurrence after radiochemotherapy plus interstitial boost. The remaining patient with complete remission was treated outside, and the specimen was free of tumor.

Two patients had required a colostomy prior to treatment because of tumor stenosis of the anus. Both had a histologically complete response, but one of them refused the replacement of the anus preternatural.

Table 1. Treatment results for anal canal carcinoma, treated with radiochemotherapy (RCT) with or without interstitial implants or salvage surgery (University of Erlangen, June 1985 through July 1989). Median follow-up 18 months (n = 30)

Results after radiochemotherapy:	
Overall complete remission	23/30 (77%)
Histologically complete remission	20/25 (80%)
Lethal complications (sepsis)	1/30 (3%)
Results after 18 months median follow-up:	
Local control after RCT without implant	17/20 (85%)
Local control after RCT with implant	6/ 9 (67%)
Local control after salvage surgery	27/28 (96%)
Survival	26/30 (87%)
Disease-free survival	25/30 (83%)
Disease-free survival without colostomy	19/30 (63%)
Sphincter preservation in disease-free survivors	19/25 (76%)

3.5 Survival and Sphincter Function

After a median follow-up of 18 months, four patients have died: two of distant metastases with local control, one with local tumor progression, and one because of a treatment-related complication (catheter sepsis). A total of 26 patients are alive, 25 of them free of disease and 19 of them (76%) without colostomy. Two patients with preserved anal aphincter are partially incontinent since the initial biopsy. The overall treatment results are shown in Table 1.

4 Discussion

Simultaneous radiochemotherapy has become the primary treatment of choice in anal canal cancer in many institutions (Ajlouni et al. 1984; Beahrs 1985; Cantril et al. 1983; Cummings 1984; Cummings et al. 1984; Dunst et al. 1987; Eby and Sullivan 1969; Flam et al. 1983; Leichman et al. 1985; Meeker et al. 1986; Michaelson et la. 1983; Newman and Quan; Nigro et al. 1974, 1983; Nigro 1984; Shank 1985; Sischy 1985; Svenson and Montague 1980; Wanebo et al. 1981). Several clinical studies suggest that long-term survival is as good as after radical surgery with or without irradiation (Beahrs 1979; Bomann et al. 1984; Frost et al. 1984; Golden and Horsley 1976; Hager and Hermanek 1986; Sugarbaker et al. 1983) with the chance of sphincter preservation in the majority of patients. In small tumors, however, good results have been achieved with irradiation alone (interstitial and/or external) without chemotherapy (Fenton et al. 1986; Papillon 1986; Papillon et al. 1983; Pipard 1989). In large tumors (T3–4), additional chemotherapy seems to increase the local control rate (Pipard 1989; Puthawala et al. 1982). Nevertheless, the general use of additional chemotherapy irrespective of tumor size has several advantages: Nearly all patients are accessible for these modalities, the treatment is easy to perform, and even large tumors have a good chance of complete regression (Fig. 3).

Our preliminary data confirm the effectiveness of simultaneous radiochemotherapy. Like other authors (Cummings 1983; Cummings et al. 1984; Pipard 1989), we have used two courses of chemotherapy with 5-FU and mitomycin C in combination with a relatively high external radiation dose of 50 Gy. The histological complete response rate of about 75% is in accordance with the literature. Long-term data are lacking, but we consider this treatment as our primary treatment of choice. In this on-going protocol, open questions at the moment concern the need of a rebiopsy for histological verification of response and the value of an interstitial boost treatment.

4.1 Value of Interstitial Implants

The data in the literature indicate that an interstitial implant increases local control if irradiation alone is used as conservative treatment (Papillon 1986; Papillon et al. 1983; Pipard 1989; Puthawala et al. 1982; Syed et al. 1978; Thiel 1989). However, the need for an interstitial boost after radiotherapy plus simultaneous chemotherapy is not clear. The best results have been obtained in series with interstitial boosts (Cummings 1983; Cummings et al. 1984; Pipard 1989), but it seems reasonable to restrict this boost treatment to those patients who are at high risk of local failure.

Patients with a histologically proven complete remission seem to be at low risk of recurrence. During the past few years we have used implants in these patients if the initial tumor was large (T3–4). In former years, however, we did not do so and have not seen any difference, although the number of patients is small. So far, our own results do not indicate that there is a benefit from inter-

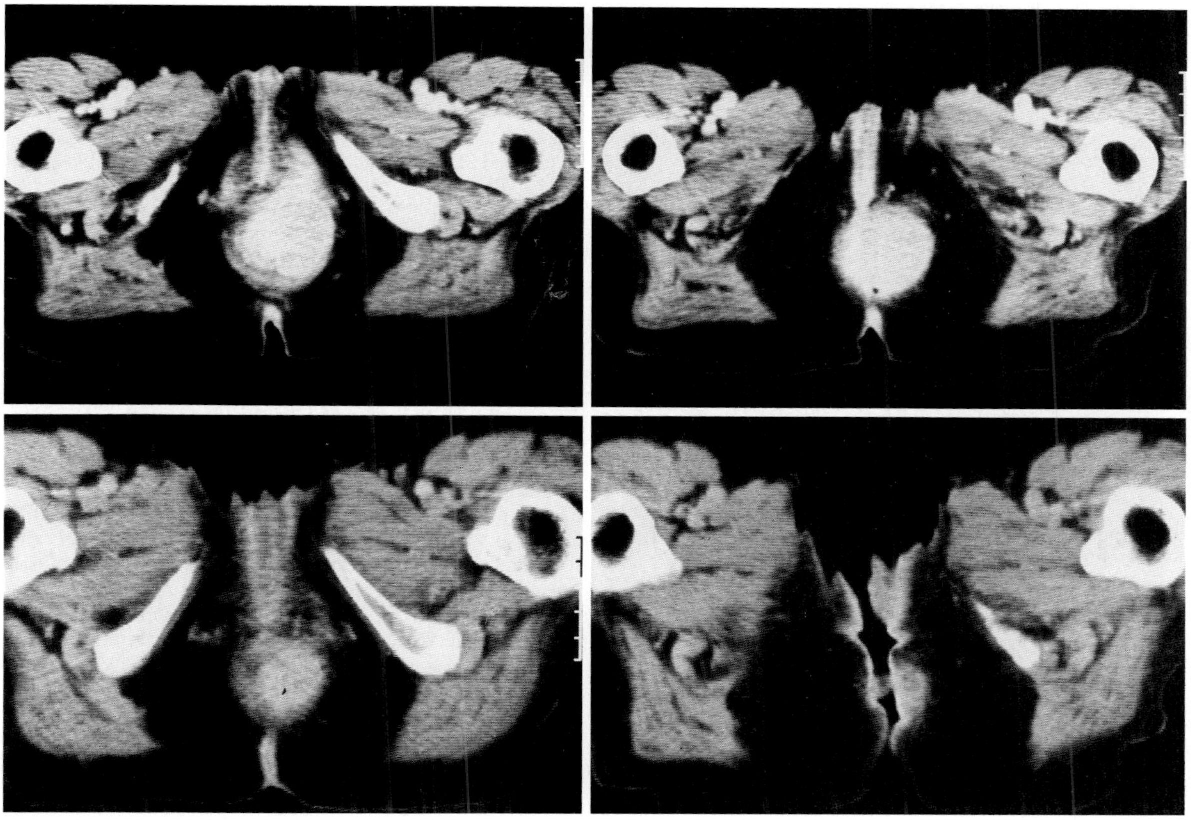

a b c d

Fig. 3a–d. Anal canal carcinoma in a 65-year-old woman with complete stenosis of the anal canal and infiltration of the vagina and gluteal muscle prior to treatment (**a**, **b**). Three weeks later after 30 Gy external beam irradiation and one course of chemotherapy, there was a significant tumor regression (**c**, **d**). Reevaluation 6 weeks after the end of treatment (50 Gy external beam plus two courses of chemotherapy) showed a clinically and histologically complete remission. The patients received no boost treatment and died 18 months later because of sytemic metastasis (paraaortic and mediastinal nodes, lung metastasis)

stitial implants after a histologically proven complete remission.

Patients with histological evidence of residual tumor might benefit from an interstitial implant. We have only limited experience, but one of three patients so treated is disease-free with a preserved sphincter. Interstitial boost treatment for this subgroup of patients is highly interesting for the radiation therapist. It would be desirable to have these patients referred to a cancer center for further conservative treatment. Thereby, it should be possible to increase the rate of sphincter preservation.

At the moment, we prefer to use interstitial implants in patients with a clinically good response but with foci of tumor cells and in complete responders if the initial tumor was large (T3–4). Our current treatment concept is shown in Fig. 4.

4.2 Value of Rebiopsies After Treatment

The clinical relevance of a rebiopsy is difficult to judge. The objective is to identify patients who require immediate additional treatment. The correlation between clinical and histological response is relatively high, especially in the case of a clinically complete remission. Furthermore, it is not proven that persistent foci of tumor cells will always cause a local recurrence. Several therapists with excellent results do not routinely use rebiopsies (Cummings et al. 1984). A rebiopsy seems unnecessary if an interstitial boost is given in all patients.

Our treatment protocol is based on external irradiation and simultaneous chemotherapy. We have up to now used interstitial implants for boost

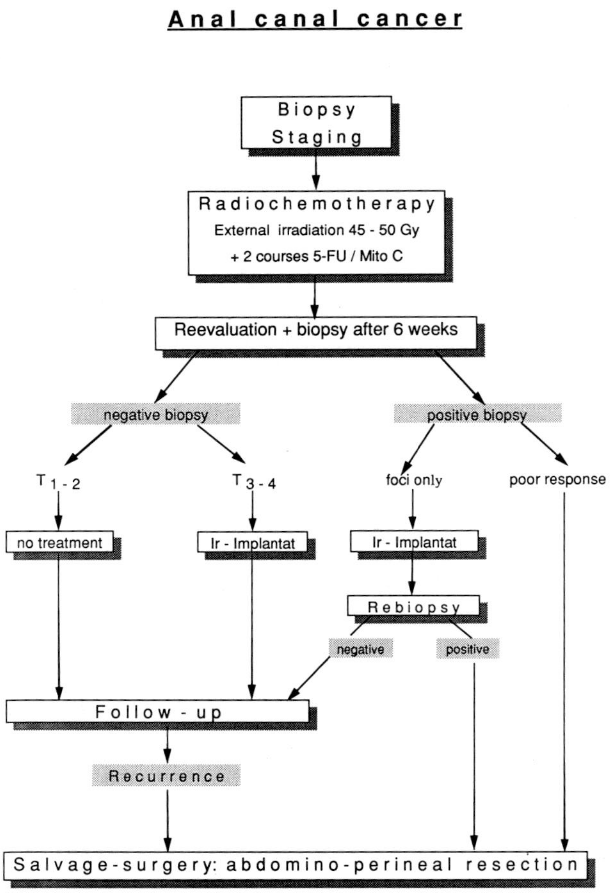

Fig. 4. Treatment schedule for anal canal carcinomas at the University of Erlangen

treatment only in a subgroup of patients (about 50%), and salvage surgery was restricted to a limited number of them. We do favor rebiopsies in most of our patients and think that histological reevaluation is helpful in further treatment planning (Fig. 3).

In conclusion, we prefer radiochemotherapy as the primary treatment for anal canal cancer. After external irradiation with 50 Gy and simultaneous chemotherapy with 5-FU/mitomycin C, clinically and histologically complete response rates in the range of 70% or more are achieved. Interstitial implants are indicated if irradiation alone without chemotherapy is used. Implants seem to improve local control in the case of incomplete remission or if the response is assessed only clinically. In the case of a histologically confirmed complete remission after radiochemotherapy, the value of an interstitial boost remains questionable.

References

Ajlouni M, Mahrt D, Milad MP (1984) Review of recent experience in the treatment of carcinoma of the anal canal. Am J Clin Oncol 7: 687–691

Beahrs OH (1979) Management of cancer of the anus. AJR 133: 791–795

Beahrs OH (1985) Management of squamous cell carcinoma of the anus and adenocarcinoma of the lower rectum. Int J Radiat Oncol Biol Phys 11: 1741–1742

Boman BM, Moertel CG, O'Connell MJ, Scott M, Weiland LH, Beart RW, Gunderson LL, Spencer RJ (1984) Carcinoma of the anal canal. Cancer 54: 114–125

Cantril ST, Green JP, Schall GL, Shaupp WC (1983) Primary radiation therapy in the treatment of anal canal carcinoma. Int J Radiat Oncol Biol Phys 9: 1271–1278

Cummings BJ (1983) Carcinoma of the anal canal – radiation or radiation plus chemotherapy. Int J Radiat Oncol Biol Phys 9: 1417–1418

Cummings BJ, Keane T, Thomas G, Harwood A, Rider W (1984) Results and toxicity of the treatment of anal canal carcinoma by radiation therapy or radiation therapy and chemotherapy. Cancer 54: 2062–2068

Dunst J, Reichard U, Wolf N, Sauer R (1987) Funktionserhaltende Therapie des Analkarzinomas durch simultane Radio-Chemotherapie. Dtsch Med Wochenschr 112: 1201–1205

Eby LS, Sullivan ES (1969) Current concepts of local excision of epidermoid carcinoma of the anus. Dis Colon Rectum 12: 332–337

Fenton J, Cutuli B, Rousseau J, Labib A, Salmon RJ, Mathieu G (1986) Anal canal carcinoma: survival and sphincter preservation after radiotherapy (195 cases). 5th Annual Meeting of the European Society for Therapeutic Radiology and Oncology, Baden-Baden

Flam MS, John M, Lavolvo MJ, Mills RJ, Romalho LD, Prother C, Mowry PA, Morgan DR, Lau BP (1983) Definite non-surgical therapy of epithelial malignancies of the anal canal. Cancer 51: 1378–1387

Frost DB, Richards PC, Montague ED, Giacco GG, Martin RG (1984) Epidermoid cancer of the anorectum. Cancer 53: 1285–1293

Golden GT, Horsley JS (1976) Surgical management of epidermoid carcinoma of the anus. Am J Surg 131: 275–280

Hager T, Hermanek P (1986) Maligne Tumoren der Analregion. In: Gall FP, Hermanek P, Tonak P (eds) Chirurgische Onkologie. Springer, Berlin Heidelberg New York, pp 581–589

Hermanek P, Sobin LH (eds) (1987) TNM classification of malignant tumors, 4th edn. Springer, Belin Heidelberg New York

Leichman L, Nigro N, Vaitkevicius VK, Considine B, Buroker T, Bradley G, Seydel HG, Olchowski S, Cummings G (1985) Cancer of the anal canal. Am J Med 78: 211–215

Meeker WR, Sickle-Sontonello BJ, Philpott G, Kennedy D, Bland KI, Hill GH, Pupp MB (1986) Combined chemotherapy, radiation and surgery for epithelial cancer of the anal canal. Cancer 57: 525–529

Michaelson RA, Magill GB, Quan SHQ, Leaming RH, Nikrui M, Stearns MW (1983) Prospective chemotherapy and radiation therapy in the management of anal epidermoid carcinoma. Cancer 51: 390–395

Newmann HK, Quan SHQ (1976) Multimodality therapy for epidermoid carcinoma of the anus. Cancer 37: 12–19

Nigro ND (1984) An evaluation of combined therapy for squamous cell cancer of the anal canal. Dis Colon Rectum 27: 763–766

Nigro ND, Vaitkevicius VK, Considine B (1974) Combined therapy for cancer of the anal canal. Dis Colon Rectum 17: 354–356

Nigro ND, Seydel HG, Considine B, Vaitkevicius VK, Leichman L, Kinzie JJ (1983) Combined peroperative radiation and chemotherapy for squamous cell carcinoma of the anal canal. Cancer 51: 1826–1829

Papillon J (1986) Current therapeutic concepts of management of cancer of the anal canal. 5th Annual Meeting of the European Society for Therapeutic Radiology and Oncology, Baden-Baden

Papillon J, Mayer M, Montbarbon JF, Gerard JP, Chassard JL, Bailly C (1983) A new approach to the management of epidermoid carcinoma of the anal canal. Cancer 51: 1830–1837

Pipard G (1989) Cancer of the anal canal: experience and results of conservative radiotherapy in Geneva. In: Wolf N, Matzel K (eds) Fortschritte in der Proktologie. Zuckschwerdt, Munich

Puthawala AA, Syed AMN, Gates TC, McNamara C (1982) Definitive treatment of extensive anorectal carcinoma by external and interstitial irradiation. Cancer 50: 1746–1750

Shank B (1985) Treatment of anal canal carcinoma. Cancer 55: 2156–2162

Sischy B (1985) The use of radiation therapy combined with chemotherapy in the management of squamous cell carcinoma of the anus and marginally resectable adenocarcinoma of the rectum. Int J Radiat Oncol Biol Phys 11: 1587–1593

Sugarbaker PH, Gunderson LL, Macdonald JS (1983) Cancer of the anal region. In: DeVita V, Hellman S, Rosenberg A (eds) Cancer Principles and practice of oncology. Lippincott, Philadelphia

Svenson EW, Montague ED (1980) Results of treatment of cloacogenic carcinoma. Cancer 46: 828–830

Syed AMN, Puthawala A, Neblett D, George FW, Myint US, Lipsett JA, Jackson BR, Flemming PA (1978) Primary treatment of the lower rectum and anal canal by a combination of external irradiation and interstitial implant. Radiology 128: 199–203

Thiel HJ (1989) Prinzipien der Kombinationstherapie des Analkarzinoms. In: Wolf N, Matzel K (eds) Fortschritte in der Proktologie. Zuckscherdt, Munich

Wanebo HJ, Futrell W, Constable W (1981) Multimodality approach to surgical management of locally advanced epidermoid carcinoma of the anorectum. Cancer 47: 2817–2826

radiotherapy only (CUMMINGS et al. 1984). The 5-year survival time was identical in both groups (70%), but radiochemotherapy improved the primary tumor control while at the same time preserving the sphincter. Pretreatment resulted in complete local tumor elimination in 60% of patients in the radiotherapy group compared with 93% in the combined therapy group. The combined procedure, however, inflicted additional gastrointestinal toxicity and hematologic side effects. The risks and benefits of radiochemotherapy therefore have to be weighed carefully, in particular since the value of chemotherapy in the multimodal therapy concept has not yet been clarified. Nigro, a supporter of radiochemotherapy in the treatment of anal carcinomas, has admitted that according to several reports the favorable results seen at present might be due to new radiotherapeutic techniques and equipment (e.g., fractionation) and, to a lesser or only minor extent, adjuvant chemotherapy. Another open question is whether the radiation dose can be reduced on account of the adjuvant chemotherapy. Furthermore, cytostatic drugs or drug combinations other than 5-fluorouracil and mitomycin C might be more effective. These problems will have to be tackled prospectively in the future, for instance, by investigating a radiochemotherapy program in which optimum radiotherapy alone is employed in treating squamous cell carcinoma of the anus in order to evaluate the toxicity and responsiveness as well as the rate of local tumor control and survival time. Such a design is at present under investigation in an EORTC study (BARTELINK and ROELOFSEN 1987).

5 Surgical Treatment in Multimodal Therapy Concepts

In the case of biopsy evidence of residual tumor tissue found after radio(chemo)therapeutic pretreatment, which further procedure to employ is still disputed and undecided. The question arises whether primarily rectotomy is indicated or rather a boost radiation which can be either exogenously applied or interstitially as an iridium implant (DUNST et al. 1987; MÜLLER and PÖTTER 1987; PAPILLON et al. 1983). Based on retrospective results, an advantage of iridium implantation has been postulated but not proven. In general, the necessity for a control biopsy after radio(chemo)-

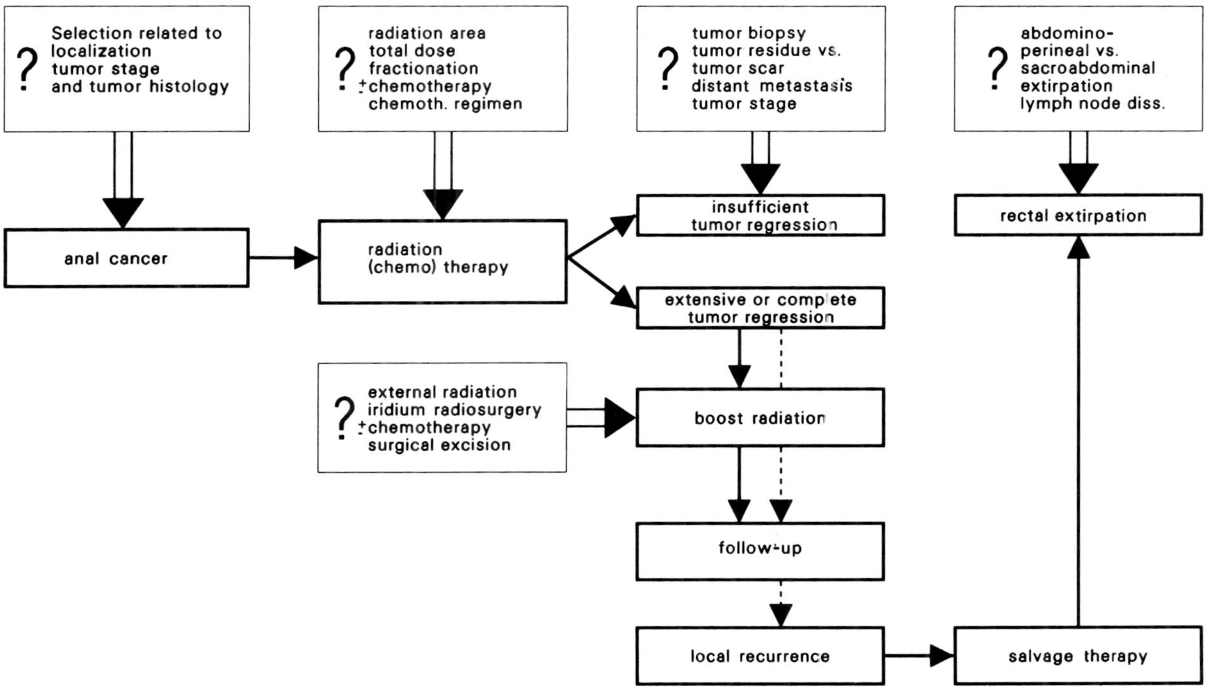

Fig. 1. Decision making and open questions on the therapy of anal cancer

therapy is becoming more and more controversial (Dunst et al. 1987), because it might not be representative or lead to misinterpretation when tissue is taken too early after the pre-radiation treatment. The biologic relevance of histologically evident small amounts of tumor tissue is also still unknown. For these reasons and because too deep biopsy taken after preradiation might result in a chronic anal fistula, histologic examination of findings after radio(chemo)therapy is in part rejected, and subsequent boost radiation is generally recommended instead. In the case of advanced local tumors the problem remains whether primary measures to preserve the sphincter are feasible or whether pre-radiation should be used only to improve the local tumor control in combined surgical and radiotherapeutic treatment. The best method to be used to treat potential metastases in the lymph nodes is also still under discussion (Schlag 1986; Schraut et al. 1983). Here, too, prospective investigations are necessary to define an optimum therapeutic concept, in which a conservative expectant approach, diagnostic dissection, and preventive radiation have to be compared.

Although primary radio(chemo)therapy is to be the first therapeutic approach in the treatment of anal carcinoma, a large number of questions are left open in the definition of this treatment concept as well as in the selection of patients (Fig. 1). To examine these problems, multi-institutional approaches will be needed because of the rare occurrence of this tumor type.

References

Bartelink H, Roelofsen I (1987) EORTC Cooperative Group of Radiotherapy and Gastrointestinal Group, protocol 22861: Radiotherapy alone or with concomitant chemotherapy in the treatment of anal carcinoma. (Study protocol available by the EORTC Data Center, Brussels)

Beahrs OH, Wilson SM (1976) Carcinoma of the anus. Ann Surg 184: 422–428

Cantril ST, Grenn JP, Schall GL, Schaupp WC (1983) Primary radiation therapy in the treatment of anal carcinoma. In J Radiat Oncol Biol Phys 9: 1271–1278

Clark J, Petrelli N, Herrera L, Mittelman A (1986) Epidermoid carcinoma of the anal canal. Cancer 57: 400–406

Cummings B, Kean T, Thomas G, Harwood A, Rider W (1984) Results and toxicity of the treatment of anal canal carcinoma by radiation therapy or radiation therapy and chemotherapy. Cancer 54: 2062–2068

Dunst J, Reichard U, Wolf N, Sauer R (1987) Funktionserhaltende Therapie des Anal-Carcinoms durch simultane Radio-Chemotherapie. Dtsch Med Wochenschr 112: 1201–1205

Flam MS, John M, Lovalo LJ, Mills RJ, Ramalho LD, Prather C, Mowry PA, Morgan DR, Lau BP (1983) Definitive nonsurgical therapy of epithelial malignancies of anal canal – a report of 12 cases. Cancer 51: 1378–1387

Hermanek P, Scheibe O, Spiessl B, Wagner G (eds) (1987) TNM – Klossifikation maligner Tumoren, 4th edn. Springer, Bolin Heidelberg New York

Leichman L, Nigro N, Vaitkevicius VK, Considine B, Buroker T, Bradley G, Seydel GH, Olchowski S, Summings G, Leichman C, Baker L (1985) Cancer of the anal canal – model for preoperative adjuvant combined modality therapy. Am J Med 78: 211–215

Meeker WR, Sickle-Santanello BJ, Philpott G, Kenady D, Bland KI, Hill GH, Popp MB (1986) Combined chemotherapy, radiation, and surgery for epithelial cancer of the anal canal. Cancer 57: 525–529

Michaelson RA, Magill GB, Quan HQ, Leaming RH, Mikrui M, Stearns MW (1983) Preoperative chemotherapy and radiation therapy in the treatment of anal epidermoid carcinoma. Cancer 51: 390–395

Müller R-P, Pötter R (1987) Strahlentherapie beim Anal- und Analkanal-Carcinom. Chir Prax 38: 3–9

Nigro ND (1987) Multidisciplinary management of cancer of the anus. World J Surg 11: 446–451

Papillon J, Mayer M, Montbarbon JF, Gerard JP, Chassard JL, Bailly C (1983) A new approach to the management of epidermoid carcinoma of the anal canal. Cancer 51: 1830–1837

Pipard G, Widgren S (1983) Le traitement conservateur du cancer de l'anus. Rev Med Suisse Romande 103: 141–146

Puthawala AA, Nisar Syed AM, Gates C, McNamara C (1982) Definitive treatment of extensive anorectal carcinoma by external and interstitial irradiation. Cancer 50: 1746–1750

Salem PA, Habboubi N, Anaissie E, Brihl E, Issa P, Abbas J, Khalyl M (1985) Cis-Dichlorodiamminerplatinum (II) is Effective in the Treatment of Anal Squamous Cell Carcinoma Proc Am Soc Clin Oncol 4: 78

Schlag P (1986) Aspekte operativer und multimodaler Therapie beim Anal-Carcinom. Chirurg 57: 488–492

Schraut WH, Wang CW, Dawson PJ, Block GE (1983) Depth of invasion, location, and size of cancer of the anus dictate operative treatment. Cancer 51: 1291–1296

Sischy B, Remington JH, Sobel SH, Savlov ED (1980) Treatment of carcinoma of the rectum and squamous carcinoma of the anus by combination chemotherapy, radiotherapy and operation. Surg Gynecol Obstet 151: 369–371

7 Prostatic Cancer

7.1 Interstitial Irradiation in Prostatic Cancer: Report of 10-Year Results

Basil S. Hilaris, Zvi Fuks, Dattatreyudu Nori, William A. Fair, and Whillet F. Whitmore

CONTENTS

1 Introduction 235
2 Material and Methods 235
2.1 Brachytherapy Technique 236
3 Treatment Results 237
4 Discussion 239
References 240

1 Introduction

Cancer of the prostate is the second most common cancer in men in the USA, representing about 20% of all male cancer deaths (American Cancer Society 1988). The incidence of prostatic carcinoma increases significantly with age, involving about one-half of the male population over the age of 70 years. There are certain puzzling national and racial differences in the incidence of this cancer. It is extremely rare in oriental people, for instance, there are only 3.9 per 100000 Japanese men and 3.7 per 100000 Chinese men in Hong Kong, as compared with an incidence of about 60 per 100000 white and 95 per 100000 black men in the USA.

The initial cancer extent and its histologic grade are the main factors influencing prognosis and treatment. A new and very promising technique for determining the local extent of prostate carcinoma and monitoring local response to therapy is transrectal sonography. The combination of physical examination, ultrasound, and transrectal biopsy is currently the single most accurate approach for the diagnosis of prostatic carcinoma; for assessing lymph node involvement, lymph node biopsy is still essential.

One of the greatest current controversies in urological oncology is the optimal management of localized stage prostatic carcinoma. Radiation therapy and radical surgery are the standard treatments in early prostatic cancer. The respective roles of the two modalities, however, have varied through the years. Interstitial brachytherapy alone or combined with external radiation is a radiotherapeutic option available for the management of early cancer of the prostate. The decision between surgery, external radiation, and brachytherapy is usually governed by the experience of the surgeon or radiation therapist involved, as well as by assessment of the side effects of each modality.

2 Material and Methods

Between 1970 and 1985, 1119 patients with stage B or small C prostatic carcinoma were treated at the Memorial Sloan-Kettering Cancer Center by pelvic lymph node dissection and interstitial iodine 125 brachytherapy (Fig. 1).

Patients with stage A lesions were excluded because it was assumed that the residual prostate following a transurethral prostatic resection was not adequate for interstitial brachytherapy. Stage C tumors were usually judged unsuitable for implantation if there was significant extension outside the prostate, laterally, apically, or into the seminal vesicles. Lesions which might ordinarily be considered too extensive for radical surgery, however, were often judged suitable for brachytherapy. The clinical staging work-up involved a complete history and physical examination, biopsy, complete blood count, a screening profile which included serum acid and alkaline phosphatase, chest radiography, and a bone scan. Skeletal radiographs, which had been used in earlier years,

Basil Hilaris, M.D., FACR, Professor and Chairman, Department of Radiation Medicine, New York Medical College, Valhalla, NY 10595, USA

Zvi Fuks, M.D., Dattatreyudu Nori, M.D., Department of Radiation Oncology; William A. Fair, M.D., Willet F. Whitmore, M.D., Department of Surgery, Memorial Sloan-Kettering Cancer Center, New York, NY, USA

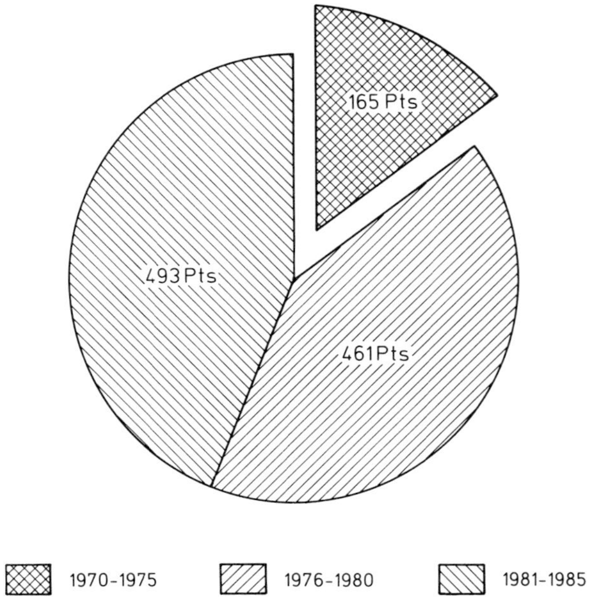

Fig. 1. Chronological distribution of 1119 patients with stage B or small C prostatic cancer treated during the period under study (1970–1975) as well as during subsequent periods of 1976–1980 and 1981–1985 at the Memorial Sloan-Kettering Cancer Center by iodine 125 brachytherapy

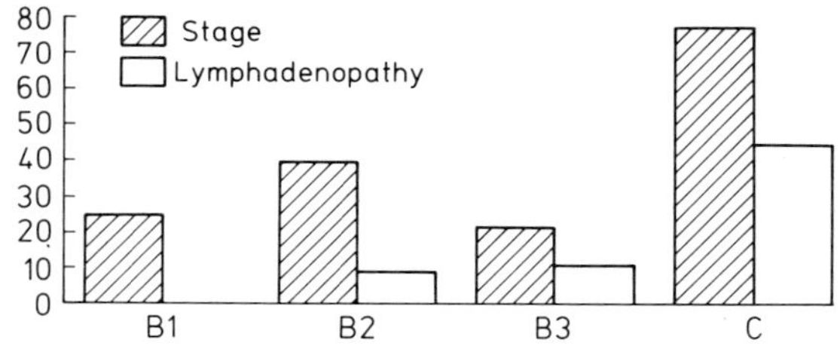

Fig. 2. Frequency of pelvic lymph node metastases according to the stage for patients treated from 1970 through 1985

are currently done only when indicated for further clarification of equivocal radionuclide scan results. Cystoendoscopy under anesthesia with prostate palpation was the final step in the selection process. Patients with bladder neck obstruction were either excluded or subjected to a partial transurethral prostatic resection 6–8 weeks prior to implantation.

The grading system utilized at the Memorial Sloan-Kettering Cancer Center considers both the pattern of the neoplasm and the degree of anaplasia, grade I being well differentiated, grade II moderately well differentiated, and grade III poorly differentiated (Barzell et al. 1977).

The staging classification used during the period under study was the American Urological Association Staging System for prostate cancer. The B2 category has been slightly modified to include nodules larger than 2 cm in one lobe of the prostate. A B3 category was added to include the remaining intraprostatic lesions. A comparison of the AUA system with the TNM system is included in the recent edition of the *Manual for Staging Cancer* (American Joint Committee on Cancer 1986).

Patients generally were followed up by periodic digital rectal examinations, roentgenograms, and serum acid and alkaline phosphatase evaluations at intervals of 3–6 months; bone scans and intravenous urograms were done at 1–2-year intervals. Elevations in the serum acid phosphatase alone were not considered evidence of progression.

This report will present long-term results on 165 patients who were treated up to December 1975 and, therefore, have been followed for a minimum period of 13 years.

Some 25 patients had a clinical stage B1 (T2a) tumor, 40 patients had a stage B2 (T2b, single lobe) tumor, 22 patients had a stage B3 tumor (T2b, both lobes), and 78 patients had a clinical stage C (T3) tumor. As stage increased, the percentage of lymph node-positive patients also increased (Fig. 2). Of the 165 patients 65 (39%) had metastatic cancer in the resected nodes (pathologic D1): 9 in stage B2, 11 in stage B3, and 45 in stage C. Of these 65 patients, 13 had only one lymph node station involved; the remaining 52 patients had extensive nodal metastases, more frequently bilateral than unilateral.

2.1 Brachytherapy Technique

The technique of implantation has been described before (Hilaris et al. 1972, 1974, 1977, 1978; Whitmore et al. 1972, 1974). The basic steps are illustrated in Fig. 3. At the end of the extraperitoneal staging lymphadenectomy, the prostate is mobilized and inspected to insure adequate exposure for the radiation oncologist. The planning of the iodine-125 implant is performed according to the New York system of dosimetry guidelines. Following the determination of the prostate volume, the required number of iodine-125 seeds, the spacing between needles and seeds, and the required number of needles is calculated using the permanent iodine-125 implant nomogram (Hilaris et al. 1988).

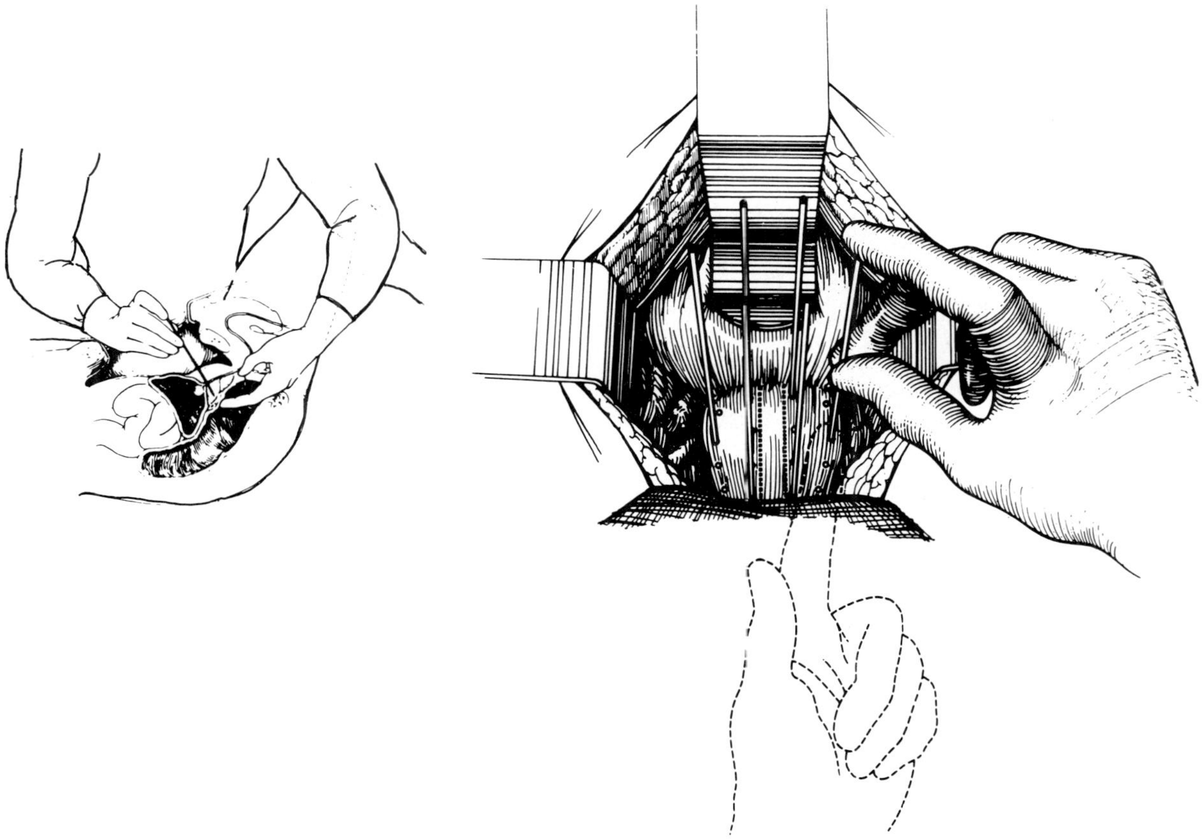

Fig. 3. Technique for needle implantation in the prostate. *Left* Pre-rectal needle insertion. *Right* Para-urethral needle insertion.Care is taken to avoid the rectal wall by keeping the index finger inside the rectum during insertion of the needles so that their tips can be sensed before entering the rectum. Care is also taken to avoid the lumen of the prostatic urethra by avoiding penetration of the urethral catheter with needles

The implant dosimetry is done 24–48 h after the brachytherapy procedure is performed and before the removal of the bladder catheter. Conventional dosimetry includes orthogonal anterior and lateral films or a stereo-shift method. The latter method is used when it is difficult to visualize the iodine-125 seeds on the lateral film. The tumor dose is reported as matched peripheral dose (MPD). Computer calculations of dose distribution and isodose contours are superimposed and displayed preferably on CT cross sections or more conventionally on the orthogonal films (Hilaris et al. 1987).

3 Treatment Results

The minimum follow-up observation period of these patients is 13 years, the maximum is 18 years, and the median follow-up period is 15 years.

Survival and disease-free survival rates were calculated from the date of the brachytherapy treatment. The survival rates were estimated using the Kaplan-Meier method.

Treatment failures by site were classified as local or distant. Local failure refers to either persisting disease after treatment or local recurrence during the period of follow-up. Distant failure refers to subsequent distant metastases, usually detected in bones. Local failure was present as the only site of failure in 24 patients; metastatic disease as the only site of failure in 36 patients; and local and metastatic disease were present in 70 patients.

The overall 15-year actuarial survival is 37% (Fig. 4) and the local disease-free survival, 29% (Fig. 5). Patients with true solitary nodules of 1.5 cm or a little more (B1), a criterion used by Jewett (1980) for radical prostatectomy, have a

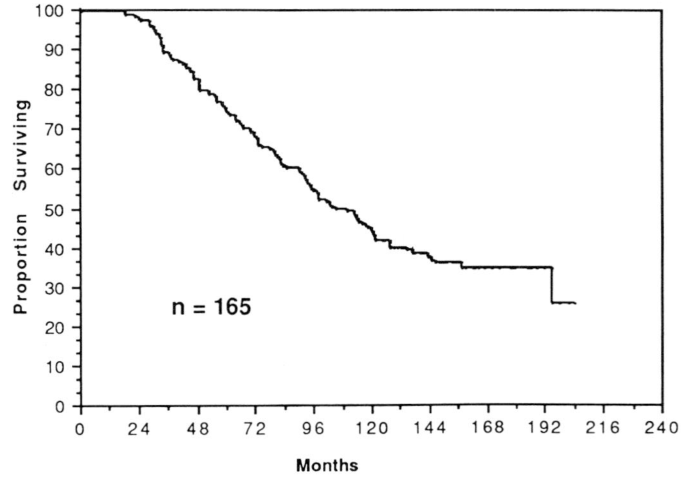

Fig. 4. Overall actuarial 15-year survival

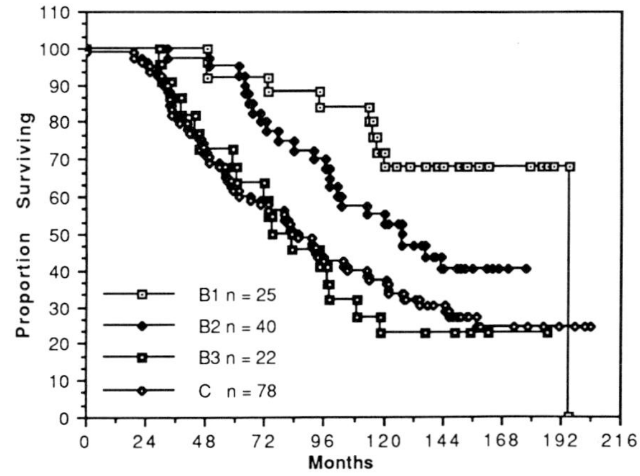

Fig. 6. Actuarial 15-year survival by stage. Note that the 15-year survival for patients with a small nodule (B1) is 70%

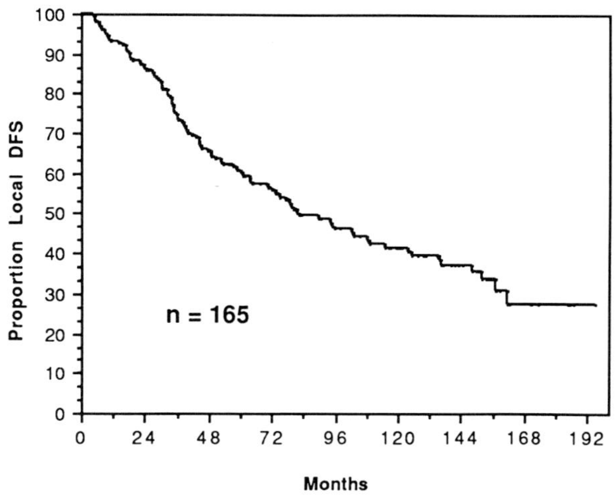

Fig. 5. Overall actuarial local disease-free 15-year survival

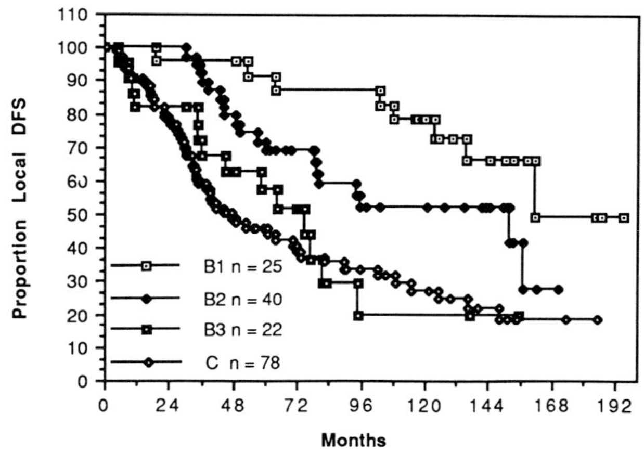

Fig. 7. Actuarial local disease-free 15-year survival (DFS)

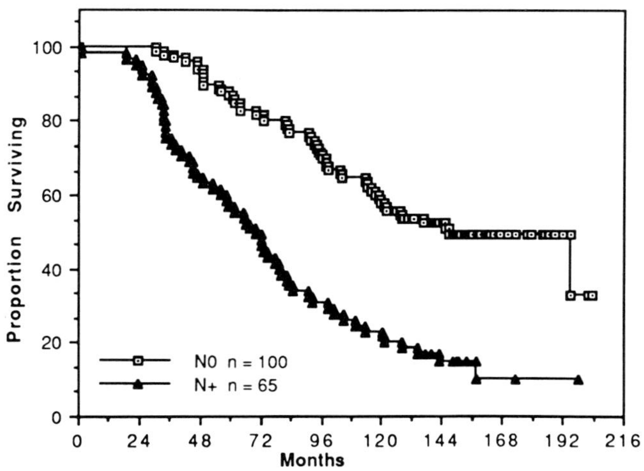

Fig. 8. Actuarial 15-year survival of patients with negative (N0) and positive (N+) nodes

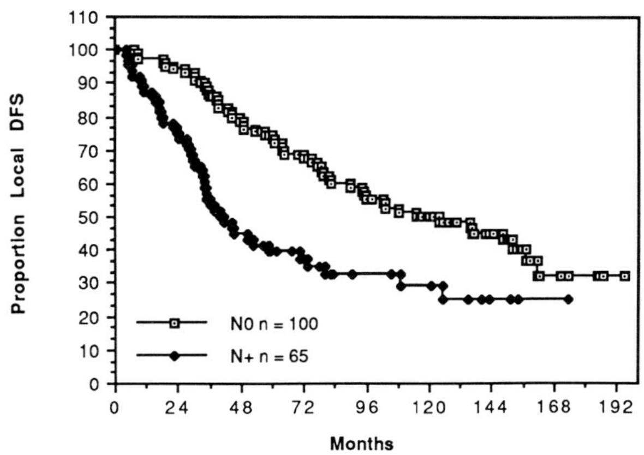

Fig. 9. Actuarial local disease-free 15-year survival (DFS) of patients with negative (N0) and positive (N+) nodes

70% survival at 15 years. It is 41% for patients with larger nodules (B2). Patients with tumor in both lobes (B3) or extraprostatic extension of the tumor (C) have 23% and 22% survival, respectively (Fig. 6). The 15-year local disease-free survival is 50%, 27%, 18%, and 18%, respectively (Fig. 7). The survival is 50% in patients with negative lymph nodes, and 10% in those patients with positive nodes (Fig. 8); the local disease-free survival is 32% and 26%, respectively (Fig. 9).

4 Discussion

Radical prostatectomy and external beam radiation have been the traditional choices for treatment of localized prostatic adenocarcinoma. Permanent interstitial I-125 implantation and pelvic lymph node dissection are another option.

Survival of the 165 patients treated by iodine-125 implantation and followed for at least 13 years (median 15 years) for B1, B2, B3, and C lesions with negative nodes is 50%, and for B2, B3, and C with positive nodes it was 10%. The 15-year local disease-free survival in patients with stage B1 disease is essentially similar to the one achieved by external beam radiation (Bagshaw et al. 1985) and surgery (Jewett 1980; Paulson et al. 1982).

Our experience indicates that the presence of pelvic lymph node metastases not only identifies a tumor which has the potential for systemic dissemination but also indicates that this tumor can no longer be contained by a treatment which has only a locoregional impact.

No severe complications were seen as a result of iodine-125 implantation. The duration of the rectal symptomatology appears to rise with increased maximum rectal dose. All rectal and urinary complications resolved spontaneously (Sogani 1983). Potency was followed by in a prospective manner and was maintained in 90% of patients who were potent preoperatively. This is in marked contrast to potency rates of 0%–10% formerly seen with radical prostatectomy although potency has been apparently well preserved in patients subjected to nerve-sparing radical prostatectomy. The incidence of potency with external radiation therapy ranges from 60% to 70% depending on the author.

Our early experience in the period 1970–1975 may not be representative of our subsequent experience for two reasons: the substitution of a modified pelvic lymph node dissection for the conventional one in July 1978 which led to a reduction in postoperative complications, without altering the accuracy of nodal stage, and modification of the technique of implantation in July 1978, designed to improve dosimetry. To test this hypothesis, DeBlasio et al. (1988) analyzed the results of treatment of 99 patients undergoing interstitial iodine 125 brachytherapy at the Memorial Sloan-Kettering Cancer Center in 1981, and he indeed demonstrated an increased local tumor control rate with the full iodine-125 implant dose of 16000 cGy.

Recent advances in ultrasound technology have made possible the development of a transrectal ultrasound unit (Bruel & Kjar, Demark), which permits simple and accurate needle guidance for percutaneous interstitial brachytherapy.

A high resolution display image provides great detail of the prostate. A probe stepping unit allows accurate volume determination of the prostatic gland. The brachytherapy procedure is performed under spinal anesthesia with the patient in a modified lithotomy position. A template grid attached to the unit improves the accuracy of the procedure. Several modifications of the existing equipment have been devised to improve the stability of the apparatus and the accuracy of the implantation. Transrectal scanning has the potential of excellent tumor localization, prostate volume determination, improved brachytherapy planning, and accurate brachytherapy dose evaluation (Hilaris et al. 1988).

References

American Cancer Society (1988) Cancer. Facts and figures. American Cancer Society, New York

American Joint Committee on Cancer (1986) Manual for staging cancer, 3rd ed. Lippincott, Philadelphia

Bagshaw MA, et al. (1985) Radiotherapy of prostatic carcinoma: long or short-term efficacy (Stanford University experience). Urology [Suppl] 25: 17–2

Brazell W, et al. (1977) Prostatic adenocarcinoma: relationship of grade and local extent of the pattern of metastases. J Urol 118: 278–282

DeBlasio D, et al. (1988) Permanent interstitial implantation of prostatic cancer in the 80s. Endocuriether Hyperthermia Oncol 4: 193–201

Hilaris BS, et al. (1972) Radical radiation therapy of cancer of the prostaste: a new approach using interstitial and external sources. Clin Bull 2(3): 94–99

Hilaris BS, et al. (1974) Radiation therapy and pelvic node dissection in the management of cancer of the prostate. AJR 121(4): 832–838

Hilaris BS, et al. (1977) Behavioral patterns of prostate adenocarcinoma following an I-125 implant and pelvic node dissection. Int J Radiat Oncol Biol Phys 2: 631–673

Hilaris BS, et al. (1978) I-125 implantation of the prostate: dose response considerations. Front Radiat Ther Oncol 12: 82–90

Hilaris BS, et al. (1987) Brachytherapy treatment planning in the radiation therapy of cancer. Front Radiat Ther Oncol 21: 94–106

Hilaris BS, et al. (eds) (1988) Atlas of brachytherapy. Macmillan, New York

Jewett HJ (1980) Radical perineal prostatectomy for palpable, clinically localized, non-obstructive cancer. Experience at the Johns Hopkins Hospital 1901–1963. J Urol 124: 492–494

Paulson DF, et al. (1982) Radical surgery versus radiotherapy for adenocarcinoma of the prostate. J Urol 128: 502–504

Sogani PC (1983) Pelvic lymphadenectomy: techniques and complications. In: Hilaris BS, et al. (eds) Brachytherapy oncology – advances in prostate and other cancers. Memorial Sloan-Kettering Cancer Center, New York, pp 79–82

Whitmore WF, et al. (1972) Retropubic implantation of iodine-125 in the treatment of prostatic cancer. J Urol 108: 918–920

Whitmore WF, et al. (1974) Implantation of I-125 in prostatic cancer. Surg Clin North Am 54(4): 887–895

7.2 Interstitial Radiotherapy: The Freiburg Experience

H. SOMMERKAMP and MICHAEL WANNENMACHER

CONTENTS

1 Introduction 241
2 Patients and Methods 241
3 Results 242
4 Discussion 243
References 244

1 Introduction

Interstitial radiotherapy of early prostatic cancer with iodine-125 is an accepted treatment alternative for patients in which either radical operation or external beam irradiation is indicated. Pioneer work with this method has been done by HILARIS, WHITMORE et al. since 1972 (WHITMORE et al. 1972). Their therapeutic concept is based on the ideal radiation properties of iodine-125 as a low-energy γ-ray-emitting radionuclide with a long half-life. These characteristics provide for a sustained continuous irradiation of the tumor with a low dose to the adjacent tissues. Due to the low depth of penetration of the radiation, precise positioning of the seeds is a prerequisite for perfect dose distribition and good local tumor control.

Iodine-125 implants for prostatic carcinoma were not in clinical use in the FRG until 1978. At that time our Depts. of Urology and Radiotherapy decided to try this method from the USA (WANNENMACHER et al. 1979).

H. SOMMERKAMP, Prof. Dr. med., Abteilung Urologie, Universitätsklinikum, Hugstetter Str. 55, 7800 Freiburg, Germany

MICHAEL WANNENMACHER, Prof. Dr. Dr., Radiologische Universitätsklinik, Abteilung Klinische Radiologie (Schwerpunkt Strahlentherapie), Im Neuenheimer Feld 400, 6900 Heidelberg, Germany

2 Patients and Methods

We have treated 77 patients since and would like to report on our experience.

In our first series of 60 patients we performed the original Memorial Sloan-Kettering Cancer Center (MSKCC) procedure wth one-stage pelvic lymphadenectomy and retropubic seed implantation irrespective of the nodal status (Table 1); 2 years ago we switched to a two-stage operation with preceding lymph staging and transperineal implantation of node-negative patients only. This was done to improve the precision of the implantation and to exclude patients unsuited for this procedure.

Patients were selected for interstitial radiotherapy on the basis of the following criteria: age under 70 years, good physical condition, tumor stages T1–3, and no evidence of distant metastases or pelvic lymph node involvement as judged by computerized tomography (CT) study. Stages A1 or A2 were considered unsuited for this treatment. There were six patients with local tumor persistence after external beam radiotherapy included in our series. Poor results in stage C tumors with low differentiation induced us to limit presently the indication to low-volume, node-negative tumors.

Complications were of considerable extent in our first series, with complete lymph node dissection in conjunction with simple drains and low-dose heparin (Fig. 1). The introduction of the "limited pelvic lymphadenectomy" as published by HERR (1982) only reduced our complication

Table 1. Stage and incidence of nodal metastasis in 60 patients with prostatic cancer (pelvic lymphadenectomy and brachytherapy) (pN+ = 32%!)

Stage	pN0	pN1	pN2
T2 (B)	19	2	3
T3 (C)	22	8	6

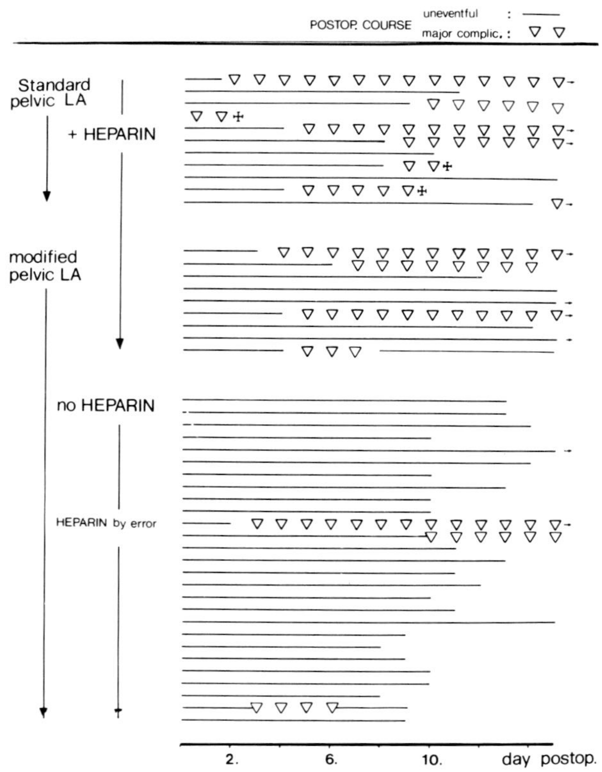

Fig. 1. Early postoperative major complications after pelvic lymphadenectomy (LA) and brachytherapy: Influence of operative technique and heparin on incidence of complications

rate – mostly hematomas and lymphoceles – to 50% when done with heparin. Not until we discontinued heparin completely did our postoperative complication rate drop to 7.7% (SOMMERKAMP 1986). Late morbidity was lower than in patients treated by external beam radiotherapy. We registered 8% of patients with sexual impotence after retropubic iodine 125 implantation.

Another specific feature of the interstitial radiotherapy with iodine-125 seeds was investigated by us (SOMMERKAMP et al. 1988): The loss of seeds after implantation. We found that an average of 8% of the seeds implanted are lost, mostly within the first 6 months (Fig. 2). This occurs in 92% of the patients. It may lead to uneven local dose distribution or underdosage and deserves attention. In some cases we had to add an external beam dose to guarantee an adequate total dose.

3 Results

Our results with respect to overall survival and disease-free patients are as follows: the 5-year crude survival was 87.5% with a disease-free percentage of 70%. Of 64 patients 8 died within this period. There is an obvious correlation between tumor differentiation and survival, with good results in highly and moderately highly differentiated tumors and poor survival rates with a great incidence of bone metastasis after treatment in G3 tumors (Table 2). This observation is in accordance with data from other authors and

Table 2. Tumor differentiation and bone metastases on follow-up in patients after brachytherapy with iodine 125

Grade	n	Metastases
G1	10	0 (0%)
G2	38	4 (10%)
G3	12	4 (33%)
Total	60	8 (13%)

Fig. 2. Seed loss after retropubic iodine 125 implantation

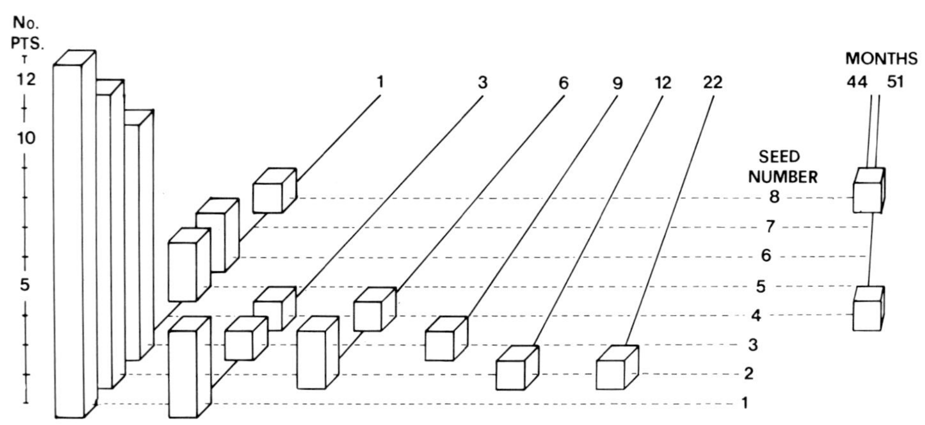

reflects two problems associated with the original MSKCC procedure. First, a large number of node-positive patients is treated by a method designed for local tumor control only; and second, G3 tumors are mostly in stage C with a large volume in which a homogeneous seed distribution is difficult to achieve by this manual retropubic technique.

A comparison with patients after external beam radiotherapy in our department yielded better results for low-grade tumors. Results in G1 and G2 tumors were slightly better after interstitial treatment. As selection criteria were different for both groups and this was not considered a randomized study, conclusions can only be drawn with caution. Most readers will be familiar with Bagshaw's comparison of his data with the MSKCC results in which no significant difference could be found. More recent publications question the superiority of interstitial over external beam radiotherapy when utilizing iodine-125 implants only.

Local tumor control should be 100% on a theoretical basis when adequate total dose and optimum dose distribution are achieved with the implants. Clinical results, however, are significantly less impressive. A complete tumor regression in our patients followed up for more than 1 year was 48% only.

Our local failure rate averaged 15% and was especially high in tumors with poor differentiation. Similar to the MSKCC experience it can be as high as 30% (Sommerkamp et al. 1987; Whitmore et al. 1985). We have to bear in mind that in radical prostatectomy the local recurrence rate in comparable stages ranges between 20% and 30%.

4 Discussion

Taking into account our treatment results and the information from large centers in the USA we modified our treatment protocol, as mentioned, to a two-stage procedure. This serves two purposes: First, it excludes patients with lymph node involvement, and second, it makes use of the superior ultrasound-controlled implantation technique as published by Holm et al. (1983).

Their technique with transrectal ultrasound monitoring was successfully employed by us. To improve radiation protection during implantation we designed a special cartridge instead of pre-

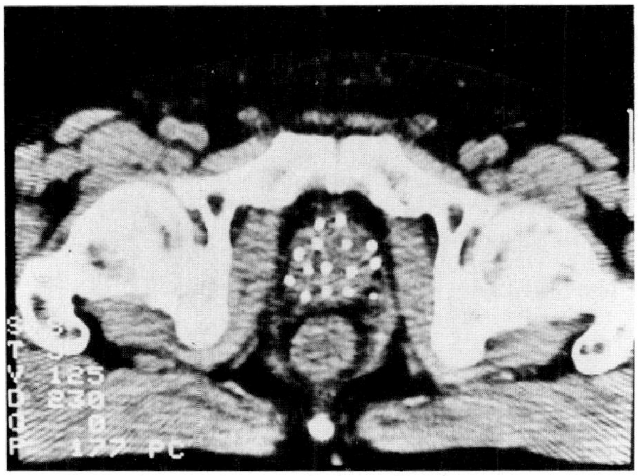

Fig. 3. Seed distribution after ultrasound-monitored, perineal, percutaneous seed implantation

charged needles, which can be used like the Mick applicator. Seed distribution is better, as shown by the postimplant isodose curves. CT-controlled distribution (Fig. 3) is precise, so this implantation can be considered optimal.

In 17 patients implanted by the perineal route since 1986 we observed a better dose distribution than in patients implanted by the retropubic manual technique. We hope that this will lead to a significant improvement in local tumor control rates. Our preliminary results are encouraging.

Consequences must be drawn from the rather discouraging results obtained with the standard protocol using iodine-125 implants only. Suggestions are:

- Limiting the indication to T1 and T2 stages
- Improving the implantation technique
- Combining interstitial and external beam irradiation in stage T3
- Preference of other radionuclides like iridium-192 for large volume T3 tumors
- Seed loss has to be taken into account with an addition of 8%–10% to the calculated dose requirement

A problem under discussion is the question of how to treat node-positive (D1) patients. Exclude them from interstitial radiotherapy or treat them in combination with adjuvant hormonal treatment? Recent data from the Mayo Clinic (Zincke et al. 1987) suggest encouraging results for early hormonal treatment in conjunction with control of the primary tumor (surgery) in patients with lymph node involvement.

References

Bagshaw MA (1989) Stellenwert der Strahlentherapie in der Behandlung des Prostatakarzinoms. In: Sommerkamp H, Altwein JE (eds) Karger

Herr HW (1982) Pelvic lymphadenectomy and iodine-125 implantation. In: Johnson DE, Boileau MA (eds) Genitourinary tumors. Grune and Stratton, New York, pp 63–73

Holm HH, Juul N, Pedersen JF, Hansen H, Stroyer I (1983) Transperineal 125-iodine seed implantation in prostatic cancer guided by transrectal ultrasonography. J Urol 130: 283

Sommerkamp H (1986) Pelvic lymphadenectomy and brachytherapy for prostatic cancer: analysis of early postoperative morbidity. Eur Urol 12: 265–269

Sommerkamp H, Knüfermann H, Wannenmacher M (1987) Grenzen der Strahlentherapie beim undifferenzierten Prostatakarzinom. II. Interstitielle Strahlentherapie. Tumordiagn Ther 8: 22–27

Sommerkamp H, Rupprecht M, Wannenmacher M (1988) Seed loss in interstitial radiotherapy of prostatic carcinoma with I-125. Int J Radiat Oncol Biol Phys 14: 389–392

Wannenmacher M, Sommerkamp H, Knüfermann, Kuphal K (1979) Die interstitielle Strahlentherapie in der Behandlung des Prostatakarzinoms. Dtsch Ärztebl 76: 1371–1378

Whitmore WF, Hilaris B, Grabstald H (1972) Retropubic implantation of iodine 125 in the treatment of prostatic cancer. J Urol 108: 918–920

Whitmore WF Jr, Hilaris B, Batata M, Sogani P, Herr H, Morse M (1985) Interstitial radiation: short-term palliation or curative therapy? Urology [Suppl] 25: 24–29

Zincke H, Utz DC, Tuhle PM, Taylor WF (1987) Treatment options for patients with stage D1 (To-3, N1-2, Mo) adenocarcinoma of prostate. Urology 30: 307–315

7.3 Cancer of the Prostate: Brachytherapy Techniques

LUTHER W. BRADY, CHRISTOPHER KOPROWSKI, ANNE MARIE BOROFSKY, and DAVID A. LIGHTFOOT

CONTENTS

1 Introduction 245
2 Biopsy Technique: Effect on Survival 245
3 Lymph Node Involvement 246
4 Bone Metastases 246
5 Techniques of Prostatectomy 246
6 External Beam Irradiation 247
7 Iodine-125 Seed Implantation 248
8 Comparison of the Efficacy of External Beam Irradiation and Implantation Techniques 253
9 Conclusions 254
References 254

1 Introduction

Prostate cancer is the most common as well as the most frequently irradiated genitourinary malignancy. In addition, it is among the most successfully treated malignancies of the genitourinary tract, both in definitive and metastatic presentations.

It is most commonly a disease of glandular origin, often peripheral and multifocal (in 75% of prostatectomy specimens) (JEWETT 1963; BYAR and MOSTOFI 1973). Early tumors are usually asymptomatic, with more advanced tumors causing urinary obstructive symptoms or pain due to metastatic disease. Most commonly, patients have advanced disease at the time of diagnosis with 75% presenting with stage B, C and D lesions (MURPHY et al. 1982). A recent trend toward earlier diagnosis has been identified. The staging of prostate cancer varies with different institutions, the most common systems being shown in Table 1.

In 1989, the AMERICAN CANCER SOCIETY anticipates that there will be 103000 new cases of carcinoma of the prostate diagnosed in the USA.

Table 1. Prostate cancer staging systems

		TNM
A_1	Incidental finding < foci, grade I	T1A
A_2	Incidental finding > foci, grade II, III	T1B
B_1	Palpable nodule < 1.5 cm	T2A
B_2	Palpable nodule > 1.5 cm or diffuse involvement of gland	T2B
C_1	Extracapsular extension	T3
C_2	Seminal vesicle, bladder neck, or lateral pelvic involvement	
D_1	Extensive pelvic involvement or pelvic nodal disease	N1–2
D_2	Bone, extrapelvic nodes, or soft tissue metastases	M1

It represents the third major cause of death from cancer in man. However, about 70% of patients receive no treatment following the diagnosis. It has often been stated that no treatment can effectively alter the overall survival from a disease that is often metastatic on presentation (JOHANSSON et al. 1989). However, JEWETT et al. (1972) identified about 5% of all patients who would benefit from radical prostatectomy. BAGSHAW (1980) showed the advantage of radical external beam radiation therapy in achieving local control without the morbidity or mortality of radical prostatectomy. The dramatic resolution of prostate cancer by estrogen therapy led to enthusiasm for hormonal manipulation, but it has been shown that this treatment regimen does not affect long-term survival (WHITMORE 1956).

2 Biopsy Technique: Effect on Survival

The ideal biopsy technique to identify the diagnosis has not been established. The use of transurethral resection has been noted to be associated with decreased survival and adverse outcome by a number of authors (MCGOWAN 1980; HANKS et al.

LUTHER W. BRADY, M.D., CHRISTOPHER KOPROWSKI, M.D., ANNE MARIE BOROFSKY, M.D., DAVID A. LIGHTFOOT, M.A., Department of Radiation Oncology and Nuclear Medicine, Hahnemann University, Mail Stop 200; Broad and Vine, Philadelphia, PA 19102–M92, USA

1983; ROSEN et al. 1984). It has been postulated that this is caused by dissemination of tumor cells at the time of the transurethral resection. However, this may also represent a selection bias since those patients with bulky disease or high grade tumors are the ones most likely to require transurethral resection. The identification of the adverse impact of transurethral resection was identified specifically in moderately or poorly differentiated stage C tumors by HANKS et al. (1985), and it was in this group that the recommendation was made for 5 Gy irradiation prior to the transurethral resection to diminish the potential of metastatic dissemination. However, PEREZ et al. (1980) found no adverse effect of transurethral resection in patient groups corrected for clinical stage and degree of differentiation. Needle biopsy, either via the transperineal or transrectal approach, is most commonly used to establish the diagnosis.

3 Lymph Mode Involvement

Involvement of regional lymph nodes is common in prostatic carcinoma with the risk of positive disease reflecting the stage and grade of the tumor. The periprostatic and obturator nodes are the primary drainage sites, followed by external iliac, internal iliac, common iliac,and periaortic nodes. GLEASON and MELLINGER (1974) have defined a grading system based on the major and minor histologic patterns of the tumor with the combined score accurately predicting the risk of lymph node metastases and survival. BAGSHAW (1986a) observed no lymph node metastases in 449 surgically staged patients with a Gleason score of 5 or less.

Since lymph node positivity correlates adversely with survival (75% distant metastasis at 5 years in positive lymph node patients; WHITMORE et al. 1979) it would seem important to identify those patients with occult nodal spread in order to offer them more extensive therapy. Lymphangiographic studies are poor indicators of nodal disease due to the inconsistent visualization of the obturator nodes on bipedal lymphangiograms. Laparotomy for lymph node biopsy is able to detect occult nodal disease (40% in localized disease, 7% in stage B1, 43% in stage B2, and 60% in stage C; FOWLER and WHITMORE 1981) but is associated with significant morbidity including a 20%–25% chance of lower extremity edema (PEREZ 1983). Since lymphadenectomy is only useful as a prognostic and diagnostic procedure and not as a therapeutic procedure, it is difficult to justify this morbidity risk.

4 Bone Metastases

The pattern of bony metastatic disease has been postulated to reflect the venous drainage of the prostate and therefore favor the vertebral bodies and pelvic bones (BATSON 1942). However, the distribution of metastases in the periphery of the body is now better appreciated due to whole body technetium 99 m bone scanning, and the predominance of central metastasis may only reveal the increased regional blood flow at these sites.

5 Techniques of Prostatectomy

The therapeutic options for localized prostatic carcinoma include radical surgery, implantation of radioactive materials into the prostate itself, or external radiation therapy. Radical prostatectomy, either by a transperineal or retropubic approach, was found by WALSH and JEWETT (1980) to achieve a 51% 15-year survival in patients with stage B1 disease, while patients with stage B2 disease had only a 26% survival. It is often not possible to determine accurate staging preoperatively.

Radical prostatectomy is associated with substantial chronic morbidity with 5%–10% long-term incontinence, 15%–20% short-term incontinence, and nearly 100% erectile impotence. Because of these complications, WALSH (1986) developed a modified technique designed to preserve the neurovascular bundle that provides for erectile potency. This has resulted in less morbidity in preliminary studies, with potency being maintained in 69% (20/29 patients) who had unilateral preservation of the neurovascular bundle. Overall, in 250 patients with a minimum 1-year follow-up, WALSH (1986) found a 69% potency rate (147/214 patients). Reportedly, the incontinence rate has also been decreased to less than 5%. However, long-term survival and local control rates are still not known. The knowledge that prostate carcinoma is a disease of the peripheral subcapsular gland and that it often involves the distal end of the gland periurethrally makes the intentional transection of the gland to preserve potency and continence a worrisome approach (Breslow et al. 1977). Additionally, the widespread application of a surgical technique without information on its 5-year efficacy needs further study, especially since the selection criteria are being expanded to include more advanced disease.

6 External Beam Irradiation

The use of external megavoltage radiation therapy was first popularized by Bagshaw. The most recent results from Stanford as well as three other, large, recent reports are summarized in Table 2 (Rosen et al. 1984; Bagshaw 1986a; Perez et al. 1986; Hanks et al. 1987). These studies indicate that prostatic cancer has a dose-response curve; in the local disease extent a dosage of at least 60–65 Gy over 6–7 weeks is required for early lesions and 65–70 Gy over 7–8 weeks for stage C disease utilizing a shrinking field technique. Dosages higher than 70 Gy over 7–8 weeks have not been shown to increase local control or survival and are associated with increased morbidity.

The morbidity associated with external beam radiation therapy of prostate cancer is shown in Table 3 (Pilepich et al. 1984). Many of these patients had surgical evaluation of the lymph nodes prior to treatment, which is known to increase morbidity when combined with radiation

Table 2. Definitive radiation treatment results of prostate cancer

Author/Institution	Year	*n*	Stage	Results		
				5 years	*8 years*	
Rosen et al./Harvard Jt Center	1984	229	A 25	96%	82%	
			B 85	77%	63%	
			C 88	61%	38%	
				5 years	*10 years*	
Perez et al./Mallinckrodt	1986	327	A_2 10	100%		
			B 113	72%	60%	
			C_1 107	60%	55%	
			C_2 97	40%	35%	
				5 years	*10 years*	*15 years*
Bagshaw/Stanford	1986	898	A/B 491	81%	60%	35%
			C 407	61%	36%	18%
				5 years	*10 years*	
Hanks et al./Patterns of Care	1987	668	A 60	85%	61%	
			B 312	75%	46%	
			C 296	58%	38%	

Table 3. Summary of treatment-related morbidity with prostate irradiation (Radiation Therapy Oncology Group 75-06; 526 patients)

	Grade					Total	
	1	2	3	4	5	No.	(%)
Cystitis	30	28	8			66	(12%)
Diarrhea	20	30	1			51	(9.7%)
Proctitis	14	20	6	1		41	(7.8%)
Genital and/or leg edema	7	13	5			25	(4.8%)
Melena	14	6	2		1	23	(4.4%)
Hematuria	4	4	8			16	(3.0%)
Urethral stricture			12			12	(2.3%)
Suprapubic fibrosis edema	8	1				9	(1.7%)
Rectal-anal stricture	4	2	1			7	(1.3%)
Vesical neck contracture			2	1		3	(0.6%)
Rectal ulcer		1	2			3	(0.6%)
Bowel Obstruction				2		2	(0.4%)

therapy. In those patients who had not been staged surgically, none developed genital or leg edema. In the Patterns of Care review of 668 patients treated from 1973 to 1975 with a 10-year follow-up, data indicated a reduction in complications from 6% to 3% by changing from a 2- to 4-field treatment technique (HANKS 1986).

The addition of concomitant hormonal therapy to radiation therapy has not been found to benefit local control or the delay of distant metastatic disease, and it is not recommended (PEREZ et al. 1986). Hormonal therapy, either by orchiectomy or supplemental estrogens, is useful in relieving bone metastatic pain and should be reserved for the treatment of metastatic disease.

7 Iodine-125 Seed Implantation

In 1975, HILARIS popularized the technique of implantation of radioactive iodine-125 seeds into the prostate following pelvic lymphadenectomy in patients with localized disease, giving high-dose localized radiation to the prostate only. This technique gained wide acceptance in the USA with results equivalent to those achieved by radical prostatectomy (HILARIS 1986; BATATA et al. 1980; GILES and BRADY 1986).

Iodine-125 seed implantation is performed at the time of a retropubic operative intervention preceded by lymph node dissection or lymph node biopsy. The theoretical advantage allows for the concentration of the radiation dose in the precise area that requires treatment at a higher dose level than would be possible if external beam radiation therapy techniques were used. The limitations of the procedure are that it requires a surgical approach, does not irradiate the regional lymphatics, and delivers the radiation dose at a very low dose rate, which may be less efficacious than that at which external beam radiation therapy is delivered. HILARIS (1986) and coworkers (HILARIS et al. 1975; BATATA et al. 1980) have achieved a 5-year survival in early tumors of 96%, 76%, and 69% in T1, T2, and T3 tumors, respectively. Delivering an adequate dose to the entire gland (180 Gy as a minimum peripheral dose) has led to results equal or superior to those of other therapeutic modalities. It is, however, a technically difficult procedure and is not uniformly applicable in all centers. In those patients who are found to have positive nodes at the time of surgery, the addition of external radiation therapy does not improve results but does increase significantly the complications from the combined treatment regimen (HILARIS 1975). In the data from the Hahnemann University experience, the diagnosis of adenocarcinoma of the prostate was made by either needle biopsy or transurethral resection of the prostate. During the megavoltage era for Hahnemann University (from 1960 to 1988), 983 patients were seen for evaluation of prostate cancer. The patients treated by radioactive iodine-125 implantation were compared with a similar group of patients treated by external beam radiation therapy. All patients were assessed as to stage of the disease by history, physical examination, negative bone scan, computerized tomography (CT) scans of the pelvis, as well as prostatic acid phosphatase tests. Lymphangiograms were not performed routinely because of the disappointing accuracy of 75%. The clinical stage of the primary tumor and the pathologic stage of the removed nodes were reported using the Whitmore modification of the American Joint Commission system. The tumors were graded according to the degree of anaplasia and glandular formation into well differentiated, moderately differentiated, and poorly differentiated (KOPROWSKI et al. 1990).

Those individuals receiving iodine-125 implantation had no overt obvious evidence of disease beyond the prostate at the time of clinical assessment. At the time of operation, lymph node dissection or biopsies were carried out, and frozen sections were made. If the lymph nodes were positive, the patient was rejected as a candidate for iodine-125 implantation. Only those individuals who had limited localized disease involving the prostate were accepted.

The implantation procedure pursued was that described by the Memorial Sloan-Kettering Cancer Center using the Mick applicator (HILARIS 1975). The total activity of the seeds to be implanted was calculated using the dimension averaging system tables with the activity of each seed generally in the range of 0.5 mCi. The desired minimum peripheral dose was 180 Gy. Computer calculation of the achieved dose for total decay of the iodine 125 was performed after the implant using stereo-shift films. The patients were monitored following implantation by careful attention to the urine and feces. One patient lost eight seeds, and several patients lost less than five seeds, but the remaining dosimetry was considered adequate after recalculation. The centrally placed seeds within the higher dose area were the ones that tended to be lost.

In the early stage of the study, four patients revealed positive lymph node disease on permanent sections and were treated by external beam radiation therapy as a supplemental treatment, delivering 40 Gy over 4–5 weeks using a linear accelerator. Because of the severe complications associated with this combined treatment technique, it was abandoned. In patients not undergoing implantation of the prostate gland, definitive external beam radiation therapy was given with 50 Gy to the pelvis over 5–6 weeks and an additional 20 Gy cone down boost to the prostate gland through a 4-field summated technique (Giles and Brady 1986).

Of the 133 patients considered suitable for the procedure, 5 elected to have external beam radiation therapy only, and 6 were referred with recurrent disease which had been previously diagnosed and treated. These 11 patients are excluded from the analysis. Therefore, there were 122 patients submitted to staging pelvic lymphadenopathy who had newly diagnosed localized prostate adenocarcinoma. The histologic differentiation and clinical and pathologic stages are shown in Tables 4 and 5. Some 36 patients (30%) were upstaged to pathologic stage D1 because of pelvic lymph node involvement; 15% of well-differentiated tumors and 13% of A2 tumors were upstaged to pathologic stage D1. Four of the 5 stage C patients had nodal disease. For the 36 stage D1 patients, various treatment options were used. Twenty patients underwent implantation only, 9 also received external radiation therapy, and 7 had external radiation therapy alone. The treatment option was influenced by negative frozen section on subsequent histology (22 of the 106 negative frozen sections), by the preference of the physician and patient, and by the extent or site of extracapsular involvement (10 had only periprostatic involvement). The patients were followed at regular intervals with history and physical examination including rectal examination, serum prostatic acid phosphatase, and bone scans. Routine biopsies were avoided unless there was evidence of regrowth of the tumor.

Failure of treatment was identified by:

1. An expanding nodule or increasing enlargement of the prostate on rectal examination.
2. Positive biopsy more than 15 months after implant or less than 15 months if an increased histologic grade was part of the primary assessment.
3. Elevated acid phosphatase level with suspicious clinical examination.
4. Bone scan or radiographic evidence of bone metastases.

The results from the Hahnemann University experience were analyzed by life table analysis on a computer system and also by multivariate statisti-

Table 4. Distribution of histological grade and pathological stage in carcinoma of the prostate ($n = 122$)

	Pathological stage							
Histology	A1	A2	B1	B2	B	C	D1	Total
Well differentiated	7	12	18	8	6	1	9	59
Moderate	–	7	9	8	1	–	15	40
Poorly differentiated	–	–	1	1	1	–	10	13
Unspecified	–	1	6	1	–	–	2	10
Total	7	20	34	18	6	1	36	122

Table 5. Change in clinical stage in the 36 pathological stage D1 patients

	Clinical stage						
Histology	A1	A2	B1	B2	B	C	Total
Well differentiated	–	1	2	3	1	2	9/59 (15%)
Moderate	–	–	3	9	2	1	15/40 (37%)
Poorly differentiated	–	2	1	7	–	–	10/13 (77%)
Unspecified	–	–	1	–	–	1	2/10 (20%)
Total	–	3	7	19	3	4	36/122 (30%)
Percentage		13%	17%	50%	33%	80%	

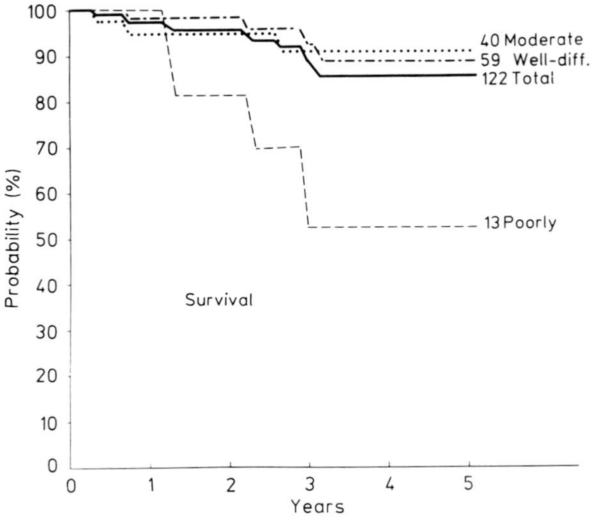

Fig. 1. Crude survival and histological grade of 122 patients submitted to lymphadenectomy

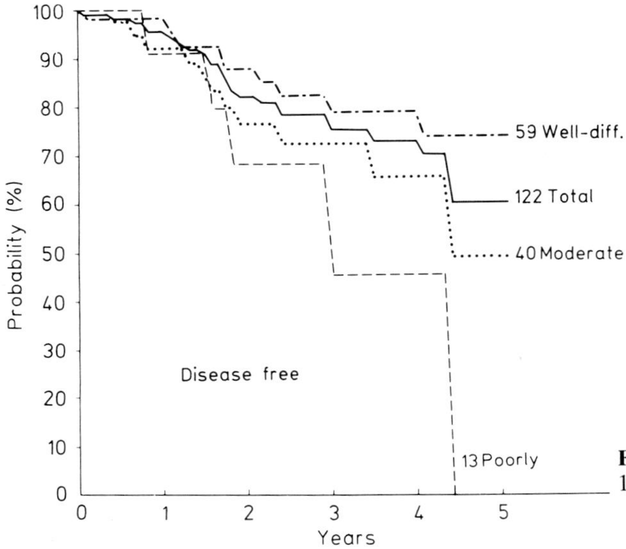

Fig. 2. Disease-free survival and histological grade of 122 treated patients

Table 6. Clinical outcome in the 114 patients (all stages) treated by implantation

	6 months	6–11 months	12–17 months	18–23 months	24+ months
No. of patients at risk	114	106	101	92	85
Complete resolution	22%	55%	69%	74%	
Incomplete or no change			27%	22%	
Local failure	–	1%	2%	6%	10%
Distant metastases	1%	1%	4%	5%	7%
Local and distant metastases	–	–	2%	3%	6%

cal techniques where necessary to control for differences in grade, stage, and length of follow-up. Response to treatment in the recurrence pattern is given in Table 6 and shows a 22% complete response to palpable tumor by 6 months, 55% by 1 year, and 74% by 2 years, with 22% showing no change or incomplete response. Two patients received a repeat iodine-125 implantation for local

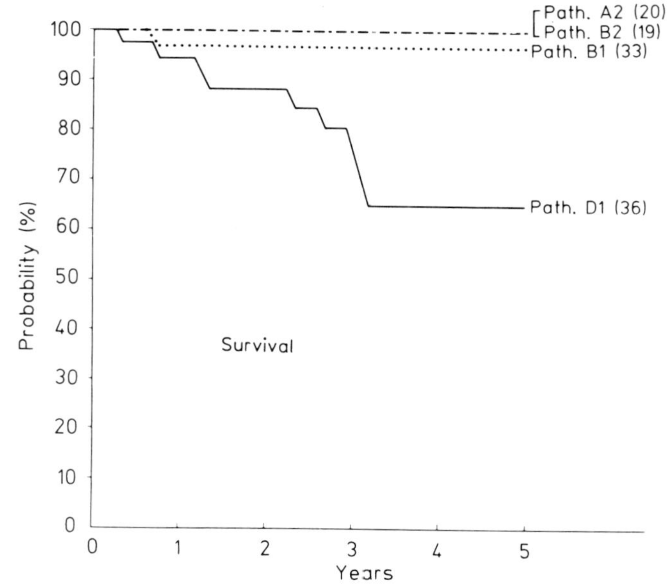

Fig. 3. Crude survival of 122 treated patients by stage of disease

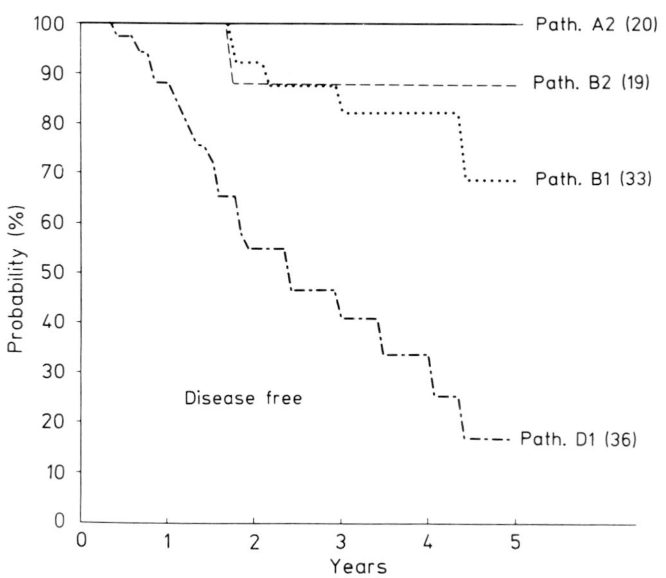

Fig. 4. Disease-free survival of 122 treated patients by stage of disease

recurrence at 18 and 42 months, and both are still alive, one with disease at 33 months and the other disease-free at 18 months after reimplantation with no significant short- or long-term morbidity.

Figures 1 and 2 indicate overall survival, probability of freedom from relapse, and histologic differentiation for all patients. The overall 5-year probability of survival is 85%, with 60% probability of no evidence of disease for all stages. Well and moderately differentiated tumors do significantly better than poorly differentiated tumors. Figures 3 and 4 incidate the survival and disease-free probability for the different stages. All patients with stage A1 lesions remained disease-free. All patients with pathologic stage A2 lesions are alive and disease-free. Of the unspecified stage B patients, one of six has shown local recurrence at 52 months.

Table 7 compares the stage groups and latency to local failure. Five of the 35 stage B1, 2 of the 19 stage B2, and 1 of the stage B patients developed recurrent disease locally, none having demonstrated distant metastatic disease. Of the 10 recurrent stage D1 patients, only 1 has local failure

Table 7. Comparison by stages in 114 implanted patients of latency to local failure

Pathological stage	Follow-up interval		
	<24 months	24–48 months	>48 months
A	0/27	0/21	0/10
B	4/58	2/36	3/16
D1	7/29	1/14	2/3

alone, and none have developed distant metastases in the absence of proved or suspicious local recurrence. There was no significant difference in survival or recurrence for pathologic stage D1 patients who had either a positive or false-negative frozen section at lymphadenectomy. The site of the metastatic disease in the pelvis (in the immediate periprostatic area or in the more distant iliac nodes) did not influence survival. There was no effect of transurethral resection of the prostate versus needle biopsy for diagnosis nor of prior transurethral resection in the past for any stage. Twelve patients were diabetic, and 75 patients were taking antihypertensive or cardiovascular medication without these concurrent medical problems affecting survival.

The reactions to the implantation procedure are shown in Table 8 and are related to the form of definitive treatment. When external radiation therapy is added to the lymphadenectomy and implant, 80% of patients experienced significant morbidity with one patient dying disease-free following colostomy for a rectal stricture. One patient in the lymphadenectomy and implant group developed a urethral stricture, and another required a penile prosthesis for impotence. Subsequent malignancies occurred in three patients (hepatoma, colon, and lung), which is comparable to REED's report with 3 of 189 patients developing subsequent rectal carcinoma after external radiation therapy.

Those patients with well-differentiated stage A1 lessions were disease-free at follow-up after implantation and lymphadenectomy. It is now recognized that such patients do not require treatment after transurethral resection other than follow-up with repeat transurethral resection 3–6 months later.

FREIHA et al. (1979) reported only 7% positive pelvic nodes in patients with stage A2 and B lesions but 90% in patients with stage C lesions with elevated acid phosphatase studies. KLEIN (1979) found 8%–45% positive nodes in patients with stage A2 and B lesions. The data from several nonrandomized reports of iodine 125 implantation in clinically localized carcinoma of the prostate have shown that there is a definite failure of local control in the pelvis as well as of distant metastases, with a continuing downward trend in the probability of being disease-free (BATATA et al. 1984; EL-MANDI et al. 1985). The Hahnemann University series show equivalent patient survival but earlier local and distant failure, perhaps due to more rigorous criteria for freedom from disease.

HILARIS et al.'s (1978) early report gave the results of survival and freedom from disease over more than 20 months in 112 patients (stages B and C). These data, however, disregard the continuing recurrences that appear thereafter, as the series' later results show (BATATA et al. 1984; GROSSMAN

Table 8. Early and late complications in 122 patients who received surgery, comparing treatments used

	Implant only	External radiotherapy only	Combined implant and external radiotherapy
No. of occurrences/No. at risk	30/106	6/7	9/9
Early			
Death	–	–	–
Would infection and/or delayed healing	2	–	1
Pulmonary embolus	1	–	–
Moderate cystitis and/or proctitis	14 (13%)	6 (86%)	8 (90%)
Mild transient edema	2	1	–
Late			
Lymphedema of pubis, groin, or lower extremity	7	1	3
Bowel, bladder, or bleeding	1	2	3 (33%)
Transurethral prostatic resection for benign prostatism	4 (4%)	–	2 (22%)
Urethral stricture	1	–	–
Rectal stricture and colostomy	–	–	1
Death			

et al. 1982). The first communications from the Memorial Sloan-Kettering Cancer Center showed that the presence of minimal nodal metastases might not adversely affect the 5-year prognosis for bone metastasis or distant failure. However, this assessment was not sustained by a longer follow-up (GROSSMAN et al. 1982). More recent reports by WHITMORE (1988) giving 10-year follow-up data on the early series from the Memorial Sloan-Kettering Cancer Center indicate a poor rate of survival without disease, probably influenced by less than rigorous criteria for the selection of patients during the early phases, many with positive nodal disease being included, and the changing evolution of the criteria used to calculate the radiation dosage, as well as the increase in the radiation dose from 140 Gy in 1965 to 238 Gy as the minimal peripheral dose in 1979 without a significant increase in morbidity or survival. There were also, during the same time period, changes in the specific γ-ray consonant for iodine 125 and the dosimetric methods used for calculating the dose (HILARIS et al. 1978).

It has been frequently stated that transurethral prostatic resection increases the incidence of failure (LEIBEL et al. 1984; PEREZ 1983). This has not been found in the Hahnemann University series; perhaps the hypothesis derives from those patients with more advanced stage disease and with poorly differentiated tumors.

8 Comparison of the Efficacy of External Beam Irradiation and Implantation Techniques

The implantation technique allows for preservation of prior sexual function in 93% of patients as reported by HILARIS et al. (1978), in 92% as reported by SCHELLHAMMER and EL-MAHDI (1983), and in 92% in the Hahnemann University series.

PILEPICH et al. (1981) have pointed out that 65–70 Gy of external beam radiation therapy over 6.5–7 weeks can be well tolerated, although 25% developed leg or genital edema when combined with lymphadenectomy. There was no increased morbidity in the Radiation Therapy Oncology Group 75-06 (PILEPICH et al. 1982) trial in which para-aortic irradiation was added, except in the lymphadenectomy patients.

In the most recent review of the Hahnemann University experience (KOPROWSKI et al. 1990) with 276 patients treated definitively by external beam radiation therapy or iodine 125 implantation, the groups were comparable for mean age and proportion diagnosed by transurethral resection, with mean follow-ups of 5 years or longer. In analysis of the overall survival, recurrence-free survival, local failure, and complication rates, a multivariable statistical technique was used to control for differences in grade, stage, and length of follow-up. There were striking differences between the implantation patients and the external beam radiation therapy patients in local failure rates and relapse-free survival, most of the latter attributable to poor local control.

Of those patients treated by iodine 125 implantation, 42% developed local failure, whereas in those patients treated by external beam radiation therapy only 10% developed local failure. Those patients treated by implantation who fared well had clinical stage A2 disease; none experienced local failure, and only 18% developed distant failure. Complication rates were marginally lower in patients treated by the implantation technique (10%) than in those treated by external beam radiation therapy (16%).

This comparison of a single institutional set of patients has led to serious doubts about the efficacy of iodine 125 implantation in maintaining local control. This appears to be satisfactory only in patients with stage A2 lesions. The complication rates were no different between the two treated groups.

It was from these data that the current recommendation was made to limit the implantation procedure to carefully selected patients and to centers with experience.

Other groups have used gold-198 grains, and more recently, transperineal temporary implantation of iridium-192 seeds in combination with external beam radiation therapy has been advocated in an attempt to improve local control further in patients with larger tumors. (PUTHARWALA et al. 1985; BAGSHAW 1986b).It is too early yet to analyze the results for meaningful comparison with the known expected outcome in such patients. Transperineally implanted iridium 192 as a temporary interstitial treatment has been employed in many institutions as a supplement to external beam radiation therapy. The iridium is implanted directly into the prostate in patients with stage C tumors after 50 Gy over 5 weeks have been delivered by external beam radiation therapy. The implant provides a 30-Gy boost to the primary tumor. These techniques have been designed to improve the potential for local control in patients with large bulky tumors. Other booster techniques under study include high-energy protons, neutron irradiation, negative pi meson irradiation, and heavy ion irradiation. Attempts are also being made to investigate hyperthermia in combination with external beam radiation therapy or interstitial implantation to improve local control in a technique pioneered by Yerushalmi et al.

9 Conclusions

The scarcity of suitable patients for radioactive implantation techniques, the technical difficulties associated with the procedure, and the considerable doubt relative to the radiation dose delivery at the low dose rate of iodine 125 indicate that it is difficult to justify a widespread use of this procedure. The results do show, however, that it gives rise to excellent long-term survival and local control in those individuals who have stage A2 lesions with no evidence of extension of the disease process beyond the prostate, as determined by lymph node biopsy or lymphadenectomy. In patients with other stages of disease, implantation techniques should be reserved to boost the local radiation dosage to the prostate using temporary implantation procedures in combination with external beam radiation therapy.

The treatment of prostate carcinoma with radiation therapy is a highly successful, well-tolerated modality. In general, external beam radiation therapy techniques are the most generally applicable program for the management of these patients. This is based upon the results that have accrued in a review of the literature as well as the high probability for microscopic disease involvement in the regional lymph node drainage of the prostate. Permanent implantation with iodine-125 seeds as well as temporary transperineal implantation by iridium-192 seeds have recognizable and difficult problems in their general application and should be restricted to those patients with pathologically staged limited local disease involving the prostate.

References

American Cancer Society (1989) Cancer. Facts and figures – 1989. American Cancer Society, New York

Bagshaw MA (1980) Eternal radiation therapy of carcinoma of the prostate. Cancer [Suppl 2] 45: 1912–1921

Bagshaw MA (1986a) Current conflicts in the management of prostatic cancer. Int J Radiat Oncol Biol Phys 12: 1721–1727

Bagshaw MA (1986b) Radiation therapy, the Stanford experience, at current controversies in the management of prostate carcinoma. American Cancer Society, Baltimore

Batata MA, Hilaris BS, Chu FCH, et al. (1980) Radiation therapy in adenocarcinoma of the prostate with pelvic lymph node involvement on lymphadenectomy. Int J Radiat Oncol Biol Phys 6: 149–153

Batata MA, Hilaris BS, Chu FCH, He S, Genest P, Jain P, Whitmore WF (1984) External beam versus brachytherapy in localized prostatic cancer. Int J Radiat Oncol Biol Phys [Suppl 2] 10: 116

Batson OU (1942) The role of the vertebral veins in metastatic processes. Ann Intern Med 16: 38–45

Breslow N, Chan CW, Dhom G, et al. (1977) Latent carcinoma of prostate at autopsy in seven areas. Int J Cancer 20: 680–688

Byar DP, Mostofi FK (1973) Carcinoma of the prostate: prognostic evaluation of certain pathologic features in 208 radical prostatectomies. Cancer 30: 5–13

El-Mahdi AM, Kuban DA, Schellhammer PF (1985) The treatment of choice for localized poorly differentiated adenocarcinoma of the prostate (Abstr). 67th Meeting of the American Radium Society

Fowler JE, Whitmore WF (1981) The incidence and extent of pelvic lymph node metastases in apparently localized prostatic cancer. Cancer 47: 2941–2945

Freiha S, Pistenmaa DA, Bagshaw MA (1979) Pelvic lymphadenectomy for staging prostatic carcinoma: is it always necessary? J Urol 122: 176–177

Giles GM, Brady LW (1986) Iodine 125 implantation after lymphadenectomy in early carcinoma of the prostate. Int J Radiat Oncol Biol Phys 12: 2117–2125

Gleason DF, Mellinger GT (1974) Veterans Administration Cooperative Urologic Research Group: Prediction of prognosis for prostatic adenocarcinoma by combined histologic grading and clinical staging. J Urol 111: 58–64

Groosman HB, Batata M, Hilaris B, Whitmore WF (1982) ^{125}I implantation for carcinoma of the prostate. Urology 20(6): 591–598

Hanks GE (1986) The RTOG patterns of care study, at current controversies in the management of prostate carcinoma. American Cancer Society, Baltimore

Hanks GE, Leibel S, Kramer S (1983) The dissemination of cancer by transurethral resection of locally advanced prostate cancer. J Urol 129: 309–311

Hanks GE, Leibel SA, Krall JM, et al. (1985) Patterns of care studies: dose response of observations for local control of adenocarcinoma of the prostate. Int J Radiat Oncol Biol Phys 11: 153–157

Hanks GE, Diamond JJ, Krall JM, et al. (1987) A 10 year follow-up of 682 patients treated for prostate cancer with radiation therapy in the United States. Int J Radiat Oncol Biol Phys 13: 499–505

Hilaris BS (ed) (1975) Handbook of interstitial brachytherapy. Publishing Sciences, Boston

Hilaris BS (1986) Brachytherapy, the memorial experience, at current controversies in the management of prostate carcinoma. American Cancer Society, Baltimore

Hilaris BS, Whitmore WF, Batata MA, et al. (1975) Cancer of the prostate. In: Hilaris BS (ed) Handbook of interstitial brachytherapy. Publishing Sciences, Acton

Hilaris BS, Whitmore WF, Batata MA, Barzell W, Tokita N (1978) ^{125}I implantation of prostate: dose-response considerations. Front Radiat Ther Oncol 12: 82–90

Jewett HJ (1963) Radical perineal prostatectomy for palpable, clinically localized, nonobstructive cancer: experience at the Johns Hopkins Hospital 1909–1963. J Urol 124: 492–494

Jewett HJ, Eggleston JC, Yawn DH (1972) Radical prostatectomy in the management of carcinoma of the prostate: probable causes of some therapeutic failures. J Urol 107: 1034–1040

Johansson J-E, Andersson S-O, Krusemo UB, Adami H-O (1989) Natural history of localized prostatic cancer. Lancet 1: 799–803

Klein LA (1979) Prostatic cancer. N Engl J Med 300(15): 824–833

Koprowski C, Berkenstock K, Barofsky AM, Ziegler JC, Lightfoot DA, Brady LW (1990) External beam irradiation versus iodine 125 implantation in the definitive treatment of prostatic cancer. (in press)

Leibel SA, Hanks GE, Kramer S (1984) Patterns of care outcome studies: results of the national practice in adenocarcinoma of the prostate. Int J Radiat Oncol Biol Phys 10: 401–409

Leiber MM (1987) Surgery vs radiation for localized prostate cancer. Oncology 1: 61–68

McGowan DG (1980) The adverse influence of prior transurethral resection on prognosis in carcinoma of prostate treated by radiation therapy. Int J Radiat Oncol Biol Phys 6: 1121–1126

Murphy GP, Natarjan N, Pontes JE, et al. (1982) The national survey of prostate cancer in the United States by the American College of Surgeons. J Urol 127: 928–934

Perez CA (1983) Presidential address of the 24th annual meeting of the American Society of Therapeutic Radiologists: carcinoma of the prostate, a vexing biologic and clinical engima. Int J Radiat Oncol Biol Phys 9: 1427–1438

Perez CA, Walz BJ, Zivnuska FR, et al. (1980) Irradiation of carcinoma of the prostate localized to the pelvis: analysis of tumor response and prognosis. Int J Radiat Oncol Biol Phys 6: 555–565

Perez CA, Pilepich MV, Zivnusta F (1986) Tumor control in definitive irradiation of localized carcinoma of the prostate. Int J Radiat Oncol Biol Phys 12: 523–531

Pilepich MV, Perez CA, Walz BJ, Zivnuska FR (1981) Complications of definitive radiotherapy for carcinoma of the prostate. Int J Radiat Oncol Biol Phys 7: 1341–1348

Pilepich MV, Satish MD, Prasad C, Perez CA (1982) Computed tomography in definitive radiotherapy of prostatic carcinoma. II. Definition of target volume. Int J Radiat Oncol Biol Phys 8: 235–240

Pilepich MV, Krall JM, George FW, et al. (1984) Treatment-related morbidity in phase III RTOG studies of extended-field irradiation for carcinoma of the prostate. Int J Radiat Oncol Biol Phys 10: 1861–1867

Putharwala AA, Syed AMN, Tansey LA, et al. (1985) Temporary iridium 192 implant in the management of carcinoma of the prostate. Endocuriether Hyperthermia Oncol 1: 25–34

Reed NS External radiation for early carcinoma of prostate.

Rosen EM, Cassady JR, Connolly J, et al. (1984) Radiotherapy for localized prostate carcinoma. Int J Radiat Oncol Biol Phys 10: 2201–2210

Schellhammer PF, El-Mahdi AM (1983) Pelvic complications after definitive treatment of prostate cancer by interstitial or external beam radiation. Urology 21(5): 451–457

Walsh PC (1986) Surgery, The Johns Hopkins experience, at current controversies in the management of prostate carcinoma. American Cancer Society, Baltimore

Walsh PC, Jewett HJ (1980) Radical surgery for prostatic cancer. Cancer 45: 1906–1911

Whitmore WF (1956) Symposium on hormones and cancer therapy. Am J Med 21: 697

Whitmore WF (1988) Clinical management of prostate cancer: an overview. In: Raymond JP, Harvey M, Ojasso T (eds) Proceedings of a satellite symposium of the 3rd International Congress on Hormones and Cancer. Raven, New York, pp 88–97

Whitmore WF, Butata MA, Hilaris BS (1979) Prostate irradiation: iodine 125 implantation. In: Johnson DE, Samuels ML (eds) Cancer of the genitourinary tract. Raven, New York, pp 195–205

7.4 Percutaneous Iodine-125 Implantation Guided by Ultrasound

JOSEF HAMMER and MARKUS RICCABONA

CONTENTS

1	Introduction	257
2	Selection of Patients	257
3	Treatment Strategies	257
3.1	Suprapubic Implantation	258
3.2	Transperineal Implantation	258
3.2.1	Volumetry and Prostate Mapping by Ultrasound	258
3.2.2	Seed Preparation and Implantation	259
4	Physical Aspects	259
5	Results, Complications, and Future Aspects	260
6	Conclusions	261
	References	262

1 Introduction

The reports by HILARIS (1968), HILARIS and BATATA (1981), and HILARIS et al. (1972, 1978) and by the working group from Freiburg, FRG (KNÜFERMANN et al. 1981; KUPHAL et al. 1979), led to the introduction of iodine-125 seeds in the treatment of prostate cancer as an alternative to radical prostatectomy. In particular, the lower percentage of long-term side effects stimulated our interest in seed implantations: The rate of incontinence and impotence is lower compared with the complications after the radical surgical approach. We started interstitial management in patients with prostate cancer in February 1981. This contribution presents the preliminary experience with suprapubic implantation and the reasons for changing to the transperineal approach. The method is described, and the short-term results and complications are given.

JOSEF HAMMER, Dr. med., Institut für Radiotherapie, and MARKUS RICCABONA, Dr. med., Urologische Abteilung, Krankenhaus der Barmherzigen Schwestern, Seilerstätte 4, 4010 Linz, Austria

2 Selection of Patients

Patients are clinically suitable for iodine-125 seed implantation in stages T1-2 pN0-1 and small grade I and II T3 stages. T3 grade III and other advanced stages are not an absolute contraindication, particularly in patients who refuse radical prostatectomy. We included patients with T3N2 tumors and two patients with small local recurrences after radical prostatectomy who preoperatively presented with T4 stages, in a palliative attempt to reduce symptoms, accepting that there may be no impact on survival. A limited transurethral prostatic resection is allowed and necessary in patients with urethral obstruction (12 of 54 patients). The basic examinations are listed in Table 1.

3 Treatment Strategies

From February 1981 to Januray 1987, 54 patients with prostate carcinoma underwent local interstitial radiotherapy using iodine-125 seeds. The median age was 62.8 years (range 45–73), and the median follow-up time was 45 months (range

Table 1. Patient selection and staging procedures

T	category
	rectal palpation
	transrectal ultrasound
	volumetry
	ultrasound-guided biopsy
N	category
	computerized tomography
	diagnostic lymphadenectomy
M	category
	bone scan
	prostate specific antigen
	prostate acid phosphatase

Table 2. Staging in 54 patients after lymphadenectomy

Stage	Retropubic	Perineal	Total	Percentage
T0(A2)N0	0	4	4	7.4
T0(A2)N1	0	1	1	1.8
T2N0	10	20	30	55.6
T2N1	1	1	2	3.7
T2N2	1	1	2	3.7
T3N0	3	3	5	9.3
T3N1	1	0	1	1.8
T3N2	5	2	7	13.0
LRN0	0	2	2	3.7
Total	20	34	54	100
	37%	63%		

LR, local recurrences after radical prostatectomy.

90–19), a very short period which allows only preliminary statements.

3.1 Suprapubic Implantation

From February 1981 until August 1984, 19 clinically N0 and 1 N+ patients were treated by diagnostic pelvic lymphadenectomy and the conventional retropubic technique in the same operating session (group 1, 20 patients). Eight of the 20 operated patients presented with pelvic lymph node metastases (Table 2) and underwent postoperatively percutaneous radiotherapy with 5000 cGy to the pelvis, excluding the implanted area.

3.2 Transperineal Implantation

Since September 1984, the implantation procedure has been modified (group 2, 34 patients). After proof of histology by biopsies and after confirmation of the clinical stage, modified (nonradical) pelvic lymphadenectomy is performed. If there is evidence of negative lymph nodes or minor metastatic disease with 1 or 2 positive nodes in one or both iliac regions, the patient is selected for iodine 125 implantation.

3.2.1 Volumetry and Prostate Mapping by Ultrasound

To plan local therapy, the size and volume of the prostate is measured endosonographically. Rectal sonography offers an exact volumetric measurement. The prostate is mapped out in 5-mm steps, the cross-sectional areas are planimetried, and the volume can then be exactly calculated. Prostate mapping is done with the aid of a multichannel puncture attachment. The channels serve as needle guides to ensure particular geometric patterns and are indicated by numbers and capital letters (Fig. 1) similar to a chessboard. This grid can also be seen on the monitor screen of the ultrasound device. Multiple biopsies are taken by means of a trucut needle from defined areas of the prostate. The biopsy spots are marked on the screen and documented by Polaroid pictures. According to the data from volumetry and pathology, the brachytherapist calculates the number of seeds and determines their localization (Fig. 2). This first step is performed to clarify all parameters which enable an accurate dose distribution.

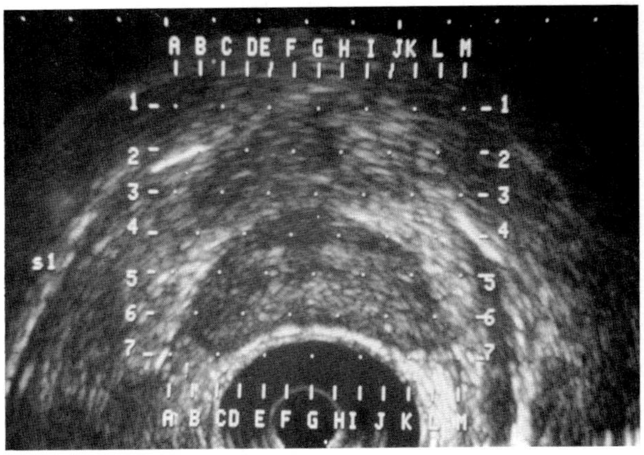

Fig. 1. Transrectal sonography with display of the template matrix used for biopsies and seed implantations

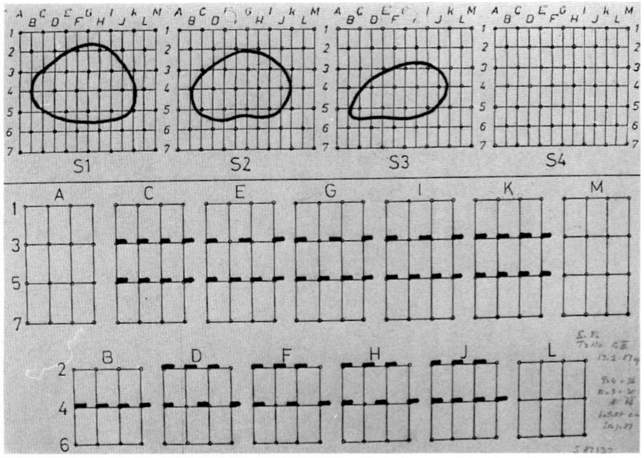

Fig. 2. Map of cross sections and preplanned allocation of seeds

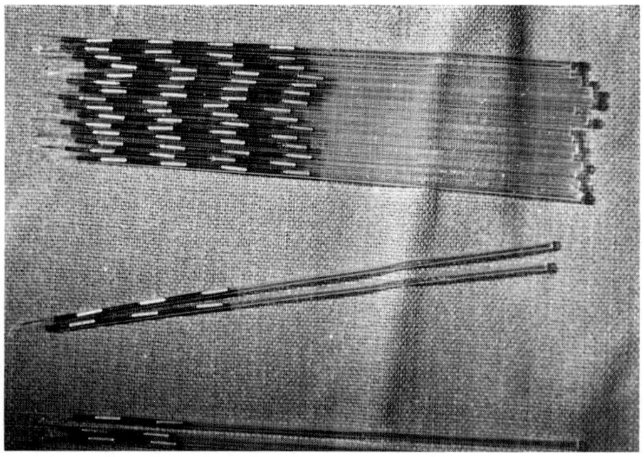

Fig. 3. Teflon carrier tubes filled with seeds and spacers

3.2.2 Seed Preparation and Implantation

Some 3–4 weeks after lymphadenectomy the second step is performed. The radioactive seeds are filled into small Teflon tubes as indicated in Fig. 3. To achieve the desired distance between the seeds inside one needle, specially hardened absorbable suture material (poly-*p*-dioxanone suture, PDS) is used. After changing to this kind of material we never observed obstructions inside the tubes or the needles (HAMMER et al. 1989). The implantation is performed in lithotomy position under general or spinal anesthesia. The transrectal probe is again fixed to the template in the same manner as during the planning procedure, and can be moved in 5 mm steps. Care should be taken to ensure that the distance from the transducer to the prostate is the same. The unloaded implantation needles are inserted into the prostate under sonographic guidance in accordance with the tumor map. The prefilled carriers are fixed into the needles, and the seeds are pushed into the prostate by means of a stylet. A similar implantation technique is described by HOLM et al. (1983). Of these 34 patients, 5 presented with positive pelvic lymph nodes after lymphadenectomy (N1-2), with a maximum of two infiltrated nodes per iliac region, and were treated with adjuvant chemotherapy (estramustine) postoperatively for a period of 6–7 months. In group 2 no additional percutaneous radiotherapy was performed.

4 Physical Aspects

Iodine-125 seeds with an activity of 0.55 mCi per seed were used. The physical parameters, dose distribution around a seed, and implant calculations are described by many authors (KRISHANSWAMY 1978, 1979; LING et al. 1983; KUPHAL and GENGNAGEL 1985; SCHELL et al. 1987; ANDERSON 1976; ANDERSON et al. 1981; GLASGOW and PEREZ 1987). The number of seeds is calculated according to the prostate's volume, tumor grade, and tumor distribution. In group 1 (open laparotomy approach) we used the Anderson nomogram and the average dimension method to determine the number of seeds. The seeds were inserted in a manual free-hand technique. The seed distribution was excellent in only a few patients; in most patients it was unsatisfactory, resulting in an insufficient dose distribution (Fig. 4). The underdosage to parts of the prostate may increase the risk of local tumor progression (HILARIS et al. 1978; SOGANI 1979). This formed the impetus to look for methods to optimize the seed distribution. Since September 1984, therefore, the transperineal approach has been employed. Compared with the Anderson nomogram we calculate 15%–25% more seeds to achieve a better dose distribution, particularly at the prostate borders. By this method some seeds are located in the prostate capsule or just outside the prostate. X-ray films of the implanted seeds give proof of the better seed distribution. The real seed distribution after implantation is not that of the preplanned ideal but much more acceptable when compared with the retropubic method (Fig. 5). The transfer of the

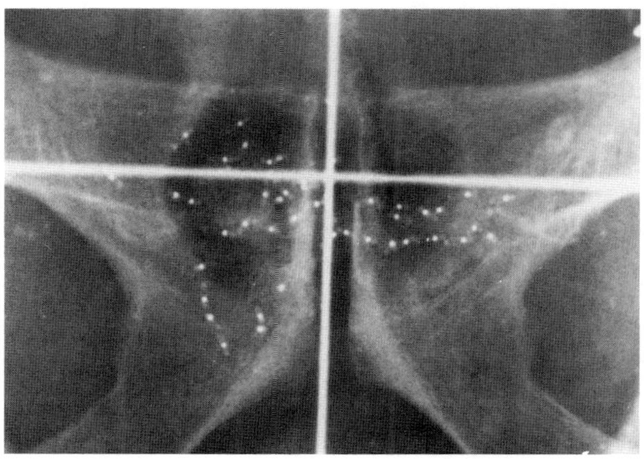

Fig. 4. Inadequate distribution of gold-marker seeds after retropubic implantation

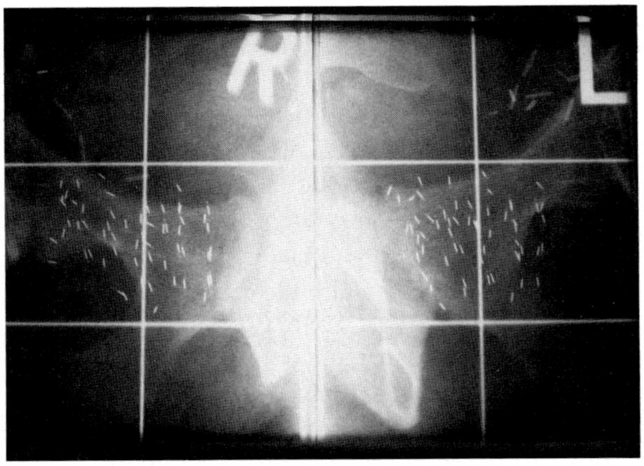

Fig. 5. Adequate distribution of seeds after transperineal implantation (orthogonal X-ray films)

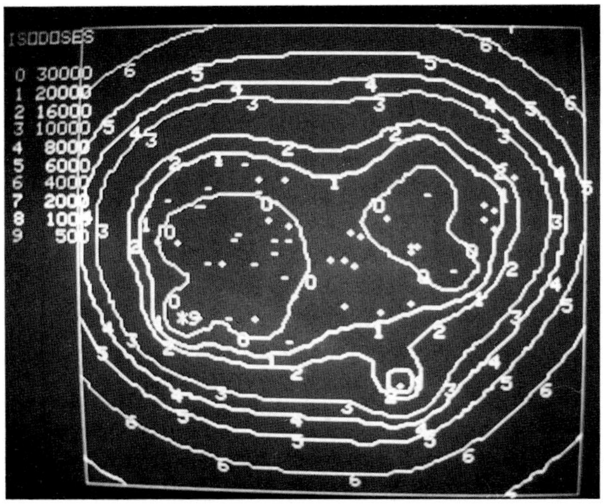

Fig. 6. Anteroposterior view of isodose distribution

seed localization into the CMS-Modulex computer is done by means of simulator films. The calculation program follows the rules and recommendations formulated by Glasgow et al. (1981) according to the data published by Krishnaswamy (1978) and Anderson et al. (1981). The "matched peripheral dose" resulted in 10000–16000 cGy and the dose peaks reached up to 30000 cGy or more (Fig. 6). Radiation exposure for medical personnel during the implantation procedure and at the ward is very low. Due to the low energy of about 30 kV radiation, exposure to personnel and family members can easily be reduced (Kuphal et al. 1979; Liu and Edwards 1979) by means of a protective flap which can be attached to the underwear.

5 Results, Complications and Future Aspects

The aim of this report is to describe the method of percutaneous, transperineal, iodine-125 seed implantation into the prostate and also the reasons for abandoning the retropubic approach. There is clear evidence that endosonographically guided seed implantation results in a better seed distribution and therefore in a better homogeneity of the dose distribution. In prostate cancer patients long-term survival data are necessary to evaluate the outcome of different treatment regimens. From the clinical point of view only preliminary results can be presented. Table 2 shows the number of patients in clinical T and pathological N stages after lymphadenectomy, separated according to the retropubic and transperineal approaches. More than 60% of patients presented with T2 tumors, and 75% were node negative. Table 3 shows the short-term results. For the first group, with implantation by laparotomy, there was a median follow-up time of 68 months. There was one local failure. The overall survival is 85% and the disease-free survival, 75% so far. Five patients developed nodular and/or distant disease. Two patients died of disease, one patient died of cardiovascular disease without evidence of tumor. In the second group, treated by the transperineal method, there was a median follow-up period of 32 months. There was one local failure and one

Table 3. Preliminary results of treatment by retropubic (group 1) and by endosonographically guided (group 2) seed implantation

	Retropubic	Transperineal
n	20	34
Median age (years)	62.7 (45–72)	62.9 (46–73)
Median follow-up (months)	68 (90–50)	32 (46–19)
Local failure	1 (50%)	1 (3%), Em-NED[b]
Distant disease (N,M)	2 (10%)	0
Free of disease	15 (75%)	34 (100%)
Overall survival	17 (85%)[a]	33 (97%)
Death	3 (10%)[a]	1 (3%) Mi-NED[c]

[a] Two patients died of distant metastases, one disease-free patient died of a cerebral stroke.
[b] After estramustine treatment there is no evidence of disease.
[c] One disease-free patient died of myocardial infarction.
Em, Estramustine Phosphate.
NED, no evidence of disease.
Mi, myocardial infarction.

Table 4. Complications and side effects in 54 patients after pelvic lymphadenectomy and iodine-125 seed implantation

	Retropubic (n = 20)	Transperineal (n = 34)
Lymphocele (puncture)	3[a]	2
Lymphedema	9 (7[a])	0
Cystitis proctitis	2[a]	0
Urethra-rectal fistula	1	0
Lung infarction	1[a]	0
Erectile impotence	1[a]	1
Perioperative mortality	0	0
Urinary incontinence	0	1[b]

[a] N-positive patients with additional percutaneous pelvic irradiation.
[b] Seed implantation into a local recurrence 2 years after radical prostatectomy

patient died of myocardial infarction. Distant metastases have not been observed in any patient so far. The only local recurrence or progression, 20 months after seed implantation, was verified by an increase of prostate volume by transrectal sonography, by histology, and by positive tumor markers. After receiving treatment with estramustine phosphate, the tumor now shows a decrease in volume, the tumor markers are negative, and the patient is clinically free from disease. The overall survival is 97%. Table 4 gives information about complications. The method involving radical lymphadenectomy and simultaneous retropubic seed implantation is connected with a relatively high incidence of side effects, particularly after combination with external radiotherapy to the iliac regions, as was performed in node-positive patients: More than 40% of the patients presented with edema of the legs and genital region. One major local complication was a very painful urethra-rectal fistula. Since September 1984 a limited lymph node dissection has been carried out as a diagnostic staging laparotomy. Systemic therapy using estramustine phosphate was done only in the case of positive nodes. In this group without pelvic irradiation, the complication rate was very low. In particular, no cystitis and no proctitis was observed. In general, the percentage of urinary incontinence and sexual impotence was very low. This is the principle goal of the prostate-preserving treatment method described above. Only one patient presented with impotence, and another one developed urinary incontinence.

Since March 1987, a multiplane ultrasound scanner has been available in our hospital. During needle implantation the cross section can be switched from transversal imaging to longitudinal scanning. We can therefore watch each needle during insertion. The relation of the needles to surrounding anatomical structures like the urethra, prostate capsule, seminal vesicles, urinary bladder, and rectal wall can be observed in detail. This approach allows an exact placement of the needle parallel to the urethra into prostate tissue and, in particular, of the needle tip into the prostate borderline at its base. Reflux of urine indicates that the needle has to be drawn back an appropriate distance. In this manner we are able to place each seed on that point which has been preplanned. During the procedure we can watch each seed leaving the needle. This method improves dose distribution. It may also improve local control, but further investigations are necessary. For this reason, patients with limited transurethral resection (TUR) need no longer be excluded.

6 Conclusions

The percutaneous, transperineal, and sonographically guided iodine-125 seed implantation method enables us, as an alternative procedure to radical prostatectomy and to external radiotherapy, to expect greater tumor control and survival in prostate cancer patients with minimal side effects. The ideal application of interstitial radiotherapy is to the early stage carcinoma, the T1, T2, and small T3 tumor with negative nodes and negative bone scan. Node-positive patients have to be treated by additional systemic therapy.

Seed implantation is easy and, in comparison with the conventional operative free-hand technique, much more precise under ultrasonic guidance and can be carried out according to the individual tumor size. Because of the very short follow-up period, no comments can be made about the contribution of optimized seed and dose distribution to long-term local tumor control. A higher local control rate compared with group 1 is to be expected.

In more advanced inoperable patients iodine-125 implantations may be used with a palliative intent. Small local recurrences after radical prostatectomy are suitable for implantation treatment. The use of transrectal ultrasound allows seed implantations after limited transurethral prostatic resection.

Acknowledgements. Sincere and grateful appreciation is expressed to Monika Habersberger and Luise Mitterbauer for their support in preparation of this manuscript, to Werner Labeck for his assistance as physicist, and to the Kretz-Technik Company for its assistance in realizing ideas and innovations, particularly in developing a multiplane ultrasound scanner.

References

Anderson LL (1976) Spacing nomograph for interstitial implants of I-125 seeds. Med Phys 3: 48

Anderson LL, Kuan HM, Ding IY (1981) Clinical dosimetry with I-125. In: George FW (ed) Modern interstitial and intracavitary radiation cancer management. Masson, New York, pp 9–16

Glasgow GP, Perez CA (1987) Physics of brachytherapy. In: Perez CA, Brady LW (eds) Principles and practice of radiation oncology. Lippincott, Philadelphia, pp 213–251

Glasgow GP, Harms WB, Purdy JA, Liu YY (1981) Implant dose calculations with the Modulex-RTP. 23rd Annual Meeting of the American Association of Physicists in Medicine, Boston

Hammer J, Hawliczek R, Kärcher KH, Riccabona M (1989) A new spacing material for interstitial implantation of radioactive seeds. Int J Radiat Oncol Biol Phys Vol 16, 1: 259–260

Hilaris BS (1968) Techniques of interstitial and intracavitary radiation. Cancer 22(4): 745–751

Hilaris BS, Batata MA (1981) Interstitial brachytherapy in the treatment of prostatic cancer. In: Wannenmacher M (ed) Kombinierte chirurgische und radiologische Therapie maligner Tumoren. Urban and Schwarzenberg, München, pp 54–62

Hilaris BS, Whitmore WF, Grabstald H, O'Kelly PJ (1972) Radical radiation therapy of cancer of the prostate: a new approach using interstitial and external sources. Clin Bull 2: 94–99

Hilaris BS, Whitmore WF, Batata MA, Brazell W, Tokita N (1978) 125-I implantation of the prostate: dose-response considerations. Oncol Front Radiat Ther 12: 82–90

Holm H, Juul N, Pedersen J, Hansen H, Stroyer I (1983) Transperineal 125-I seed implantation in prostatic cancer guided by transrectal ultrasonography. J Urol 130: 283–286

Knüfermann H, Bruggmoser G, Wannenmacher M (1981) Die interstitielle Strahlentherapie in der Behandlung des Prostatakarzinoms. In: Wannenmacher M (ed) Kombinierte chirurgische und radiologische Therapie maligner Tumoren. Urban and Schwarzenberg, München, pp 63–69

Krishnaswamy V (1978) Dose distribution around an 125-I seed source in tissue. Radiology 126(2): 489–491

Krishnaswamy V (1979) Dose tables for 125-I-seed implants. Radiology 132: 727–730

Kuphal K, Gengnagel W (1985) Dosisleistungskonstante und radiale Dosisfunktion einer Jod-125-Strahlenquelle in verschiedenen Stoffen. Strahlentherapie 161: 414–420

Kuphal K, Knüfermann H, Bruggmoser G (1979) Erste Strahlenschutzerfahrungen bei der interstitiellen Therapie mit Jod-125 Seeds. In: Reich H (ed) Medizinische Physik Dr. Alfred Hüthig Verlag, Heidelberg, pp 115–120

Ling CC, Yorke ED, Spiro IJ, Kubiatowicz D, Bennett D (1983) Physical dosimetry of 125-I seeds of a new design for interstitial implant. Int J Radiat Oncol Biol Phys 9: 1947–1952

Liu J, Edwards FM (1979) Radiation exposure to medical personnel during iodine-125-seed implantation of the prostate. Radiology 132: 748–749

Schell MC, Ling CC, Gromadzki ZC, Working KR (1987) Dose distributions of model 6702 I-125 seeds in water. Int J Radiat Oncol Biol Phys 13: 795–799

Sogani PC (1979) Carcinoma of the prostate: treatment with pelvic lymphadenectomy and iodine-125 implants. Clin Bull 9: 24–31

7.5 Transperineal ^{125}I Implantation Guided by Ultrasound: Preliminary Results of 150 Cases

GERHARD SCHLEGEL and KLAUS LUTZ

CONTENTS

1 Introduction . . . 263
2 Technique . . . 263
2.1 History . . . 263
2.2 Selection of Patients . . . 263
2.3 Radiotherapy Planning . . . 264
3 Patients and Methods . . . 264
3.1 Additional Therapy . . . 265
4 Results . . . 265
5 Discussion . . . 266
6 Conclusions . . . 267
References . . . 267

1 Introduction

According to the AMERICAN CANCER SOCIETY (1987), 63% of all prostate cancers are discovered while still localized within the general region of the prostate, and 83% of all patients whose tumors are diagnosed at this stage survive more than 5 years after treatment. This success is based on early diagnosis as well as on therapeutic means. Treatment alternatives include: (1) radical prostatectomy, (2) external beam radiation therapy alone, (3) interstitial implantation of the prostate gland with ^{125}I, (4) combination of interstitial implantation and external beam radiation therapy. The indications, advantages, disadvantages, and complications of each method are summarized in Table 1.

The safest method of eradication of the malignant tumor is radical prostatectomy with total excision of all malignant tissue. Selection of patients for this operation has been very restrictive, however, with only 10% being suitable candidates. The best alternative method with curative intention and without significant contraindication is the transperineal, sonographically controlled, ^{125}I seed implantation technique.

GERHARD SCHLEGEL, Dr. med., Radiologische Klinik, KLAUS LUTZ, Dr. med., Urologische Klinik, Katharinenhospital, Kriegsbergstr. 60, 7000 Stuttgart 10, Germany

This form of treatment is characterised by minimal invasiveness and thus is ideal for older patients as well as patients with increased operative risk. By this local treatment the 20000 cGy isodose encloses the total gland. This very high tumor dose may be applied to the prostate cancer without risk to surrounding organs. Table 2 describes the physical and radiobiological data of ^{125}I seed therapy.

2 Technique

2.1 History

In 1917 YOUNG and FRONZ reported a series of patients treated with radium needles implanted transperineally, transrectally, and transurethrally. Even with this proceeding (BARRINGER 1942), several long-term disease-free survivors were described. In 1970 Whitmore and Hilaris initiated interstitial irradiation with ^{125}I seeds at the Memorial Sloan-Kettering Cancer Center (WHITMORE et al. 1972). The transperineal, ultrasonographically controlled, seed implantation of ^{125}I seeds was introduced in 1981 by HOLM et al.

2.2 Selection of Patients

The implantation of ^{125}I seeds in the prostate gland is indicated if only very limited extracapsular tumor extension is found. This is best evaluated by transrectal ultrasonic scanning. Also, ultrasonically guided prostatic biopsy from various defined locations (mapping) is the safest method for obtaining the true histologic grade, which is the most important factor in the overall prognosis (BAGSHAW 1986).

Chest X-ray and bone scans were performed in order to confirm the absence of distant metastases

Table 1. Indications, advantages, disadvantages, and complications of four treatment alternatives to localized prostate cancer

	Radical prostatectomy	Interstitial ^{125}I implantation		External beam radiotherapy
		Operative	Guided by ultrasound	
Indications	T1-2N0M0 G1–2 <70 years	T1-2N0M0 G1–3 <70 years	T1-3N0M0 G1–3 <80 years	T1-4N0-2 G1–3 No limitation
Performance status (Karnofsky Index)	>80	>80	>60	>40
Advantages	Total excision of all malignant tissue	High local tumor dose No operative risk Hospitalization for 2 days		No need for radiation protection
Disadvantages	Hospitalization for 4–6 weeks	Measure of radiation protection required		Duration of treatment 6–8 weeks for local tumor control (not in all cases)
Complications	6%–23%[a]	20%[b]	0%	10%[c]

[a] van der Werf-Messing (1982). [b] Wannenmacher et al. (1979). [c] Scardino and Bretas (1987).

Table 2. Physical and radiobiological data for ^{125}I-seed therapy

Energy level:	27.4 and 31.4 keV
$t_{1/2}$:	59.6 days
Activity/seed:	0.5 mCi
Half-value layer (tissue):	1.3 cm
Mean peripheral dose:	20 000 cGy
Biologically equivalent dose:	ca. 7000 cGy
Cumulative dose in 31 days:	6000 cGy
Dose rate in 31 days:	ca. 8 cGy/h

and lymph node staging, by means of pelvic lymphadenectomy or computerized tomography (CT) of the minor pelvis in case of increased operative risk. Prostate specific antigen (PSA) serves as an aid in detecting the extension of the disease beyond the prostatic capsule, i.e., to the regional lymph nodes or distant metastasis.

2.3 Radiotherapy Planning

To obtain the sectional images for radiotherapy planning and volumetry, transrectal scanning is necessary. According to the data the radiologist calculates the number and precise localization of the seeds to ensure a total tumor dose of 20 000 cGy encircling the gland. This procedure and the technique of implantation were described by Holm et al. (1983). Figure 1 demonstrates the principle of the insertion of the ^{125}I seeds.

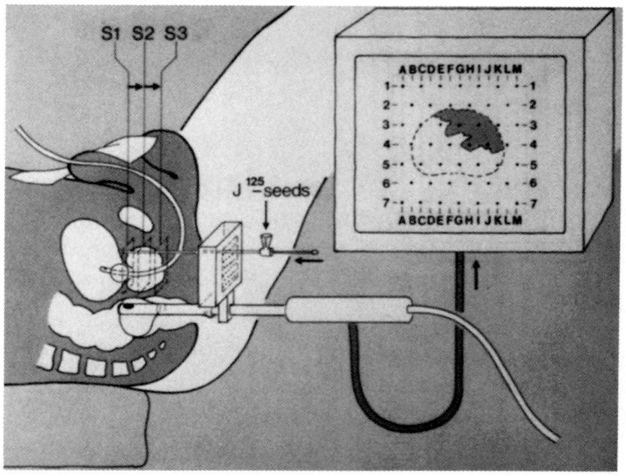

Fig. 1. Principle of ultrasonically guided ^{125}I seed insertion

3 Patients and Methods

A total of 150 patients with carcinoma of the prostate were treated between May 1986 and September 1988 at the Clinic of Radiology in collaboration with the Clinic of Urology, Katharinenhospital, Stuttgart. All had undergone ultrasound controlled biopsy using mapping technique for histological confirmation.

One patient died 13 months after the implantation from cardiac infarction without any evidence of tumor activity. Another patient with tumor stage T3N0M0G2 relapsed 15 months after the interstitial radiotherapy. The interstitial applica-

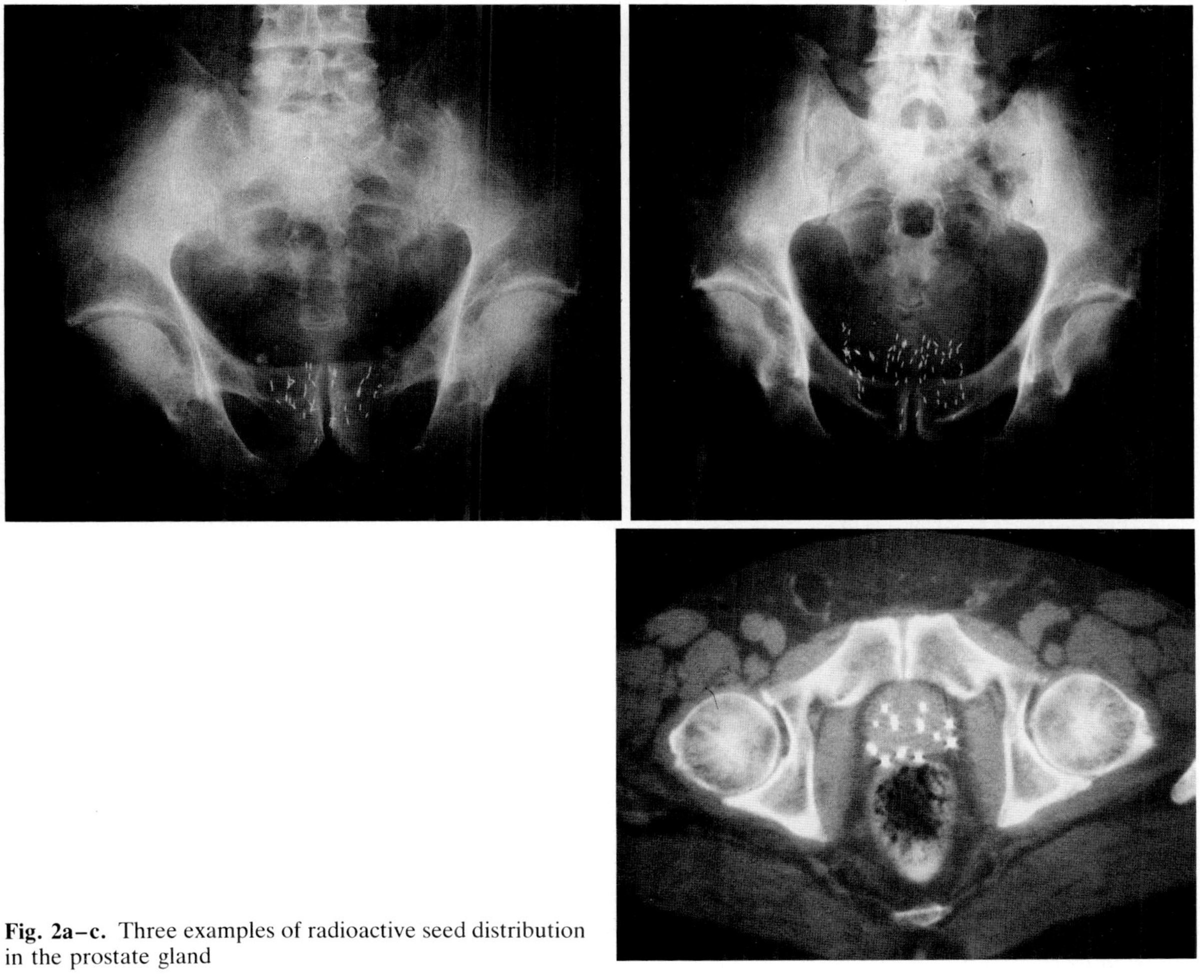

Fig. 2a–c. Three examples of radioactive seed distribution in the prostate gland

tion was performed using a Mick applicator with the help of transrectal sonography with a 5-MHZ probe. Figure 2 shows three examples of radioactive seed distribution in the prostate gland.

With the exception of a mild transient dysuria for some weeks, no intraoperative or postoperative side effects were observed. In accordance with federal regulations on radio-protection all the patients were discharged 48 h after seed application. Beginning 6 weeks after treatment, patients were seen at 3–6 month intervals. Patients in contact with children or pregnant women were instructed to wear a lead rubber shield during the first 6 months.

3.1 Additional Therapy

Prior to radiological treatment 134 (89.3%) patients underwent orchiectomy and 35 (23.3%) patients with T3 or N+ received chemotherapeutical treatment with flutamide or estramustine phosphate. Table 3 gives the patients' characteristics and the distribution of tumor stages according to the TNM classification (Hermanek et al. 1987) at the beginning of interstitial radiotherapy.

4 Results

With the exception of the patient with progressive disease, 6 months following interstitial radiotherapy all PSA levels dropped into the normal range. Rectal examination after 18 months showed complete regression of the tumor in 69%. Furthermore, transrectal sonographic findings were normal in 84%. Regression of the mean tumor volume is demonstrated in Fig. 3.

Needle biopsies were not done before 18

Table 3. Patients' characteristics and distribution of tumor stages according to the TNM classification (HERMANEK et al. 1987) at the beginning of interstitial radiotherapy

Stage or grade[a]		n
T1N0M0		8
T2N0M0		59
T3N0M0		73
T1-3N1M0		4
T1-3N2M0		6
G1 well differentiated	26 (17.3%)	
G2 moderate "	73 (48.7%)	
G3 poorly "	41 (27.3%)	
G2–3	10 (6.7%)	
Total number:	150	
Age:	52–79 years (mean: 65.2)	
Volume:	17–114 cm^3 (mean: 42.9)	
Number of seeds per patient:	23–102 (mean: 44.7)	

[a] According to HERMANEK et al. (1987).

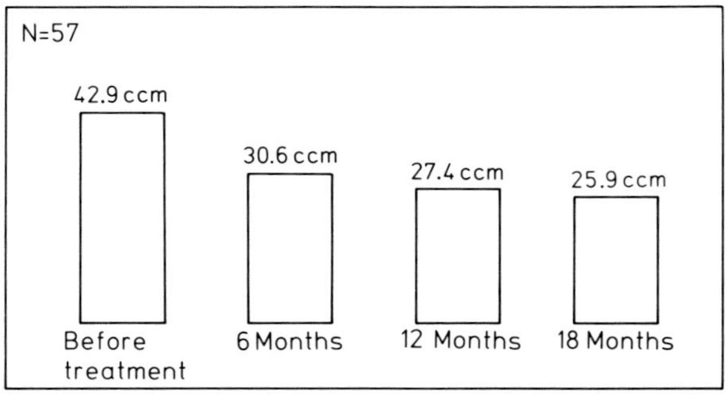

Fig. 3. Regression of the mean tumor volume after interstitial radiotherapy

months following interstitial treatment. Preliminary results are shown in Table 4. All patients treated with orchiectomy and ^{125}I seed implantation showed complete tumor regression. Patients treated by ^{125}I seed implantation alone without orchiectomy showed positive biopsy in 2 of 4 (nos. 1 and 2) after 18 months. Two other patients (T3N0M0G3) had complete tumor regression (nos. 5 and 11).

5 Discussion

A total of 150 patients with localized adenocarcinoma of the prostate were treated with ultrasonically guided, transperitoneal, ^{125}I seed implantation. No intraoperative or postoperative side effects were observed. In accordance with federal regulations on radio-protection, all patients were discharged from the hospital 48 h after seed application. Implantation guided by ultrasound is applicable in patients able to receive some form of anesthesia. Patients with high operative risks or over 70 years of age underwent orchiectomy to improve the prognosis. Staging lymphadenectomy was considered unnecessary in the face of other noninvasive staging methods (eg., CT). In our opinion, lymphadenectomy is indicated only in patients selected for radical prostatectomy. The ^{125}I seed implantation and orchiectomy were performed in 10 patients with N1 or N2 lymph

Table 4. Results of biopsies done 18 months after interstitial treatment

Stage	Treatment			Local clinical findings		PSA[b]	Biopsy
	^{125}I	Orchiectomy	Med.[a]	Sonogram	Palpation		
1 T3N0M0G1	+	–	–	+	+	++	+
2 T3N0M0G2	+	–	–	–	–	–	+
3 T3N0M0G2–3	+	+	–	–	–	–	–
4 T2N0M0G1	+	+	–	–	–	–	–
5 T3N0M0G2	+	–	–	–	–	–	–
6 T2N0M0G3	+	+	+	–	+	–	–
7 T3N0M0G2	+	+	+	–	+	–	–
8 T3N0M0G3	+	+	+	+	+	–	–
9 T3N0M0G2–3	+	+	+	–	–	–	–
10 T3N0M0G1	+	+	–	–	–	–	–
11 T3N0M0G2	+	–	–	–	–	–	–
12 T2N0M0G2	+	+	–	–	–	–	–
13 T3N0M0G2	+	+	–	–	–	–	–

[a] Chemotherapeutical treatment with flutamide or estramustine phosphate.
[b] PSA, prostate specific antigen.

node metastasis. After a mean follow-up of 1 year, none showed progression. These results are comparable with those of Zincke et al. (1987) on radical prostatectomy and orchiectomy in node-positive cases.

Our concept of treatment without lymphadenectomy does not permit either exact staging in prostate carcinoma or definitive assessment of the success of ^{125}I seed implantation. Nevertheless, these preliminary results, even in patients with locally extensive tumors, encourage us to continue our treatment concept. A randomized, multicenterstudy would be helpful to solve some still open questions.

6 Conclusions

The radiosensitivity of adenocarcinoma of the prostate is widely accepted. Tumors localized to the gland can be treated with excellent results by interstitial ^{125}I seed application guided by transrectal ultrasound without major side effects. The combination of interstitial, ultrasonically guided radiotherapy with orchiectomy is a very effective and extremely well tolerated treatment. This has been confirmed in our experience on 150 patients treated over 2 years.

References

American Cancer society (1987) Cancer. Facts and figures-1987. American Cancer Society, New York, p 12

Bagshaw MA (1986) Current conflicts in the management of prostatic cancer. Int J Radiat Oncol Biol Phys 12: 1721–1727

Barringer BS (1942) Prostatic carcinoma. J Urol 47: 306–308

Hermanek P, Scheibe O, Spiessl B, Wagner G (1987) TNM Klassifikation malgner Tumoren, 4th edn. Springer, Berlin Heidelberg New York, pp 134–136

Holm HH, Strøyer I, Hansen H, Stadil F (1981) Ultrasonically guide percutaneous interstitial implantation of iodine 125 seeds in cancer therapy. Br J Radiol 54: 665–670

Holm HH, Juul N, Pedersen JF, Hansen H, Strøyer I (1983) Transperineal 125-iodine seed implantation in prostatic cancer guided by transrectal ultrasonography J Urol 130: 283–286

Scardino PT, Bretas F (1987) Interstitial radiotherapy. In: Bruce AW, Tarchtenberg J (eds) Adenocarcinoma of the prostate. Springer, Berlin Heidelberg New York, pp 145–158

Van der Werf-Messing B (1982) Radiation therapy of carcinoma of the prostate. In: Jacobi GH, Hohenfellner R (eds) Prostate cancer. Williams and Wilkins, Baltimore, pp 195–211

Wannenmacher M, Sommerkamp H, Knüfermann H, Kuphal K (1979) Die interstitielle Strahlentherapie in der Behandlung des Prostatakarzinoms. Dtsch Ärztebl 76: 1371–1378

Whitmore WF Jr, Hilaris BS, Grabstald H (1972) Retropubic implantation of iodine-125 in the treatment of prostatic cancer. J Urol 108: 918–920

Young HH, Fronz W (1917) Some new methods in the treatment of carcinoma of the lower genitourinary tract with radium. J Urol 1: 505–536

Zincke H, Utz DC, Vahle PM, Taylor WF (1987) Treatment options for patients with stage D1 (T0-3 N1-3 M0) adenocarcinoma of prostate. Urology 30: 307–315

8 Gynecological Malignancies

8.1 Interventional Radiation Therapy Techniques in Gynecology

HANS-PETER HEILMANN

CONTENTS

1 Introduction 271
2 Main Radionuclides Used and Their Physical Characteristics 271
3 Radium Treatment Techniques of Cervix Carcinoma 271
3.1 The Paris Technique 272
3.2 The Stockholm Technique 272
3.3 The Manchester Technique 273
3.4 The Munich Technique 273
3.5 The Hamburg Technique 274
3.6 The Houston Technique 274
3.7 Interstitial Technique 275
4 Intracavitary Treatment Techniques of Carcinoma of the Endometrium 275
5 Treatment Techniques of Carcinoma of the Vagina 275
6 Interstitial Methods 276
7 Afterloading 276
8 Future Developments 277
References 277

1 Introduction

In intracavitary and interstitial treatment of gynecologic tumors, innumerable techniques have been devised, utilizing intracavitary or interstitial radium therapy, transvaginal therapy, external irradiation, or combinations thereof. The design of the applicators, the number and fractionation of the applications, and the dose rate (hourly in tensive) have been the main sources of variation in intracavitary radium techniques.

Dose distribution in space and time of radium-treatment in different gynelogical or radiological departments never has been standardized. In spite of great differences in methodology in different hospitals there were no treatment related differences in cure rates. Dose distribution in space and time, too, was not a consequence of theoretical considerations but of the low dose rate of radium and of the poor technical standard of external radiotherapy by the sixties (FRISCHKORN 1985).

So, before megavolt therapy was available, cervix carcinoma (and larynx cancer) were the only sites of curative treatment with radiation therapy. This may be an explanation for the enormous efforts made by different hospitals in radium treatment of cervix cancer.

In the meantime, not only gynecological tumors but a variety of different tumors are effectively treated by intracavitary or interstitial radiation techniques. Therefore, brachytherapy methods are integrated or should be integrated into every radiation therapy department. At least afterloading devices should be available. This chapter gives an overview on different techniques of intracavitary and interstitial treatment of gynecological cancers with special respect to the classic techniques of treatment of cervix carcinoma.

2 Main Radionuclides Used and Their Physical Characteristics

The main radionuclides used in intracavitary and interstitial treatment and their physical characteristics are shown in Table 1. Main radionuclides are radium and, for afterloading devides, cesium 137, cobalt 60, and iridium 192. Californium 252 is relatively seldom used and not further mentioned in this paper.

3 Radium Treatment Techniques of Cervix Carcinoma

Radium treatment of cervix carcinoma is one of the oldest techniques of radiotherapy. DOMINICI (1902, cited from HEYMAN 1918) was one of the first to use radium salt in a glass tube for intra-

HANS-PETER HEILMANN, Prof. Dr. med., Hermann-Holthusen-Institut für Radiotherapie, Allgemeines Krankenhaus St. Georg, Lohmühlenstr. 5, 2000 Hamburg 1, Germany

Table 1. Properties of radionuclides suitable for intracavitary techniques

Radionuclide	Physical half-life	Specific γ-ray constant ($R\,m\,Ci^{-1}\,hr^{-1}$ at 1 cm)	Photon energy (MeV)	Transmission through 2 cm Pb
Radium	1620 years	8.25	0.19–2.43	30%
Cesium 137	30 years	3.3	0.66	12%
Cobalt 60	5.3 years	13.1	1.17; 1.33	39%
Iridium 192	74.4 days	4.8	0.296–0.613	
Gold 198	2.7 days	2.31	0.41	
Californium 252	2.65 years		0–13 (neutrons, 60%) 0.5–1.0 (photons, 40%)	

uterine application. Later he constructed a special applicator for intrauterine application using different metals, e.g., silver and gold. Another special apparatus for the treatment of cervix cancer was build by WICKHAM (1913, cited from HEYMAN 1918).

In the first two decades of this century, in different hospitals, the technique of radium treatment of cancer of the cervix was developed. The Stockholm (HEYMAN 1918) and the Paris techniques (REGAUD 1926) have evolved as the most successful ones. Many institutions throughout the world use them, either in their original form or with modifications. They represent two distinctly different techniques, both in the physical properties of the applicators and in the duration and intensity of irradiation.

The intrauterine application of radium was to treat the cervix tumor; the aim of the intravaginal application was not only to treat the vagina but to get as much dose as possible to the pelvic wall because of the limited possibilities of external irradiation.

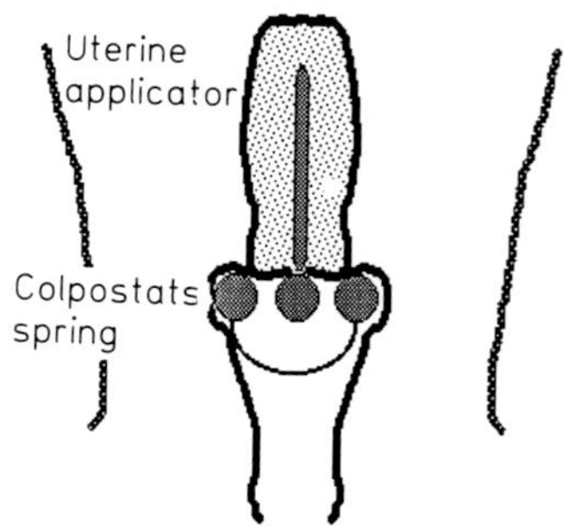

Fig. 1. The Paris method

Table 2. Dosage levels of the Paris method

mg Ra	Time	Dose	Dose rate
33.0	5 days	vagina 3960 mg h = 8000 R	1.1 R/min
24.0	5 days	uterus 2880 mg h = 14 000 R	1.95 R/min

3.1 The Paris Technique

The Paris technique (REGAUD 1926) is characterized by a low intensity radium irradiation over a relatively long period, mostly 1 week (Fig. 1, Table 2). The intravaginal sources are placed in a colpostat comprising two impermeable cork cylinders, one in each vaginal fornix, banded by a metal spring. A central source overlying the cervix is also used in selected patients. The number of units and loading factors are selected according to the length of the uterine canal (10–30 mg) but never exceeded 40 mg. The vaginal loading is usually 30 mg depending upon the size of the colpostats (corks of 2, 2.5, and 3 cm in diameter). The treatment is normally continued for 5 days. This provides for a total treatment of 7000–8000 mg h (REGAUD 1934).

3.2 The Stockholm Technique

The Stockholm technique is characterized by short applications of high intensity radium irradiation repeated two or three times (Fig. 2, Table 3). This method was introduced by FORSELL (1917) together with HEYMAN (1918, 1935) at the Radiumhemmet Stockholm.

The original Stockholm method has been further developed and considerably modified by KOTTMEIER (1958; KOTTMEIER and FURSSNER 1955; WALSTAM 1954) since 1948 by greater individualization, radicalization of radium therapy, and more elaborate external irradiation.

The vaginal applicators take the form of boxes or a tapered collar to cover the vaginal component

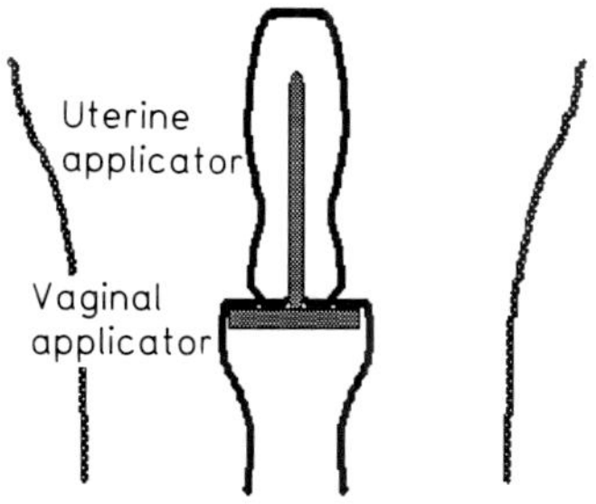

Fig. 2. The Stockholm method

Table 3. Dosage levels of the Stockholm method

mg Ra	Time	Dose	Dose rate
		vagina	
85.8	3 × 20 h in 4 weeks	3 × 1716 = 5148 mg h 3 × 3900 = 11700 R	3.2 R/min
		uterus	
33.0	3 × 20 h in 4 weeks	3 × 660 = 1980 mg h 3 × 3300 = 9900 R	2.75 R/min

of the tumor and contain up to 100 mg of radium. There is no link between uterine and vaginal applicators.

The applicators are retained during treatment by gauze packing to avoid the rectal or bladder tissues coming into contact with them. In order to minimize the normal tissue reactions and achieve maximum tumor control, treatment is divided into two fractions each lasting between 20 and 24 h, separated by an interval of 3 weeks. (Originally three applications at intervals of 1–3 weeks were given.) The total treatment provides between 7000 and 8000 mg h but is reduced when external beam therapy is added. "The most fundamental and important principle is individualized treatment and dedicated expertise" (HEYMAN 1974a).

The diameter of the intrauterine applicators is 7 mm. The uterine canal is completely filled with the applicator, the lower 2 cm, however, without radium in order to avoid hot spots. Dosimetry in bladder and rectum is done in order to calculate application time. The radium load of the vagina is 60–80 mg, of the uterus 53–74 mg.

3.3 The Manchester Technique

The most well-known development of the Paris method is the Manchester technique originated by TOD and MEREDITH (1938) and later modified (Fig. 3). Hard rubber ovoids replaced the colpostats, which are held apart by a washer or

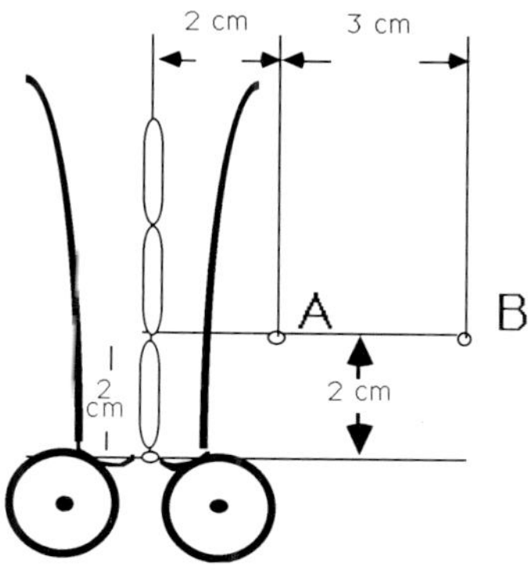

Fig. 3. The Manchester method

spacer. These ovoids were shaped to produce a radiation dose uniform to within 10% over the surface of the ovoids.

A section of three sizes for both the intrauterine tubes and vaginal ovoids, with different source loading, are available to meet the problems of different sizes of uterus and vagina. Treatment is separated into two fractions of 70 h each over a total time of 10 days to deliver 8000 roentgens to the Manchester point A (PATERSON 1952).

One important principles of the Manchester system is to prescribe treatment in terms of a dose of radiation to defined points of interest within the paracervical tissues and the pelvic wall. Accurate source positioning in relation to the normal anatomical structures is strongly stressed, and check X-rays following insertion of the sources should always be carried out. The Manchester point A (TOD and MEREDITH 1938) came into use as a prescribing point because of the importance of limiting the dose of radiation within the paracervical tissues, where the uterine artery and ureter are closely related and where the least variation in dose occurs.

Point A was originally defined in a frontal plane 2 cm lateral to the uterine canal and 2 cm up from the fornix. In 1953 TOD and MEREDITH redefined its location as 2 cm from the end of the lowest uterine source (level of internal os) and 2 cm laterally in the frontal plane.

The radium load was 20–35 mg to the uterine applicator and 17.5–22.5 mg to the "Manchester ovoids".

3.4 The Munich Technique

This method uses a fixed combination of intrauterine and vaginal applicator (Fig. 4). The uter-

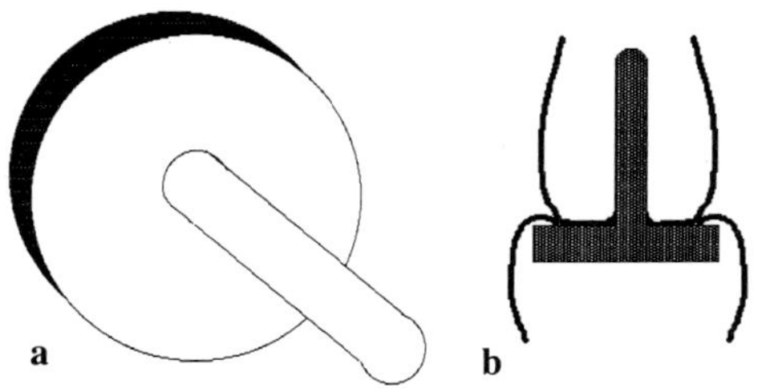

Fig. 4a,b. The Munich method

ine tube is linked to the vaginal applicator by a screw fixture. The vaginal applicator has the form of a circle (EYMER and RIES 1934, cited from EYMER and RIES 1941, EYMER and RIES 1941). There are different size of combinations. Treatment consists of three applications with 2000 mg h each in intervals from 1–2 weeks.

This technique has been very popular in the FRG, especially with gynecologic departments. A potential disadvantage is a high dose to the ureter in cases where the uterus is not centrally located. The technique requires skillful application because of the difficulty of seeing the uterine canal while bringing the applicator into the vagina.

3.5 The Hamburg Technique

As mentioned before, the philosophy of the Paris and the Stockholm techniques varies. The Paris technique uses one application with a low dose rate over a long period; the Stockholm method two or three applications with a relatively high amount of radium and limited application time.

The Hamburg method (HAMANN et al. 1934) uses the advantages of both techniques (Fig. 5): for intrauterine application, a tube with only 13–33 mg radium is used with an application time of 5 days (low dose rate, one application: so-called Paris tube). Simultaneously with the intrauterine tube, a vaginal applicator of the Stockholm method (a so-called Stockholm plate) with 60–80 mg radium is used for 24 h and then removed. The applicators are fixed by gauze packing, which also gives distance from the rectum. This is renewed after removal of the vaginal applicator. The vaginal treatment is repeated twice at intervals of 1–2 weeks.

By the Paris tube, only one intrauterine application under general anesthesia is required. Using the Stockholm plates, size of the applicator can be reduced with shrinking tumor size in order to avoid a high dose to the rectum and bladder. With this method, the dose in point A (according to the Manchester system) was 6000–9000 rad.

3.6 The Houston Technique

The Houston technique is characterized by a special applicator, the tandem (FLETCHER et al. 1952, 1953, 1958), and a combination of radium treatment with external irradiation (FLETCHER 1973) (Fig. 6). The tandem consists of two vaginal applicators, the ovoids, and a curved uterine applicator. The radium is brought into the tandem by manual afterloading procedures. Principles of the Houston method are: (a) Treatment method is chosen according to the tumor volume: "treat the tumor, not the stage!" (b) The dose of external irradiation and that of intracavitary treatment depend upon each other. Total doses of external and intracavitary treatment can be as high as 90–100 Gy in the tumor area. The Fletcher tandem has been modified by ERNST (1949), HENSCHKE (1960), and SUIT et al. (1963).

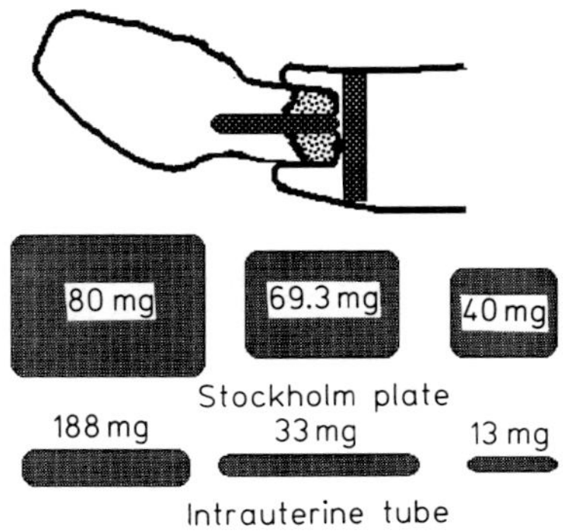

Fig. 5. The Hamburg method

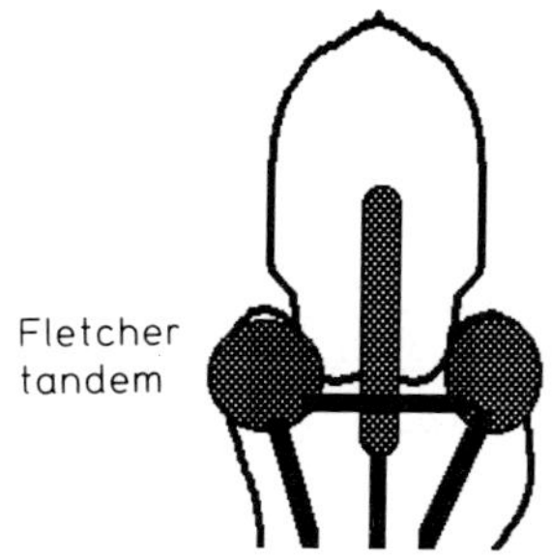

Fig. 6. The Houston method

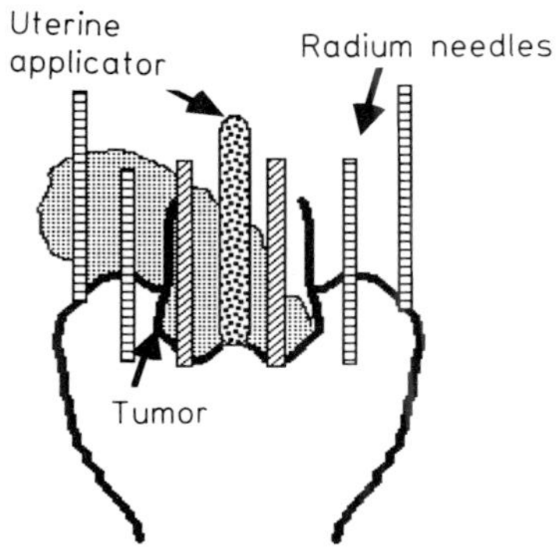

Fig. 7. The Interstitial technique

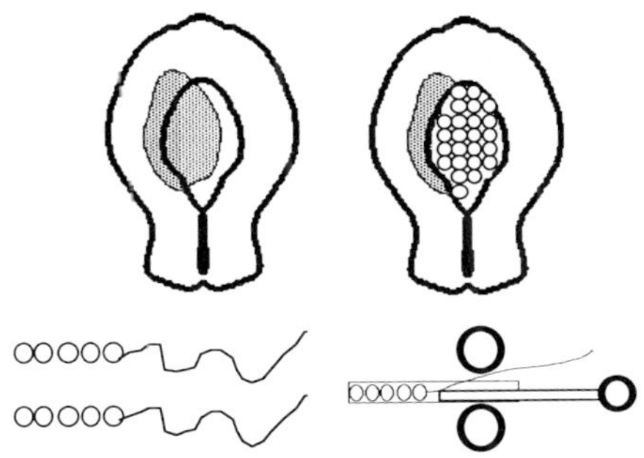

Fig. 8. Coblat-pearl technique for treatment of carcinoma of the endometrium. *Above left*, uterus with tumor; *above right*, cavum uteri filled with cobalt-60 pearls after curettage of the tumor; *below left*, wires with cobalt-60 pearls; *below right*, special applicator for cobalt-60 pearls

3.7 Interstitial Technique

Interstitial radiation therapy of carcinoma of the cervix was proposed by CORSCADEN et al. (1948, 1954) (Fig. 7). Radium needles are brought to the cervix and the parametria in order to give a high dose (10000 to 14800 "gamma r") to the tumor. This method has not become very popular.

4 Intracavitary Treatment Techniques of Carcinoma of the Endometrium

In carcinoma of the endometrium the techniques of intracavitary irradiation are quite different to those in carcinoma of the cervix. The cavum uteri has to be irradiated uniformly with a high dose. Three techniques, relatively similar to each other, have been developed:

(a) Heyman capsules (HEYMAN 1936, 1947b): Radium capsules are brought to the uterine cavity in order to fill it and give a uniform dose distribution.
(b) Ries eggs (RIES 1944): Like the Heyman capsules, the Ries eggs are radium applicators.
(c) Cobalt pearls (BECKER and SCHEER 1952; GAUWERKY, personal communication): This method is slightly different. Pearls of cobalt 60 are lined up on a wire (5 pearls on each) and brought to the cavum uteri by a special applicator (Fig. 8). A dose of 3000 cGy is given within a 1-cm distance of the applicators, and treatment is repeated after an interval of 3 weeks (total dose 6000 cGy). External irradiation with 3000 cGy is performed after intracavitary treatment with blocking of the preirradiated region. Knowledge of the extent of the cavum uteri and of the tumor before treatment is mandatory. Therefore, hysterography is recommended (GAUWERKY 1953).

5 Treatment Techniques of Carcinoma of the Vagina

Treatment of carcinoma of the vaginal must take into account the precise site of origin of the lesion, its possible maximal vaginal extent, and the presence or absence of involvement of adjacent

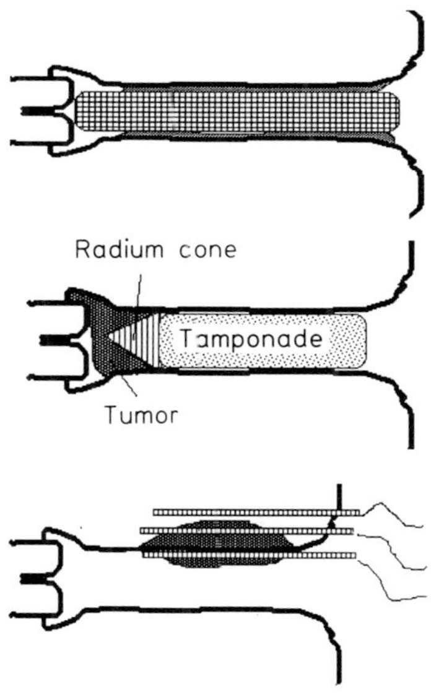

Fig. 9. Intracavitary and interstitial techniques for treatment of carcinoma of the vagina. *Above*, cylindrical applicator for radium or cesium 137; *Middle*, cone applicator with gauze packing; *below*, radium needles

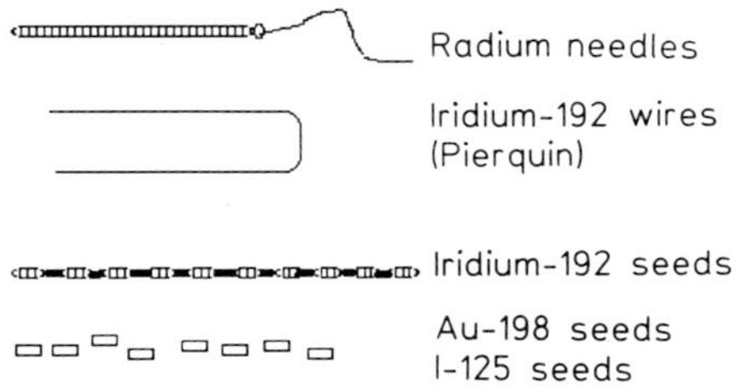

Fig. 10. Needles, wires, and seeds for interstitial therapy

structures such as the cervix, rectum, vulva, urethra or bladder. Therefore, a wide variety of applicators, flat and curved boxes as well as cylindrical devices, for radium therapy are used (Fig. 9) (WHELTON and KOTTMEIER 1962). In special situations, interstitial treatment with radium needles is possible (HOLTHUSEN and GAUWERKY 1949). For cylindrical vaginal applicators, cesium 137 can be used instead of radium (GAUWERKY 1958). In most cases, a combination with external irradiation is necessary.

6 Interstitial Methods

The eldest technique of interstitial irradiation, as mentioned before, involves radium needles (Fig. 10). PATERSON and PARKER (1938) developed a dosage system which has in the meantime become somewhat classic. Later on, cesium-137 needles were constructed. Up to now, a variety of isotopes and techniques of application have been developed, such as seeds of gold 198 (SINCLAIR 1952) and iodine 125 (HILARIS et al. 1976), and seeds and wires of iridium 192 (HILARIS 1975; SYED and FEDER 1977; SYED et al. 1978; PIERQUIN et al. 1978; DELCLOS 1984). Permanent interstitial implantation has been used since 1917 (FAILLA, cited from HILARIS 1975). Single-seed inserters have been replaced by "guns" (HODT et al. 1952) which allow the insertion of more than one seed. HENSCHKE (1956) replaced the old permanent implantation technique with one based on the afterloading principle (manual afterloading). This is characterized by the insertion of unloaded needles and then afterloading with radioactive sources. The most well-known instrument today is the Mick applicator (Mick Radio-Nuclear Instruments, New York; SCHULZ and BUSCH 1981).

Iridium-192 wires or chains with seeds are used for temporary implants in different sites (PIERQUIN et al. 1978; HILARIS 1975; POTISH and WILLIAMSON 1984). These highly sophisticated techniques are described in other chapters.

7 Afterloading

Today, in the intracavitary treatment of carcinoma of the cervix, afterloading techniques are standard in most departments. Many of them have developed from the classic techniques of radium treatment previously mentioned. There are three different types: (a) manual afterloading, (b) remote low-dose afterloading, (c) remote high-dose afterloading.

Manual afterloading has been described before (see Houston technique). Another method, derived from the Paris technique, is the Creteil method (PIERQUIN and MARINELLO 1986). Some low-dose remote afterloading techniques are derived from the Paris and the Manchester systems; others are influenced by the Stockholm method (WALSTAM 1965). The Institut Gustave-Roussy's method (GERBAULET et al. 1986) and the Saint Cloud method (DELOUCHE and GEST 1986) use a machine called the Curietron (DELOUCHE et al. 1967). The Selectron is a further development of the Manchester system (WILKINSON et al. 1983; JONES et al. 1987). In low-dose systems, the sources (mostly cesium 137) are brought to the applicators and stay there during therapy. Dose distribution is changed by employing a different

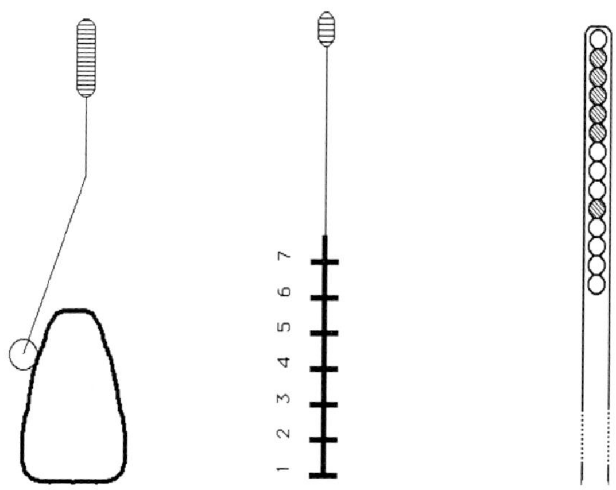

Fig. 11. Principles of remote afterloading. *Right*, Buchler principle: cycling source with varied speeds by scanning a disc, *middle*, Gamma-Med principle: paces with varied time in each position: *Left*, Selectron principle: active and inactive sources of cesium 137 are mixed

number and different activities of the sources (Fig. 11).

High-dose afterloading machines (Cathetron, Buchler, Gamma-Med) use sources of cobalt 60 (O'CONNELL et al. 1967), cesium 137, or iridium 192. To achieve different dose distributions, two techniques are used (Fig. 11): cycling of the source and movement of the source pace by pace.

Cycling sources in remote afterloading techniques are described by HENSCHKE et al. (1966). This technique is used with the Buchler machine: A cycling cesium-137 source is guided by several different discs. By scanning the contour of the disc with a steering wheel the speed of oscillation is altered, and through this the dose distribution is influenced. With this machine, a comparison was made between radium therapy and high-dose afterloading treatment (ROTTE 1985).

In the Gamma-Med machine (Sauerwein), the iridium source is moved pace by pace. Dose distribution is influenced by the time the source remains in each position. Because of the small size of the iridium source, the Gamma-Med can be used with small catheters for intracavitary treatment of the bronchi, esophagus, and bile duct as well as for interstitial therapy (Fig. 12). Further details are given in other chapters of this book.

8 Future Developments

As has been shown, brachytherapy methods in gynecology were of the greatest importance when external radiotherapy was underdeveloped. With the introduction of cobalt machines and linear accelerators, their standing has diminished slightly, but even today intracavitary techniques play a significant role in the therapy of gynecologic malignancies. Because of radiation protection, the use of radium is practically omitted, and remote afterloading machines are used in most departments. There are still a lot of problems, especially with high-dose machines. Single doses and total doses in relation to cure and side effects have to be studied thoroughly, as well as the best combinations with external irradiation. Therefore, the situation today is somewhat similar to the time when intracavitary therapy with radium was started.

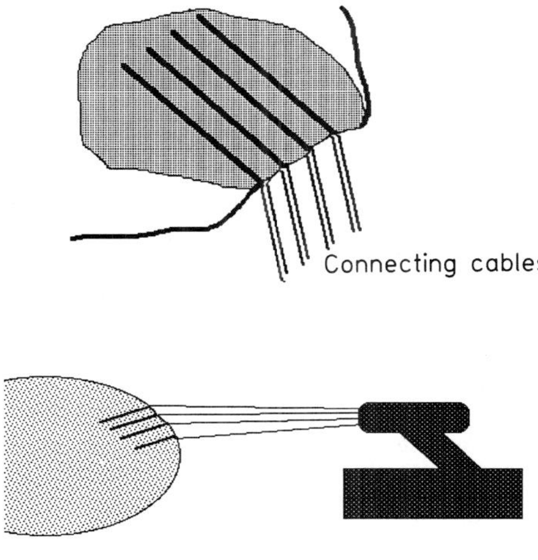

Fig. 12. Interstitial remote afterloading. *Above*, tumor with needles and cables to the afterloading machine, *below*, machine with cables and needles

References

Becker J, Scheer KE (1952) Strahlentherapeutische Anwendung von radioaktivem Kobalt in Form von Perlen. Strahlentherapie 86: 540–547

Corscaden JA, Gusberg SB, Donlan CP (1948) Precision dosage in interstitial radiation of carcinoma of the cervix. AJR 60: 522–534

Corscaden JA, Gusberg SB, Kosar W (1954) Interstitial radium treatment of cancer of the cervix uteri. AJR 72: 278–283

Delclos L (1984) Interstitial irradiation techniques. In: Levitt SH, Tapley NV (eds) Technological basis of radiation therapy: practical clinical applications. Lea and Febiger, Philadelphia, pp 55–84

Delouche G, Gest J (1986) Saint-Cloud method. In: Chassagne D (ed) Teaching course in brachytherapy in gynecology. European Society for Therapeutic Radiology and Oncology. Baden-Baden, p 9

Delouch G, Milhaud F, Gest J (1967) La curiethérapie gynécologique endocavitaire par césium 137 (appareillage à préparation non radio-active). J Radiol Electrol 48: 229–242

Ernst EC (1949) Probable trends in irradiation treatment of carcinoma of the cervix uteri with improved expanding type of radium applicator. Radiology 52: 46–58

Eymer H, Ries J (1941) Die Ergebnisse der Strahlenbehandlung der Gebärmutterhalskrebse an der Münchner Universitäts-Frauenklinik im Jahre 1934. Strahlentherapie 69: 12–16

Fletcher GH (1973) Textbook of radiotherapy. Lea and Febiger, Philadelphia

Fletcher GH, Shalek RJ, Wall JA, Bloedorn FG (1952) A physical approach to the design of applicators in radium therapy of carcinoma of the cervix. AJR 68: 935–947

Fletcher GH, Wall JA, Bloedorn FG, Shalek RJ, Wooton P (1953) Direct measurements and isodose calculations is radium therapy of carcinoma of the cervix. Radiology 61: 885–90

Fletcher GH, Brown TC, Rutledge FN (1958) Clinical significance of rectal and bladder dose measurements in radium therapy of cancer of the uterine cervix. AJR 79: 421–450

Forsell G (1917) Übersicht über die Resultate der Krebs-behandlung am Radiumhemmet Stockholm 1910–1915. Fortschr Geb Röntgenstr 25: 142–149

Frischkom R (1985) Erfahrungen bei der Umstellung der Radiumtherapie auf ein Afterloadingverfahren unter besonderer Berücksichtigung der räumlichen Dosisverteilung. Strahlentherapie 161: 281–285

Gauwerky F (1953) Zur Strahlenbehandlung des Korpuskarzinoms des Uterus unter besonderer Berücksichtigung hysterographischer Befunde. Fortschr Geb Rontgenstr [Sonderbd 36] 79: 51–58

Gauwerky F (1958) Erfahrungen mit der Verwendung von Caesium 137 bei der intrakavitären Curietherapie gynäkologischer Karzinome. Strahlentherapie 105: 107–118

Gerbaulet A, Haie C, Chassagne D (1986) The modern afterloading techniques derived from classical low dose rate systems: Paris technique. In: Chassagne D (ed) Teaching course in brachytherapy in gynecology. European Society for Therapeutic Radiology and Oncology, Baden-Baden, p 1

Hamann A, Göbel A, Englmann K (1934) Die Strahlenbehandlung der Gebärmutterkrebse im Allgemeinen Krankenhaus St. Georg in Hamburg (Juni 1929–Dezember 1931). Strahlentherapie 50: 529–556

Henschke UK (1956) Interstitial implantation with radioisotopes. In: Hahn PF (ed) Therapeutic use of artificial radioisotopes. Wiley, New York, pp 375–397

Henschke UK (1960) Afterloading application for radiation therapy of carcinoma of the uterus. Radiology 74: 834

Henschke UK, Hilaris BS, Mahan GD (1966) Intracavitary radiation therapy in cancer of the uterine cervix by remote afterloading with cycling sources. AJR 96: 45–51

Heyman J (1918) Die Radiumbehandlung des Uteruskrebses. Arch Gynakol 108: 229–474

Heyman J (1935) The so-called Stockholm method and the results of treatment of uterine cancer at the Radiumhemmet. Acta Radiol (Stockh) 16: 129–148

Heyman J (1936) The radiumhemmet method of treatment and results in cancer of the corpus uteri. J Obstet Gynecol 43: 655–666

Heyman J (1947a) Improvement of results in the treatment of uterine cancer. JAMA 135: 412–416

Heyman J (1947b) The radiotherapeutic treatment of cancer of the corpus uteri. Br J Radiol 20: 85–91

Hilaris B (1975) Handbook of interstitial brachytherapy. Publishing Sciences, Acton

Hilaris B, Kim JH, Tokita N (1976) Low energy radionuclides for permanent interstitial implantation. AJR 126: 171–178

Hodt JJ, Sinclair WK, Smithers DW (1952) A gun for interstitial implantation of radioactive gold grains. Br J Radiol 25: 419–421

Holthusen H, Gauwerky F (1949) Ergebnisse der Strahlentherapie gynäkologischer Karzinome im Krankenhaus St. Georg. In: Holthusen H (ed) Aktuelle Probleme der Pathologie und Therapie. Thieme, Stuttgart, pp 153–167

Hunter RD, Cowie VJ, Blair V, Cole MP (1986) A clinical trial of two conceptually different radical radiotherapy treatments in stage III carcinoma of the cervix. Clin Radiol 37: 23–27

Jones DA, Notley HM, Hunter RD (1987) Geometry adopted by Manchester radium applicators and selectron afterloading applicators in intracavitary treatment for carcinoma of the cervix uteri. Br J Radiol 60: 481–485

Kottmeier HL (1958) Current treatment of carcinoma of the cervix. Am J Obstet Gynecol 76: 243–251

Kottmeier HL, Forssner E (1955) Die Entwicklung der Therapie des Kollumkarzinoms am Radiumhemmet. MMW 97: 1019–1020, 1028–1030

O'Connell D, Joslin CAF, Howard N, Ramsay NW, Liversage WE (1967) The treatment of uterine carcinoma using the cathetron. I. Br J Radiol 40: 882–887

Paterson R (1952) Studies in optimum dosage. Br J Radiol 25: 505–516

Paterson R, Parker HM (1938) A dosage system for interstitial radium therapy. Br J Radiol 11: 252–266, 313–340

Pierquin B, Marinello G (1986) The Creteil method. In: Chassagne D (ed) Teaching course in brachytherapy in gynecology. European Society for Therapeutic Radiology and Oncology, Baden-Baden, p 8

Pierquin B, Chassagne D, Chahbazian CM, Wilson JF (1978) Brachytherapy. Green, St Louis

Potish RA, Williamson JF (1984) Interstitial iridium template techniques. In: Levitt SH, Tapley NV (eds) Technological basis of radiation therapy: practical clinical applications. Lea and Febiger, Philadelphia, pp 85–100

Regaud C (1926) Traitement des cancers du col de l'uterus par les radiations: idée sommaires des méthodes et des résultats; indications thérapeutiques. 7th Congress of the Société Internationale de la Chirurgie 1: 35–146

Regaud C (1934) Considérations sur la radiothérapie des cancers cervico-uterus d'après l'expérience et les résultats acquis à L'Institut du Radium des Paris. Radiophys Radiother 3: 155–170

Ries J (1944) Neues zur Behandlung des Gebärmutterkrebses: Die totale Radiumtamponade des Uterus. Nachr Krebsbek 12: 94–97

Rotte K (1985) Klinische Ergebnisse der Afterloading-Kurzzeittherapie im Vergleich zur Radiumtherapie. Strahlentherapie 161: 323–328

Schulz U, Busch M (1981) Ein neuer Applikator zur interstitiellen Therapie mit Au-198 und J-125 seeds. Strahlentherapie 157: 104–105

Sinclair WK (1952) Artificial radioactive sources for interstitial therapy. Br J Radiol 25: 417–419

Suit HE, Moore EB, Fletcher GH, Worsnop R (1963) Modification of Fletcher ovoid system for afterloading, using standard-sized radium tubes (milligram and microgram). Radiology 81: 126–131

Syed AMN, Feder BH (1977) Technique of after-loading interstitial implants. Radiol Clin (Basel) 46: 458–475

Syed AMN, Feder BH, George FW III, Neblett D (1978) Iridium-192 afterloaded implant in the retreatment of head and neck cancers. Br J Radiol 51: 814–820

Tod MC, Meredith WJ (1938) A dosage system for use in treatment of cancer of the uterine cervix. Br J Radiol 11: 809–824

Tod MC, Meredith WJ (1953) Treatment of cancer of the cervix uteri – a revised "Manchester method". Br J Radiol 26: 252–257

Walstam R (1954) The dosage distribution in the pelvis in radium treatment of carcinoma of the cervix. Acta Radiol (Stockh) 42: 237–249

Walstam R (1965) Remote-controlled afterloading radiotherapy apparatus. Phys Med Biol 7: 225–228

Whelton J, Kottmeier HL (1962) Primary carcinoma of the vagina: a study of a Radiumhemmet series of 145 cases. Acta Obstet Gynecol Scand 41: 22–40

Wilkinson JM, Moore CJ, Notley HM, Hunter RD (1983) The use of Selectron afterloading equipment to simulate and extend the Manchester system for intracavitary therapy of the cervix uteri. Br J Radiol 56: 409–414

8.2 Technical Aspects of Bladder Dosimetry in Intracavitary Irradiation of Carcinoma of the Cervix

A.J. Subandono Tjokrowardojo and A.G. Visser

CONTENTS

1 Introduction 279
2 Materials and Methods 279
2.1 Technique of Insertion 279
2.2 Dosimetric Results 282
5 Conclusions 283
References 283

1 Introduction

In the treatment of the carcinoma of the cervix, high dose rate afterloading (HDR-AL) brachytherapy was started in Rotterdam in 1978. From the viewpoint of dosimetry high dose rate applications are preferable because the short application times allow for complete fixation of the applicator in combination with immobilization through spinal or epidural anesthesia. Stable geometrical conditions are obtained. Direct measurements in the bladder appear very difficult to realize. The main problem is to ensure that the probe is positioned at the area of the bladder base where late radiation reactions occur most frequently.

The shape of the bladder is dependent on the quantity of its contents and on the positions of the uterine cervix and the anterior vaginal wall. The place of the bladder base is not changed by filling or emptying the bladder, but it can shift markedly by inserting an applicator into the vagina.

In a filled bladder three compartments can be distinguished, i.e., two lateral pouches and a central section. In the central section, the bladder bulges in the ventral direction. When the bladder is empty, the lateral walls move in the medial direction, and the vault comes down in plies and folds until it reaches the bladder base. Under such conditions, any probe inserted for measurements at the bladder base will easily be caught between those structures and deviate in a ventral direction, thus precluding the discovery of any relation between complications and measured dose (Fig. 1).

Late radiation reactions following brachytherapy mostly occur near the median plane in the area where the bladder base passes over into the posterior wall. The balloon of a Foley catheter, which is usually applied for dose determination, does not visualize this area but that of the bladder neck. Therefore, in Rotterdam, a thin metal chain is temporarily inserted into the bladder and used as a marker of the area of interest.

A.J. Subandono Tjokrowardojo, M.D., A.G. Visser, Ph.D., Dr. Daniel den Hoed Cancer Center and Rotterdam Radio-Therapeutic Institute. Groene Hilledijk 301, 3075 EA Rotterdam, The Netherlands

2 Materials and Methods

2.1 Technique of Insertion

The following necessary tools have to be sterilized: a pair of gloves, a chain (silver or stainless steel) with a length of 40 cm, a 17-cm long tube, and a 21-cm long rod with a diameter of 2.5 mm, this being equal to that of the chain.

The outer diameter of the tube is 3.5 mm, and its wall thickness is 0.4 mm. It is preferable to use a chain which can easily be bent in one plane only.

Before starting the insertion the bladder is emptied and filled with a standard quantity of sterile saline (200 cc). In Asian countries, 150 cc appear to be endured much better.

It is important to prevent unwanted displacements of the applicator during the after insertion of the chain because the lateral bladder pouches are often asymmetric. In such a case a slight movement of the applicator will easily cause slipping of the chain into the deepest lateral bladder pouch.

Before starting the insertion, the chain is introduced into the tube as far as its upper end. The lower end of the tube and the chain are firmly held between thumb and index finger. The remaining part of the chain is kept in the hollow of the same

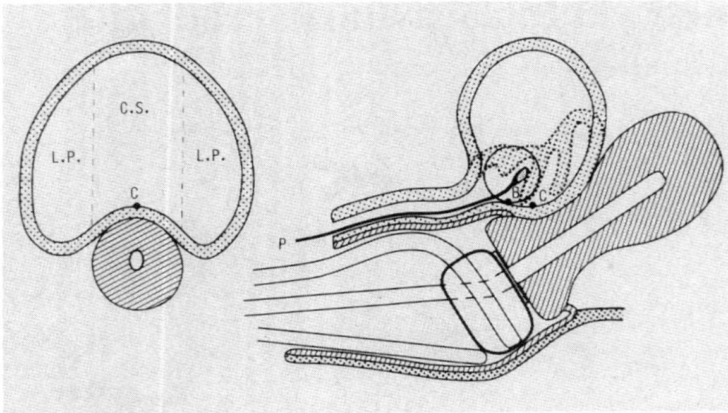

Fig. 1. Bladder, uterus, and vagina with applicator and rectal rectractor. *Left*, cross section of filled bladder where the chain intersects the median plane. *Right*, longitudinal view with vault positions when bladder is full or empty. *LP*, Lateral pouch; *CS*, central section; *B*, balloon, posterior point; *C*, point where chain crosses the median plane; *P*, probe inserted into empty bladder

hand in order to keep it sterile (Fig. 2). The upper end of the tube is cautiously inserted into the urethra at an angle of about 20° in a ventral direction, i.e., inclined upwards (Fig. 3). If the lateral pouches are not equally deep, which is often the case, the deepest one should be chosen for insertion of the first section of the chain. When 5–6 cm of the tube has been introduced, the upper end is laterally moved over an angle of 30°–40° and inserted further until it touches the posterior bladder wall. After that, the tube is retired about 2–3 cm in order to make room for the chain which is then pushed forward by the other hand into the tube until 3–5 cm are positioned in the lateral pouch. One should be careful to keep the upper end of the tube at such a level that the entire chain cannot disappear into the bladder. Then the upper end of the tube is moved in a median and ventral direction. In the meantime two fingers should firmly close the lower end of the tube, otherwise too much saline will flow out.

After crossing the median line, the tube is inserted into the other bladder pouch, again until the posterior wall is touched. Then the tube is retired about 2–3 cm, and the chain is pushed forward again over 3–5 cm. The next section of the chain is placed at the distal part of the bladder base while cautiously retracting the tube.

As soon as the lower end of the chain has disappeared into the tube, the rod is placed against the chain, and for removal the tube is pulled out, downward over the rod, taking care that the whole chain cannot slip into the bladder, and its lower end is kept protruding out of the external uretha orifice (Fig. 4a).

The position of the chain is checked by means of fluoroscopy, avoiding any movement either of the applicator or of the patient. If the position is wrong (e.g., the chain lies on only one part of the lateral pouch), the chain has to be removed and inserted again. If much saline has been lost during

2 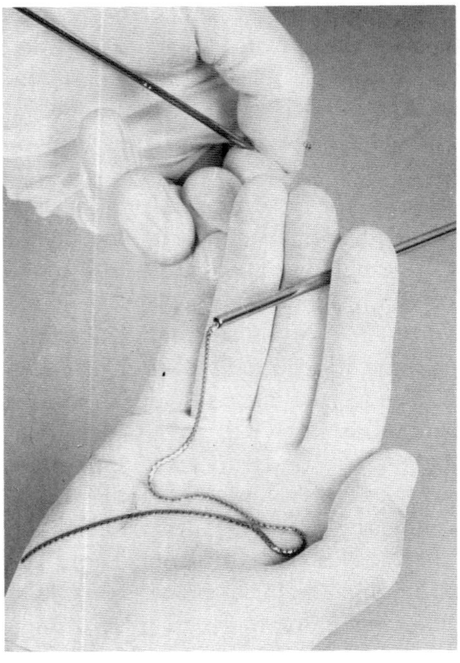

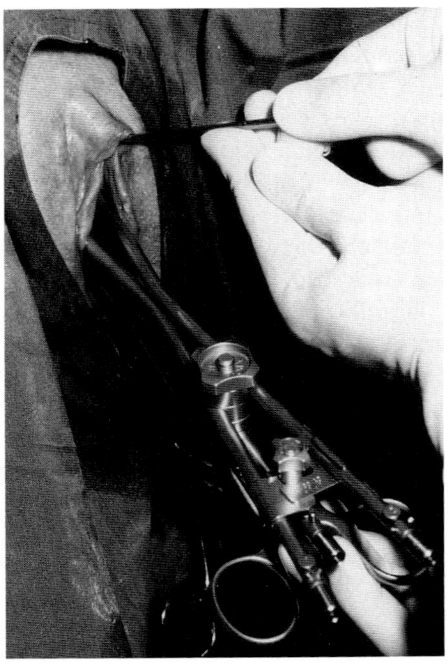 3

Fig. 2. The tools: Tube with chain in right hand, rod in left hand

Fig. 3. Insertion of the upper end of the tube into the urethra at an angle of about 20° in a ventral direction, i.e., inclined upwards

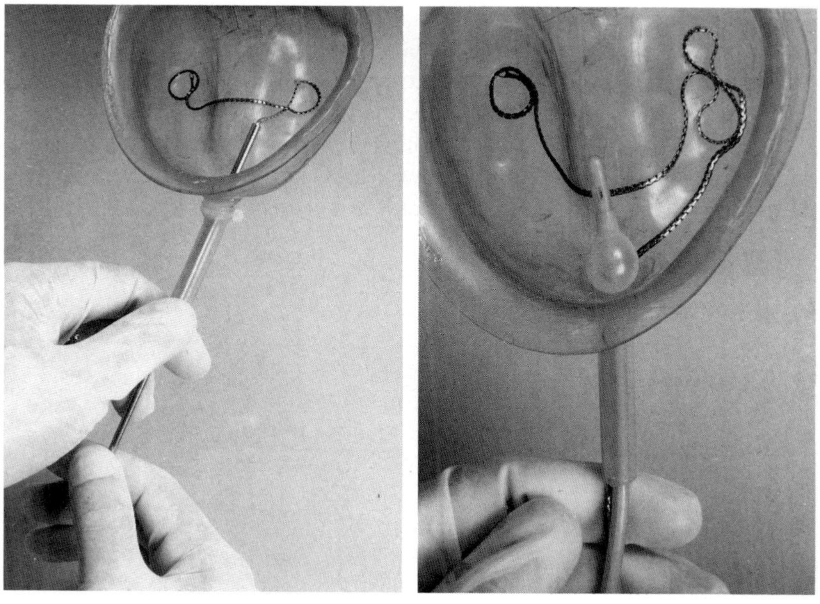

Fig. 4. a Chain on the bladder base from one bladder pouch crossing the median line to the other bladder pouch. b The position of the chain on the bladder base and a small balloon catheter (12 or 14 ch) filled with 5 cc contrast medium, against the urethral ostium

the first insertion (50 cc or more) then another 50–100 cc has to be added into the bladder before the second attempt.

In Rotterdam, intracavitary irradiation is performed with legs stretched and with the applicator fixed to the table top. Any movement of the applicator can result in displacement of the chain. Therefore, after having verified the correct position of the chain by fluoroscopy, its free end can be fixed with tape to the vulva (never to the leg).

A small catheter (12 or 14 ch) is inserted very carefully and, after filling the balloon with 5 cc with contrast medium, is gently pulled down as far as possible (Fig. 4b).

Then the applicator is fixed to the table, taking care that no unwanted movement of the apparatus can change the position of the bladder base in order not to disturb the position of the chain. After checking again by means of fluoroscopy the position of the applicator and of the chain, stereographic lateral and anteroposterior radiographs are taken for dosimetry. This can be performed without moving the patient, as the application takes place on a dedicated simulator (Fig. 5).

Even taking all the precautions mentioned above, the technique of insertion of the chain appeared to be difficult 10% of the time: 9% due to a wrong position, namely, the chain lies on only one part of the lateral pouch; and 1% disappearance of the entire chain into the bladder. In the

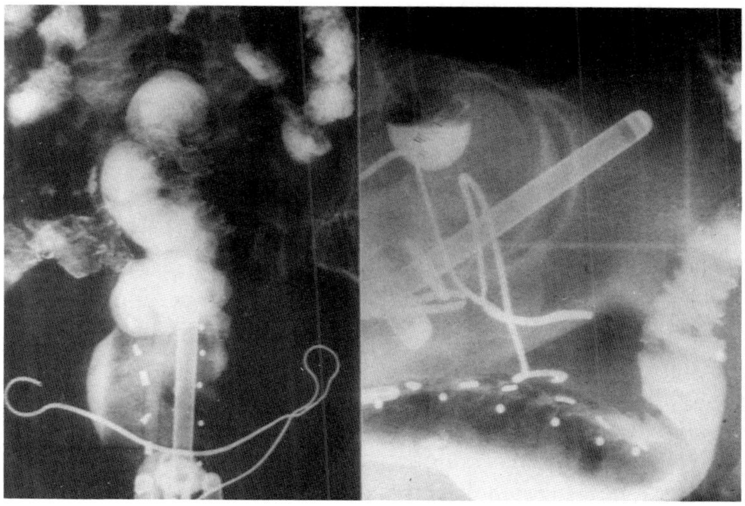

Fig. 5. Anteroposterior (*left*) and lateral (*right*) radiographs. Applicator with intrauterine tube and ovoids. Chain and balloon in filled bladder, foam plastic plug with markets in rectum, contrast-filled gut and sigmoid colon. Isodose pattern from intrauterine sources on lateral radiograph

beginning failures may certainly be encountered. In more than 300 insertions, disappearance of the entire chain into the bladder occurred in only 4. With aid of stone-grasping forceps, the chain can easily be removed under cystoscopic control. A thin nylon thread, fixed at the lower end of the chain, appeared to be useful for preventing this.

2.2 Dosimetric Results

In a comparative study, bladder dosimetry with the aid of the chain was combined for more than 2 years with dose determination by means of the balloon of a Foley catheter. The point at which the chain crosses the median line is used for dose calculation. On lateral radiographs this point can usually be recognized easily. The lateral radiograph is taken with a constant magnification factor of 1.4. With transparent overlays containing sets of isodose curves plotted with the same magnification factor, the dose in the median plane at point C of the chain can be estimated.

In order to gain information on the relative contribution of the intrauterine and vaginal sources, one set of isodose curves corresponding to the intrauterine sources and another set corresponding to those of both ovoids together are used. In Rotterdam, this type of dosimetry is always followed by stereo X-ray photogrammetry performed with the aid of stereographically exposed radiographs (Kuipers 1982). This method is more time-consuming and is therefore routinely used as a control following treatment. Only if the method with isodose curves over the lateral radiographs appears not to give reliable results are the doses determined by means of stereoradiographs.

The dose measured at the posterior edge of the balloon corresponds with the point at which this touches the bladder base. It should be remarked that in the case of an asymmetric position of the balloon, small inaccuracies may occur.

From 140 applications the ratio D_B/D_C between the dose determined with isodose overlays placed on the lateral radiographs with aid of the balloon (D_B) and the chain (D_C), is represented as a histogram in Fig. 6a. It appears that the average ratio was 0.89 with a standard deviation of 0.22.

Point C of the bladder base used in dose calculations based on the position of the chain is usually found more cranial than point B at the posterior edge of the balloon. This is illustrated in Fig. 1 in which the geometrical relationships

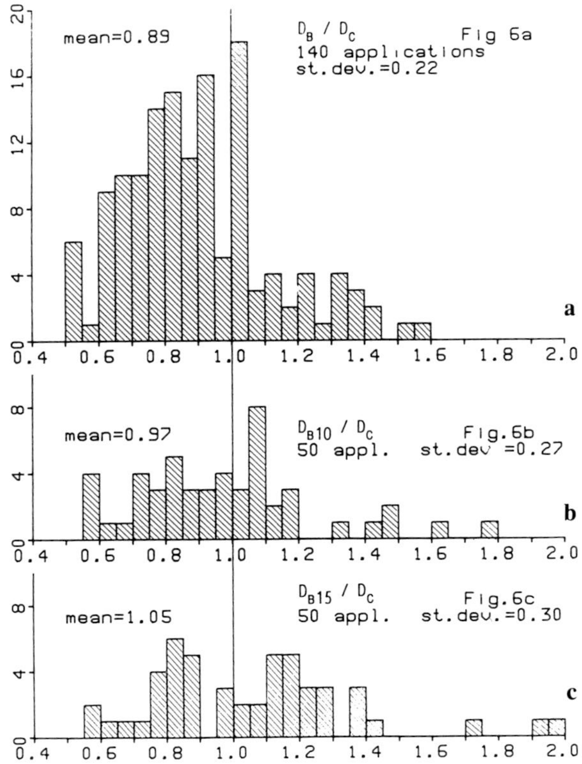

Fig. 6. Histograms of the ratio D_B/D_C (**a**), of the ratio D_{B10}/D_C (**b**), and of the ratio D_{B15}/D_C (**c**)

between points B and C and the center of the flange of the intrauterine applicator are represented according to the mean values of measurements made in 120 applications.

With the uterus in 30°–40° anteversion the contribution of the intrauterine sources increases steeply when moving from point B in the cranial direction. This is the cause of a higher dose at point C than at point B. In clinical practice, it appears simpler to insert the balloon than the chain for dosimetry. Therefore, it has been proposed (A. Gerbaulet, Paris) to simulate point C by taking a point 10 mm away from point B in the cranial direction, which might be a convenient simplification. This point will be denoted as B10. However, whether for dosimetry point C can be replaced by point B10 should still be verified. The choice of a distance of 10 mm may be somewhat arbitrary; possibly 15 mm might fit as well. Thus, a similar point called B15 is introduced here for comparison.

The histograms in Figs. 6b and c show the distribution of the dose ratios D_{B10}/D_C and D_{B15}/D_C, respectively. Both distributions show large deviations from 1.0 with standard deviations of

0.27 and 0.30. This means, in general, a poor correlation between the measured doses at point C of the chain and the simulated reference points B10 and B15. With point B10 deviations larger than 20% were observed in more than one-third of the applications (38%); deviations larger than 10% occurred in about two-third (64%). At point B15 these deviations are even larger, 68% and 88%, respectively. Although the average ratios are fairly close to 1 (0.97 and 1.05, respectively) the large deviations in individual applications preclude the use of these simulated reference points for reliable bladder dosimetry.

Concerning the results mentioned before, it appears that the bladder dose calculated at the chain was on average 11% higher than that derived from the position of the balloon. However, the large range of individual deviations precludes using the balloon and adding 11% to the result.

3 Conclusions

In intracavitary irradiation of carcinoma of the cervix, bladder dosimetry performed with the aid of a small chain temporarily inserted into the bladder appeared to give a more reliable indication of the dose at the area where late reactions most frequently occur than the usual method with the balloon of a Foley catheter. Therefore, the data achieved with the chain enable a more efficient prevention of bladder complications.

Replacement of point C of the chain by means of a simulated point B10 or B15 derived from the position of point B of the balloon by shifting over 10 or 15 mm, respectively, in a cranial direction do not seem to give reliable results.

If a higher dose is expected, one or more of the following measures can be taken:

a) Change position of the applicator.
b) Modify the positions of the sources.
c) Modify dose contribution from the intrauterine and ovoid sources.
d) Diminish the total dose from brachytherapy and compensate by some extra fractions of external beam irradiation.

However, if the anatomy is quite inappropriate for intracavitary treatment, a simple hysterectomy is indicated.

References

Chassagne D, Horiot JC (1977) Prépositions pour une définition commune des points de références en curiethérapie gynécologique. J Radiol Electrol 58: 371–373

Joslin CAF, Smith CW, Mallik A (1972) The treatment of cervix cancer using high activity 60 Co sources. Br J Radiol 45: 257–270

Kuipers TJ (1982) Stereo X-ray photogrammetry applied for prevention of sigmoid-colon damage caused by radiation from intrauterine sources. Int J Radiat Oncol Biol Phys 5: 1011–1017

Kuipers TJ, Visser AG (1986) Technical aspects of bladder dosimetry in intracavitary irradiation of cervix carcinoma. Radiother Oncol 7: 7–12

Wilkinson J, Moore CJ, Notly M, Hunter RD (1983) The use of the Selectron afterloading equipment of simulate and extend the Manchester system for intracavitary therapy of the cervix uteri. Br J Radiol 56: 409–414

8.3 Intracavitary High Dose Rate Afterloading Therapy with Iridium-192: Basic Physical Measurements, Dosimetry, and Localisation

NORFRIED THESEN

CONTENTS

1 Characteristics of Radionuclides Used with Intracavitary Afterloading 285
2 Source Specification and Check of Source Position 285
2.1 Dosimetry in Air 285
2.2 Dosimetry in a Phantom 286
2.3 Check of the Source Positions 287
3 Localisation of the Applicator and Organs at Risk 287
4 Determination of Absorbed Dose to Organs at Risk 288
4.1 Measurement 288
4.2 Calculation and Graphic Method 288
4.3 Comparison of Measured and Calculated Doses to Organs at Risk 289
References 290

1 Characteristics of Radionuclides Used with Intracavitary Afterloading

The γ-ray emitting isotopes like radium 226, cobalt 60, iridium 192, and cesium 137 are used with intracavitary afterloading techniques. Though their physical parameters are quite different (Table 1), the isodose distribution does not change very much if therapy-relevant distances of 1–5 cm are respected. The dose decrease around a point source in water for the above-named isotopes, with small deviations, follows the inverse square law and is therefore independent of isotope and photon energy. This results from a nearly complete compensation of absorption by scattering in the vicinity of the source. Depending on the isotope, the correction F(r) for absorption and scattering up to distances of 5 cm ranges between 0% and 8% (Fig. 1) (DUTREIX et al. 1982; YOUNG 1983; MEISBERGER et al. 1968; MELI et al. 1988).

NORFRIED THESEN, Dr. rer. nat., Klinik und Poliklinik für Strahlentherapie, der Universität zu Köln, Joseph-Stelzmann-Str. 9, 5000 Köln 41, Germany

The choice of isotope therefore has consequences for:

- The radiation protection (shielding of the storage safe and the treatment room).
- The achievable activity per unit volume and the size of the source (influence on treatment time and number of patients treated per day).
- The cost of a new source, related to the different half-lives of the isotopes.

2 Source Specification and Check of Source Position

Intracavitary dose distributions are individually calculated by computer or supplied by the manufacturer. The absorbed dose in the tumor is, however, only in agreement with these data if activity and position of the source are identical in calculation and treatment. Both parameters can be checked by measurement.

2.1 Dosimetry in Air

Instead of the activity the reference kerma rate to air in air at 1 m at the transverse axis of the source, including source filtration and corrected for air attenuation and scattering, has to be determined (ICRU 1985). To achieve a sufficient dose rate it will be necessary to measure over a shorter distance than 1 m. In this case, the reference air kerma rate for a point source can be calculated by the following equation:

$$\mathring{K}_{a,1} = r^2 \, \mathring{K}_{a,r}$$

where $\mathring{K}_{a,1}$ is the reference kerma rate (in mGy/min) to air in air at 1 m, $\mathring{K}_{a,r}$ is the kerma rate in r m and r is the distance in m.

Linear sources can be regarded as point sources

Table 1. Physical parameters of the radionuclides used with intracavitary afterloading (Dutreix et al. 1982; Young 1983)

Isotope	Half-life (years)	Mass per activity $\left(\frac{\text{mg}}{100\,\text{MBq}}\right)$	Source Dimensions	Air kerma rate constant $\left(\frac{\mu\text{Gy m}^2}{\text{h MBq}}\right)$	Exposure constant $\left(\frac{\text{R m}^2}{\text{h Ci}}\right)$	Energy (MeV)	Half value layer (cm Pb)
radium 226	1600	2735	big	0.197	0.825	0.1–2.4	1.6
cesium 137	30	31.2	middel	0.076	0.335	0.662	0.66
cobalt 60	5.27	2.39	small	0.309	1.31	1.17 1.33	1.2
iridium 192	0.203	0.294	very small	0.116	0.490	0.32–0.61	0.57

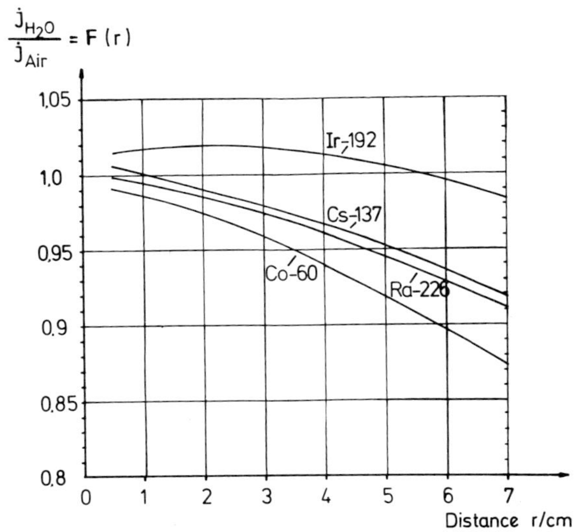

Fig. 1. Correction F(r) for attenuation and scattering in water as a function of distance r for intracavitary radionuclides according to Meisberger et al. (1968)

Fig. 2. Experimental set-up for measurement of the reference kerma rate to air in air

Fig. 3. Lucite phantom for source calibration

if the measuring distance is at least 5 times the largest source or chamber dimension. For dosimeters still calibrated in exposure rate, $\dot{J}$, the air kerma rate neglecting the bremsstrahlung can be calculated by the exposure-to-dose conversion factor:

$$\dot{K} = 8.73 \cdot \dot{J}$$

$$\frac{\text{mGy}}{\text{h}} \quad \frac{\text{mGy}}{\text{R}} \quad \frac{\text{R}}{\text{h}}$$

If the measurement is done in air, it will be helpful to consider the following conditions: ionisation chamber diameter (4–7 mm); measuring distance in air 4–10 cm; (a shorter distance yields a higher measuring signal, but errors in distance are more critical); precisely defined measuring distance (Fig. 2); lack of scattering (walls, floor).

2.2 Dosimetry in a Phantom

The measurement for source specification can be done in a solid phantom if absorption and scat-

2
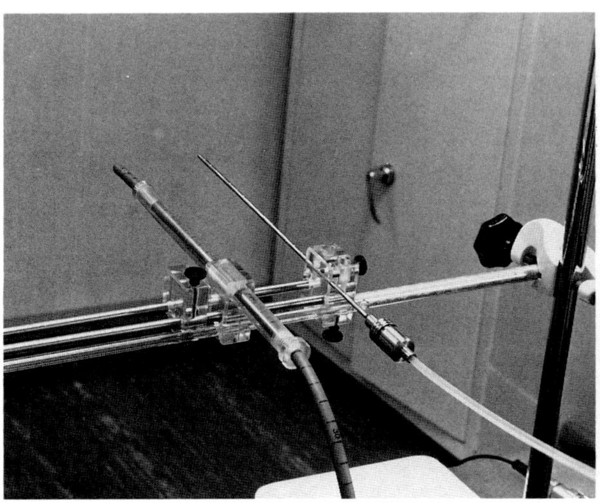

3

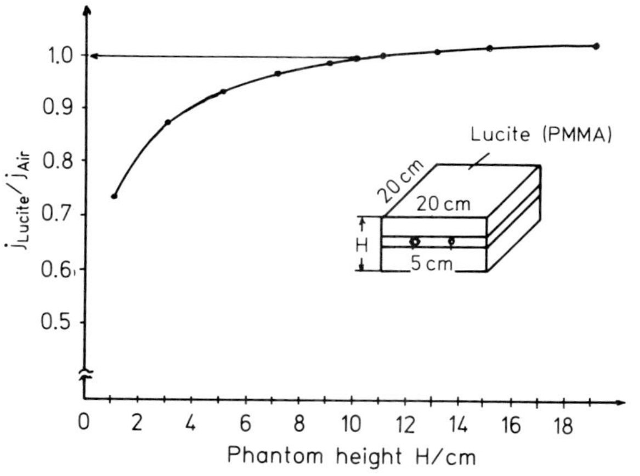

Fig. 4. Comparison of exposure rate of an iridium 192 point source in air to exposure rate in a lucite phantom at a constant measuring distance of 5 cm as a function of phantom height

tering are corrected for (MEISBERGER et al. 1968; MELI et al. 1988). Compared with the measurement in air, a solid phantom guarantees fixed and stable geometrical conditions (Fig. 3). The measured dose rate is dependent on the dimensions of the surrounding phantom. For iridium 192 and a distance of 5 cm, the error of the measured value in the phantom, compared with air, is smaller than 0.5%, if phantom dimensions of 20 × 20 × 10 cm are used. If the phantom height is less than 10 cm, deviations increase up to 37% for a height of 1 cm (Fig. 4).

2.3 Check of the Source Positions

Deviations of the expected source position in the applicator can be detected by contact radiography (Fig. 5) with single packed verification films or photographic papers. The expected source position can either be marked with a pin in advance or can be documented during exposure by thin lead wires between film and source. The applicator can be documented by X-radiography. Because of the high contact dose rate, the photographic material should cover a range of doses from 1 to 5 Gy, allowing the dwell time of the source at the selected position to be large compared with the moving time.

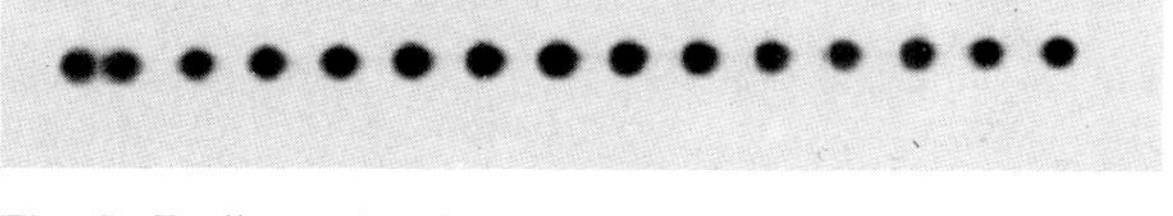

Fig. 5. Radiograph of a linear applicator with 15 point sources; source no. 1 is mispositioned

3 Localisation of the Applicator and Organs at Risk

The quality of the treatment and the complication rate depend highly on exact information about the actual applicator position and its distance to organs at risk. This enables the radiotherapist to detect and correct a mispositioned applicator before treatment and to predetermine the maximal absorbed dose to the organs at risk.

With intracavitary afterloading treatment of carcinoma of the cervix and the vagina, X-radiographs in two perpendicular planes (anteroposterior and

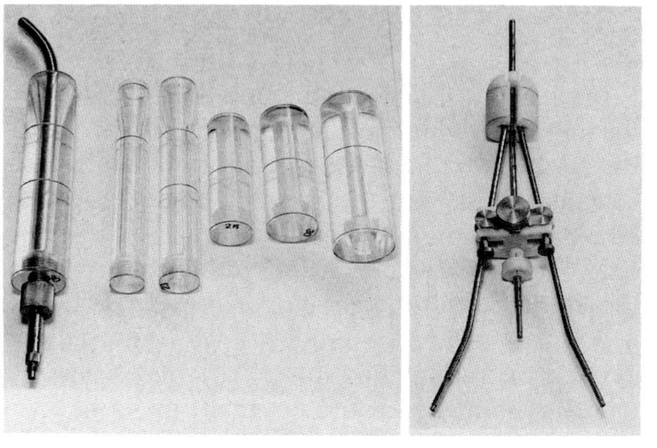

Fig. 6. One- (*left*) and three-way (*right*) applicators for treatment of carcinoma of the uterine cervix. The applicator surface is marked with lead rings

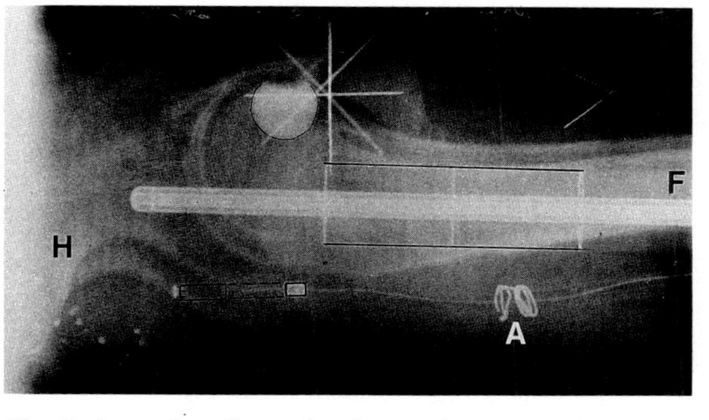

Fig. 7. Lateral radiograph of a patient treated for carcinoma of the cervix with an one-way applicator. Applicator surface, portio, anus, and position of the rectal probe are visible. The base of the bladder is marked with a balloon catheter. The definite distance of the lead rings allows an evaluation of the magnification factor

lateral) can give sufficient information if the applicator and important structures are made visible by lead marks (0.3-mm wire) or contrast medium (Figs. 6, 7). The special marks help to identify:

- The vaginal wall and the applicator surface (lead rings on the applicator surface lie at a definite distance from which the magnification factor can be determined).
- The portio (marked distal end of the applicator, portio flange, or clip).
- The base of the bladder (balloon catheter with contrast medium or a metal chain; (Kuipers and Visser 1986).
- The ionization chamber (lead marks at beginning and end of the chamber).
- The anus (lead ring on the detector cable).

4 Determination of Absorbed Dose to Organs at Risk

From its concept brachytherapy is an irradiation technique with high local contact doses and a very steep dose gradient to the surrounding tissue. Organs at risk like the rectum, bladder, ureter, and sigmoid colon are in the direct vicinity of the target volume and receive a considerable part of the absorbed dose to the target volume (Fig. 8).

The severity and frequency of complications depend on the applied doses, the observance of tissue tolerance, and the optimal time-dose fractionation (Crook et al. 1987; Glaser et al. 1985).

Absorbed dose to the organs at risk can be determined by three methods: (a) in vivo measurements (ionization chamber, solid-state detector, TLD, (b) calculation, (c) superimposition of a lateral radiograph with a set of isodose curves, plotted with the correct magnification factor.

4.1 Measurement

Ionization or semiconductor dosimeters with probe diameters of 4–7 mm are used for in vivo measurements in the rectum and bladder. For reasons of radiation protection, direct manual measurements at several points in the organs at risk are not recommended in high dose rate (HDR) afterloading. Dosimeters with a single detector only give information at one point of the organ. In order to be independent of this arbitrary point, dosimeters with several linearly arranged probes have been designed, which allow simultaneous measurements at 5–6 different positions. Linearly arranged TLD probes in a flexible tube can give a comparable result (Planskoy and Lim 1980). The maximum dose along the introduced tube can be determined graphically (Busch 1978) or by computer evaluation. In vivo measurements in general lead to an underestimation of the maximal absorbed dose in the rectum and bladder. The reasons are discussed in Sect. 4.3.

4.2 Calculation and Graphic Method

With the aid of radiographs, maximal doses to organs at risk can be calculated by computer using the documented coordinates on the films or can be determined by superimposing a sagittal plane of the isodose distribution on the lateral radiograph.

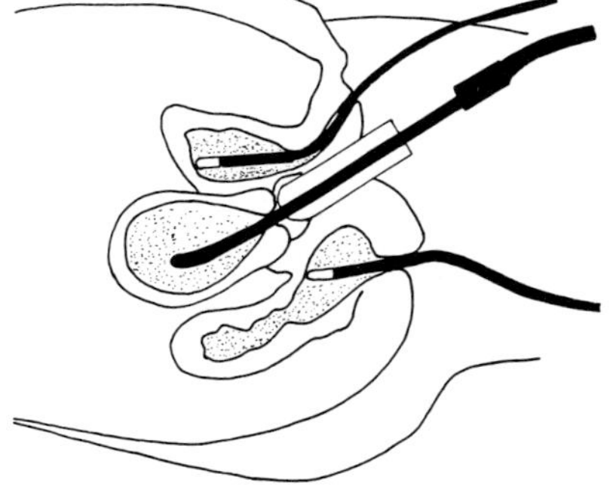

Fig. 8. Sagittal plane of the female pelvis with an one-way applicator. The ionisation chambers in the rectum and bladder demonstrate the typical errors in position

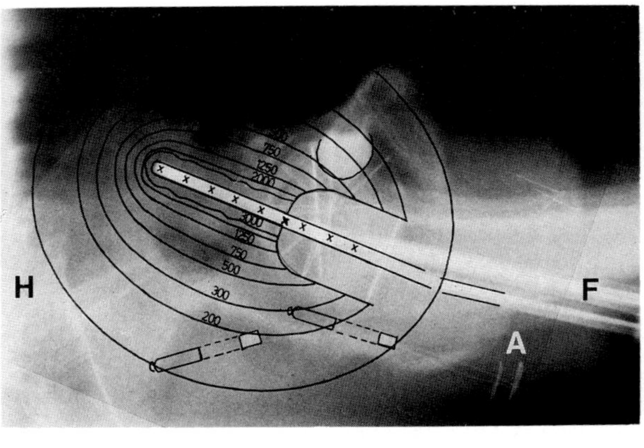

Fig. 9. Superimposition of two lateral radiographs from two treatments of the same patient, take with different craniocaudal positions of the rectal chamber. The difference in measured dose can be seen from the isodoses

Table 2. Minimal distances from portio surface to rectum and bladder evaluated from computerized tomography according to HIMMELMANN et al. (1983)

	Minimal distance (mm)	Mean value (mm)
Cervix–bladder	3–5	4
Cervix–rectum	3–6	4
Diameter of the cervix	27–41	29

The magnification factor for the isodose plot changes with the position of the applicator in the patient. It must therefore be recorded on the film (Figs. 8, 9).

In addition to the magnification factor, the distance between the applicator surface and the organs at risk must be known for a direct determination of the doses. These distances can be determined individually by intracavitary transvesical and transrectal ultrasound and simultaneous radiography (MAK et al. 1987) or individual computerized tomography (CT) scans (COLTART et al. 1987) or can be taken as mean values from published data. For carcinoma of the cervix, Table 2 gives values for these distances which have been evaluated from CT scans (HIMMELMANN et al. 1983).

4.3 Comparison of Measured and Calculated Doses to Organs at Risk

If in vivo measurements and calculations with intracavitary treatments are compared, the measured values always prove to be much lower than the calculated doses, and therefore the maximal absorbed dose to the rectum and bladder is always underestimated. There are three errors responsible for these discrepancies:

- The craniocaudal detector position in the rectum and bladder is not optimal (this error can be reduced by using a dosimeter with multiple probes or by premarking the inserted length on the detector cable according to the isodose distribution of the applicator).
- The detector has no direct contact with the anterior wall of the rectum or the base of the bladder (The detector position depends on the anatomical situation. In most cases the detector is situated in a variable distance to the walls of the rectum and bladder).
- The measuring point in the chamber is always at 2–3 mm distance from the organ wall, depending on the detector radius, even if direct contact to the wall is achieved.

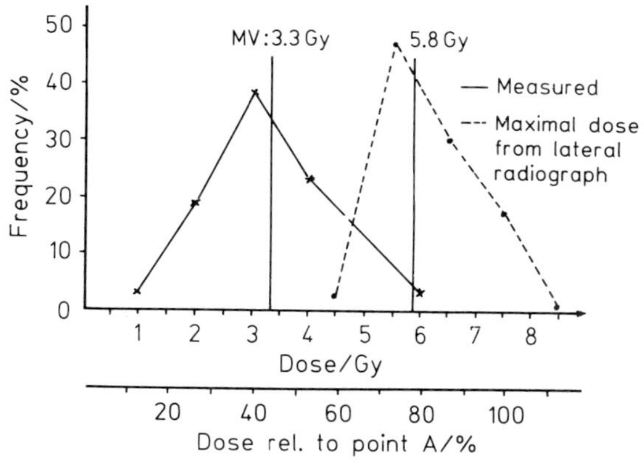

Fig. 10. Comparison of measured and graphically determined absorbed dose to the rectum from 71 high dose rate afterloading treatments (23 patients) of carcinoma of the cervix with a three-way applicator. From measurements with an ionization chamber, a mean value of 3.3 Gy per treatment was estimated, while the calculation led to a mean dose of 5.8 Gy per treatment at the anterior wall of the rectum, i.e., 44% and 78% of the fractionated dose to point A

From a comparative study of 71 afterloading treatments of carcinoma of the cervix with a three-way applicator (Fig. 6), a mean dose of 3.3 Gy per treatment to the rectum resulted from in vivo measurements, i.e., 44% of the fractionated dose of 7.5 Gy to point A (Fig. 10). From a graphical evaluation of the lateral radiographs a mean absorbed dose of 5.8 Gy per treatment was calculated, i.e., 78% of the fractionated dose to point A. Thus, the in vivo measurement under-estimated the absorbed dose to the rectum by an average factor of 1.77.

Errors of in vivo measurements in the bladder were larger. Exposure of the bladder, however, seems not to be associated with the same complication rates seen for the rectum (CROOK et al. 1987; GLASER et al. 1985). Because of the large error in positioning and the additional risk of infections, the dose to the bladder was determined by graphic evaluation. From the graphical evaluation of the same treatments a mean absorbed dose of 6.4 Gy was found, i.e., 86% of the fractionated dose to point A. Table 3 shows comparable published data.

From these results the radiotherapist must conclude that in vivo measurements in the rectum and bladder alone are not sufficient to observe

Table 3. Measured and calculated doese in the rectum and bladder (BATES and BERRY 1980; GLASER et al. 1985; HIMMELMANN and RAGNHULT 1983)

Reference	Dose absolute and relative to point A		Total dose to point A (Gy)	Method
	Bladder	Rectum		
WARD et al. (1980) London		65% 6.2 Gy	3 × 9.5 = 28.5	Ionization chamber
DALE (1980) London		40% 2.4 Gy	9 × 6 = 54	Scintillation detector
PLANSKOY and LIM (1980) London	60%	40–50% 3–3.8 Gy	3 × 7.5 = 22.5	TLD in a flexible tube
SNELLING et al. (1980) London		65%	3 × 7.5 = 22.5	Measurement
HIMMELMANN et al. (1980) Göteborg		39% 3.4 Gy	5 × 8.8 = 44	Measurement
	76% 6.8 Gy	73% 6.4 Gy		Calculated 5 mm from applicator
THESEN (1985) Köln		44% 3.3 Gy	3 × 7.5 = 22.5	Ionization chamber
	86% 6.4 Gy	77% 5.8 Gy		Calculated 5 mm from applicator

tissue tolerance in organs at risk precisely enough, especially if afterloading units are operating with treatment limits which are directly taken from this in vivo dosimetry.

References

Bates TD, Berry RJ (1980) High-dose-rate afterloading in the treatment of cancer of the uterus. Br J Radiol [Special Rep] 17

Busch M (1978) Die Messung der Strahlenbelastung von Blase und Rektum bei der gynäkologischen Kontakttherapie. Strahlentherapie 154: 681–685

Coltart RS, Nethersell ABW, Thoma S, Dixon AV (1987) A CT based dosimetry system for intracavitary therapy in carcinoma of the cervix. Radiother Oncol 10: 295–305

Crook JM, Esche BA, Isturiz CCJ, Sentenac I, Horiot JC (1987) Dose-volume analysis and the prevention of radiation sequelae in cervical cancer. Radiother Oncol 8: 321–332

Dale RG (1980) Dosimetry of the Charing Cross Cathetron. Br J Radiol [Special Rep] 17: 38–42

Dutreix A, Marinello G, Wambersie A (1982) Dosimétrie en curiethérapie. Masson, Paris

Glaser FH, Grimm D, Hänsgen G, Rauh G, Schuchardt V (1985) Klinische Erfahrungen bei der Afterloading-Kurzeittherapie im Vergleich zur konventionellen Brachytherapie bei der Behandlung gynäkologischer Tumoren. Strahlentherapie 161: 459–475

Himmelmann A, Ragnhult I (1983) High dose-rate afterloading treatment in carcinoma of the uterine cervix using an individual planning and reconstruction system. Acta Radiol [Oncol] (Stockh) 22: 263–271

Himmelmann A, Karlstedt K, Ragnhult I (1980) Early experience with high dose-rate (Ralstron) treatment in Göteborg. Br J Radiol [Special Rep] 17: 106–110

Himmelmann A, Bjurstam N, Ragnhult I (1983) Computed tomography of the cervix and distances to the bladder and rectum in intracavitary radiation treatment of gyneological cancer. Strahlentherapie 159: 198–202

International Commission on Radiation Units and Measurements (ICRU) (1985) Dose and volume specification for reporting intracavitary therapy in gyneology. ICRU, Bethesda (Report no 38)

Kuipers T, Visser AG (1986) Technical aspects of bladder dosimetry in intracavitary irradiation of cervix carcinoma. Radiother Oncol 7: 7–12

Mak ACA, van't Riet A, Ypma AFGM, Veen RE, van Slooten FHS (1987) Dose determination in bladder and rectum during intracavitary irradiation of cervix carcinoma. Radiother Oncol 10: 97–100

Meisberger LL, Keller RJ, Shalek RI (1968) The effective attenuation in water of the gamma rays of gold-198, iridium-192, cesium-137, radium-226 and cobalt-60. Radiology 90: 953–957

Meli JA, Meigooni AS, Nath R (1988) On the choice of phantom material for dosimetry of Ir-192 sources. Int J Radiat Oncol Biol Phys 14-587–594

Planskoy B, Lim A (1980) Cathetron dosimetry at the Middlesex Hospital, London. Br J. Radiol [Special Rep] 17: 45–49

Snelling MD, Lambert HE, Yarnold L (1980) Clinical results and complications following treatment of carcinoma of the cervix and endometrium using Cathetron at Middlesex Hospital. Br J Radiol [Special Rep] 17: 33–37

Thesen N (1985) Bestrahlungstechnik, Dokumentation und individuelle Dosimetrie bei der intrakavitären Kurzzeit-Afterloadingtherapie. Strahlentherapie 161: 476–486

Ward AJ, Stubbs B, Dixon B, Firth LA (1980) A shedule of radiotherapy, including the use of the Cathetron for advanced carcinoma of the cervix. Br J Radiol [Special Rep] 17: 17–23

Young ME (1983) Radiological physics, 3rd edn. Lewis London

8.4 Present Status and Perspectives of High Dose Rate Afterloading in Gynecologic Malignancies

HEINER ANNWEILER and MANFRED BUSCH

CONTENTS

1 Introduction 291
2 Dose Reference Points and Dose Distribution 291
3 Dosage and Fractionation 292
4 Dose Rate Factor 293
5 Results from the Literature 295
6 Discussion and Perspectives 295
References 296

1 Introduction

In 1903, 5 years after the discovery of radium by M. Curie, M. Cleaves and A. Döderlein (OESER 1954) were the first who proposed using it for the intracavitary irradiation of cervical cancer. K. Strebel described a manual afterloading treatment in order to minimize the radioactive exposure for the medical staff. In 1910, the surgeon R. Abbe proposed an afterloading technique with celluloid tubes. During the era of the Stockholm and Paris techniques, the afterloading method was forgotten until Fletcher rediscovered it in 1952: He preloaded the cervical canal with empty tubes, fixed them, and afterloaded them with radioactive sources. In 1964, Henschke described a treatment with a single cobalt-60 source, which could be moved in 1-cm steps under manual remote control and by this guaranteed total radioprotection.

This was the beginning of a revolutional development in many countries and led to the production of more than ten different afterloading devices (important ones are listed in table 1). Today, more than one hundred afterloading devices are clinically used in the FRG. Most of them are high dose rate (HDR) devices with radioactive sources of up to 20 Ci, which deliver dose rates of about 0.5–2.5 Gy/min at point A. Cobalt 60, cesium 137, and iridium 192 are the isotopes which are mainly employed. The remote-controlled afterloading technique is performed either electromechanically (oscillation or stepwise with one active source at the top of a wire) or pneumatically (arrangement of active sources and inactive spacers). There are devices which deliver only standard dose distributions and others with individually adapted isodoses. There are gynecological applicators for intracavitary and interstitial techniques. Intracavitary applicators with three channels and two ovoid-like spacers in the vaginal vault are able to produce pear-shaped dose distributions. Isodose lines of Fletcher applicators can be simulated in order to reduce the dosage to organs at risk.

HEINER ANNWEILER, Dr. med., MANFRED BUSCH, Prof. Dr. med., Universitätsklinikum der Gesamthochschule Essen, Radiologische Klinik und Poliklinik, Hufelandstr. 55, 4300 Essen 1, Germany

2 Dose Reference Points and Dose Distribution

In 1938 Tod and Meredith were the first who defined dose reference points in the female pelvis for intracavitary brachytherapy. The dose was prescribed at the new defined points A (2 cm lateral and cranial of the external os of the uterus) and B (3 cm lateral to point A). Point A is near to the ureter, which is an organ at risk. Point B is near to the pelvic wall and represents tumor infiltration or involved lymph nodes. After the introduction of afterloading devices more definitions of dose reference points followed: X1, X2, AF, VA1–5, MY, X, and B2 (BATES and BERRY 1980, p 77). Some authors prefer the dose reference line A (line B) which is located parallel and 2 cm (5 cm) lateral to the central axis (VAHRSON and RAUTHE 1988, p 139). Investigators of low dose rate (LDR) afterloading advices prefer dose reference points near the organs at risk: the rectum and bladder (ESCHE et al. 1987). Nevertheless, points A and B remain the most frequently used reference points for HDR afterloading.

Most of the devices have standard dose distributions for their applicators; some of them can

Table 1. Low (LDR) and high dose rate (HDR) afterloading devices (Bates and Berry 1980; Rotte 1981)

Manufacturer	Author/dose rate	Source/safe capacity	Principle of movement	Number of channels	Experience since
Buchler FRG	Rotte HDR	^{192}Ir 10 Ci	oscillating	1	1971
Decatron GDR	Glaser HDR	^{192}Ir 20 Ci	stepped	3	1974
Brachytron Canada	v. Essen HDR	^{60}Co 20 Ci	oscillating	3	1963
Ralstron Japan	Wakabayashi HDR	^{60}Co 20 Ci	stationary	6	1965
Cathetron GB	O'Connell HDR	^{60}Co 50 Ci	stationary	9	1963
Gamma-Med FRG	Mundinger HDR	^{192}Ir 120 Ci	oscillating	1	1964
GammaMed II FRG	Busch HDR	^{192}Ir 10 Ci	stepped	12	1974
Selectron NL	van't Hooft HDR LDR	^{192}Ir/20 Ci ^{60}Co/10 Ci ^{137}Cs/2.5 Ci	stationary	3–6	1977
Curietron France	Chassagne LDR	^{137}C 0.15 Ci	stationary	4	1969

optimize the dose distribution by computing the dwell times of the source when the desired dosage is given at prescribed reference points (Busch et al. 1977; Busch 1981).

It is generally recommended that orthogonal radiographs be taken from the pelvis and the fixed applicator just before irradiation (Bates and Berry 1980, p 133). After the evaluation of the radiographs, the position of the applicator can be corrected if necessary, and the dose at the bladder and rectum can be computed. The documentation of the applicator position and the knowledge concerning the dose distribution is useful for an exact adaptation to the percutaneous fields: Either a biaxial rotational technique or fixed fields technique with wedge is used. Thus, areas of over- and underdosage at the borders can be avoided.

3 Dosage and Fractionation

By changing from radium to HDR afterloading techniques, the extremely high dose rate requires diminution of the single and total doses. The treatment time for one fraction is reduced from several hours or few days to some minutes. The repair mechanisms which take place during irradiation have only a few minutes in which to work. Therefore, the radiobiological effect on tumor as well as on healthy tissue is much stronger with HDR therapy. If severe side effects are to be avoided, the single dose has to be reduced, and more fractions must be given. Early experience has demonstrated that a single dose of more than 10 Gy causes an increasing rate of early and late

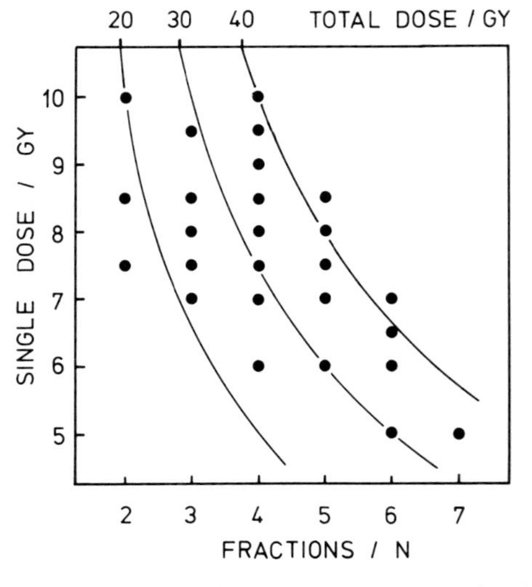

Fig. 1. Spectrum of single and total doses at point A and fractionation of high dose rate afterloading therapy (Bates and Berry 1980)

Table 2. Rate of severe side effects at the rectum in correlation with the high dose rate afterloading single dose at point A (TAINA 1981; VAHRSON and RAUTHE 1988; GLASER et al. 1982)

Reference	Treatment period	Tumor localization/ pretreatment	Patients (n)	Single dose (Gy)	Medium of single dose (Gy)	Fractions (N)	Rate of serious complications (%)	Gradient complications per dose (%/Gy)
TAINA		Cervic	101	7.5		4–5	20.6	8.72
1981	1969–1976	RT		10.0			42.4	
		Corpus		7.5		4–5	11.4	8.76
		RT		10.0			33.3	
BUSCH	1976–1982	Cervix	61	<9.8	8	2–3	4	8.81
1985		RT		10.4–12.4	13		40	
				>13	13		50	
GLASER		Cervix		<7.5	7	5–6	0	7.94
1982	1974–1980	Corpus	921	7.5–10	8.75		3.1	
		RT		>10	12		28.9	
KUCERA		Corpus	94	7–8.5	7.75	2	0	8.18
1988	1980–1984	RT		10	10		18.4	
SORBE		Corpus	379	7.5–9[a]	8.25	4–6	6	6.10
1988	1970–1985	OP		13.5[a]	13.5		38[b]	
HASHIMOTO		Cervix	267	2.5–7	6	3–15	1	6.8
1988		RT		7–10	8.5		18	9.2
				10–20	15		78	

[a] Dose in 5 mm tissue depth
[b] Complications grade 2
RT, radiotherapy; OP, operation.
RT = external irradiation
OP = total abdominal hysterectomy and bilateral salpingo-oopharectomy

side effects such as fistula formation or stenosis. On the other hand, a single dose below 6 Gy requires too many fractions (more than 6). The possible range of HDR single doses thus lies between 6 and 10 Gy. Most authors recommend single doses between 7 and 8.5 Gy (Fig. 1). At the beginning of the HDR afterloading era single doses of up to 15 Gy were given. An evaluation of the late side effects of six treatment schedules with different HDR single doses indicates that a change of the single dose from 7.5 to 10 Gy is followed by a rise in the rate of side effects of about 20%. If a linear ascent were given, in this range, an increase of 8% more side effects per 1 Gy additional dose is expected (Fig. 2, Table 2). In reality the increase of side effects is much stronger for high single doses.

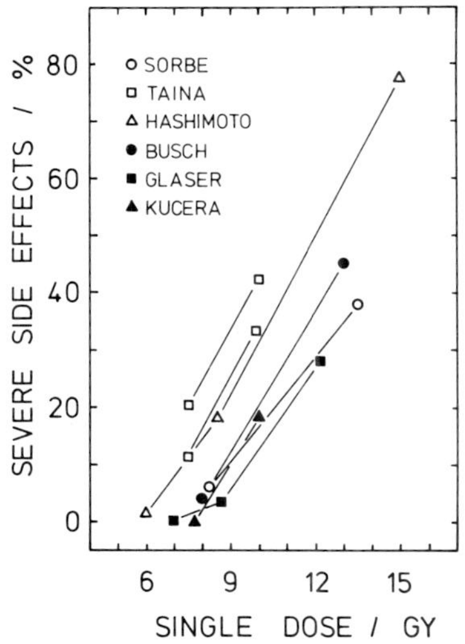

Fig. 2. Increase of severe side effects at the rectum with single dose at point A

4 Dose Rate Factor

In 1966 Liversage asked: What dose in 15 min is equivalent to a dose of X rads in 24 h? He realized that there was no simple answer to give to such a question.

Today, the answer is as difficult as 20 years ago, because there are still open questions about isoeffect doses of LDR and HDR therapy. According to experimental investigations of Trott (1978) the correction factor (CRF) should be in a range between 0.5 and 0.75, depending on the dose rate. The dose rate factor (DRF) is defined as the inverse of the correction factor

$$\text{DRF} = 1/\text{CRF}$$

$$\text{and} \quad \text{Dose (LDR)} = \text{DRF} \times \text{Dose (HDR)}$$

The DRF gives the relation between LDR dose and HDR dose; it is the proportion for lowering the total brachytherapy dose when changing from radium to HDR afterloading regimes. According to Trott's investigations, the DRF should be in the range between 1.33 and 2.0. Clinical investigations of Arai et al. recommend a DRF between 1.5 and 2.0, depending on fractionation (3–8 fractions) (Bates and Berry 1980, p 92). Rotte recommends a DRF between 1.25 and 2.0 (Rotte 1981).

Today, some authors have a long experience

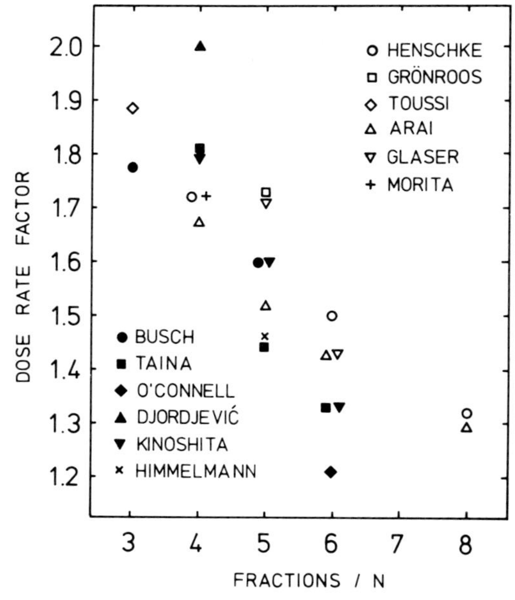

Fig. 3. Dose rate factors for gynecological high dose rate afterloading techniques

Table 3. Evaluation of different low and high dose rate (LDR/HDR) regimes and the dose rate factor (DRF) (Bates and Berry 1980; Vahrson and Rauthe 1988)

Reference		LDR		HDR		DRF
		Dose (Gy)	Fractions (N)	Dose (Gy)	Fractions (N)	
Djordjevic	1988	60	2	30	4	2.00
Morita	1988	50	4	29	4	1.72
Toussi	1980	75	2	40	3	1.88
Henschke	1980	60	2	34.9	4	1.72
				40.2	6	1.50
				45.4	8	1.32
Arai	1980	42	2	25.4	4	1.67
				27.9	5	1.52
				29.7	6	1.43
				32.5	8	1.30
Grönroos	1980	65	2	37.5	5	1.73
Busch		40	2	22.5	3	1.77
		60	3	37.5	5	1.60
Glaser	1982	53	3	35	5	1.71
				42	6	1.43
Kinoshita	1980	50	3	28	4	1.80
		48	3	30	5	1.60
				36	6	1.33
Taina	1981	54.6		30	4	1.82
			1	37.5	5	1.45
		40		30	6	1.33
O'Connell	1980	40		33	6	1.21
Himmelmann	1988	62	3	42.5	5	1.46

Table 4. Influence of high dose rate (HDR) fractionation on the dose rate factor (DRF)

HDR fractions	DRF (range)	DRF (medium)
3	>1.75	1.85
4	1.65–2.00	1.75
5	1.45–1.75	1.55
6	1.20–1.50	1.35
8	<1.35	1.30

with radium as well as with afterloading, and an evaluation of their results is possible. In general, the compared treatment schedules have a similar teletherapy dosage. Most of the authors report better or at least the same survival rates and lower side effect rates for HDR therapy. The published clinical data deliver dose rate factors between 1.2 and 2.0 (Fig. 3, Tables 3, 4).

5 Results from the Literature

In 1913 during the congress of the Deutsche Gesellschaft für Gynäkologie at Halle, Döderlein, Bumm, Krönig, and coworkers reported on curing carcinomas of the cervix with intracavitary application of radium. Since then, the cure rate has risen. Between 1944 and 1964, 5-year survival rates with radium and orthovolt X-rays/cobalt 60 are 61%–70% (stage I), 35%–42% (II), and 10%–20% (III). First results with the Cathetron between 1967 and 1974 were reported by Joslin: 94% (I), 57% (II), and 42% (III) (ALTH 1977, p 107). The annual report on results of treatment in gynecological cancer (PETTERSON 1985) gives 5-year results (treatment period 1976–1978) for conventional and afterloading intracavitary irradiation. The most spectacular difference can be found for stage III cervical carcinoma: 29.3% (conventional) and 41.0% (afterloading). Today, many 5-year results are available for HDR and LDR afterloading techniques (Table 5).

At the beginning of HDR afterloading therapy the single doses applied were too high, so that severe side effects occurred in about 20%–50% (Fig. 2). Today, many publications show that if the recommended dosage is used, the rate of serious complications lies below 10%, usually under 5% (Table 6).

6 Discussion and Perspectives

The discussion about the useful and negligible dose reference points in gynecological afterloading therapy continues: Is it sufficient for comparing results if one or two points of a dose distribution are defined or do we need more? Is it necessary to

Table 5. Five-year survival rates (%) of afterloading brachytherapy and radium therapy in cervical carcinoma (stages I-III) (BATES and BERRY 1980; VAHRSON and RAUTHE 1988; PETTERSON 1985)

Reference	Number of patients				Five-year results		
	I	II	III	I–III	I	II	III
HDR afterloading							
JOSLIN 1980	83	120	87	290	95	55	47
NEWMAN 1983	92	103	92	287	77	50	25
SHIGEMATSU 1983		43	100	143		80	53
UTLEY 1984	29	30	43	102	89	50	33
ROTTE 1985	44	97	22	143	71	72	45
GLASER 1988	150	175	76	401	77	66	57
MIZOE 1988	17	68	212	297	71	66	39
ILIC 1988	232	462	430	1124	93	73	54
DJORDJEVIC 1988	73	115		188	78	65	
ARAI 1980	86	173	212	471	83	71	51
LDR afterloading							
HORIOT 1984				1485	89	76–85	50–62
PERNOT 1985				216	78	62	31
ARAI 1980	31	125	253	409	87	75	49
Radium							
GLASER 1988				623	79	59	22
Annual Report 1985	1257	2587	2296	5140	69	54	29

Table 6. Serious complications of afterloading and radium brachytherapy in carcinoma of the uterus (Bates and Berry 1980; Vahrson and Rauthe 1988)

Reference	Patients (*n*)	Complications (%)
HDR afterloading		
Glaser 1988	1131	1.3
Rotte 1985	223	8.1
Snelling 1980	68	5.8
Kauppila 1980	217	5.5
Frankendal 1980	130	2.3
Joslin 1982	354	6.8
Newman 1983	291	4.8
Cikaric[a] 1988	140	12.1
LDR afterloading		
Horiot 1984	1485	9.8
Pernot 1985	216	3.2
Radium		
Glaser 1988	623	23.8
Cikaric 1988	187	26.2

define the dose on a line A? One must consider cigar-shaped or pear-shaped dose distributions and the different dose reference points for tumor infiltration of the cervix, corpus, or vagina. It seems nearly impossible to take more than two reference points for each region into consideration, if the dose distribution is to be described. It is much more useful for the optimization of the treatment schedule if the dose distribution for brachytherapy is well adapted to the field borders of the external irradiation.

These distributions of total doses represent only the physical doses and not the biologically effective ones. Knowing the effective doses is necessary if, e.g., the dose at the organ at risk must be computed. If HDR afterloading is used, the DRF could help to estimate the biological effective dose which corresponds to radium doses. The DRF is not a qualified tool to compute the HDR doses exactly because many other parameters (radium fractionation, length of treatment-free intervals, HDR total dose, teletherapy technique and dosage, etc.) were not taken explicitly into consideration. The DRF only gives a rough estimation of the dose reduction.

The cited 5-year results confirm the impression of many brachytherapists that the LDR and HDR afterloading techniques will surpass the radium technique. The reasons for the improvement of the results of cervical carcinoma are multiple. Most authors feel that the exactly placed and fixed applicator with a well-defined, constant, and reproducible dose distribution is a good basis for obtaining positive results. The discussion whether fractionated HDR or protracted LDR brachytherapy is more useful is still going on. Nevertheless, many authors have achieved better results with afterloading (HDR and LDR) than with radium methods.

A few important rules for carrying out HDR afterloading therapy are recommended:

- The recommended dose range for single doses is 7–8.5 Gy at point A
- Appropriate treatment-free intervals between HDR fractions
- The exactly placed applicator is the condition for a well-defined and adaptable dose distribution
- The radiographs of the applicator can help to adapt the teletherapy fields
- The total dose distribution and the doses at the organs at risk should be computed.

The main advantages for the patient of HDR afterloading compared with radium are:

- Better survival rates and lower rates of serious complications
- The physical and psychological strain is diminished
- Outpatient treatment is possible
- Anesthesia is not always necessary

and for the hospital and staff are:

- Many applications per day possible
- Only one radioprotected room is necessary
- No radioactive exposure to the staff

References

Abbe R (1910) News Note. Arch Roentg Ray 15: 74

Alth G (1977) Technik des Nachladeverfahrens. Maudrich, Vienna

Annweiler H, Roth SL, Thesen N, Sack H (1985) Vorstellung eines Therapieschemas und der Bestrahlungtechnik bei der Kruzzeit-Kontaktbestrahlung des Kollumkarzinoms. Strahlentherapie 161: 286–292

Bates TD, Berry RJ (1980) High dose rate afterloading in the treatment of cancer of the uterus. Br J Radiol [Special rep] 17: 1–89

Busch M (1981) Der therapeutische Dosierungsspielraum bei der Kurzzeitafterloading Therpie in der Gynäkologie. Strahlentherapie [Special Vol] 76: 320

Busch M, Alberti W (1985) High dose rate afterloading therapy of uterine cancer. Essen Afterloading Symposium, Radiologisches Zentrum, Universitätsklinikum Essen

Busch M, Makoski B, Schulz U, Sauerwein K (1977) Das Essener Nachladeverfahren für die intrakavitäre Strahlentherapie. Strahlentherapie 153: 581–588

Cleaves MA (1903) Radium Therapy Med. Record 64: 601

Esche BA, Crook JM, Isturiz J, Horiot J-C (1987) Reference volume, milligram-hours and external irradiation for the Fletcher applicator. Radiother Oncol 9: 255–261

Fletcher GH (1952) A physical approach to the design of applicators in radium therapy of cancer of the cervix uteri. Am J Roentg 68: 935–949

Glaser FH, Grimm D, Hänsgen G, Rauh G, Heider K-M, Kraft M, Salewski D, Schuchardt V (1982) Fraktioniertes Kurzzeit-Afterloading mit hohen Dosisraten. Radiobiol Radiother (Biol) 23: 481–496

Glaser FH, Grimm D, Hänsgen G, Rauh G, Schuchardt V (1985) Klinische Erfahrungen bei der Afterloading-Kurzzeittherapie im Vergleich zur konventionellen Brachytheapie bei der Behandlung gynäkologischer Tumoren. Strahlentherapie 161: 459–475

Hammer J, Kärcher KH (1988) Fortschritte in der interstitiellen und intrakavitären Strahlentherapie. Zuckschwerdt, Mūnich

Henschke UK, Hilaris B, Mahan D (11964) Remote afterloading with intracavitary applicators. Radiology 83: 344

Horiot JC (1985) Results of radiotherapy alone in the treatment of 1485 cases of cancer of the cervix. In: Tungsubutra K (ed) Diagnosis and treatment of carcinoma of the cervix in development areas. Proceedings of the international working party meeting in Thailand, 1985. A. Hilger pp 33–39

Liversage WE (1966) The application of cell survival theory to high dose- rate intracavitary therapy. Br J Radiol 39: 338–349

Newman H, James KW, Smith CW (1985) Treatment of cancer of the cervix with a high dose rate afterloading machine. Int J Radiat Oncol Biol Phys 11: 931–937

Oeser H (1954) Strahlenbehandlung der Geschwülste. Urban & Schwarzenberg, München p 145

Pernot M, Bey P, Stines J, Hoffstetter S (1985) Devenir à long terme des complications recto-sigmoïdiennes chez les malades traitées par irradiation exclusive pour cancer du col uterin. J Eur Radiother 6: 207–217

Petterson F (1985) Annual report on the results of treatment in gynecological cancer, vol 19. Radiumhemmet, Stockholm

Rotte K (1981) Ferngesteuerte Afterloadingverfahren. Strahlentherapie [Special vol] 76: 313

Rotte K (1985) Klinische Ergebnisse der Afterloading-Kurzzeittherapie im Vergleich zur Radiumtherapie. Strahlentherapie 161: 323–328

Strebel K (1903) Vorschiläge Zur Radiotherapie Dtsch Med Z 24: 1145–1146

Shigematsu Y, Nishiyama K (1983) Treatment of carinoma of the uterine cervix by remotely controlled afterloading intracavitary radiotherapy with high dose rate. Int J Radiation Oncol Biol Phys 9: 351–356

Taina E (1981) High versus low dose-rate intracavitary radiotherapy in the treatment of carcinoma of the uterus. Acta Obstet Gynecol Scand [Suppl] 103: 12–13

Tod MC, Meredith WI (1938) Dosage system for use in treatment of cancer of uterine cervix. Brit J Radiol 11: 809

Trott KR (1978) Der Einfluß der Dosisleistung auf die therapeutische Wirkung von Co-60–Gammabestrahlung beim Adenokarzinom der Maus. Strahlentherapie 154: 656–658

Utley JF, von Essen CF, Horn RA, Moeller JH (1984) High dose rate afterloading brachytherapy in carcinoma of the uterine cervix. Int J Radiat Oncol Biol Phys 10: 2259–2263

Vahrson H, Rauthe G (1988) High dose rate afterloading in the treatment of cancer of the uterus, breast and rectum. Strahlenther Onkol [Suppl] 82

8.5 Five- to Seven-Year Results in High Dose Rate Radiation of Cancer of the Cervix

GERHART ALTH

CONTENTS

1 Introduction 299
2 Technical Conditions 299
3 The Physical Part 300
4 The Radiobiological Part 301
5 Late Complications 302
6 The 5-Years Survival Results 302
7 The Protocol 304
References 304

1 Introduction

In 1977, the plate method in the afterloading technique using a high dose rate (HDR), 3-channel, remote afterloading system (TEM, Cathetron) was presented for the first time. The integration of the plate method has several advantages:

a) Plate method is a familiar method, which has been used for more than 50 years at this place.
b) Optimal dose distribution for a (large) tumor in the region of point P.
c) Inhibition of a breakthrough of an endometrical cancer via the cervical region into the portio.
d) Standardization of the applicators and dose distribution.
e) Possibility of choice of an applicator suitable for an exact volume adaptation in the vagina.
f) Colpostatic effect by using the applicator and a short-time, high dose rate technique.

The goals were to decrease the complications regarding the rectum and other intestinal parts, bladder and ureter, and vaginal tissue; and total eradication or inactivation of malignant cells with the chance of local cure.

Contrary to other doctrines the fractionation number was increased from the usual 2–4 up to 8–14 in the space of 9–10 weeks including additional percutaneous teletherapy.

In this chapter the observations of short and late complications in a 10-year period following this HDR therapy are reported, as well as the results of gynecological treatment with 5–7 years' survival time.

Before employing this kind of brachytherapy, three technical conditions should be considered: (a) it is a high dose rate technique, (b) using higher fractionation than usual and (c) a 3-channel remote afterloading device.

2 Technical Conditions

In 1976, I started with the 3-channel remote afterloading device, the Cathetron, integrated to a 9-chamber tresor and equipped with varying length cobalt-60 sources of different activities. The average charge is between 2 and 3.5 Ci per channel.

In 1982, a 3-channel remote afterloading unit, "Buchler" K 50, with three cesium-137 sources (2 Ci) and an oscillating middle source was installed.

For this modification of the method a new kind of applicator was developed. By carefully inflecting the end of the tubes and allowing for a corresponding radius for untroubled movement of the sources, the tubes were put into one, two, or three calculated (computed) drill holes of a cubic plastic material, whereby the distance to the anterior rectal wall by elongation of the plastic plate from the plane of the tubes was variable and could be individually formed (Figs. 1, 2).

This has several particular features:

a) Ideal planar contact to the tumor.
b) Sufficient distancing of source from the rectal wall or bladder through the possibility of variable placement (intermediate plastic material or lead shielding).

GERHART ALTH, Prim. Prof., M.D., Allgemeines Krankenhaus der Stadt Wien-Lainz, Abteilung Strahlentherapie, Wolkersbergenstr. 1, 1190 Wien, Austria

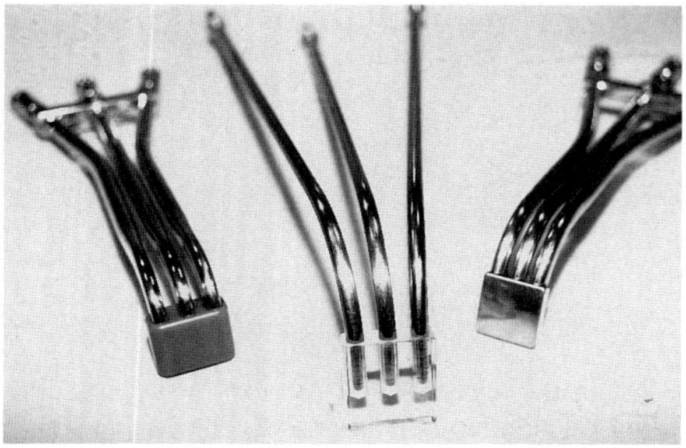

Fig.1. Three-channel Alth applicators for cobalt-60 and cesium-137 high dose rate afterloading devices developed in 1976 and 1981 in Vienna

c) Standardization of the dose distribution by development of
d) Applicator types, which are sufficient for the common gynecological brachytherapy techniques.

3 The Physical Part

The dose surrounding the plate applicator with three standing sources of 2 Ci per channel over 5 min is 15 Gy at 2 mm distance from the applicator surface, 1.7 Gy (11.7%) in the region of point A, and 0.5 Gy (3.3%) at point B, compared with point P (Fig. 3a).

The intrauterine tube (cobalt 60 or cesium 137) delivers a dose of 6 Gy to point A, 10 Gy to point P, and 1.3 Gy to point B (Fig. 3b).

There are three possibilities to reduce radiation exposure to the anterior wall of the rectum. Elongation of the plate applicator, leaving the posterior vaginal speculum in the vagina during the time of treatment, tamponade of the vagina. The average dose at the anterior wall of the rectum amounts to 2.0–2.5 Gy at a single dose of 15 Gy at 2 mm from the line or plane P. The dose limitation (50 Gy, according to Kottmeier) during

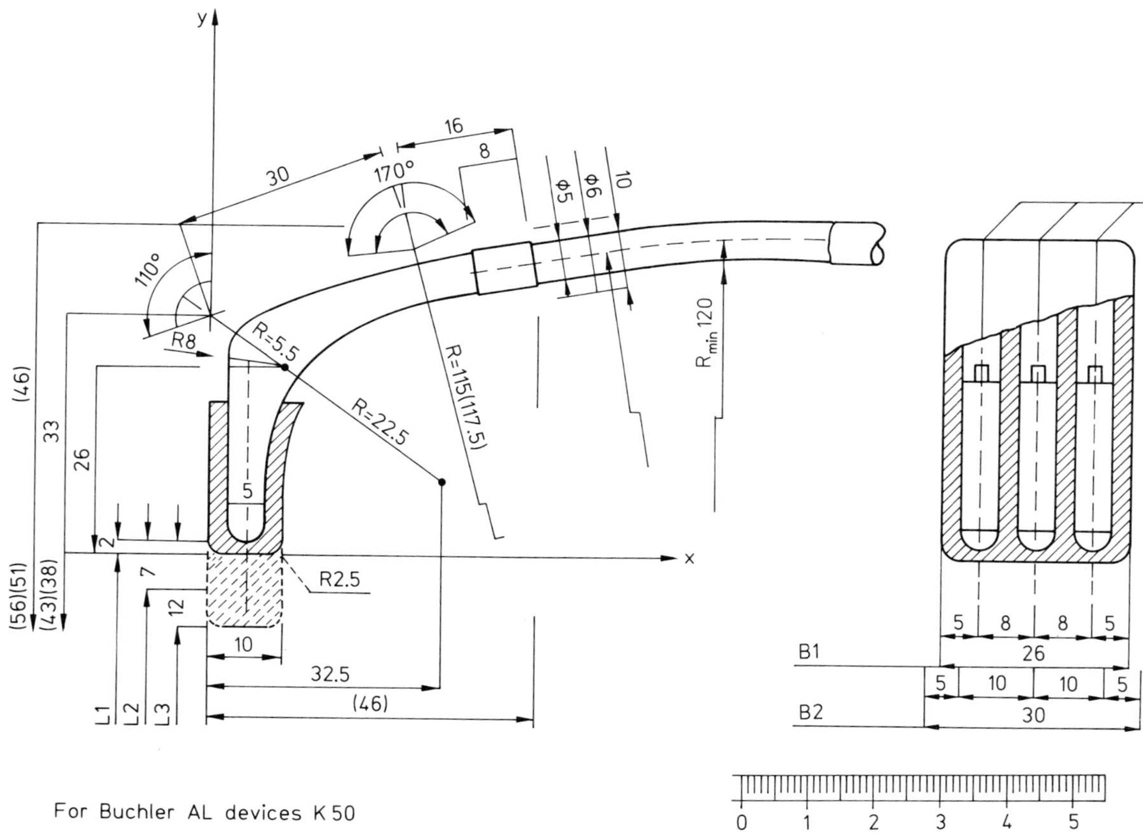

Fig. 2. Construction sketch of the Buchler afterloading 3-channel applicator

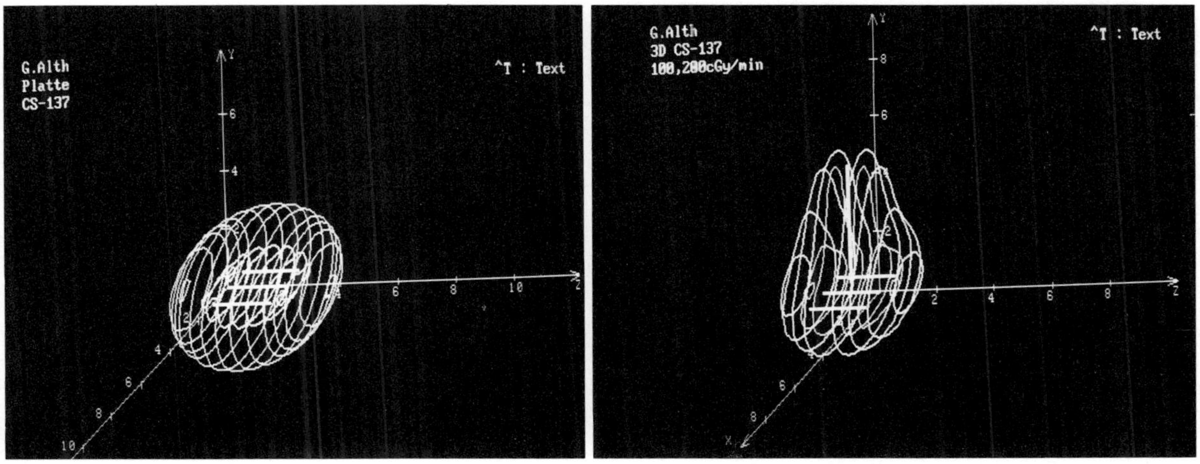

a b

Fig. 3.a,b. Three-dimensional target volumes for two different isodose levels by Kallinger (level 1, 100 cGy/min; level 2, 20 cGy/min) for the treatment of cervical cancer. **a** Plate, **b** plate plus tube

total treatment has never been reached. Experience shows that only random tests with one to three patients and tumor-related measurements are necessary to check the quality for rectal exposure.

Protection of the bladder is achieved by the correct positioning of the applicator and its angle to the bladder. When delivering a dose of 20 Gy to point P, the measured exposure in the bladder is less than 2 Gy; the same can be said for the anterior wall of the rectum.

4 The Radiobiological Part

When treating cancer of the cervix, the following dose distribution should be achieved by brachytherapy: 180 Gy to point P, 60 Gy to point A, 11 Gy to point B during a treatment time of approximately 9 weeks (63 days).

At the same time the brachytherapy is reinforced by teletherapy; 2 Gy/day up to 45–60 Gy in total, depending on the staging, are administered (Table 1): stage I FIGO, 45 Gy; IIa, 50 Gy; IIb, 55–60 Gy; and III, 60 Gy. The region of point P is totally shielded by lead; point A is 40%–50% shielded to a maximal dose of 85 Gy; and point B receives 60–65 Gy.

Table 1. Point- and stage-related dosage of the high dose rate afterloading plate method.

	Time (weeks)	Total dose			Time (days)	Fractionation	Single dose (Gy)		
		Point P	Line A	Line B			Point P	Line A	Line B
Plate applicator and intrauterine tube	9	6060 ret	2020 ret	370 ret	63	8–14			
Plate (3 × 74 GBg) (2 mm depth)	9	120 Gy	14 Gy	4 Gy	63	8	15	17	0.5
Intrauterine tube	9	60 Gy	36 Gy	7 Gy	63	6	10	6	1.3
Teletherapy (2 Gy/day)	6	180 Gy shielded 100%	85 Gy shielded 50% 25 Gy	65 Gy 50 Gy	42	20–30 30			
Teletherapy with lead absorber	6	shielded 100%	shielded 40% 1055 ret	 1758 ret	42 30–42	30 22–30			
Stage Ia 45 Gy			3075 ret	2128 ret					

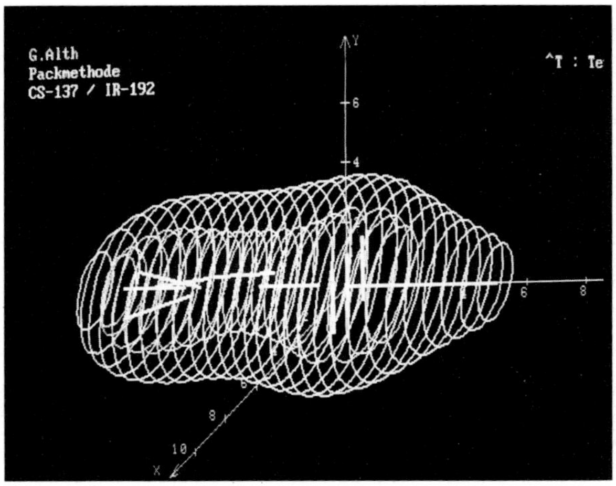

Fig. 4. Three-dimensional target volumes for two different isodose levels by Kallinger (level 1, 100 cGy/min; level 2, 20 cGy/min) for the treatment of endometrical cancer. Iridium-192 packing, cobalt-60 vaginal cylinder, and cesium-137 plate

The number of fractions of the primary radiotherapy amounts to about 8–10 intravaginal plate applications and 6–8 sessions with intrauterine tubes, in all 12–18 brachytherapy plus 20–30 teletherapy treatments, giving a maximum of 50 treatments over 9–10 weeks.

Since 1985, the primary therapy for endometrical cancer has been iridium-192 wire tube packing (30 h over 14 days, then later 20 h, leading to an average dosage of 260 Gy at 5 mm depth) supplemented 6 times by the plate applicator and 2 times by vaginal cylinder. The single dose of the plate therapy is equivalent to that at the cervix uteri (15 Gy in 2 mm depth at point P). The single dose of the vaginal cylinder is calculated to be 15 Gy at 2 mm depth beyond the anterior wall (urethra) of the vagina, used to prevent neoplastic anterior breakthrough of the cervix and portio (Fig. 4).

After trying the conventional afterloading method with single or double intrauterine tubes I returned to the earlier method of radiopacking, because an increase of the recurrence rate from 10% with conventional radium therapy to 26% was observed despite using Fletcher's superposition method and the usual HDR afterloading tubes for intracavitary brachyradiotherapy.

Table 2. Complications of the plate method over the past 10 years (n = 1811)

	n	Percentage
No complications	1745	96.4
Proctitis I, II,	9	0.5
III, IV (ulcer)	4	0.22
Cystitis I, II,	17	0.94
III, IV (fistulation)	3	0.17
Early reaction of the cervical vaginal vault tissue	17	0.94
Late reaction of the cervical vaginal vault tissue	14	0.77
Ureter fibrosis	0	0
Sigmoiditis (and stenosis)	1	0.25
Complications	65	3.59

5 Late Complications

The complication rate was distinguished by a significant decrease of proctitis I–IV and cystitis I–IV. In summary, among all patients (n = 1811) from 1976 to 1986, only 1% of complications at the rectum and 1% of complications at the bladder were noted; the fistulation incidence decreased to 0.3% at the rectum, and treatment-related complications involving the bladder did not appear.

The same results were found with the treatment of corpus carcinoma, with 1.5% complication rate at the rectum and none at the bladder. This is contrary to the results of radium therapy which show fistulation in 6% and proctitis II-IV in 7% (Table 2).

Late mucosal reactions of the portio, cervix uteri, and vaginal tissue were seen over a period of 3–6 months, when occasionally in a protocol of a prospective study the single dose of a plate application was 30 Gy a week in 2 mm depth of the plane P. Some 23 instances of a late reaction of the vaginal and cervical tissue were attributed to this high single dose irradiation (of 40 = 57%).

Of the 23, 5 were observed as midlate or early reaction. This group of patients consists of alcoholics and heavy smokers.

6 The 5-Years Survival Results

The following results were collected from January 1976 until December 1979. In all, 239 patients were treated for primary carcinoma of the cervix

Table 3. Results of primary high dose rate plate and tube treatment (n = 239)

Stage	n	1a	2a	3a	4a	5a	
I	56	92.8	89.3	87.3	85.0	82.7	
IIa	50	94.0	85.8	75.6	71.7	65.8	
IIb	48	83.3	66.2	59.8	57.6	59.2	
III	72	70.8	54.9	51.2	46.8	46.8	+11.7
IV	13	61.5	43.9	21.9	21.9	21.9	
$\bar{x}$%		82.2	70.7	63.9	60.0	57.2	
% primary radiotherapy: postoperative radiotherapy		−11.5	−15.9	−13.1	−17.0	−18.5	

Table 4. Actuarial 5-year survival results after postoperative treatment (n = 172)

Stage	n	1a	2a	3a	4a	5a+	
I	97	97.9	94.8	92.5	92.5	92.5	
IIa	41	95.1	92.5	78.5	78.5	74.9	
IIb	18	94.4	89.6	68.8	68.8	68.8	
III	16	81.2	56.2	35.1	35.1	35.1	+11.7
IV		93.7	86.6	77.0	77.0	75.7	
$\bar{x}$%		+11.5	+15.9	+13.1	+17.0	+18.5	
% primary radiotherapy: postoperative radiotherapy							

Table 5. Results from October 1976 to December 1980 of primary high dose rate treatment by plate-tube method of cervical carcinoma (n = 239)

Stage	n	Died of local relapse	Died of Distant metastases	Lost pat. met.	Survival 5A Specific cause (%)	Survival 5A Patients lost (%)	Died intercurr. Ned	Crude survival (%)	Actuarial survival (Kaplan-Meyer) (%)
(Figo)									
I	56	3 (2[a])	2	1	91.1	89.3	10	71.4	82.7
II	98	12 (9[a])	3	3	84.7	81.6	19	62.2	62.5
III	72	31 (18[a])	–	7	56.9	47.2	2	44.4	46.8
IV	13	10	–	2	23	7.7	1	0	21.9

[a] Pelvic wall

uteri. The actuarial 5-year survival results achieved by primary radiotherapy with the plate method are stage I (Figo), 82.7%; IIa, 65.8%; IIb, 59.2%; III, 46.8%; and IV, 21.9%.

The average age of the patients with stages I and II is higher than normal, because in these stages there in a strong influence of negative selection. This fact reduces distinctly the results (Manchester method n = 49, plate method n = 190) (Table 3).

On the other hand improved results with postoperative radiotherapy are evident in 172 patients with the plate-only method: Figo stage I, 92.5%; IIa, 74.9%; IIb, 68.8%; III, 35.1%; only in stage III (35.1%) did the primary therapy offered 10% better results (Table 4).

The last results from October 1976 until December 1980 with and without Kaplan-Meyer's factors referring to primary therapy for cervical carcinoma by the plate-tube method (excluding all other modifications) in percentage were: Figo stage I, 90.3% (78); II, 81% (69); III, 46% (41); IV, 8.4% (0) (Table 5).

If the tumor is not completely removed at surgery, the patient has a worse condition due to surgeon-caused tumor cell propagation than a primary-treated patient without dissemination risks or other difficulties, for example, an involved

Table 6. Protocols for irradiation treatment of gynecological carcinoma

	Plate	Intrauterine tube	Teletherapy
Primary therapy			
Ca. portionis uteri	9–14 ×	6 ×	
Ca. cervicis uteri	6–8 ×	6–8 ×	
Ca. corporis uteri	6 ×	2 × Vaginal cylinder	Plus ^{192}Ir wire packing of 260 Gy at 5 mm depth in myometrium (2 treatments totalling 50 h) (line 11)
Postoperative therapy			
Ca. portionis uteri	6–8 ×		Teletherapy obligatory
Ca. cervicis uteri	6–8 ×		
Ca. corporis uteri	6 ×	2 × Vaginal cylinder	Teletherapy dependent on infiltration of myometrium

ureter. Similar results were obtained for primary and postoperative radiotherapy of carcinoma of the endometrium.

The actuarial 5-year survival results are 84.4% for stage I (41 patients) in comparison with 92.1% for stage I and 64.4% for stage II in 195 patients treated by operation. The average age of 69 years in the stage I patient group receiving primary therapy suggests that it is a collective of senior citizens at high risk.

7 The Protocol

The protocols of irradiation treatment for gynecological carcinoma are given in Table 6.

Line M is the formerly imaginary, now measureable uterus-surrounding line in the myometrium at a depth of 5 mm deep to the surface.

All these brachytherapy-treated patients underwent supporting teletherapy.

The indication for teletherapy depended on infiltration of the myometrium. A tumor cell invasion of more than 3 mm depth demanded teletherapy.

References

Alth G (1977a) Die Einführung der Plattenmethode in die Nachladetechnik. Deutscher Röntgenkongreß, Münster

Alth G (1977b) Technik des Nachladeverfahrens. Maudrich, Vienna, pp 33, 34, 89

Alth G (1978) Curietherapie mit dem Nachladeverfahren unter besonderer Berücksichtigung der Plattenmethode. Maudrich, Vienna, p 20

Alth G, Koren H, Fucik F, Steyrer K (1979) Afterloading High-dose-rate-Verfahren. Biomed 11: 10–16

Ellis F (1971) Nominal standard dose and the ret. Br J Radiol 44: 101–108

Fletcher GH (1952) A physical approach to the design of applicators in radium therapy of cancer of the cervix uteri. AJR 68: 935–949

Gauwerky F (1957) Standardisierung und individuelle Anpassung bei der Strahlenbehandlung der Gebärmutter- und Scheiden-karzinome. Strahlentherapie 103: 16–47

Gauwerky F (1975) Über das Zielvolumenkonzept der strahlentherapeutischen Planung. Radiologe 15: 217–223

Glaser FH, Grimm D, Heider K-M (1984) Zum Einfluß der zeitlichen Dosisverteilung bei protrahierter und fraktionierter Brachytherapie gynäkologischer Tumoren. Mathematisch formulierte Modellvorstellungen und klinische Erfahrungen beim fraktionierten Kurzzeitafterloading mit hohen Dosisraten. Radiobiol Radiother (Berl) 25: 231–240

Joslin CAF, O'Conell D, Howard N (1967) The treatment of uterine carcinoma using the Cathetron. III. Clinical considerations and preliminary reports on treatment results. Br J Radiol 40: 895–904

Joslin CAF, Smith CW, Mallik A (1972) Treatment of cervix cancer using high activity Co-60 sources. Br J Radiol 45: 257–270

Joslin CAF, Smith W (1974) Radiobiological implications of cathetron therapy. In: Simon NM, Snelling TD (eds) Cancer of the cervix. DHEW, Washington, p 372 (Publication no (FDA) 74-8021)

Koren H, Freud R (1987) Dreidimensionale Bestrahlungsplanung in der Brachytherapie mit einem PC. In: Fortschritte in der interstitiellen und, intrakavitären Strahlentherapie. Zuckschwerdt, Münich, pp 30–31

Kottmeier HL (1952) Die Therapie des Collumcarcinomas. Oncologia 5: 243–259

Kottmeier HL (1953a) Carcinoma of the female genitalia. Williams and Wilkins, Baltimore

Kottmeier HL (1953b) Die Behandlungstechnik und Erfolge der Strahlentherapie des Collumkarzinoms am Radiumhemmet in Stockholm. Arch Gynäkol 183: 430–443

Paterson R (1952) Studies in optimum dosage. Br J Radiol 25: 505–516

Paterson R (1954) Radiotherapy in cancer of cervix: rising cure rates follow improvement in technique. Acta Radiol [Suppl] (Stockh) 116

Paterson R, Parker HM, Spiers FW (1936) A system of dosage for cylindrical distributions of radium. Br J Radiol 9: 487–508

Rotte K, Linka F, Felder D (1973) Intrakavitäre Bestrahlung des Uteruskarzinoms durch ein Afterloading-Gerät mit punkförmiger Iridium-192-Quelle. Strahlentherapie 145: 523

Strebel K (1903) Vorschläge zur Radiotherapie. Dtsch Med Z 24: 1145–1146

Tod M, Meredith WJ (1938) Radium treatment of cancer of the uterine cervix by the "Manchester method". Br J Radiol 809

Tod MC, Meredith WJ (1941) Optimum dosage in treatment of carcinoma of uterine cervix by radiation. Br J Radiol 14: 23–29

Trott KR (1976) Der Einfluß der Dosisleistung in der Strahlentherapie. 57. Tagung der Deutschen Röntgengesellschaft, Essen

8.6 Prospective Clinical Trial Concerning High Dose Rate Afterloading Therapy in Cancer of the Cervix and Endometrium

DIETER KOB and KARL-HEINZ KLOETZER

CONTENTS

1 Introduction 307
2 Materials and Methods 307
3 Results 308
4 Discussion 309
References 309

1 Introduction

The present trend in brachycurietherapy for gynecological malignancies, from conventional methods to the afterloading technique, offers undeniable advantages with respect to labor-saving and radiation protection, particularly when applying high dose rates, and permits exact localization and application. Improvements in the spatial dose distribution are possible due to movable radiation sources and applicator shields. Considering the biological aspect, however, short-term afterloading therapy has disadvantages, since both for repair processes and synchronization effects the long-term irradiation has proven to be more favorable than short-term irradiation, and the influence of the oxygen factor diminishes with decreasing dose rate.

It is, therefore, understandable that now as before the high dose rate (HDR) afterloading therapy meets with a certain skepticism even though the negative consequences of high-dose fractions must be less serious with a small volume dose distribution than with greater volume doses.

In the GDR the afterloading technique was introduced on a countrywide scale at the end of the 1970s and was influenced by the positive experiences of Glaser (GLASER 1979; GLASER et al. 1977). At the time of the installation of the Decatron machine in this clinic no definite dose recommendations could be given, and in addition it was necessary to make considerable technical changes on the machine (WALTER et al. 1983). Therefore, I decided to undertake a prospective randomized study in order to re-examine the fractionation recommendations.

DIETER KOB, Prof. Dr., KARL-HEINZ KLOETZER, Dr. Sc., Klinik für Radiologie der Friedrich-Schiller-Universität, Bachstr. 18, 6900 Jena, Germany

2 Materials and Methods

Given the recommendations from radiation therapists (JOSLIN et al. 1972, 1977; JOSLIN 1978; ROTTE 1975a, b, 1978, 1985) on the maximum single dose of 10 Gy and total dose of 40 Gy to the references points the dosage regimes 4 × 10 Gy, 5 × 8 Gy, and 8 × 5 Gy with external radiation therapy of carcinoma of the cervix and endometrium were examined in a study started in 1981 (dose rate, 2–4 Gy/min). Dosage is expressed at points A (cervical cancer) and My (endometrial cancer). The single doses of 10 Gy and 8 Gy were administered once a week, the single dose of 5 Gy twice a week. Brachytherapy was combined with percutaneous radiotherapy of 45 Gy in the area of the pelvic wall by means of biaxial telecobalt pendulum therapy (pendulum angle 10°–170°, field size 6 × 16 cm^2, distance of the pendulum axes 10 cm, focus–axis distance 60 cm). The spatial dosage distribution was adapted to conventional brachytherapy following the Stockholm method and the endometrium packing method by Heyman, respectively, standardized according to the length of the cervical canal. The randomization of the group of subjects was made according to the entry of the reference letters. For reasons of medical ethics the number of patients had to be limited to a minimum. Therefore, the study ended with a total of 105 patients with cervical cancer and 37 patients with endometrial cancer in the middle of 1986.

Table 1. Age distribution in patients studied

	4 × 10 Gy	5 × 8 Gy	8 × 5 Gy
Cervical cancer			
n	36	33	36
$\bar{x}$ (years)	65.2	64.9	65.1
± σ n	12.3	12.0	10.7
± σ n− 1	12.5	12.1	10.9
Endometrical cancer			
n	14	12	11
$\bar{x}$ (years)	66.6	70.8	70.6
± σ n	11.6	8.1	8.4
± σ n − 1	12.0	8.4	8.8

Table 2. Stage distribution of patients studied

Stage	4 × 10 Gy	5 × 8 Gy	8 × 5 Gy
Cervical cancer			
Ib	6/36	7/33	6/36
IIa	1/36	4/33	7/36
IIb	11/36	6/33	5/36
IIIa	2/36	–	–
IIIb	16/36	16/33	18/36
Endometrial cancer			
I	10/14	8/12	7/11
II	4/14	4/12	4/11

3 Results

The achieved uniform sorting of the patients into the investigation groups with respect to age distribution is shown in Table 1. A nearly identical age distribution was obtained with cervical cancer. In the case of endometrial cancer such a distribution could not be achieved. This difference is of no importance for the overall evaluation.

The distribution of disease stages (Table 2) is nearly uniform without consideration of a subgrouping into "a" or "b", so one can assume that a reliable uniform distribution of important prognostic factors exists.

The complications observed in the period from 3 months to 5 years after radiation therapy are more frequent and serious compared with my experiences with conventional brachytherapy (Table 3). The trend is visible that in the 8 × 5 Gy group severe intestine complications occur less frequently. The vesicovaginal fistula which appears once in this group was thought to exist before the radiation treatment as a consequence of conization and completely developed within one year. In the case of endometrical cancer the adverse reactions are by far less frequent or not detectable in the 8 × 5 Gy group.

Table 3. Radiation complications (>3 months)

Complications	4 × 10 Gy	5 × 8 Gy	8 × 5 Gy
Cervical cancer			
Chronic cystitis	1/36	3/33	2/33
Bladder ulcer	2/36	2/33	2/36
Vesicovaginal fistula	–	–	1/36
Proctitis	4/36	5/33	5/36
Intestinal ulcer	–	1/33	3/36
Inflammatory performation	–	1/33	–
Intestine stenosis (ileus)	1/36	2/33	1/36
Rectovaginal fistula	3/36	3/33	1/36
Endometrical cancer			
Bladder	–	–	–
Proctitis	1/14	–	–
Inflammatory stenosis (ileus)	1/14	1/12	–
Inflammatory perforation	–	–	1/11

Dosimetry at the urinary bladder and rectum was estimated via indirect determination by X-radiograms taken at two levels and, since 1984, in parallel with the AM 6 dosimeter. Here it appeared that the calculated values were on average 8% lower than the measured doses, so I must state retrospectively that the tolerance limits were partly exceeded. The relatively rigid rectum probe of the AM 6 proved to be the reason for this. In spite of equal irradiation conditions average dose deviations of up to 31% appeared in the urinary bladder and 26% in the rectum, which underlines the demand for permanent measurement in the bladder and rectum. Local tumor

Table 4. Local tumor control (>6 months to 60 months)

	4 × 10 Gy	5 × 8 Gy	8 × 5 Gy
Cervical cancer			
No tumor	16/36 (44.4%)	21/33 (63.6%)	24/36 (66.7%)
Status progressus	16/36 (44.4%)	5/33 (15.2%)	6/36 (16.7%)
Recidivation	4/36 (11.1%)	7/33 (21.2%)	6/36 (16.7%)
Endometrical cancer			
No tumor	10/14 (71.4%)	8/12 (66.7%)	10/11 (90.9%)
Status progressus	3/14 (21.4%)	2/12 (16.7%)	
Recidivation	1/14 (7.1%)	2/12 (16.7%)	1/11 (9.1%)

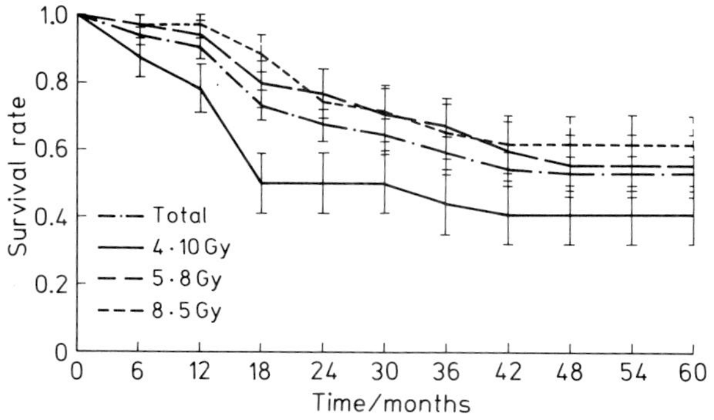

Fig. 1. Survival of patients with cervical cancer

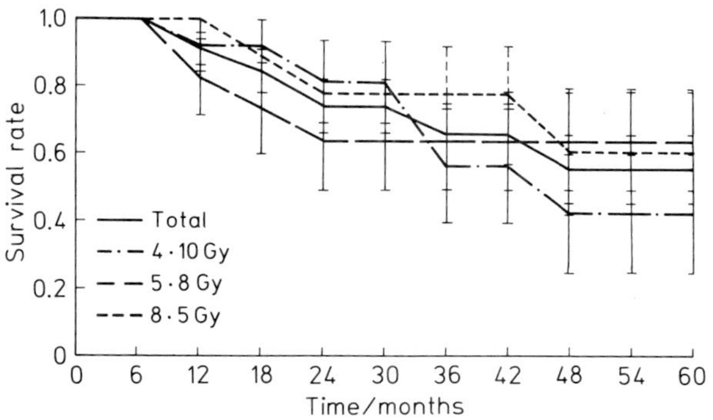

Fig. 2. Survival of patients with endometrial cancer

control as the actual parameter of the therapy efficency (Table 4) shows that for cervical cancer the 4 × 10 Gy group suffered significantly worse results. For endometrial cancer the trend to better results through a higher fractionation number can be recognized.

The results of local tumor control and radiation complications are given in the form of the survival achieved (Figs. 1, 2). For this purpose a life-table chart after a modified log-rank test according to PETO and PIKE (1973) was chosen. Thus, significant differences can be derived from the graphic presentation. With cervical cancer, the resuslts of the 8 × 5 Gy group and the 5 × 8 Gy group are identical. The 4 × 10 Gy group has a significantly worse survival. With endometrial cancer, the best results were obtained with 8 × 5 Gy.

The survival achieved with by employing afterloading therapy, with the exception of the 4 × 10 Gy group with cervical cancer, is as a whole distinctly better than the results gained in this clinic with conventional brachytherapy in the period before 1981 (KLOETZER et al. 1985; SOMMER et al. 1984). With the conventional method a 5-year survival of 54% for patients with stages I to III cervical cancer and of 49% for patients with stages I to II endometrial cancer was obtained.

Table 5. Total doses of high dose rate afterloading therapy expected to be equivalent to conventional brachytherapy (60 Gy/7 days = 30 × 2 Gy/6 weeks) with respect to normal tissue tolerance

Single dose	5 Gy	8 Gy	10 Gy
According to NSD concept	45 Gy	30 Gy	36 Gy
LQ early reaction	48 Gy	40 Gy	36 Gy
LQ late reaction	37 Gy	27 Gy	23 Gy

4 Discussion

For the interpretation of the results and the derivation of practical conclusions I re-examined the single doses with iso-effect relations applied in this study on the basis of the Nominal Standard Dose (NSD) concept and the linear-quadratic (LQ) cellular survival model. The NSD calculation was made according to the formula for the partial tolerance of healthy tissues put forward by GABRIEL-JÜRGENS et al. (1976) in accordance with ELLIS (1963, 1969, 1975).

For the relation of the constants α/β of the LQ model the calculation was based, in analogy to BARENDSEN (1984), on a value of 10 Gy for early reactions and of 2.5 Gy for late reactions. Compared with conventional brachytherapy of 60 Gy over 7 days or percutaneous radiation therapy of 30 × 2 Gy over 42 days, the maximum doses for healthy tissues given in Table 5 can be calculated for the single doses administered in this study. The calculated equivalent total doses for LQ late reactions show that on the assumption of the validity of the LQ model, all the fractionation regimes were dosed too high.

References

Barendsen GW (1984) Differences among tissues with respect to isoeffect, relations for fractionated irradiation. Strahlentherapie 160: 667–669

Ellis F (1963) Dose-time relationship in clinical radiotherapy. In: Raven RW (ed) Cancer progress. Butterworth, London, pp 163–176

Ellis F (1969) Dose, time and fractionation: a clinical hypothesis. Clin Radiol 20: 1–7

Ellis F (1975) Dose modification for low – to high dose-rate intracavitary radiation (dose reduction factor). WHO, Geneva, pp 129–131 (Report RAD 75)

Gabriel-Jürgens P, Gremmel H, Wendhausen H (1976) Die Entwicklung und Anwendung der Nominal Standard Dose für die Toleranzdosis des gesunden Gewebes in der Strahlentherapie. Strahlentherapie 151: 99–112

Glaser FH (1979) Methodische, strahlenbiologische und klinische Aspekte der Kontakt-Curie-Therapie nach dem Afterloading-Prinzip mit hohen Aktivitäten. Dissertation, University of Halle

Glaser FH, Rauh G, Grimm D, Salewski D, Muth CP, Heider KM, Kraft M (1977) Das Decatron-remote-afterloading mit hoher Dosisleistung in der Kontakt-Curie-Therapie. Radiobiol Radiother (Berl) 18: 707–716

Joslin CAF (1978) The Cathetron as part of the radical management of cervix cancer. Br J Radiol [Special Rep] 17: 11–16

Joslin CAF, Smith CW, Mallik A (1972) The treatment of cervix cancer using high activity 60-Co sources. Br J Radiol 35: 257–270

Joslin CAF, Vaishampayan GV, Mallik A (1977) The treatment of early cancer of the corpus uteri. Br J Radiol 50: 38–45

Kloetzer KH, Sommer H, Goetze B, Kob D (1985) Klinik und Therapieergehnisse beim Endometriumkarzinom am Bereich Medizin der Friedrich-Schiller-Universität Jena in den Jahren 1964 bis 1976. Zentralbl Gynäkol 107: 732–737

Peto R, Pike MC (1973) Conservation of the approximation Σ (O − E)^2/E in the Logrank test for survival data or tumor incidence data. Biometrics 29: 579

Rotte K (1975a) Technik, Strahlenbiologie und Ergebnisse der Afterloading-Behandlung gynäkogischer Karzinome. RöntgenBerichle 4: 251–264

Rotte K (1975b) Übersicht über die Afterloadingverfahren in der gynäkologischen Strahlentherapie. Strahlentherapie 150: 237–242

Rotte K (1978) A randomized clinical trial comparing a high dose-rate with a conventional dose-rate technique. Br J Radiol [Special Rep] 17: 75–85

Rotte K (1985) Klinische Ergebnisse der Afterloading-Kurzzeittherapie im Vergleich zur Radiumtherapie. Strahlentherapie 161: 321–328

Sommer H, Kloetzer KH, Kob D, Stech D, Nöschel H (1984) Ergebnisse der Behandlung des Zervixmalignoms am Bereich Medizin der Friedrich-Schiller-Universität Jena in den Jahren 1965 bis 1976. Zentralbl Gynäkol 106: 463–472

Walter W, Günther R, Kob D (1983) Jenaer Steuereinheit für das Afterloading-Gerät DECATRON. Radiobiol Radiother (Berl) 24: 591–602

8.7 Comparison Between High Dose Rate Afterloading and Conventional Radium Therapy

KARSTEN ROTTE

CONTENTS

1 Introduction 311
2 Treatment Planning 311
3 Radiobiological Considerations 311
4 Clinical Results 314
5 Conclusions 316
References 316

1 Introduction

Although there is a certain rivalry in primary treatment of cancer of the cervix between surgical procedures and radiotherapy, intracavitary brachytherapy will continue to play an important role. As an example, in Fig. 1 is shown to what extent the different treatment modes were applied and the shifts that occurred in the Women's Hospital of the University of Würzburg since 1950. On the other hand there is no doubt that the radium application carried out over decades will be completely replaced by remote-controlled afterloading procedures in the near future. Remote-controlled afterloading machines and the necessary applicators have now reached such a level of technical perfection that they meet all requirements even for individual computerized tomography (CT)-based treatment planning (BAUER et al. 1981; BUSCH et al. 1977; FRISCHKORN 1976; GAUWERKY 1977; HENSCHKE et al. 1963; HERBOLSHEIMER 1989; HIMMELMANN and RAGNHULT 1986; LADNER 1989; ROTTE 1981, 1983; VAHRSON 1989; WALSTAM 1977).

2 Treatment Planning

For an exact dose calculation in brachytherapy it is essential to know the position of the applicators within the patient. Therefore, either biplane or stereo X-radiographs have to be taken after each application as shown in Figs. 2 and 3.

Modern therapy planning units with the help of such digitilized X-radiographs enable one to calculate the dose distribution for each patient. This procedure is quite time-consuming though. Therefore, in the daily routine it is advisable to use so-called optimized standard therapy. First, a collection of dose distributions which are used most frequently for the different localizations of gynecological tumors is put together. From this collection the radiotherapist then chooses the applicators which seem most suitable for the patient in question and inserts them into the target volume. In a next step the approximate dose distribution of the applicators is laid over the X-radiographs of these inserted applicators and the dose computed within the target and the critical organs. If necessary, it can be optimized then with the help of a planning system (Figs. 4, 5) (LÖFFLER and van der LAARSE 1988).

3 Radiobiological Considerations

Whereas with classic radium therapy only radioactive sources with limited activity can be applied

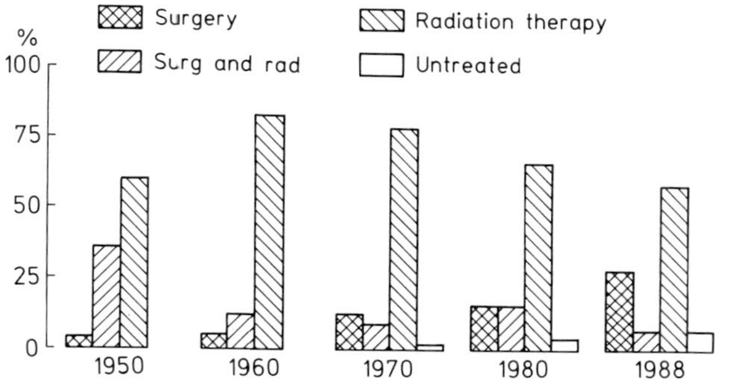

Fig. 1. Percentage of sorts of primary treatment used for carcinoma of the cervix at the Women's Hospital of the University of Würzburg

KARSTEN ROTTE, Prof. Dr. med., Strahlenabteilung, Universitäts-Frauenklinik Würzburg, Josef-Schneider-Str. 4, 8700 Würzburg, Germany

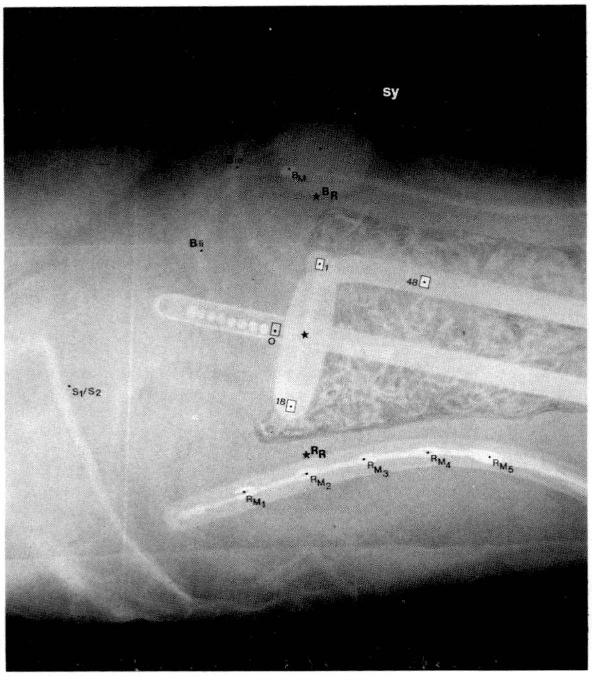

2

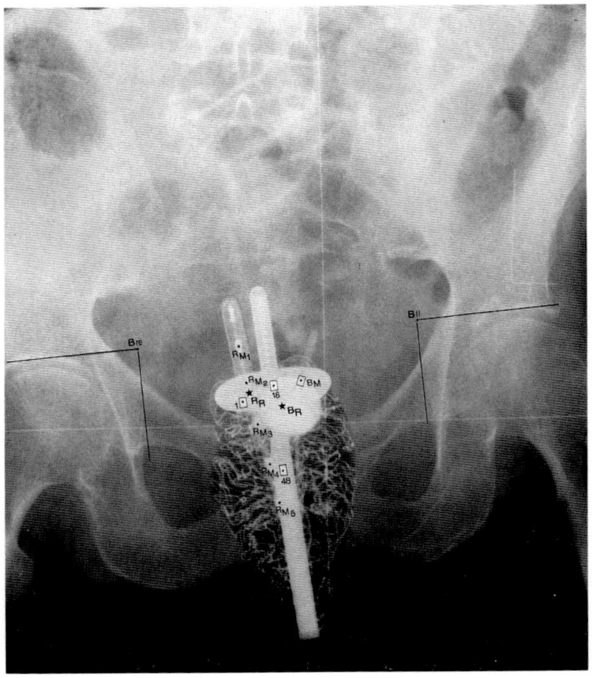

3

Fig. 2. Biplane X-radiograph of applicators in situ. Reference points for calculating the dose distribution according to ICRU report no. 38 are marked on the film

Fig. 3. Biplane X-radiograph of applicators in situ. Reference points for calculating the dose distribution according to ICRU report no. 38 are marked on the film

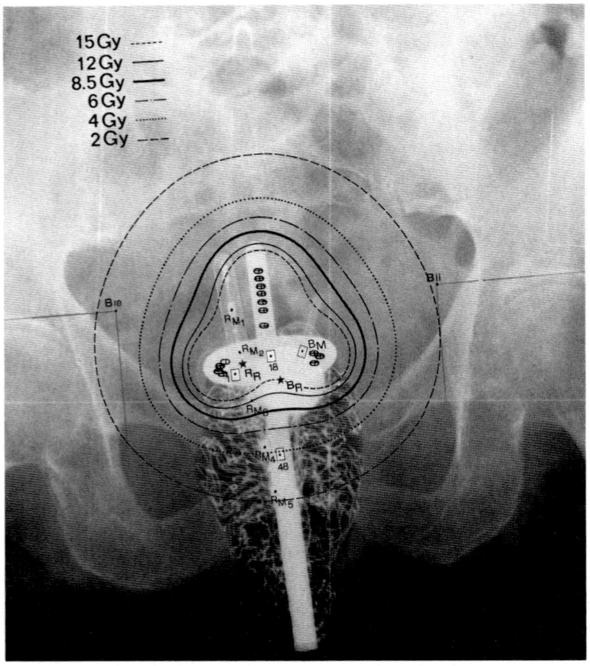

Fig. 4. Biplane X-radiograph of applicators in situ with superimposed dose distribution

for radioprotective reasons (maximal dose rate within the target volume of about 60 cGy/h), the remote-controlled afterloading technique permits the use of radioactive sources with very high activity because of the absolute safety (radioprotection) to staff. A dose rate of 200 cGy/min and more in the tumor can easily be achieved. Radiotherapist thus have a large spectrum of variation in the choice of dose rate at their disposal with which to treat the tumor. From the radiobiological point of view there are three variations possible for remote-controlled afterloading procedures, as shown in Table 1. Although the nomenclature as yet is not uniform in the literature an attempt has been made by the ICRU report no. 38 (Wyckoff 1985). Radiobiologically the medium dose rate presents the least problems since only another radioactive isotope is utilized under *otherwise identical radiobiological conditions* to those in classic radium therapy. The problems encountered with low dose rate procedure *are not of radiobiological nature* because one can draw upon the experiences gained with protracted radium irradiation over many decades, e.g. as carried out in France. They concern rather radiophysics, e.g. geometrical stability of the applicators during treatment. The high dose rate (HDR) procedures, in contrast, have only become possible with the introduction of remote-controlled afterloading machines. It is thus not surprising that despite ex-

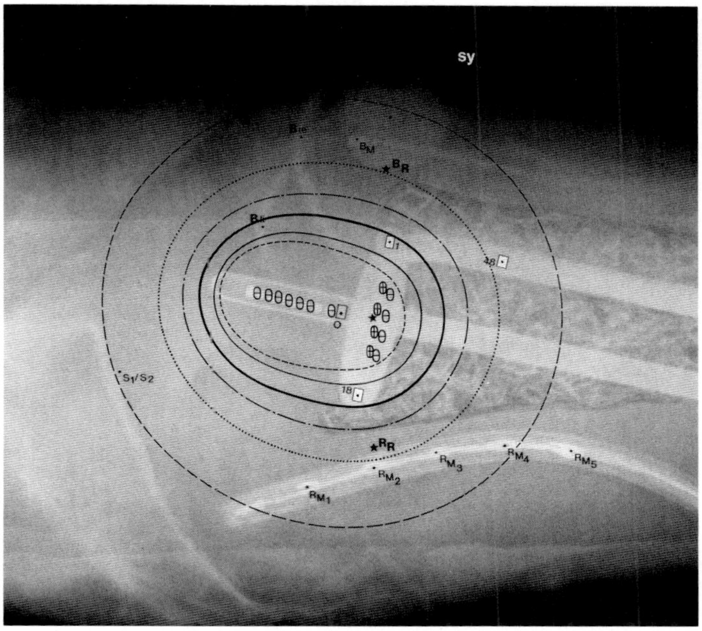

Fig. 5. Biplane X-radiograph of applicators in situ with superimposed dose distribution

Table 1. Definitions of the terms high dose rate, medium dose rate, and low dose rate in brachytherapy

High dose rate:	Fractionated irradiation with a dose rate of 200–300 cGy/min
Medium dose rate:	Fractionated irradiation with a dose rate of 1–2 cGy/min
Low dose rate:	Single protracted irradiation with a dose rate of 0.5 cGy/min or less

tensive radiobiological investigations by, amongst others, COHEN (1980), DALE (1985), ELLIS (1971, 1980), FOWLER (1989), LIVERSAGE (1980), ORTON (1989),TURESSON and NOTTER (1980), and TROTT (1975, 1978), not all questions, particularly for everyday clinical purposes, concerning the comparison of isoeffects between irradiation with high and low dose rates have been finally answered. According to TROTT (1978) the total dose when using HDR irradiation of 2 or more Gy/min should be given in as many fractions as possible. In order to achieve an isoeffect between protracted irradiation with 0.01 Gy/min and fractionated irradiation with 3 Gy/min, a correction factor of 0.5–0.75 should be used, based on his investigations.

KINOSHITA and ONCHO (1980) and ARAI et al. (1980) came to the same conclusion based on treatment results from several radiotherapy centers in Japan from a collective of over 1000 patients (Fig. 6).

LIVERSAGE (1980) has compared the radiation modalities from eight institutions having more than 5-years' experience in intracavitary HDR irradiation with the isoeffect doses from conventional radium treatment as can be calculated from the equations of KIRK et al. (1971), ELLIS (1971), or LIVERSAGE (1969, 1971) (Table 2). It was shown that the total doses irradiated by the majority of these clinics were lower than the equivalent doses calculated from the equations. However, doses calculated by the three radiobiological systems in question differ by 60%–70%. COHEN (1980) came to a similar conclusion from his calculations.

Table 2. Comparison of predicted dose (at high dose rate) vs. equivalent dose (from clinical experience) (from LIVERSAGE 1980)

	Cumulative Radiation Effect (CRE)	Liversage	Time-Dose-Fraction (TDF)
HENSCHKE (a)	1.73	1.27	1.11
(b)	1.69	1.27	1.09
(c)	1.67	1.25	1.07
O'CONNELL	1.60	1.17	1.01
GLASER	2.33	1.58	1.32
ROTTE	1.37	0.91	0.75
GRÖNROOS	2.31	1.59	1.38
SNELLING	1.88	1.39	1.23
KINOSHITA	1.78	1.31	1.09
TOUSSI	1.85	1.29	1.09
MEAN	1.82	1.30	1.11

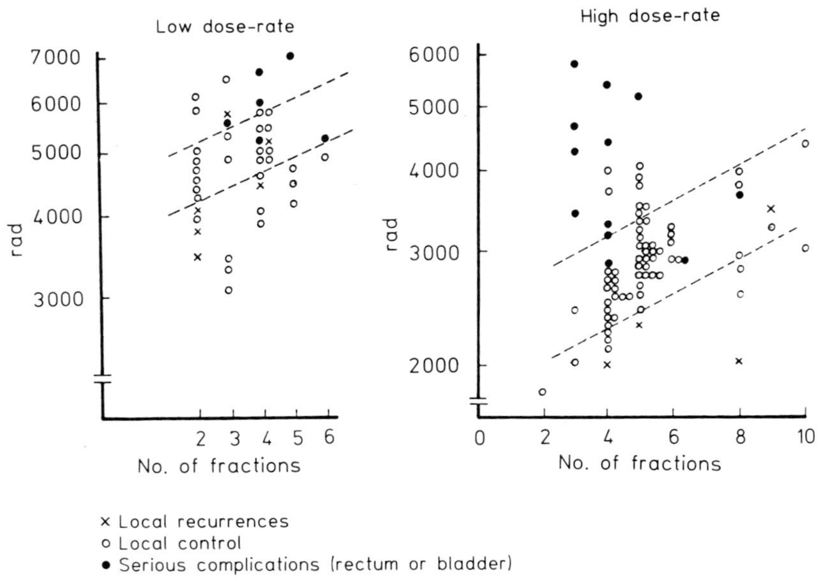

Fig. 6. Results of treatment in various fractions, times, and dosages at point A, comparing high dose rate (*right*) with low dose rate (*left*) (from KINOSHITA and ONCHO 1980)

Despite these discrepancies, today it can be considered certain that the therapeutic index involving HDR treatment is smaller than with a low or medium dose rate therapy (radium). Thus, much more sophisticated treatment planning is necessary. Hereby, special attention must be paid to the balance between intracavitary irradiation and external beam therapy.

4 Clinical Results

In the following, treatment results and side effects encountered of intracavitary HDR afterloading treatment from 1973 until 1988 will be compared with an intracavitary radium treatment carried out in this same clinic during the same time interval. This evaluation comprises altogether 1512 patients. First it should be stressed that every effort was made to keep technique and dose distribution as well as single and over all dosage applied equal in both groups. With intracavitary brachytherapy 54 Gy were applied to point A with radium and 42.5 Gy with HDR afterloading, i.e., a factor of biological dose correction of 0.8 was used based on investigations by LIVERSAGE (1969), ORTON (1989), TROTT (1978) and others. Also, in the beginning, the overall doses in both groups was split into three equal fractions. Although the original equivalent doses to be given were calculated by the equation of LIVERSAGE (1969), it became obvious, especially in stage III tumors, that three fractions of HDR afterloading did not give the same results as three fractions of radium therapy. Therefore, I changed to five equal fractions in HDR afterloading therapy (see Fig. 9).

Percutaneously 50 Gy were given to the parametrium with a linear accelerator or a telecobalt unit using biaxial, bisegmental arc therapy. With this kind of therapy the dose in a region from the midline of the pelvis to point A on both sides was reduced to 25%, i.e., only 12.5 Gy altogether were delivered to this region.

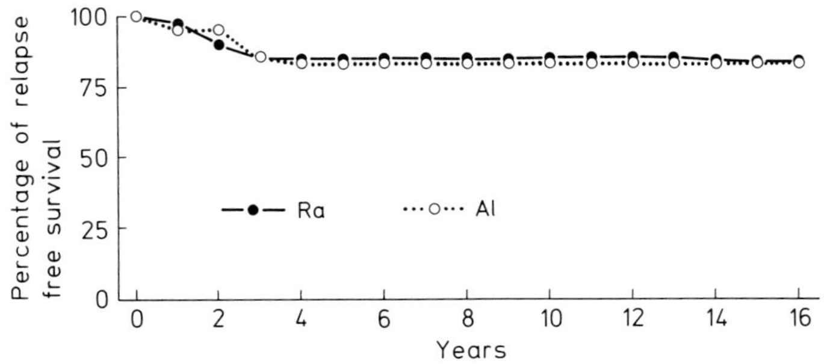

Fig. 7. Radium (*Ra*) vs. high dose rate afterloading (*Al*): actuarial survival of patients with carcinoma of the cervix uteri stage I. Observation time 1973–1988 (n = 462; 238 Ra/224 Al)

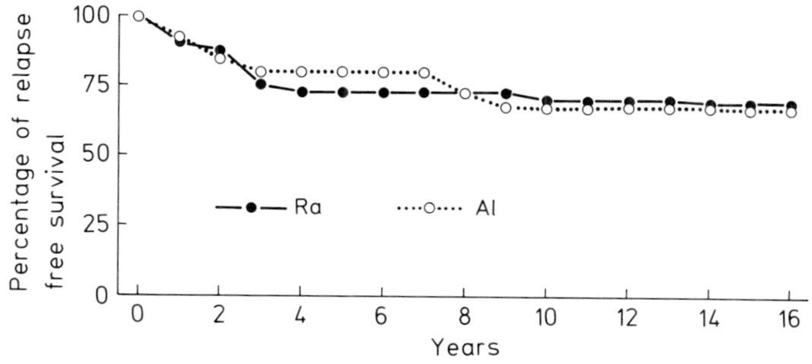

Fig. 8. Radium (*Ra*) vs. high dose rate afterloading (*Al*): actuarial survival of patients with carcinoma of the cervix uteri stage II. Observation time 1973–1988 (n = 560; 308 Ra/252 Al)

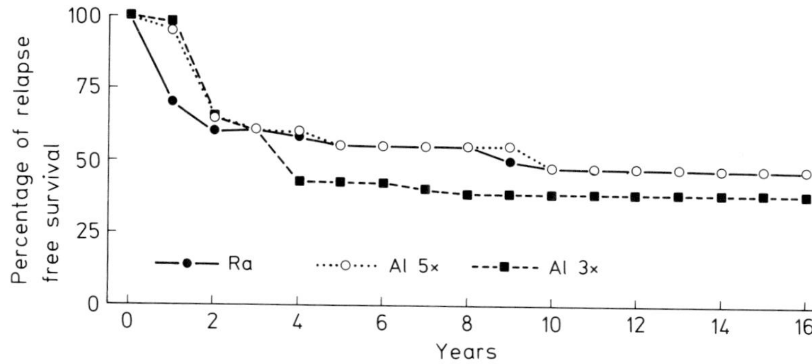

Fig. 9. Radium (*Ra*) vs. high dose rate afterloading (*Al*): actuarial survival of patients with carcinoma of the cervix uteri stage III. Observation time 1973–1988 (n = 490; 224 Ra/168 Al 5× fractions/98 Al 3× fractions)

As can be seen in Figs. 7 and 8, there is no significant difference between radium therapy and HDR afterloading treatment in stages I and II. In addition, no effect of the varied fractionation schemes of HDR afterloading on the results could be observed. On the other hand, stage III tumors showed a distinct difference in the cure rate after a threefold fractionated afterloading treatment compared with a threefold fractionated radium or a fivefold fractionated afterloading therapy (Fig. 9).

I interpret this as follows: The proportion of the hypoxic cell population in the tumor increases with tumor size. The radioresistance of hypoxic tumor cells, however, is greater with HDR than with low dose rate irradiation treatment. The investigations by TROTT (1978) have shown that this relative radioresistance of hypoxic tumor cells can be compensated for by an increased fractionation. The clinical experiences here confirm this.

All patients were examined cystoscopically and rectoscopically prior to and at the end of treatment as well as at 6-monthly intervals over the first 2 years. I observed no increased frequency of side effects with HDR afterloading procedures compared with classic radium treatment. Furthermore, side effects in HDR afterloading decrease if elaborate treatment planning and optimization is carried out (Table 3).

Table 3. Radium (*RA*) vs. high dose rate afterloading (*AL*): percentage of side effects in patients with all stages of carcinoma of the cervix uteri. Observation period 1973–1979: no individual treatment planning was carried out. Observation period 1980–1988: individual treatment planning was carried out for each patient

	RA		AL	
	73–79	80–88	73–79	80–88
Bladder				
Ulcer	2.4%	2.2%	5.7%	1.1%
Flstula	0.0%	0.0%	0.0%	0.0%
Rectum				
Ulcer	11.6%	5.8%	6.9%	1.4%
Flstula	1.2%	0.2%	1.5%	0.0%
All lesions	15.3%	8.3%	14.4%	2.5%

The following authors have also evaluated large patient collectives: GLASER (1986) DDR, JOSLIN et al. (1972), SNELLING et al. (1980), England, FRANKENDAL (1980), Sweden, KAUPPILA and KIVINIITY (1980), TAINA (1981a, b) Finnland, KINOSHITA and ONCHO (1980), ARAI et al. (1980), Japan. They reported similar treatment results and side effects as were encountered in this evaluation.

5 Conclusions

Based on the results of this evaluation and the data published by the other investigators mentioned above I feel that the advantages and problems HDR intracavitary therapy are as follows:

Advantages

1. Excellent reproducibility of the beam geometry in the target volume.
2. Lower risk of embolism for patients at risk by means of short irradiation times and thus only short immobilization.
3. Treatment can be carried out on an outpatient basis.
4. The subjective burden to the patient is less compared with prolonged irradiation with radium.
5. A combined treatment with radiosensitizers is possible.

Problems: At present a number of radiobiological models (LIVERSAGE 1969; KIRK et al. 1971; ELLIS 1971; FOWLER 1989; ORTON 1989) are used to calculate effects based on radiation. One of the most favored models is the so-called linear quadratic model, because only one parameter is to known to describe early or late effects. However, the calculation of isoeffects between HDR treatment and low or middle dose rate therapy is as yet not solved satisfactorily at least from the clinical standpoint. Two phenomena in particular are not clear. First, in HDR afterloading, there is a critical radiobiological potential for overdosing with respect to late complications or underdosing with respect to tumor sterilization by significant ratios of 20%–30% if five large fractions are used. The discrepancy is smaller if larger numbers of smaller fractions are used, but clinical experience, *e.g. treatment results and encountered side effects* accumulated so far does not appear to have made this necessary. Second, the equations published differ in their calculations from each other up to 60%.

Therefore, the clinician using HDR afterloading techniques must keep in mind three points:

1. The therapeutic index of HDR treatment is smaller than with a low dose rate therapy.
2. Physical dose distribution must be calculated very meticuously and careful packing of, e.g., the bladder and rectum away from the vaginal and uterine sources has to be done.
3. The high radioresistance of hypoxic tumor cells to HDR irradiation in comparison with low dose rate therapy must be compensated for by a higher fractionation (at least 5–7 fractions).

References

Arai T, Morita S, Kutsutani Y, Iinuma T (1980) Relationship between total iso-effect dose and number of fractions for the treatment of uterine cervical carcinoma by high dose-rate intracavitary irradiation. Br J Radiol [Special Rep] 17: 89–92

Bauer MD, von Fournier D, Fehrentz F, Kuttig H, zum Winkel K, Neldner F (1981) Afterloading-Methode zur Simulation der intrauterinen Packmethode beim Korpuskarzinom. Strahlentherapie 157: 793–800

Busch M, Makoski B, Schulz U, Sauerwein K (1977) Das Essener Nachladeverfahren für die intrakavitäre Strahlentherapie. Strahlentherapie 153: 581–588

Cohen L (1980) Biological models and computed isoeffect tables for continuous low dose-rate and intermittent fractionated radiation therapy. Br J Radiol [Special Rep] 17: 138–145

Dale RG (1985) The application of the linear-quadratic dose-effect equation to fractionated and protracted radiotherapy. Br J Radiol 58: 515–528

Ellis F (1971) Nominal standard dose and the ret. Br J Radiol 44: 101–108

Ellis F (1980) Low to high dose-rate by the TDF system. Br J Radiol [Special Rep] 17: 146–156

Fowler JF (1989) The linear quardatic formula and progress in fractionated radiotherapy – a review. Br J Radiol 62: 679–694

Frankendal B (1980) Afterloading technique with high dose-rate irradiation of cancer of the uterus. Br J Radiol [Special Rep] 17: 99–101

Frischkorn R (1976) Gynäkologische Strahlentherapie. In: Döderlein G, Wulf KH (eds) Klinik der Frauenheilkunde und Geburtshilfe. Ein Handbuch für die Praxis, vol 2. Urban und Schwarzenberg, München, pp 501–723

Gauwerky F (1977) Kurzzeit-Afterloading-Curietherapie gynäkologischer Karzinome, Technik und Problematik. Strahlentherapie 153: 793–801

Glaser FH (1986) Comparison of HDR afterloading with 192-Ir versus conventional radium therapy in cervix cancer: 5-year results and complications. In: Vahrson H, Rauthe G (eds) High dose rate afterloading in the treatment of cancer of the uterus, breast and rectum. Urban and Schwarzenberg, Mūnich, p 106

Henschke U, Hilaris B, Mahan D (1963) Afterloading in interstitial and intracavitary radiation theapy. AJR 90: 386–395

Herbolsheimer M (1989) Intrauterine packing by remote HDR-afterloading in endometrial carcinoma. In: Rotte K, Kiffer J (eds) Changes in brachytherapy. Wachholz, Nürnberg, pp 130–138

Himmelmann A, Ragnhult I (1986) HDR brachytherapy with improved accuracy: results compared to those of a mannual radium system. In: Vahrson H, Rauthe G (eds) High dose rate afterloading in the treatment of cancer of the uterus, breast and rectum. Urban and Schwarzenberg, Münich p 137

Joslin CAF, Smith C, Mallik A (1972) The treatment of cervix cancer using high activity sources. Br J Radiol 45: 257–270

Kauppila A, Kiviniity K (1980) High dose-rate intracavitary irradiation in the treatment of cervical and endometrical carcinomas: preliminary observations. Br J Radiol [Special Rep] 17: 59–64

Kinoshita M, Oncho K (1980) The use of the Cathetron in the Sawara General Hospital: the first attempt to introduce the Cathetron into Japan. Br J Radiol [Special Rep] 17: 86–88

Kirk J, Gray WM, Watson ER (1971) Cumulative radiation effect. I. Fractionated treatment regimes. Clin Radiol 22: 145–155

Ladner HA (1989) Comparison of the results of primary radiotherapy and radical surgery in cervical carcinoma. In: Rotte K, Kiffer J (eds) Changes in brachytherapy. Wachholz, Nürnberg, pp 99–108

Liversage WE (1969) A general formula for equating protracted and acute regimes of radiation. Br J Radiol 42: 432–440

Liversage WE (1971) A critical look at the ret. Br J Radiol 44: 91–100

Liversage WE (1980) A comparison of the predictions of the CRE, TDF and Liversage formulae with clinical experience. Br J Radiol [Special Rep] 17: 182–189

Löffler E, van der Laarse R (1988) Technique and individual afterloading treatment planning simulating classic Stockholm brachytherapy for cervix cancer. In: Vahrson H, Rauthe G (eds) High dose rate afterloading in the treatment of cancer of the uterus, breast and rectum. Urban and Schwarzenberg, Münich, pp 83–89

Orton CG (1989) Biological aspects of combined radiotherapy, brachytherapy and teletherapy. In: Rotte K, Kiffer J (eds) Changes in brachytherapy. Wachholz, Nürnberg, pp 7–14

Rotte K (1981) Ferngesteuerte Afterloadingverfahren. Strahlentherapie [Special Vol] 76: 313–320

Rotte K (1983) Das ferngesteuerte Nachladeverfahren (Remote-Controlled Afterloading) für die intrakavitäre Kontakttherapie (Brachytherapie) gynäkologischer Karzinome. Radiologe 23: 20–23

Snelling TD, Lambert HE, Yarnold J (1980) Clinical results and complications following treatment of carcinoma of the cervix and endometrium using the Cathetron at the Middlesex Hospital. Br J Radiol [Special Rep] 17: 32–37

Taina E (1981a) Complications following high and low dose-rate intracavitary radiotherapy for stage I-II cervical carcinoma: a comparison of remotely afterloaded Co^{60} (Cathetron) and conventional radium therapy. Acta Obstet Gynecol Scand 103: 59–66.

Taina E (1981b) A comparison of clinical results following high dose-rate intracavitary afterloading irradiation with Co^{60} (Cathetron) and conventional radium therapy for stage I-II endometrial carcinoma. Acta Obstet Gynecol Scand 103: 51–58

Trott KR (1975) Strahlenbiologische Überlegungen bei der Wahl der Dosisleistung in der intrakavitären Strahlentherapie. Strahlentherapie 150: 261–265

Trott KR (1978) Der Einfluß der Dosisleistung auf die therapeutische Wirkung von ^{60}Co-Gammabestrahlung beim Adenokarzinom der Maus. Strahlentherapie 154: 656–658

Turesson I, Notter G (1980) An experimental study of single and fractionated high dose-rate and continuous low dose-rate irradiation on normal tissue. Br J Radiol [Special Rep] 17: 165–177

Vahrson H (1989) Clinical experience with fractionated high dose rate-afterloading brachytherapy in carcinoma of the cervix. In: Rotte K, Kiffer J (eds) Changes in brachytherapy. Wachholz, Nürnberg, pp 108–117

Walstam R (1977) Strahlenphysikalische Voraussetzungen der ferngesteuerten Afterloadbestrahlung. Strahlentherapie 153: 802–806

Wyckoff HO (1985) Dose and volume specification for reporting intracavitary therapy in gynecology. International Commission on Radiation Units and Measurements, Bethesda (ICRU report no 38)

8.8 High Dose Rate Afterloading Compared with Conventional Brachytherapy

FELIX H. GLASER

CONTENTS

1 Introduction . . . 319
2 Treatment Technique and Material . . . 319
3 Treatment Modalities . . . 320
3.1 Carcinoma of the Cervix . . . 320
3.1.1 Primary Intracavitary Treatment . . . 320
3.1.2 Postoperative Intravaginal Treatment . . . 321
3.2 Carcinoma of the Endometrium . . . 321
3.2.1 Primary Intracavitary Treatment . . . 321
3.2.2 Postoperative Intravaginal Treatment . . . 322
4 Results . . . 322
4.1 Carcinoma of the Cervix . . . 322
4.1.1 Primary Intracavitary Treatment . . . 322
4.1.2 Postoperative Intravaginal Treatment . . . 323
4.1.3 Complete Results in Primary and Postoperative HDR Afterloading . . . 324
4.1.4 Frequency of Early and Late Side Effects . . . 325
4.2 Carcinoma of the Endometrium . . . 326
4.2.1 Primary Intracavitary Treatment . . . 326
4.2.2 Postoperative Intravaginal Treatment . . . 327
4.2.3 Complete Results in Primary and Postoperative HDR Afterloading . . . 327
4.2.4 Frequency of Early and Late Side Effects . . . 328
5 Discussion . . . 329
6 Conclusions . . . 330
References . . . 330

1 Introduction

Fractionated short-term afterloading with high dose rates (HDR) prevents the risk of radiation exposure to the staff, facilitates an optimization of dose distribution in space, makes the treatment modalities easier for the patients and hospital, and allows a considerable increase of treatment capacity without additional need of staff or capital. The afterloading principle has introduced a real renaissance in brachytherapy which was unthinkable a few years ago. The field of brachytherapy seemed to become more and more confined in the era of external high-voltage therapy, and it was also questioned by well-known experts for reasons of protection from radiation exposure. However, progress in radiotherapy is impressively reflected in radiotherapy of gynecologic tumors and in particular by brachytherapy. The risk of radiation exposure to the staff stands in the way of all known advantages of brachytherapy, at least in the classic form of radium therapy as it was developed at the beginning of the century. The necessity to develop and introduce afterloading in brachytherapy arose from the need for radiation protection of the staff not from bad results or failures in conventional protracted low dose rate (LDR) brachytherapy. Remote-controlled afterloading has important advantages as well as radiobiological pecularities (Table 1) and opens up the possibility of applying higher active radiation sources with HDR. In contrast to LDR brachytherapy HDR afterloading breaks with the classic principle of protraction in brachytherapy so the fundamentally changed dose-time relation requires a higher fractionation to profit from the advantage of the time factor in repair and recovery of normal healthy tissue connected with adequate tumor control (JOSLIN et al. 1972; NOTTER 1980; SNELLING et al. 1980; GLASER 1980; GLASER et al. 1982, 1985; ROTTE 1983).

By reason of the relatively higher radiobiologic efficiency of up to 60% (JOSLIN et al. 1972; GLASER 1980; ROTTE 1983) in HDR afterloading at a nearly linearly increasing dose-time relationship, the necessary total dosage has to be reduced to the range of 30–50 Gy in contrast to long-term LDR brachytherapy with doses between 60–80 Gy at determined points of reference, e.g., point A in cervical cancer therapy.

FELIX H. GLASER, Prof. Dr. Sc., Klinik und Poliklinik für Radiologie, Medizinische Akademie Erfurt, Nordhäuser Str. 74, 5010 Erfurt, Germany

2 Treatment Technique and Material

In August 1974 HDR afterloading was started at the Clinic and Polyclinic of Radiology of the

Table 1. Advantages and problems in high dose rate afterloading

Advantages	Problems
– radiation protection	– cannot be protracted
– short treatment time	– absolutely changed dose-time relationships
– external applicator fixation	– decreased therapeutic square in contrast to low dose rate
– constant reproducible irradiation geometry	– larger radiobiological efficiency of higher single doses
– psychical and physical salvation	– higher fractionation necessary
– abolition of primary treatment mortality	– adapted overall treatment time
– irradiation planning more exact and flexible	– equivalent dosage to achieve isoeffects
– ambulatory treatment (ca. 40% of all patients)	
– proven efficiency of treatment	
– good rate of healing	
– good compatibility	
– acceptable proportion of side effects	

Martin-Luther University Halle, GRD (GLASER et al. 1977; GLASER and RAUH 1978) with remote-controlled afterloading equipment developed by us, the Decatron (Fig. 1). The Decatron works with a nearly pinpoint iridium-192 source; its source activity lies between 185 and 740 GBq (5–20 Ci), and the dose rates at point A amount up to 3 Gy/min. The isodose shaping is achieved by an electromechanically controlled step-by-step movement of the source. The size of one step ranges from 5 up to 20 mm, usually 10 mm. The metal applicators, straight or curved, are externally fixed to the treatment table, and this guarantees an exact and geometrically determined reproducible spatial dose distribution and represents the basis for the higher fractionated treatment regime. The position of the applicator is controlled by orthogonal X-ray examinations, and the position can be changed according to the treatment planning. All intracavitary applications are carried out without general narcosis, rather only under drug sedation. Due to the short irradiation times of only a matter of minutes, primary treatment lethality in this form of brachytherapy is fully eliminated. In conventional LDR brachytherapy it is about 6%–8% (FRISCHKORN 1971). Over 40% of patients can be treated with HDR afterloading on an outpatient basis, under the care and control of specialists.

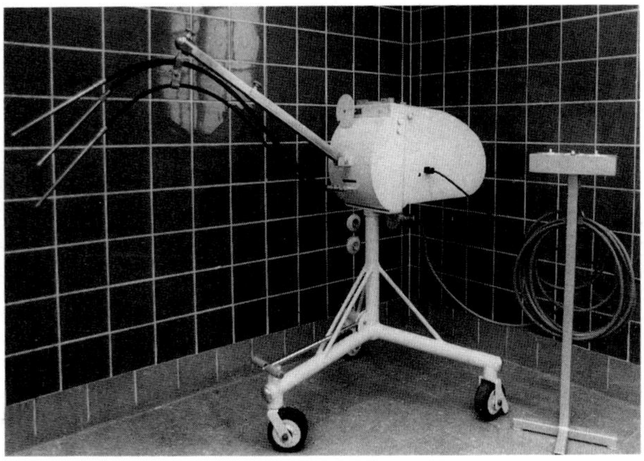

Fig. 1. Remote-controlled afterloading equipment, the Decatron, using ^{192}Ir sources with high activity up to 740 GBq (20 Ci); source transport and isodose shaping done electromechanically via step-by-step movement of the nearly pinpoint source, step size 5–10 mm

From August 1974 to December 1984 more than 3300 patients with gynecologic tumors were treated, among them 1307 carcinomas of the uterine cervix and 928 carcinomas of the endometrium. The referring hospitals were 22 gynecologic clinics and departments of the District of Halle, GDR, among them the Clinic and Polyclinic of Gynecology and Obstetrics of the Martin-Luther University. In cervical carcinoma the histologic classification in more than 90% was variously differentiated squamous cell carcinoma. The histology in endometrial carcinoma showed variously differentiated adenocarcinoma in 96% and sarcomas and other infrequently occurring malignant tumors in 4%.

3 Treatment Modalities

3.1 Carcinoma of the Cervix

3.1.1 Primary Intracavitary Treatment

The treatment regime in primarily irradiated cervical carcinoma is presented in Table 2. It is compared with the conventional methodology of LDR brachytherapy according to modified Stockholm regime. In 5–6 fractions with single doses of 7 Gy per fraction, total doses of 35–42 Gy at the A line were applied. The HDR afterloading applications were given at weekly intervals so the resulting overall treatment time comes to 28–35 days. Additional external beam therapy by cobalt-60 γ-irradiation or 9-MeV photons was carried out simultaneously on the other 4 days a

Table 2. Treatment regime in primary irradiation of cervical carcinoma: comparison of high dose rate afterloading with low dose rate brachytherapy

	Conventional technique (low activities)	Afterloading technique (high activities)
Activity	10–17 GBq ^{137}Cs	185–740 GBq ^{192}Ir
Dose rate at		
– point A	up to 5 cGy/min	up to 3 Gy/min
– point B	up to 1 cGy/min	up to 0.6 Gy/min
Fractions (*n*)	3	5–6
Single dose at		
– point A	15–20 Gy	6–7 Gy
– point B	3–4 Gy	1.2–1.5 Gy
Treatment time	8–12 h	3–30 min
Interval	1–2 weeks	1 weeks
Overall time	2–4 weeks	4–5 weeks
Total dose at		
– point A	45–60 Gy	35–42 Gy
– point B	9–1.2 Gy	6–9 Gy
Plus additional external beam therapy		
20–25 × 1.3 Gy = 26–32.5 Gy at point A		
× 2.0 Gy = 40–50 Gy at point B		
Complete dose at		
– point A	70–86 Gy	61–68 Gy
– point B	50–60 Gy	50–55 Gy

weeks. The summed doses lay between 61 Gy and 68 Gy at the A line and between 50 Gy and 55 Gy at the B line.

3.1.2 *Postoperative Intravaginal Treatment*

The treatment regime in postoperative treatment of cervical carcinoma is presented in comparison with the conventional LDR brachytherapy in Table 3. In 4 fractions single doses between 7.5 and 10 Gy fraction were applied at 10-mm tissue depth above the vaginal stump and at 5-mm tissue depth in the vaginal wall by a lead-shielded single point stand position of the iridium source to total doses of 30–40 Gy at weekly intervals. With stages II and III carcinoma, external beam mega-voltage therapy was given simultaneously in the same manner as in primary treatment. The summed doses to the vaginal stump lay between 60 and 65 Gy and in the area of the pelvic wall at 55 Gy.

Table 3. Treatment regime in postoperative irradiation of cervical carcinoma: comparison of high dose rate afterloading with low dose rate brachytherapy

	Conventional technique (low activities)	Afterloading technique (high activities)
Activity	7–10 GBq ^{137}Cs	185–740 GBq ^{192}Ir
Dose rate at 1-cm tissue depth	up to 4 cGy/min	up to 2.5 Gy/min
Fractions (*n*)	3	4
Single dose	20 Gy	7.5–10 Gy
Treatment time	8–12 h	2–40 min
Interval	1–2 weeks	1 week
Overall time	2–4 weeks	3 weeks
Total dose	60 Gy	30–40 Gy
Plus additional external beam therapy		
25 × 2 Gy = 50 Gy iliacal in stages II and III		

3.2 Carcinoma of the Endometrium

3.2.1 *Primary Intracavitary Treatment*

In primary intracavitary HDR afterloading treatment the whole uterus and the upper third of the vagina were included in the target volume for irradiation. In principle, a combination of intracavitary brachytherapy with external beam therapy is recommended and worth striving for. The spatial dose distribution was chosen in such a way that the target volume was bounded by a reference line which runs laterally at a distance of 2 cm from the middle axis of the uterus through the reference point P-My (myometrium) (Tode and Tode 1984). This spatial dose distribution is to be understood in analogy with the reference point PA in cervical carcinoma according to Todd and Meredith (1938, 1953). The isodose shaping was determined by the time the radiaton source was retained in such a way that the fundus and the edges of the uterine tubes were included in this 100% isodose. This reference line narrowed to about portio width in the caudal direction and ended in the upper third of the vagina.

The primary intracavitary treatment regime of endometrial carcinoma is given in Table 4 and compared with LDR brachytherapy.

In 4–5 fractions single doses between 8 and 10 Gy per fraction were applied to achieve total doses of 40 Gy at the P-My line in combination with external beam therapy or to 50 Gy in HDR afterloading irradiation alone.

The total doses in combined primary irradiation therapy were between 65 and 70 Gy in the target region of the primary tumor and between 50 and 55 Gy in the area of the pelvis wall. In HDR afterloading therapy alone in 5–6 fractions with the same single dosage at intervals of

Table 4. Treatment regime in primary irradiation of endometrial carcinoma: comparison of high dose rate afterloading with low dose rate brachytherapy

	Conventional technique (low activities)	Afterloading technique (high activities)
Activity	5–10 GBq ^{137}Cs 5–8 GBq ^{60}Co	185–740 GBq ^{192}Ir
Dose rate at P-My		
– myometrium	4–7 cGy/min	up to 2.4 Gy/min
– in Px-iliacal	0.9–1.5 cGy/min	up to 0.5 Gy/min
Fractions	3	4–6
Single dose at		
– P-My	13–20 Gy	8–10 Gy
– Px	2.5–4 Gy	1.6–2.0 Gy
Treatment time	8–20 h	4–20 min
Interval	1–2 weeks	1 week
Overall time	2–4 weeks	3–4 weeks
Total dose at		
– P-My	40–60 Gy	40–50 Gy
– Px	7.5–12 Gy	6.4–10 Gy
Plus additional external beam therapy		
20–25 × 1.3 Gy = 26–32.5 Gy at P-My		
× 2.0 Gy = 40–50 Gy at Px		
Complete dose at		
– P-My	70–90 Gy	60–76 Gy
– Px	50–60 Gy	50–60 Gy

3–4 days, total doses of 48–50 Gy were applied, and the overall treatment time was shortened to a matter of 18–20 days.

3.2.2 Postoperative Intravaginal Treatment

Postoperative irradiation treatment of endometrial carcinoma aims at the prevention of metastatic growth on the vaginal stump. It includes external beam therapy of the pelvic area in those patients in whom the myometrium was infiltrated beyond the inner third, lymph node involvement was proven, or metastases existed in the ovaries. Intravaginal HDR afterloading therapy was applied to cover the length of the vaginal stump. The nearly homogeneous reference isodose of 5 mm tissue depth in the vaginal mucous membrane was reached by means of a step-by-step switched source movement and the corresponding times of stay of the nearly pinpoint-like ^{192}Ir source.

It is advisable to distribute the isodose distally in such a way that the 100% reference isodose does not reach deeper than approximately 10 mm above the orificium urethrae externum. Extension towards the introitus vaginae involves the risk of

Table 5. Treatment regime in postoperative irradiation of endometrial carcinoma: comparison of high dose rate afterloading with low dose rate brachytherapy

	Conventional technique (low activities)	Afterloading technique (high activities)
Activity	7–8 GBq ^{137}Cs	186–740 GBq ^{192}Ir
Dose rate in 0.5-cm tissue depth	3–5 cGy/min	up to 2.5 Gy/min
Fractions	3	4
Single dose	1.5–2.0 Gy	7.5 Gy
Treatment time	5–20 h	6–20 min
Interval	1–2 weeks	1 week
Overall time	2–4 weeks	3 weeks
Total dose	45–60 Gy	30 Gy
Plus additional external beam therapy		
25 × 2.0 Gy = 50 Gy iliacal in stages II and III		

an increase of early and late side effects to the vagina due to its increasing radiosensitivity from cranial to caudal (Gauwerky 1971). The treatment regime of HDR afterloading in comparison with LDR brachytherapy is given in Table 5.

In 4 fractions single doses of 7.5 Gy per fraction at 5-mm tissue depth were applied to a total dose of 30 Gy at weekly intervals. The additional external beam therapy with a total dose of 50 Gy in single fractions of 2 Gy per fraction was carried out by biaxial rotation irradiation with telecobalt or with 9-MeV photon stationary fields by a linear accelerator. The radiation exposure to the middle of the pelvis including the vaginal stump was not higher than 30 Gy from external beam irradiation.

4 Results

4.1 Carcinoma of the Cervix

4.1.1 Primary Intracavitary Treatment

The results of combined treatment of 737 patients with primarily irradiated cervical carcinoma stages I–IV, can be seen in Table 6. There is a negative selection regarding age and staging. This becomes especially apparent in the comparatively low survival rates of only about 60% in stages I and II, as many patients (25 of 45 = 55.6% in stage I and 31 of 74 = 41.9% in stage II) died intercurrently and free of tumor. The efficiency of the combined HDR afterloading and megavoltage therapy

Table 6. Five-year results of primary high dose rate afterloading treatment in 737 patients with cervical carcinoma (all stages) 1974–1984 (average age = 61.5 years)

Relapse-free survival after (years)	I(n = 155)		II(n = 215)		III(n = 319)		IV(n = 48)		Total	
	%	n	%	n	%	n	%	n	%	n
1	93.5	(145/155)	91.2	(196/215)	69.6	(222/319)	0	(0/48)	76.4	(563/737)
2	80.6	(104/129)	73.6	(134/182)	55.8	(150/269)			66.9	(388/580)
3	73.5	(75/102)	66.4	(97/146)	48.4	(109/225)			59.4	(281/473)
4	64.4	(56/ 87)	55.3	(63/114)	44.6	(74/166)			52.6	(193/367)
5	57.4	(35/ 61)	61.0	(47/ 77)	42.5	(51/120)			51.6	(133/258)

Control biopsy (303/380) = 79.7% histologically negative
Autopsy (69/113) = 61.1% histologically negative

Table 7. Five-year results of primary high dose rate afterloading treatment in 587 patients with cervical carcinoma (curative treatment) 1974–1984

Relapse-free survival after (years)	I(n = 152)		II(n = 210)		III(n = 225)		Total	
	%	n	%	n	%	n	%	n
1	95.4	(145/152)	93.3	(196/210)	98.7	(222/225)	95.9	(563/587)
2	82.5	(104/126)	75.7	(134/177)	81.1	(150/185)	79.5	(388/488)
3	75.8	(75/ 99)	68.8	(97/141)	69.9	(109/156)	71.0	(281/396)
4	66.7	(56/ 84)	57.8	(63/109)	60.7	(74/122)	61.3	(193/315)
5	60.3	(35/ 58)	65.3	(47/ 72)	57.9	(51/ 88)	61.0	(133/218)

shows itself in stage III patients with a 5-year relapse-free survival rate of 42.5%. In stage IV patients only a palliative or symptomatic HDR afterloading was given to stop bleeding and to abolish pain. All of those patients died within the 1 year of follow-up. In 303 of 380 clinically relapse-free patients, a control biopsy or cytological follow-up after 6–8 months did not reveal any tumor cells. This fact corresponds to a local freedom from tumor of 79.7%, which was proven histologically. In 69 of 113 dead patients (61.1%), autopsy demonstrated local freedom from tumors, too. Up to now, 339 patients have died, among them 81 (24%) intercurrently, 69 (20.4%) free of tumor as seen at autopsy and 189 from cervical carcinoma. Eight women died from secondary tumors of the lungs, bladder, pancreas, or kidneys. In 24 (3.3%) a local relapse occurred. In 12 of these patients (50%) a therapeutic effect was achieved by a second treatment with HDR afterloading with a higher fractionation (up to 10 fractions) with reduced single doses per fraction of 5–6 Gy and at intervals of 3–4 days. A total of 150 women of all 737 could only be treated palliatively or symptomatically due to old age, poor constitution, and advanced staging. For curative aims, therefore, 587 patients could be analyzed. The results in curative HDR afterloading treatment of these 587 women in stages I–III are recorded in Table 7. The relapse-free, 5-year, stage-related survival rates are 60.3% in stage I, 65.3% in stage II, 57.9% in stage III, and 61% for the average in all stages.

4.1.2 *Postoperative Intravaginal Treatment*

The results of postoperative HDR afterloading therapy of cervical carcinoma in 570 patients in all stages I–IV are given in Table 8. The average age in this group lies visibily below that of the primarily irradiated patients. The results according to stage are as follows: 82.6% in stage I, 74.8% in stage II, 44.4% in stage III, and 74.9% average in all stages. All patients in stage IV with distant metastasis died within the 1st year of control. Some 31 women were only irradiated postoperatively with a palliative or symptomatic aim due to poor constitution and secondary diseases. Thus, 539 patients could be analyzed for curative aims. The results of curative postoperative HDR afterloading therapy of these 539 patients are presented in Table 9. The relapse-free 5-year survival rate is 84.0% in stage I, 78.6% in stage II, 60.0% in stage III, and 79.1% for the average in all stages. Up to now 130 women have died,

Table 8. Five-year results of postoperative high dose rate afterloading treatment in 570 patients with cervical carcinoma (all stages) 1974–1984 (average age = 46.6 years)

Relapse-free survival after (years)	I(n = 152)		II(n = 343)		III(n = 71)		IV(n = 4)		Total	
	%	n	%	n	%	n	%	n	%	n
1	98.0	(149/152)	94.7	(325/343)	78.9	(56/71)	0	(0/4)	93.0	(530/570)
2	94.9	(131/138)	88.1	(275/312)	55.4	(36/65)			85.8	(442/515)
3	91.0	(122/134)	81.5	(224/275)	47.1	(24/51)			80.4	(370/460)
4	87.6	(113/129)	77.3	(177/229)	36.6	(15/41)			76.4	(305/399)
5	82.6	(100/121)	74.8	(143/191)	44.4	(12/27)			74.9	(254/339)

Table 9. Five-year results of postoperative high dose rate afterloading treatment in 539 patients with cervical carcinoma (curative treatment) 1974–1984

Relapse-free survival after (years)	I(n = 150)		II(n = 333)		III(n = 56)		Total	
	%	n	%	n	%	n	%	n
1	99.3	(149/150)	97.6	(325/333)	100.0	(56/56)	98.1	(530/539)
2	96.3	(131/136)	90.8	(275/303)	73.5	(36/49)	90.6	(442/488)
3	92.4	(122/132)	84.2	(224/266)	60.0	(24/40)	84.5	(370/438)
4	89.0	(113/127)	80.5	(177/220)	55.5	(15/27)	81.6	(305/374)
5	84.0	(100/119)	78.6	(143/182)	60.0	(12/20)	79.1	(254/321)

Table 10. Complete 5-year rsults of high dose rate afterloading treatment in 1307 patients with cervical carcinoma (all stages) 1974–1984 (average age = 55.0 years)

Relapse-free survival after (years)	I(n = 307)		II(n = 558)		III(n = 390)		IV(n = 52)		Total	
	%	n	%	n	%	n	%	n	%	n
1	95.8	(294/307)	93.4	(521/558)	71.3	(278/390)	0	(0/52)	83.6	(1093/1307)
2	88.0	(235/267)	82.8	(409/494)	55.7	(186/334)			75.8	(830/1095)
3	83.5	(197/236)	76.2	(321/421)	48.2	(133/276)			69.8	(651/ 933)
4	78.2	(169/216)	70.0	(240/343)	43.0	(89/207)			65.0	(498/ 766)
5	74.2	(135/182)	71.3	(191/268)	42.9	(63/147)			65.2	(389/ 597)

18 (13.8%) intercurrently and clinically free of tumor. In 28 patients (4.9%) metastasis into the vaginal stump developed. In 8 of these (28.6%) after a second HDR afterloading treatment with a higher fractionation (up to 8 fractions) and reduced single doses of 5 Gy per fraction over 3–4 day intervals, complete local tumor control was seen.

4.1.3 Complete Results in Primary and Postoperative HDR Afterloading

The results in primary and postoperative HDR afterloading treatment of all 1307 patients with cervical carcinoma are presented in Table 10. According to the stage the relapse-free 5-year survival rate comes to 74.2% in stage I, 71.3% in stage II, 42.9% in stage III, and on average 65.2%. As mentioned above, 181 patients could be treated only palliatively or symptomatically. A total of 1126 patients were irradiated curatively, and these results are given in Table 11. The 5-year, stage-related survival rate is 76.3% in stage I, 74.8% in stage II, 58.3% in stage III, and 72.0% on average for all stages. These results were compared with those of 623 patients with cervical carcinoma, who were treated under conventional LDR brachytherapy from 1970 to 1976 at this clinic (Table 12). The grouping are similar regarding the proportion of primary and postoperative irradiation treatment given and the staging. The results in stage I are rather identical with 76.3% in HDR afterloading and 79.5% in LDR brachytherapy. The results in stage II are with 74.8 and 58.9, respectively, and especially in stage III with 58.3% and 21.6%, respectively, significantly higher, and this improvement is also

Table 11. Complete 5-year results of high dose rate afterloading treatment in 1126 patients with cervical carcinoma (curative treatment) 1974–1984

Relapse-free survival after (years)	I(n = 302)		II(n = 543)		III(n = 281)		Total	
	%	n	%	n	%	n	%	n
1	97.4	(294/302)	95.9	(521/543)	98.9	(278/281)	97.1	(1093/1126)
2	89.7	(235/262)	85.2	(409/480)	79.5	(186/234)	85.0	(830/ 976)
3	85.3	(197/231)	78.9	(321/407)	67.9	(133/196)	78.1	(651/ 834)
4	80.1	(169/211)	72.9	(240/329)	59.7	(89/149)	72.3	(498/ 689)
5	76.3	(135/177)	74.8	(190/254)	58.3	(63/108)	72.0	(388/ 539)

Table 12. Comparison of 5-year results in 1749 patients with cervical carcinoma treated by one of two irradiation techniques (1126 high dose rate afterloading; 623 low dose rate brachytherapy)

	Low dose rate brachytherapy (^{60}Co or ^{137}Cs) 1970–1976 (n = 623; 288 primary, 335 postop.)		High dose rate afterloading (^{192}Ir) Decatron 1974–1984 (n = 1126; 587 primary, 539 postop.)	
Relapse-free survival after 5 years (%)	32.6	64.5	61.0	79.1
Stage I	79.5		76.3	
II	58.0		74.8	
III	21.6		58.3	
Total	49.8		72.0	

Table 13. Early and late side effects on the urinary bladder: comparison of high dose rate afterloading with low dose rate brachytherapy

	Early reactions		Late reactions: Cystitis		Shrinking bladder		Ureter stenosis		Fistula		Total	
	%	n	%	n	%	n	%	n	%	n	%	n
Low dose rate brachytherapy (n = 623)	14.1	88	4.5	28	2.7	17	3.9	24	4.3	27	15.4	96
High dose rate afterloading by Decatron (n = 1307)	2.0	26	0.4	5	0.4	5	0.2	3	0.2	3	1.2	16

shown in the whole collection with 72.0% and 49.8%, respectively (P = 0.001).

4.1.4. Frequency of Early and Late Side Effects

In Tables 13 and 14 the early and late side effects on the bladder, rectum, and intestine in HDR afterloading of cervical carcinoma are described and compared with those observed after conventional LDR brachytherapy. Early reactions of inflammatory character with HDR afterloading were found in only 2.0% as compared with 14.1% with LDR brachytherapy. The frequency of late side effects such as ulceration, shrinking bladder, bleeding, ureter stenosis, and fistula is reduced from 15.4% with LDR brachytherapy to 1.2% with HDR afterloading. Only three vesicovaginal fistulae were observed. In Table 14 the early and late side effects on the rectum and bowel are given. Early reactions such as acute proctitis were found with LDR brachytherapy in 31.6%, which could be decreased to 2.6% by using HDR afterloading. The proportion of late side effects was diminished from 22.7% with LDR brachytherapy to 2.8% with HDR afterloading. These results and the decrease of early and late side effects with HDR afterloading are statistically significant (P = 0.01).

Table 14. Early and late side effects on the rectum: comparison of high dose rate afterloading with low dose rate brachytherapy

	Early reactions		Late reactions								Total	
			Proctitis		Ulcus; bleeding		Stenosis		Fistula			
	%	n	%	n	%	n	%	n	%	n	%	n
Low dose rate brachytherapy ($n = 623$)	31.6	197	10.0	62	2.1	13	4.7	29	6.1	38	22.7	142
High dose rate afterloading by Decatron ($n = 1307$)	2.6	34	1.5	20	0.3	4	0.5	7	0.4	5	2.8	36

4.2 Carcinoma of the Endometrium

4.2.1 Primary Intracavitary Treatment

The average age of the total patient collection with endometrial carcinoma was 62.9 years; that of the primarily irradiated group was 69.2 years, and that of postoperatively treated patients was 59.3 years. This distribution shows a negative selection. The survival rates of 375 patients with primary intracavitary treatment of endometrial carcinoma is presented in Table 15. The follow-up time ranged between 3 and more than 5 years. The absolute 5-year survival rate was 43.4%. However, the above-mentioned negative selection of patients referred for primary radiotherapy sets limits to results from combined HDR afterloading and external beam treatments for biological reasons. In this study only 50 women (13%) could undergo the combined treatment modality; 87% of patients were treated with intracavitary HDR afterloading alone. In 65 patients only palliative or symptomatic intracavitary HDR afterloading therapy with reduced dosage could be carried out. These results could not be analyzed as there was no curative aim. The 5-year survival rate was 52.2% for the 310 patients who were treated curatively with the full dosage as planned. These data alone do not explain the radiobiological effectiveness of HDR afterloading as 5-year survival rates in patients with primarily treated endometrial carcinoma must be evaluated with reservation due to the known negative selection. Many of the elderly women died within the 5-year follow-up clinically free of tumor (Alth et al. 1973; Mayer 1974; Keller et al. 1976). Local effectiveness of HDR afterloading becomes evident in the clinically and histologically proven tumor-free higher survival within the first 2 years of follow-up. After the 1st year 99% of the curatively treated patients survived in contrast to only 81.9% for the total sample. After the 2nd year 81.6% of the curatively treated patients survived in contrast to only 67.5% of all treated women.

Table 15. Five-year results of primary high dose rate afterloading treatment in 375 patients with endometrial carcinoma (all stages) and in 310 for curative treatment (1974–1984)

Relapse-free survival (%) after (years)	All stages ($n = 375$)		Curative treatment ($n = 310$)	
	n	%	n	%
1	307/375	81.9	307/310	99.0
2	253/375	67.5	253/310	81.6
3	215/375	57.3	215/310	69.4
4	156/310	50.3	156/256	60.9
5	109/251	43.4	109/209	52.2

Control bropsy: 209/310 = 67.4% histologically negative
Autopsy : 60/197 = 30.5% locally free of tumor

Patients died : 197
– from basic disease : 104
– from secondary tumors : 8
– intercurrently : 85

These differences are statistically significant ($P = 0.001$). In 209 of 310 (67.4%) curatively irradiated patients a control biopsy was carried out after 6–8 months which did not reveal any tumor cells. In 22 patients persisting tumor tissue was detected, and another intracavitary HDR afterloading therapy with the same dosage was performed. In 13 patients the progression of the tumor was stopped. Marked side effects on the bladder or rectum did not occur. In 3 patients with relapsing genital bleeding and contraindications to surgery a third HDR afterloading therapy with reduced dosage for symptomatic hemostasis was successfully carried out. A total of 197 patients died within the follow-up period. In 60 (30.5%)

Table 16. Five-year results of postoperative high dose rate afterloading treatment in 553 patients with endomentrial carcinoma (1974–1984)

Relapse-free survival (%) after (years)	I ($n = 278$)		II ($n = 239$)		III + IV ($n = 36$)		Total ($n = 553$)	
	n	%	n	%	n	%	n	%
1	(267/278)	96.0	(229/239)	95.8	(27/36)	75.0	(523/553)	94.6
2	(259/278)	93.2	(210/239)	87.9	(25/36)	69.4	(494/553)	89.3
3	(249/278)	89.6	(194/239)	81.2	(22/36)	61.1	(465/553)	84.1
4	(199/234)	85.0	(171/216)	79.2	(20/33)	60.6	(390/483)	80.7
5	(165/203)	81.3	(151/196)	77.0	(14/26)	53.8	(330/425)	77.6

autopsy demonstrated local freedom from tumor. Some 104 women died from their basic disease and 8 from secondary tumors outside the genital system; 85 patients died intercurrently and free of tumor.

4.2.2 Postoperative Intravaginal Treatment

The survival rates of postoperative intravaginal treatment of endometrial carcinoma in 553 patients are presented in Table 16. All patients were followed up from 3 to more than 5 years.

The relapse-free, 5-year survival rates lay between 81.3% and 53.8% depending on the staging (77.6% on average). The small number of only 36 patients with stages III and IV disease with ovarian metastases or metastatic spread in the pelvis cannot be evaluated statistically. So far 114 women have died; 25 of them (21.9%) died intercurrently and clinically free of tumor, and 13 (11.4%) died from a secondary carcinoma outside the genital system while clinically tumor-free as regards the endometrial carcinoma.

In 5 patients (1%) neoplasma on the vaginal stump were detected. In 3 patients with filtration of the adenocarcinoma, the metastatic spread can be assumed to be due to the basic disease. Two patients suffered from squamous cell carcinoma after 5 and 6 years of follow-up, which is considered a secondary carcinoma of the vagina. All of these 5 patients were treated again with HDR afterloading. The total doses lay between 40 and 50 Gy, the single doses between 5 and 6 Gy, with fractionation intervals of 3–4 days. The progression of metastases due to the adenocarcinoma could not be stopped. The 2 women with squamous cell tumors are clinically free of tumor in the 4th and 5th years after relapse irradiation.

4.2.3 Complete Results in Primary and Postoperative HDR Afterloading

Treatment results of the total of all 928 evaluated patients with endometrial carcinoma are recorded in Table 17. The overall 5-year survival rate lies at approximately 65%. Curative treatment vs. palliative and symptomatic irradiation alone must be regarded under the aspects of primary HDR afterloading therapy given above.

In Table 18 the results of HDR afterloading therapy are compared with those achieved by irradiation with conventional LDR brachytherapy by the packing method. This comparison is made by means of historic case material of 289 patients who were treated from 1972 to 1978 in this clinic. The two groups are comparable according to age, staging, and proportion receiving primary or postoperative therapy. The classification of patients into one of the two techniques was only a question of organization, purely accidental, without defined selection criteria, as both methods were applied parallel to each other from 1974 to 1978. Both in primary intracavitary radiotherapy with 43.4% as compared with 30.7% and in postoperative intravaginal treatment with 77.6% as

Table 17. Complete 5-year results of high dose rate afterloading treatment in 928 patients with endometrial carcinoma (all stages) and in 853 for curative treatment (1974–1984)

Relapse-free survival (%) after (years)	All stages ($n = 928$)		Curative treatment ($n = 853$)	
	n	%	n	%
1	(830/928)	89.4	(830/853)	97.3
2	(747/928)	80.5	(747/853)	87.6
3	(680/928)	73.3	(680/853)	79.7
4	(546/793)	68.9	(546/730)	74.8
5	(439/676)	64.9	(439/629)	69.8

Table 18. Comparison of 5-year results in 1217 patients with endometrial carcinoma treated by one of two irradiation techniques (928 high dose rate afterloading; 289 low dose rate brachytherapy)

	Low dose rate brachytherapy (1972–1978) (^{60}Co, ^{137}Cs) ($n = 289$)		High dose rate after afterloading (1974–1984) (^{192}Ir, Decatron) ($n = 928$)	
	postop ($n = 201$)	primary ($n = 88$)	postop ($n = 553$)	primary ($n = 375$)
Relapse-free survival after 5 years (%)	75.1	30.7	77.6	43.4
Total (%)	61.6		64.9	

Table 19. Early and late side effects on the urinary bladder: comparison of high dose rate afterloading with low dose rate brachytherapy

	Early reactions		Late reactions: Cystitis		Shrinking bladder		Ureter stenosis		Fistula		Total	
	%	n	%	n	%	n	%	n	%	n	%	n
Low dose rate brachytherapy ($n = 289$)	18.7	54	2.8	8	1.3	4	2.1	6	1.7	5	7.9	23
High dose rate afterloading ($n = 928$)	1.6	15	1.6	15	0.3	3	0	0	0.4	4	2.3	22

Table 20. Early and late side effects on the rectum: comparison of high dose rate afterloading with low dose rate brachytherapy

	Early reactions		Late reactions: Proctitis		Ulcus; bleeding		Stenosis		Fistula		Total	
	%	n	%	n	%	n	%	n	%	n	%	n
Low dose rate brachytherapy ($n = 289$)	30.8	89	10.4	30	3.8	11	2.8	8	2.8	8	19.8	57
High dose rate afterloading ($n = 928$)	3.1	29	4.6	43	0.3	3	0.6	6	0.7	7	6.3	59

compared with 75.1% the 5-year survival rates increased with HDR afterloading and showed an improvement with 64.9% as compared with 61.1% for the total sample. However, these differences are not statistically significant.

4.2.4 Frequency of Early and Late Side Effects

Early and late side effects on the bladder, rectum, and vagina which occurred during the application of HDR afterloading and the follow-up period were compared with the incidence rate of side effects in 289 patients who were treated with brachytherapy between 1972 and 1978. Early reactions on the bladder which required additional treatment or interruption of therapy occurred with LDR brachytherapy in 18.7% and decreased with HDR afterloading to 1.6% (Table 19). The proportion of late side effects was decreased from 7.9% to 2.3%, respectively, and only four fistulae arose with the latter. Early side effects on the rectum and sigmoid colon which demanded additional treatment or treatment interruption were observed with LDR brachytherapy in 30.8% of all patients, which decreased with HDR afterloading to 3.1% (Table 20). The frequency of late reactions in HDR afterloading as compared with LDR brachytherapy decreased from 19.8% to 6.3%, respectively.

In both patient groups the incidence of inflammatory reactions of the vagina was dose-dependent

Table 21. Frequency (%) of early and late side effects on the bladder and rectum according to single and total radiation exposure

Dose (in Gy)	Bladder		Rectum	
	early	late	early	late
Single exposure				
−7.5	1.0	–	0.3	–
−10	3.9	–	10.3	3.1
>10	22.9	5.4	36.2	28.9
Total exposure				
−20	–	–	–	–
−30	7.4	–	13.2	5.7
−40	16.5	3.1	23.1	10.8
>40	30.0	13.0	86.7	73.3

and determined by the chosen reference isodose from proximal to distal. It is advisable to limit the target volume in intravaginal HDR afterloading to a tissue depth of about 5 mm in the vaginal mucous membrane and not deeper than 10 mm above the orificium urethrae externum. In a treatment modality of 4 fractions with 7.5 Gy per fraction and a total dosage of 30 Gy, the incidence of early reactions on the mucosa is 1% and is influenced by individual factors, such as superinfection, diabetes, and Kraurosis vulvae. Stenosing late reactions can be avoided by local treatment with antiphlogistic agents.

A statistically significant dose dependence as regards the occurrence of early and late side effects on the bladder and rectum was observed in HDR afterloading therapy of endometrial and cervical carcinoma (Table 21). The increase was statistically highly significant for single radiation exposure of more than 10 Gy per fraction and total radiation exposures of more than 35–40 Gy ($p = 0.001$). It is obvious that the quantitatively higher incidence of reactions of the rectum in comparison with the bladder is volume-dependent (Glaser et al. 1988).

5 Discussion

The comparison of these results with those of the international statistics in the annual report, vol. 19 (results from 1976–1978) is given in Tables 22 and 23. As regards cervical carcinoma (Table 22) at the Clinic and Polyclinic of Radiology of the Martin-Luther University Halle, GDR, in stage I the 5-year survival rate is 73.9%, which is in keeping with the international average. It is 67.0% in stage II, which is above the international range, and 42.6% in stage III, which is better than the standard; the average (64.1%) for all treated patients is also high on the international scale.

In endometrial carcinoma the unselected 5-year results (Table 23) are with 439 of 676 patients (64.9%) in an international top position and comparable with those from Graz, Helsinki, Los Angeles, and Warsaw, and better than the results from Munich, Vienna, and Würzburg. The methodical advantages of fractionated HDR, short-term afterloading include external fixation of the applicators, thus creating the basis for a repeatable, fractionated, radiobiologically founded treatment regime. The exact, geometrically determined, and reproducible spatial dose distribution

Table 22. Five-year results in treatment of cervical carcinoma: comparison of high dose rate afterloading with international statistics in the annual report, vol. 19 (1976–1978)

Institution	*n*	I	II	III	IV	Total	
		(%)	(%)	(%)		(%)	(*n*)
UFK Innsbruck	318	90.4	69.6	48.9	1/14	66.4	211
IFK Vienna	497	83.0	56.0	29.5	4/33	47.1	234
Toronto	589	76.9	57.8	35.2	3/25	55.7	328
Prague	333	78.0	58.0	27.9	0/ 6	54.7	182
UFK Helsinki	327	82.0	54.5	26.8	0/ 1	62.4	204
Bordeaux	233	92.5	69.8	36.4	2/14	63.5	148
UFK Leipzig	626	85.5	48.0	19.1	1/11	67.3	421
UFK Freiburg	346	80.3	62.8	25.8	1/15	60.4	209
I.FK Munich	354	80.8	53.8	33.7	3/19	54.8	194
Manchester	1053	74.1	49.5	23.8	6/79	44.6	470
Oslo	1022	85.4	56.9	30.9	6/55	66.1	676
Minneapolis	178	83.5	53.3	5.3	1/ 7	65.2	116
Personal data (1976–1978)	245	73.9	67.0	42.6	0	64.1	157

Table 23. Five-year results in treatment of endometrial carcinoma: comparison of high dose rate afterloading with international statistics in the annual report, vol 19 (1976–1978)

Institution	*n*	All stages (%)	Survivals
Graz	146	63.0	92
Helsinki	253	67.2	170
Los Angeles	96	62.5	60
Munich, I.Fk	226	54.4	123
Oslo	678	73.2	496
Rostock	114	50.9	58
Stockholm	541	74.7	404
Warsaw	193	66.8	129
Vienna, I.Fk	312	51.6	161
Würzburg	170	53.5	91
Personal data	676	64.9	439

within the reference points from the axis of the uterus and area of the pelvis (Tode and Tode 1984) offers advantages as compared with conventional, protractedly fractionated LDR brachytherapy in combination with external beam therapy. It contributes essentially to a decrease in the incidence of early and late side effects in the critical tissues and neighboring organs of the bladder and rectum. The modifications possible to the target volume in the region of the primary tumor in the uterus which is to be treated by HDR afterloading is of decisive importance (Glaser et al. 1988). This applies to endometrial as well as to cervical cancer: In endometrial carcinoma with an increased accentuation in the fundus region and the edges of the tubes, including the whole uterus with the upper third of the vagina, the conventional "pear-shaped" isodose configuration is not longer necessary as in the era of high-voltage external beam therapy. Historically, in the era of conventional X-ray therapy the insufficient percutaneous dosage was supplemented from inside in the sense of "utilizing the available space" (Gauwerky 1971). These difficulties do not exist with external high-voltage therapy. As the isodose configuration in external beam therapy runs parallel to the body axis except in conformation irradiation, the use of the pear-shaped isodose configuration leads to zones of excessive dose summation with hot spots and to an increase in the incidence of side effects. In contrast to the pear-shaped isodose configuration, my isodose configuration proceeds in a more elongated form (Fig. 2), comparable with a corncob with the broader part at the height of the fundus region in endometrial carcinoma and at the level of the portio in cervical cancer (Glaser et al. 1985).

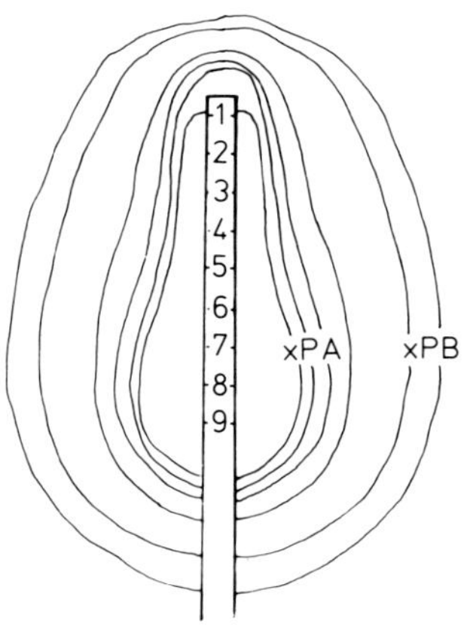

Fig. 2. Isodose configuration like a corncob in high dose rate afterloading treatment in cervical cancer

6 Conclusions

The presented long-term results of fractionated HDR afterloading therapy in 1307 patients with cervical carcinoma and 928 patients with endometrial carcinoma treated between 1974 and 1984 prove the high radiobiological effectiveness of this kind of intracavitary and intravaginal brachytherapy. The methodologic advantages, the adjustable spatially and radiobiologically founded (Glaser et al. 1984) time-dose distribution, render the significant decrease in early and late side effects possible.

References

Alth G, Schwert J, Wickenhauser I, Bremig K, Fucik F (1973) Bericht über 40 Jahre kurative Strahlentherapie des Carcinoma corporis uteri. Strahlentherapie 146: 523–529.

Frischkorn R (1971) Die Strahlenbehandlungdes Korpuskarzinoms. In: Diethelm L, Olsson O, Strnad F, Vieten H, Zuppinger A (eds) Spezielle Strahlentherapie maligner Tumoren. Springer, Berlin Heidelberg New York, p 71 (Handbuch der medizinischen Radiologie, vol 19/3)

Gauwerky F (1971) Tumoren der Vagina. In: Diethelm L, Olsson O, Strnad F, Vieten H, Zuppinger A (eds) Spezielle Strahlentherapie maligner Tumoren. Springer, Berlin Heidelberg New York, p 17 (Handbuch der medizinischen Radiologie, vol 19/3)

Glaser FH (1980) Decatron remote after-loading therapy

with high activity sources. Br J Radiol [Special Rep] 17: 51

Glaser FH, Rauh G (1978) Neue Aspekte in der Kontakt-Gammatherapie. DECATRON – ein ferngesteuertes afterloading mit hohen Aktivitäten. Dtsch Gesundheitswes 33: 2155–2157

Glaser FH, Rauh G, Grimm D, Salewski D, Muth CP, Heider KM, Kraft M (1977) Das DECATRON – remote afterloading mit hoher Dosisleistung in der Kontakt-Curie-Therapie. Radiobiol Radiother (Berl) 18: 707–716

Glaser FH, Grimm D, Hänsgen G, Rauh G, Heider KM, Kraft M, Salewski D, Schuchardt V (1982) Fraktioniertes Kurzzeit-Afterloading mit hohen Dosisraten bei der Behandlung von gynäkologischen Tumoren. Radiobiol Radiother (Berl) 23: 481–496.

Glaser FH, Grimm D, Heider KM (1984) Zum Einfluß zeitlicher Dosisverteilung bei protrahierter und fraktionierter Brachytherapie gynäkologischer Tumoren. Mathematisch formulierte Modellvorstellungen und klinische Erfahrungen beim fraktionierten Kurzzeit-Afterloading mit hohen Dosisraten. Radiobiol Radiother (Berl) 25: 231–240.

Glaser FH, Grimm D, Hänsgen G, Rauh G, Schuchardt V (1985) Klinische Erfahrungen bei der Afterloading-Kruzzeittherapie im Vergleich zur konventionellen Brachytherapie bei der Behandlung gynäkologischer Tumoren. Strahlentherapie 161: 459–475

Glaser FH, Grimm D, Hänsgen G, Rauh G, Schuchardt V (1988) High-dose-rate Afterloading in der primären und postoperativen Brachytherapie beim Endometriumkarzinom: Methodik, Ergebnisse, Begleit- und Folgereaktionen. In: Hammer J, Kärcher DH (eds) Fortschritte in der interstitiellen und intrakavitären Strahlentherapie. Zuckschwerdt, Münich p 211

Joslin CAF, Smith CW, Mallik A (1972) The treatment of cervix cancer using high activity 60-Co sources. Br J Radiol 45: 257–270

Keller H, Nöcker D, Hering K, Soweidan S (1976) Bericht über 4468 bestrahlte Uteruskarzinome. Strahlentherapie 151: 208–213

Mayer A (1974) Strahlenbehandlung der klinisch inoperablen Korpuskarzinome. Strahlentherapie 148: 53–56

Notter G (1980) Clinical comparison of different regimes of fractionation and dosage. Br J Radiol [Special Rep] 17: 190

Rotte K (1983) Das ferngesteuerte Nachladverfahren (Remote-Controlled Afterloading) für die intrakavitäre Kontakttherapie (Brachytherapie) gynäkologischer Karzinome. Radiologe 23: 20–23

Snelling MD, Lambert HE, Yarnold J (1980) Clinical results and complications following treatment of carcinoma of the cervix and endometrium using the Cathetron at the Middlesex Hospital. Br J Radiol [Special Rep] 17: 32

Todd MC, Meredith WJ (1938) A dosage system for use in the treatment of cancer of the uterine cervix. Br J Radiol 11: 809–824

Todd MC, Meredith WJ (1953) Treatment of cancer of the cervix uteri – a revised "Manchester-method". Br J Radiol 26: 252–257

Tode G, Tode D (1984) Standardisierung von Referenzpunkten für die gynäkologische Strahlentherapie. Zentralbl Gynäkol 106: 620–623

8.9 Low and High Dose Rate Afterloading in Gynecological Malignancies

D. von Fournier, H.W. Anton, H. Junkermann, and G. Wolf

CONTENTS

1 Introduction . . . 333
2 Isotopes for Afterloading Brachytherapy . . . 333
3 Applicators for LDR and HDR Afterloading . . . 333
4 Standard Treatment and Variations . . . 336
5 Comparison of LDR and HDR Afterloading Treatment . . . 336
5.1 Clinical Trials . . . 338
References . . . 339

1 Introduction

Low dose rate (LDR) afterloading with a remote-controlled afterloading system is replacing the well-tried brachytherapy with radium of the past decades. Using an equivalent geometrical set-up and an equivalent radioactive source, results equally as good as in radium therapy can be achieved. It has been possible with the afterloading system to introduce sources of a higher dose rate (HDR), that is, short-term irradiation in terms of minutes.

However, the fact still has to be established as to whether, as far as therapy results and side effects are concerned, it is as effective as LDR treatment.

2 Isotopes for Afterloading Brachytherapy

Table 1 shows the afterloading systems in use for gynecological malignancies in Heidelberg (University Women's Hospital and University Radiological Clinic) since 1975. At present the following are used: The LDR Selectron with ^{137}Cs, the HDR Selectron with ^{60}Co, and the micro-Selectron which can be used for HDR (^{193}Ir) or LDR afterloading (either ^{192}Ir or ^{137}Cs). The micro-Selectron can be used for intracavitary as well as interestitial treatment. For LDR afterloading the ^{137}Cs isotope is preferred. In the meantime very small active ^{137}Cs sources have been devised for the micro-Selectron, making interstitial therapy with LDR possible. For HDR afterloading either ^{60}Co or ^{192}Ir is used, the advantage of the latter being that the source has an extremely small diameter. A disadvantage of ^{192}Ir is its short half-life.

Figure 1 shows that by using the small geometrical setup and an equivalent activity, much the same dose distribution can be achieved with ^{137}Cs afterloading as with conventional ^{226}Ra. Figure 2 shows that in LDR afterloading with ^{137}Cs the dose rate to the points of interest in the pelvis is comparable with that of conventional radium therapy.

3 Applicators for LDR and HDR Afterloading

Figure 3 shows a selection of standard applicators used in Heidelberg. In the treatment of carcinoma of the cervix an intrauterine tube and tandem (2) or an intrauterine tube with a ring applicator (3) can be used. The ring applicator simulates the "Stockholm" method.

Figure 4 shows a vaginal applicator consisting of a template with eight peripheral needle sources. The needles can be used with HDR as well as LDR afterloading for interstitial therapy.

Figure 5 shows an "umbrella" applicator for intrauterine irradiation in the case of endometrium carcinoma. The catheter can be straightened within an outer tube. The applicator should be used with ^{192}Ir (HDR) or ^{137}Cs (LDR).

D. von Fournier, Prof. Dr. med., H.W. Anton, Dr., H. Junkermann, Dr., and G. Wolf, Dipl.-Phys., Radiologische Universitätsklinik, Abteilung Gynäkologisch-geburtshilfliche Radiologie, Vosstr. 9, 6900 Heidelberg, Germany

Table 1. Remote-controlled afterloading units used in Heidelberg (1975–1989)

Used since	Unit	Channels	Radionuclide, activity per single source	Active sources	Variability of source distribution	Transport of sources	Treatment schedule
1975	Curietron	4–8	^{137}Cs 150 mCi (5.65 BGq)	linear sources fixed	+	mechanical	LDR 3 × 20 h
1987	Gammamed II	1	^{192}Ir 10 Ci (370 GBq)	1 source stepwise movement	++	mechanical	HDR 3–6 × 0.5 h
1980	Selectron LDR	6	^{137}Cs 2.5 Ci (93 GBq)	36 pellets	+++	pneumatic	LDR 3 × 20 h
1986	Selectron	3	^{192}Ir 20 Ci (740 GBq) ^{60}Co 10 Ci (370 GBq)	36 pellets	+++	pneumatic	HDR 6–7 × 0.5 h
1986	Micro-selectron HDR and LDR	1	^{192}Ir 20 Ci (740 GBq) [or ^{137}Cs]	15 linear sources fixed	+++	mechanical	HDR 0.5 h LDR 20–40 h (for intracavitary or interstitial use)

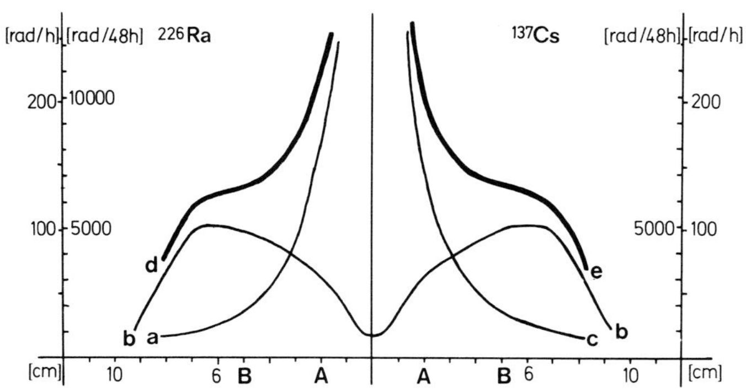

Fig. 1. Dose distribution in frontal view through the mid-pelvic area. *Left*, conventional brachytherapy with ^{226}Ra. *Right*, low dose rate remote-controlled afterloading techniques with ^{137}Cs. For both techniques as equivalent dose distribution can be achieved

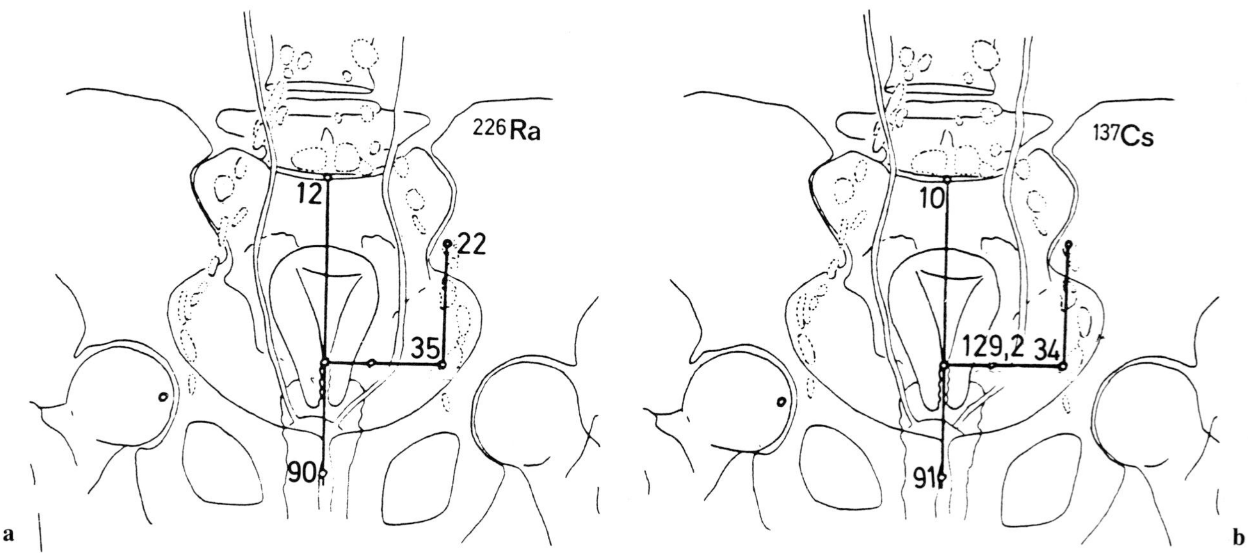

Fig. 2a,b. Frontal view of the pelvis in conventional radium brachytherapy (**a**) and in ^{137}Cs afterloading technique (**b**). Dose rates in cGy/h shown at points of interest for treatment planning (promotorium, mid-vagina, point A, B, etc.); they are very similar for both methods

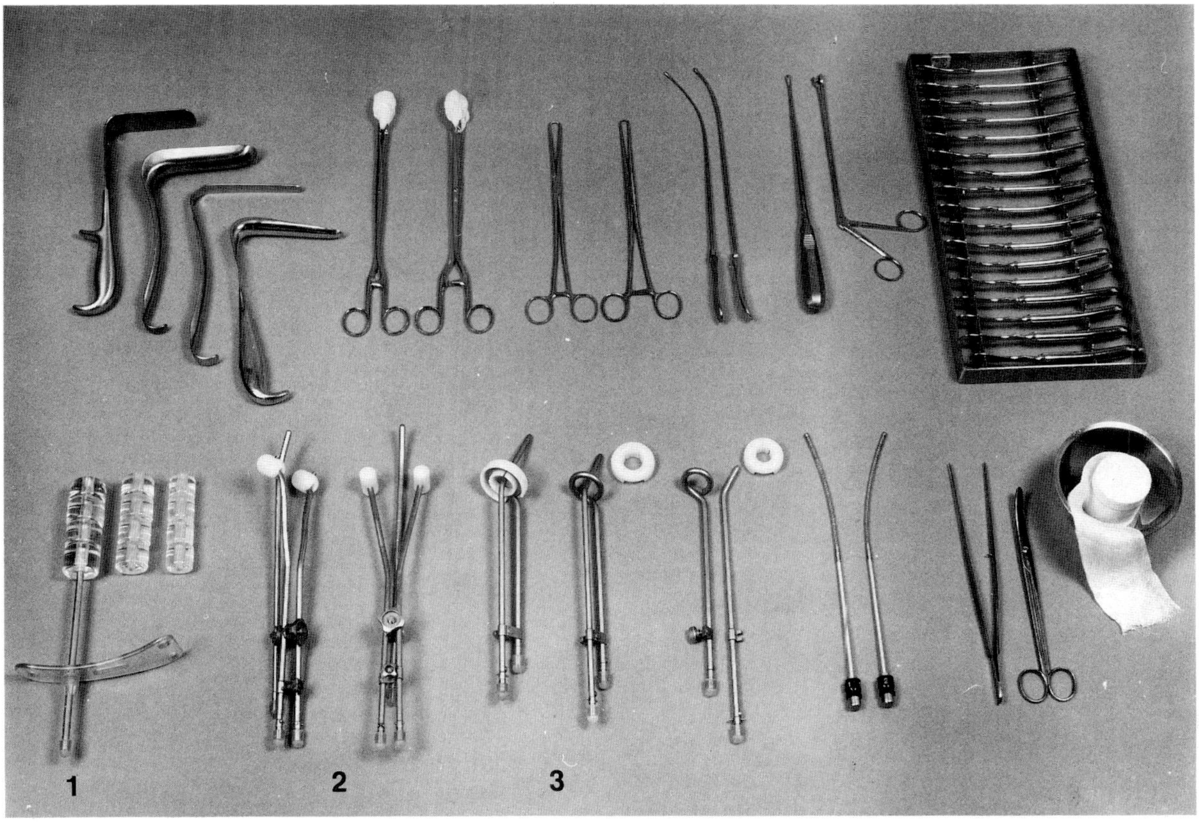

Fig. 3. Applicators used for low as well as for high dose rate afterloading: *1*, Vaginal applicator; *2*, intrauterine tube and tandem, available with rectal retractor; *3*, intrauterine tube with ring applicator

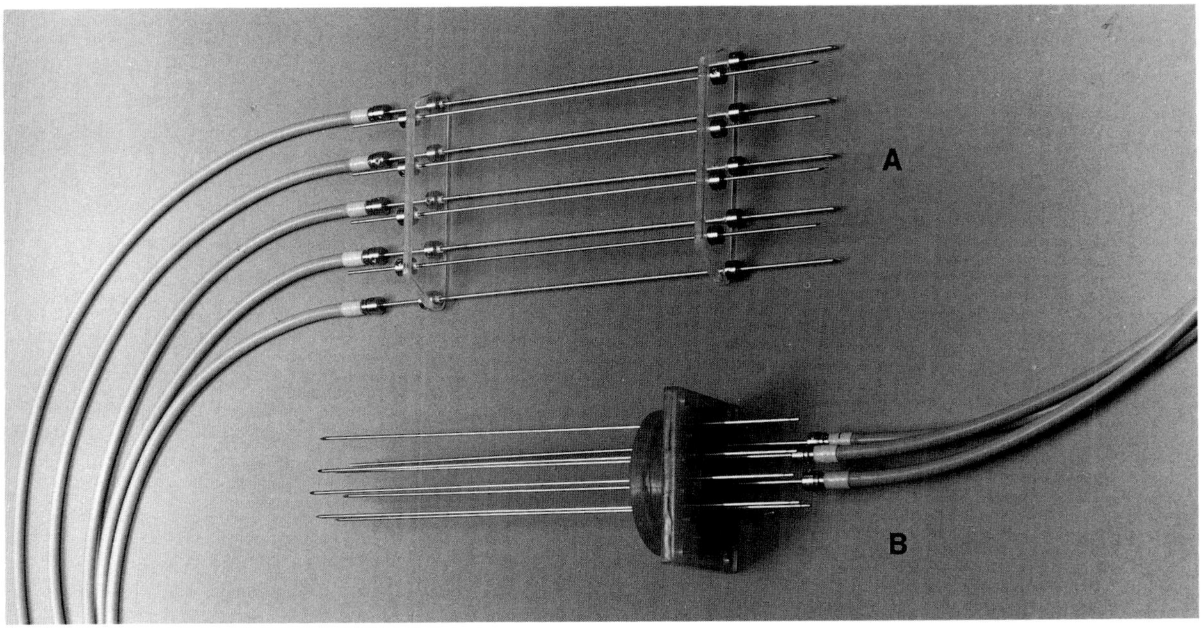

Fig. 4. Vaginal applicator: Template with eight peripheral needle sources used with the micro-Selectron, low and high dose rate techniques

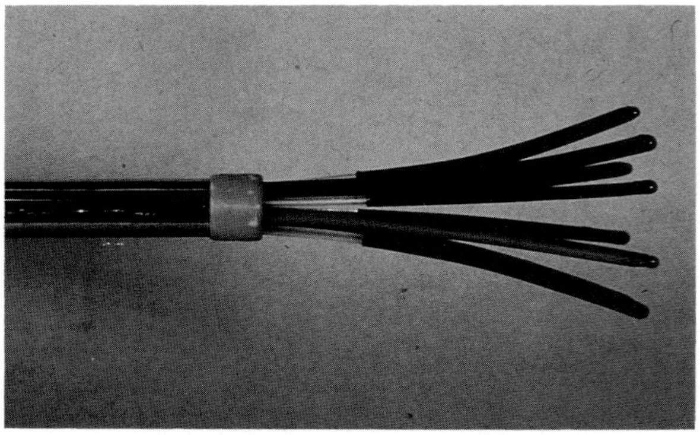

Fig. 5. "Umbrella" applicator for intrauterine irradiation in the case of endometrium carcinoma. The catheters can be straightened with an outer tube. Used with the micro-Selectron

4 Standard Treatment and Variations

The standard primary radiation therapy for carcinoma of the cervix is given in Table 2 for LDR as well as for HDR afterloading.

In some cases this typical treatment does not seem suitable because brachtherapy cannot reach the outer areas of the disease. In a series of 67 such patients with stage III an extern approach only was used, with 60–64 Gy in the center of the pelvis. The overall 5-year survival rate was 28.4%. A more favorable group of patients with stage IIIb was treated with combined brachy- and teletheraphy and achieved a 5-year survival of 42.6% (Table 3). The policy today in such cases is to give a homogeneous whole pelvis dose of 40–50 Gy followed by brachytherapy at a somewhat reduced dose.

5 Comparison of LDR and HDR Afterloading Treatment

LDR brachytherapy with irradiation over several hours or even days is based on decades of experience with ^{226}Ra and offers acceptable radiobiological qualities and a wide range of therapeutic uses. By comparison, several factors concerning HDR brachytherapy with irradiation over a short time (a matter of minutes) remain unclear.

For example, it is unclear which absolute total dose in HDR is equivalent to the LDR technique. Furthermore, the number of optimal fractions are not certain; estimates vary between 3 and 14 (or more) single fractions. It is still unclear what, in fact, the optimal total dose – not to be exceeded – per single fraction by HDR is. The most significant advantages of short-term irradiation (HDR) are:

- Irradiation on an outpatient basis is possible
- The radioactive sources can be monitored satisfactorily under short-term therapy (a matter of minutes)
- The short rest period reduces the risk of thrombosis
- More patients per unit time can be treated

On the basis of radiobiological considerations the following factors are valid. With HDR a lower total dose (in terms of Gy) shows an equal impact on the tumor and healthy tissue as with LDR since

Table 2. Standard afterloading treatment in cancer of the cervix: comparison of low (LDR) with high dose rate (HDR) techniques

Technique	Total dose at point:		Number of fractions	Time interval	Treatment time	Activity
	A	B				
LDR	70 Gy	19 Gy	3	14 days	18 h	^{137}Cs 322 mCi (=126 mg Rå equiv.)
plus teletherapy	10 Gy	50 Gy	25	5×/week		
dose rate	$130 cGy h^{-1}$	$35 cGy h^{-1}$				
HDR	48 Gy	12 Gy	6	8 days	0.4 h	^{60}Co 1320 mCi (=2080 mg Ra equiv.)
plus teletherapy	10 Gy	50 Gy	25	5×/week		
Dose rate	$2100 cGy h^{-1}$	$540 cGy h^{-1}$				

Table 3. Results of treatment in patients with carcinoma of the cervix stage IIIb. Patients considered unsuitable for typical brachytherapy because of local extension of disease were treated with external irradiation only

Treatment modality	5-year survival rate	
Radium (7200 mgeh, 2 fractions, 10-day interval) + external irradiation of pelvic wall with 40 Gy	81/190 (42.6%)	
^{60}Co monoaxial rotation to 60 Gy in the center + external irradiation of lateral pelvic wall with 40 Gy	13/41 (31.7%)	19/67 (28.6%)
^{60}Co biaxial rotation to 64 Gy in center of pelvis + 50 Gy at lateral pelvic wall	6/26 (23.0%)	

the biological effect per unit dose depends upon the dose rate of the source. Figure 6 shows the schematic changes in the area of therapeutic interest between 10 cGy/h and 6000 cGy/h (for ^{137}Cs). Significant changes in the effectiveness of the ionising irradiation take place in this area. The biological effect increases in correlation with the dose rate since the irradiation becomes increasingly lethal and cell recovery more and more difficult during the irradiation process. Likewise, the oxygen enhancement ratio also increases with the dose rate. That is to say, the higher the dose rate the lower the relative radiosensibility of the hypoxic cells. From this it follows that an

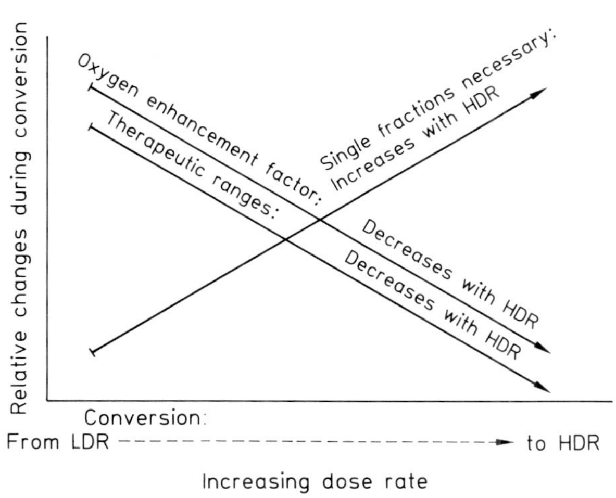

Fig. 6. Schematic changes of biological effects, radioresistance under hypoxia, and therapeutic range on conversion of low to high dose rate in the dose rate area between 10 cGy/h and 6000 cGy/h (according to LINDROP and ROTBLAT 1963; HALL and BEDFORD 1964; DENEKAMP and FOWLER 1966; FOX and NIAS 1970; HALL 1972)

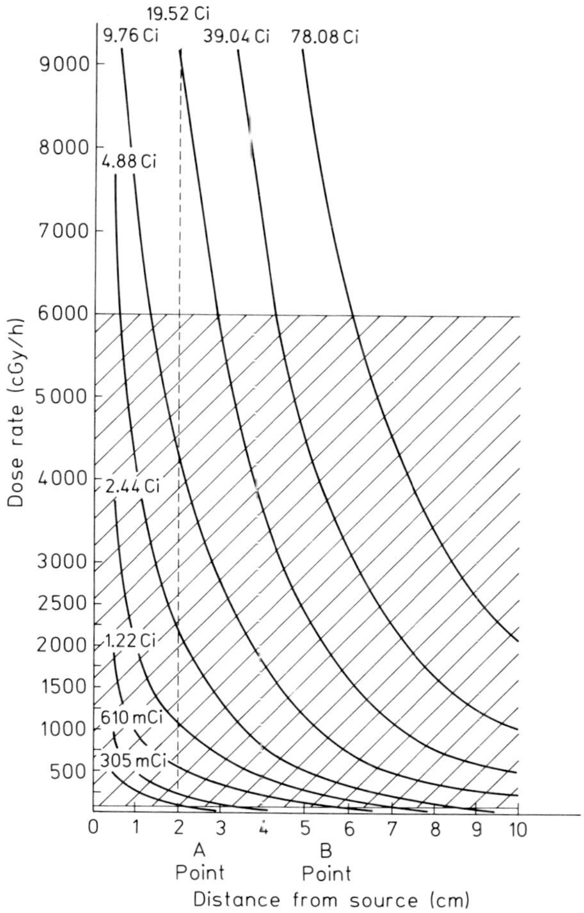

Fig. 7. Schematic chart of dose rate fall in gynecological low, middle, and high dose rate brachytherapy. Dense hatching indicates the area with the most defined radiobiological changes in relation to dose rate

increase in the dose rate results in a reduction of the therapeutic range.

Figure 7 depicts the dose rates employed in LDR and HDR gynecological brachytherapy. On the abscissa the distance from the active sources is marked in centimeters; the dense hatching represents the area in which, in correlation to the dose rate, enhanced biological changes take place as a result of the effects of the total dose given.

In contrast to teletherapy, in brachytherapy one is dealing with extreme variations in the dose rate to each single point at varying distances from the source. Close to the source position, where the dose rate is at its maximum, relative changes in the dose rate play only a minor role. In both HDR and LDR therapy the extremely high total dose to this area (2 cm, point A) always leads to total cell damage. More distance from the source (between 2 cm, point A, and 5 cm point B) the dose rate gains in significance. In protracted irradiation with LDR the cells can begin to recover from sublethal damage, while irradiation treatment is still going on. Hypoxic tumor cells show less resistance to protracted irradiation (LDR). These favorable effects can be achieved out in HDR by using a higher number of single fractions.

One problem for gynecological brachytherapy in HDR treatment lies in the fact that no more than 6–7 fractions can be clinically justified without increasing intolerably the burden to the patient and staff. According to radiobiological facts one should aim for a total HDR dose of 50–60 Gy over 10 (or more) single fractions at point A. The basis for this is that the single fraction with HDR should not exceed 9 Gy. In order to allow for the higher biological impact on both malignant and healthy tissue with HDR, the total dose in HDR treatment has to be reduced by 20%–40% if the fractionation is limited to 6 single fractions as a maximum. In summary, in order to convert LDR to HDR, an equivalent dose can be administered to single target points but not because of the variation in dose rate depending on the distance from the applicator to the entire target volume.

Because of the differing radiobiological effects, which again vary according to the distance from the source, an optimal single dose rate, number of doses, and intervals between fractions cannot be reliably calculated with any known equation (Ellis 1969; Kirk et al. 1975; Kellerer 1977a, b; Liversage 1969; Morrison 1988; Dale 1985; Orton 1987). For example, the model of Dale (1985) predicts that 7 fractions of 340 cGy per fraction are needed to control safely both tumor and late side effects. However, Orton (1987) predicts that even 2 fractions of 770 cGy per fraction (point A) will maintain tumor control and late effects. This prediction does not comply with any clinical experience. Clinical trials, comparing LDR with HDR are necessary in this situation.

5.1 Clinical Trials

Table 4 gives an overall summary of four clinical studies in which HDR and LDR were compared.

Inque et al. (1978) compare LDR treatment with 2 fractions in carcinoma of the cervix stages IIb and III with HDR treatment with 3 fractions. Marked side effects were observed in the HDR group (see Table 5). In stage IIb patients LDR came off better in terms of overall survival. In stage III patients no significant differences were discernible. The unfavorable overall survival rates in stage IIb disease with HDR are explained by the author as being the result of an almost 50% reduction in the total dose, thus reducing detrimental side effects. Dose reduction plays a less significant role in stage III patients since here high external irradiation of the entire treated volume considerably influenced survival with both methods. Taina (1981) compared LDR with 2 fractions and HDR with 3–5 fractions. The 5-year survival rate was the same. However, Taina (1981) observed an increase in early and late side effects with HDR, in particular a significant increase in late side effects (see Table 6). After increasing the

Table 4. Comparison of low with high dose rate during brachytherapy in clinical studies of carcinoma of the cervix (Inque 1978; Taina 1981; Rotte 1981, 1983)

Author	Tumor stage	Brachytherapy						Teletherapy
		High source activity			Low source activity			Dose in the reference point of the parametrium
		Isotope	Point A dose	Fractions	Isotope	Point A dose	Fractions	
Taina	I-II	^{60}Co	30–50 Gy 30–37.5 Gy	3–5	^{226}Ra	55 Gy	2	–
Rotte	I-III	^{192}Ir	60 Gy	4/6	^{226}Ra	120 Gy	3	40 Gy
Inque	Ia	^{60}Co	40–50 Gy	4–5	^{137}Cs	70 Gy	7	–
	Ib/IIa	^{60}Co	50 Gy	5	^{137}Co	70 Gy	7	30 Gy
	IIb/III	^{60}Co	30 Gy	3	^{137}Co	60 Gy	7	40 Gy
Janik	I-IV	^{226}Ra	32 Gy 20–25 Gy	1 3	^{226}Ra	60 Gy 20–25 Gy	3 3	30 Gy

Table 5. Carcinoma of the cervix stages IIb-III. Severe side effects occur with equal therapeutical effectiveness if with high dose rate (HDR) only 3 fractions are chosen (according to INQUE 1987)

	n	Number of fractions	Bleeding		
			Rectum	Bladder	Ileus
LDR	106	2	27 (25%)	7 (7%)	0
HDR	143	3	51 (36%)	3 (2%)	3

Table 6. Comparison of side effects from low (LDR) and high dose rate (HDR) treatment in carcinoma of the cervix stages Ia-IIb (Figo) according to TAINA (1981)

	Number of fractions	*n*	5-year survival	Side effects	
				early	late
LDR	2	84	78.6%	3.6%	13.1%
HDR	4–5	101	81.6%	10.9%	27.7%
				n.s.	($P<0.01$)

fractions from 3–4 to 4–5 with HDR, the side effects diminished.

In stages I and II patients ROTTE (1981) achieved the same survival rate for HDR and LDR. For stage III disease LDR (radium therapy) with more effective. As a consequence ROTTE (1983) increased the fractions from 3 to 6 with HDR at a dose rate of 10 Gy to point A. In this way, he achieved HDR results in stage III patients equal to those for LDR.

References

Becker J, Scheer KE (1952) Strahlentherapeutische Anwendung von radioaktivem Kobalt in Form von Perlen. Strahlentherapie 86: 540

Chassagne D (1973) Low-dose-rate technique of endocavitary brachytherapy. Proc R Soc Med 64: 601

Dale RG (1985) The application of the linear-quadratic dose effect equation to fractionated and protracted radiotherapy. Br J Radiol 58: 515–528

Delouch G, Harvey P, Laval C, Rambert P, Gest J (1979) Résultats de la radiothérapie de 406 cancers du col de l'utérus (T2 distaux et T3). Bull Cancer (Paris) 66: 549

Denekamp J, Fowler JF (1966) Further investigations of the response of irradiated mouse skin. International Journal of Radiation Biology 10: 435

Ellis F, Dose N (1969) Time and fractionation: a clinical hypothesis. Clin Radiol 20: 1–7

Ellis F (1971) Nominal standard dose and the ret. Br J Radiol 44: 101

Field SD, Hornsey SH (1952) Repair in normal tissues and the possivle relevance to radiotherapy. Strahlentherapie 153: 371

Fletcher GH, Rutledge FN (1967) Overall results in radiotherapy for carcinoma of the cervix. Clin Obstet Gynecol 10: 958

Fox M, Nias AHW (1970) The influence of recovery from sublethal damage on the response of cells to protracted irradiation at low dose-rate. Current topics in radiation research 7: 71

Hall EJ (1972) Radiation dose-rate: a factor of importance in radiobiology and radiotherapy. Brit J of Radiol 45: 81

Hall EJ, Bedford JS (1964) Dose rate: its effect on the survival of Hela cells irradiated with gamma-rays. Radiation Research 22: 305

Inque T, Hori S, Miyata L, Ozekio, Shigematsu D (1978) High versus low dose rate intracavitory irradiation of carcinoma of the uterine cervix. Acta Radiol [Oncol] (Stockh) 17: 277

Kellerer AM (1977) Grundlagen der Ellis-Formel. Strahlentherapie 153: 384

Kirk J, Gray WM, Watson ER (1975a) Cumulative radiation effect. IV. Normalization of fractionated and continuous therapy – area and volume correction factors. Clin Radiol 26: 77–88

Kirk J, Gray W, Watson ER (1975b) Cumulative radiation effect. V. Time gaps in treatment regimes. Clin Radiol 26: 159

Liversage WE (1969) A general formula for equating protracted and acute regimes of radiation. Br J Radiol 42: 432–440

Morrison RA (1988) The K-effect formulas for low dose rate irradiation with gamma rays. Endocuriether Hyperthermia Oncol 4: 125–128

Lindrop P, Rotbalt J (1963) In: Cellular Basis and Aetiology of Late Somatic Effects of Ionizing Radiations. Academic Press, London, p 313

Orton CG (1987) What minimum number of fractions is required with high dose-rate remote afterloading (Correspondence). Br J Radiol 60: 300–301

Pierquin B, Marinello G (1981) Plesiocuriethérapie des cancers du col de l'utérus. J Ent Radiother 4: 231

Rotte K (1981) Ferngesteuerte Afterloadingverfahren. Strahlentherapie [Special Vol] 76: 312–320

Rotte K (1983) Das ferngesteuerte Nachladeverfahren (Remote-Controlled Afterloading) für die intrakavitäre Kontakttherapie (Brachytherapie) gynäkologischer Karzinome. Radiologe 23: 20

Taina E (1981) High versus low dose rate intracavitary radiotherapy in the treatment of carcinoma of the uterus. Acta Obstet Gynecol Scand [Suppl] 103: 214

Von Fournier D, Kuttig H, Kubli F, Braun K (1975) Die Anwendung von Kobalt-60-Perlen beim corpus-Carcinom. Technik, Komplikation und Ergebnisse. Strahlentherapie 150: 273

Von Fournier D, Senf W, Kuttig H, Kubli F (1976) Verbesserung der gynäkologischen Radiumtherapie durch Afterloadingtechnik mit Cäsium-137. Strahlentherapie 151: 195

9 Interstitial Hyperthermia

9.1 Techniques and Clinical Experience of Interstitial Thermoradiotherapy

M. HEINRICH SEEGENSCHMIEDT, ROLF SAUER, LUTHER W. BRADY, and ULF L. KARLSSON

CONTENTS

1	Introduction	343
2	Biological Considerations	343
3	Interstitial Hyperthermia Techniques	344
3.1	Radiofrequency Techniques	344
3.1.1	Resistive Local Current Field Technique	344
3.1.2	Capacitive Radiofrequency Technique	346
3.2	Microwave Technique	346
3.3	Hot Source Techniques	347
3.3.1	Inductive Ferromagnetic Seed Technique	347
3.3.2	Conductive Hot Water Perfusion Technique	349
3.3.3	Conductive Laser Technique	349
3.4	Current Technical Status and Future Prospects	349
4	Clinical Experience	350
4.1	Treatment Rationale and Indications	350
4.2	Clinical Results	351
4.2.1	Interstitial Radiofrequency Hyperthermia	351
4.2.2	Interstitial Microwave Hyperthermia	351
4.2.3	Interstitial Ferromagnetic Seed Hyperthermia	351
4.2.4	Interstitial Hot Water Perfusion Hyperthermia	352
4.2.5	Personal Interstitial Microwave Hyperthermia Experience	352
4.3	Treatment Complications	353
4.4	Prognostic Treatment Factors	353
5	Conclusions	353
	References	354

1 Introduction

In recent years abundant information has been gained on various biological, technical and clinical aspects of hyperthermia in combination with radiotherapy (thermoradiotherapy). Using various heating methods, superficial and interstitial hyperthermia methods have achieved a sufficiently high standard of technology to provoke interest among clinical oncologists. Since hyperthermia is regarded as an effective radiosensitizing as well as chemosensitizing agent, it shows great potential for palliative and/or adjuvant treatment approaches when combined with radiotherapy and cytotoxic agents (BRADY and SEEGENSCHMIEDT 1987; PEREZ et al. 1987; OVERGAARD 1987, 1989).

Similar to other interventional radiotherapy techniques, such as brachycurietherapy, the use of interstitial heating techniques generally implies that the heating sources are implanted within the target tissue. In addition endocavitary and perfusional hyperthermia may be regarded as additional methods of "interventional hyperthermia"; however, these subjects will not be addressed in this review. Theoretical and experimental evidence indicates several advantages to interstitial over utilization of external heating techniques: (a) better heating uniformity within the target volume (b) extensive, reliable, invasive, 3-dimensional thermometry data, (c) improved hyperthermia treatment control, and (d) improved sparing of surrounding normal tissue.

Interstitial hyperthermia (IHT) was initially proposed by Doss and coworkers at the beginning of the modern era of clinical hyperthermia in 1975 (DOSS and MCCABE 1976); since then, further techniques and clinical applications have been developed in the early 1980s by a few institutions and groups of researchers (ARCANGELI et al. 1982; BREZOVICH et al. 1984a, b; COSSET et al. 1982; COUGHLIN et al. 1983; LE BOURGEOIS et al. 1978; JOSEPH et al. 1981; MANNING and GERNER 1983; VORA et al. 1982).

M. HEINRICH SEEGENSCHMIEDT, Dr. med., ROLF SAUER, Prof. Dr., Strahlentherapeutische Klinik und Poliklinik der Universität Erlangen-Nürnberg, Universitätsstr. 27, 8520 Erlangen, Germany

LUTHER W. BRADY, M.D., Professor and Chairman, ULF L. KARLSSON, M.D., Department of Radiation Oncology and Nuclear Medicine, School of Medicine, Hahnemann University, Mail Stop 200; Broad and Vine, Philadelphia, PA 19102-1192, USA

2 Biological Considerations

The biological interaction between radiation and heat is based on two effects (LEEPER 1985): First, heat has cytotoxic effects without specific intrinsic sensitivity to tumor cells. Environmental conditions (chronic hypoxia, nutritional deprivation, and increased acidity) which are typically found in poorly vascularized solid tumors can enhance the thermal sensitivity of cells. Thus, temperatures above 43°C, which are still tolerated by normal tissues, can destroy large proportions of many solid tumors. Second, heat offers radiosensitizing effects at temperatures above 41°C, which is mainly expressed as direct sensitization of radiation damage involving decreased repair of sublethal (SLD) and potentially lethal (PLD) radiation damage and enhanced cell killing in the radioresistant G_2 and/or S phases of the cell cycle. The almost complementary interaction of heat and irradiation provides a strong rationale for its combined use: hyperthermia is able either to increase the biological effect of a given radiation dose (hyperthermic radiosensitization) and/or to destroy normally radioresistant tumors cells (hyperthermic cytotoxicity). These effects may be expressed as the therapeutic enhancement ratio (TER).

Both phenomena have been demonstrated to occur for low-dose radiation exposure combined with hyperthermia in various in vitro and in vivo experiments (Ben-Hur et al. 1974; Miller et al. 1978; Harisiadis et al. 1978; Gerner et al. 1983; Moorthy et al. 1984; Hall 1985; Jones et al. 1989). Experimental data suggest that potentiation of low-dose irradiation with ^{192}Ir by a single heat treatment may reach a TER of 1.3–1.6 and is maximized if hyperthermia is given simultaneously with the interstitial irradiation (Jones et al. 1989).

3 Interstitial Hyperthermia Techniques

The various methods and physical agents employed for interstitial hyperthermia have been previously reviewed (Coughlin et al. 1983; Brezovich et al. 1984a; Emami et al. 1984, 1987; Neyzari and Cheung 1984; Cosset et al. 1986; Perez et al. 1987; Gautherie 1989; Seegenschmiedt and Sauer 1989) and are summarized in Table 1. Three different techinques exist: (a) resistive radiofrequency (RF), (b) radiative microwave (MW), and (c) conductive hot source (HS) techniques either employed with ferromagnetic seeds (FMS), hot water perfusion (HWP), or interstitial laser technique, which has been recently introduced (Milligan and Overholt 1989).

3.1 Radiofrequency Techniques

Arrays of metallic implants (hollow stainless steel needles or wires) that conduct radiofrequency currents can be arranged to form opposite ("active" and "passive") electrodes with the tissue as intervening medium inbetween. Depending upon the employed RF range a resistive type of heating occurs with low frequencies at 0.5–1.0 MHz compared with a capacitive type of heating with higher frequencies at 8–27 MHz.

3.1.1 Resistive Local Current Field Technique

The interstitial "resistive type" radiofrequency (R-RF) hyperthermia operates with metallic implants (rigid needles or flexible wires) and low frequency electric currents, which are driven between electrically connected pairs or arrays of implants creating a localized current field (LCF). Slabs of tissue between the respective pairs of adjacent needle or wire arrays are simultaneously energized and resistively heated. The technique was first proposed by Doss and McCabe 1976. For human tissues 0.5–1.0 MHz RF electric currents are used, which produce mainly resistive heating, whereas capacitive reactance is negligible. Moreover, this low RF range reduces "electrical noise" interference with other equipment components, allows good impedance matching, and prevents potentially hazardous neuronal depolarization.

The RF power deposition is dependent and proportional to the square of the electrical current density and the specific resitivity of the various tissues; therefore, maximal power deposition occurs close to the metallic implants and diverges and decreases inbetween, which can cause broad variations of heating. However, due to the high power deposition and high temperature region immediately encompassing the implants conductive heat transfer can contribute to the heating of the intervening regions of lower temperatures. In general, the magnitude of resistive and conductive heating depends upon the specific geometry of the implant volume, electrical tissue properties, and the "thermal clearance" through differences in tissue perfusion.

Various types of amplifiers and independent power control for each electrode pair have been developed and successfully applied in human tumors (Aristizabal and Oleson 1984; Cosset et al. 1984, 1985, 1986; Emami et al. 1984, 1987;

Table 1. Survey on interstitial hyperthermia techniques

Radiofrequency (RF) technique	Microwave (MW) technique	Ferromagnetic (FM) technique	Hot perfusion technique
Principle: resistive/capacitive heating – resistive between alternative needle pairs and/or arrays at 0.1–1.0 MHz, or capacitive system at 8/27 MHz.	*Principle:* radioactive heating – emitted by semirigid coaxial MW antennae at 300–2450 MHz, depending on tissue quarter wavelength.	*Principle:* inductive/conductive heating – contactless induction of FM particles with 0.5–2 MHz RF external magnetic induction coil.	*Principle:* conductive heating – "over compensating" hot liquid perfusion system with hot water and 150–200 mW cm^{-1}°C^{-1} heat deposition.
Features: Perfect contact between tissue and needle electrodes required; Perfect geometry of the implant required (template technique); Flexible and perfect arrangement of connected electrode arrays required; Multipoint-feedback software, specific thermometry and power control.	*Features:* Use of conventional brachytherapy afterloading plastic tubes; Independent heating of each individual antenna in tissue; Ellipsodial heating pattern, with intrinsic "hot spot" in the "junction" plane and "cold spot" at the antenna tip; Variable "active" heating length with insertion depth and frequency.	*Features:* Biocompatible silver-gold electroplated alloys, such as 70%Ni–30%Cu; Curie point at therapeutic temperatures between 45°C–55°C; Narrow thermal margin between the ferro- and paramagnetic status; Parallelity of patient, FM seeds, and magnetic induction field axis required.	*Features:* Metallic needles or flexible plastic tubes, usual brachytherapy technique; Uni- or bidirectional tubes with hot and cold water influx channels and variable central temperature mixing chamber, turbulent and laminar flow; high water flow of 2.5 ml s^{-1}, high intraluminal water temperatures of 45°C–55°C required.
Advantages: No limitation of implant dimension, homogeneous heating along needles.	*Advantages:* Wider spacing of antennae possible, usual brachytherapy techniques.	*Advantages:* Noninvasive automatic thermal regulation; usual brachycurie technique.	*Advantages:* Easy handling and low cost of the system, great variability.
Disadvantages: Rigidity of metal needle electrodes; specific needle power steering and high precision implant required.	*Disadvantages:* Limited heating length depending on insertion depth, frequency, tissue; uncontrolled power reflection.	*Disadvantages:* Migration of permanent seeds; difficulties with thermal regulation and biocompatibility of seeds.	*Disadvantages:* Close spacing and exact probe implantation geometry required, difficulties at high perfusion.

Frazier and Corry 1984; Gautherie et al. 1989; Joseph et al. 1981; Lilly et al. 1983; Linares et al. 1986; Manning et al. 1982; Oleson et al. 1984; Vora et al. 1982, 1988, 1989; Yabumoto and Suyama 1984; Yabumoto et al. 1989). RF heating has some drawbacks: The rigidity of the implanted needles can cause discomfort during the extended hyperthermia and brachytherapy treatment sessions. Skin entry and exit points as well as tissue interfaces cannot be effectively shielded from excessive heat exposure. These problems can be more pronounced when heating sessions are conducted in OR under general anaesthesia without the patient's sensation.

Phantom studies in homogeneously perfused (Stohbehn 1983) and nonperfused materials (Oleson and Cetas 1982) have shown the large variations in power deposition that might influence the heating pattern due to heterogeneous implant geometry, electrical tissue properties, and perfusion characteristics. Perfect contact between metallic implants and tissue, optimal implant geometry with strict parallelism, equidistant "active length", and close spacing of electrodes (≤1 cm) are strictly required.

Improved R-RF hyperthermia systems have been proposed which should reduce the heterogeneity of interstitial RF heating. Using the suggestion of Astrahan 1982, a multipoint feedback control (MFC) system with computer software has been developed, which can sequence power to various combinations of implants and regulate the individual dwell times and power levels of each electrode pair. A sophisticated computer software controls each needle pair using a thermometry feedback with at least 32 thermocouple sensors (Rude et al. 1987, Prionas et al. 1989).

Cosset et al. put forth the use of electrically insulating material (plastic tubes) around the metallic implants in order to prevent undesired power deposition and to tailor the individual heating volume. He implemented several customised "metallic-plastic" tubes with insulated plastic ends and a metallic central secion of variable length, thus allowing individually shaped heating volumes in head and neck tumors. The implanted needles were used both for intersitial heating and radiotherapy (Cosset et al. 1984, 1985, 1986).

Permanently implanted needle electrodes together with a large external surface electrode have been tested in phantoms and animals (Lilly et al. 1983) and in intrathoracic lesions (Corry and Barlogie 1982; Frazier and Corry 1984). A miniature localized current field (LCF) system at 0.5 MHz has been developed to be used in conjunction with an ophthalmic ^{125}I plaque for treatment of choroidal melanomas; successful testing in phantoms and animals has been reported (Finger et al. 1986; Astrahan et al. 1987).

Most recently, a novel design of interstitial RF electrodes has been proposed, consisting of multiple, electrically independent aperture elements in order to allow longitudinal and axial control of power deposition. Both four-element rigid and flexible RF electrodes have been designed with the option to power each element independently at different levels. The rigid design of the electrodes consists of four coaxial layers each of electrically conducting (Cr/Al/Ni) and electrically insulting (SiO_2) materials successively placed on a hollow stainless stell trocar, whereas the flexible electrode is molded in flexible epoxy material consisting of short sections of stainless steel tubes, which are connected to an enamel-coated copper conductor and twisted around an inner hollow core (Prionas et al. 1989).

3.1.2 Capacitive Radiofrequency Technique

Recently a variation of the LCF heating system using either 8 MHz (Akuta et al. 1989) or 27 MHz (Marchal et al. 1989; Duerloo et al. 1989; Visser et al. 1989) radiofrequency has been proposed by several groups. In contrast to the low frequency LCF resistive technique, the direct galvanic contact is replaced by a capacitive coupling between the metallic implants (wires or needles) and tissue. In physical terms, the heating within the target volume is induced by an asymmetric capacitive coupling with currents passing between the "active" implants (coated wires or needles inserted in plastic catheters) through the dissipative medium (phantom or living tissue) to a "passive" external ground plane placed under the heated object. Separate ground planes and isolation transformers are implemented to avoid crosstalk between different electrodes, and variable tuning coils are included for appropriate impedance matching. With this technique there are no constraints to the electrode design: any size, number, and shape (e.g. looping technique) of electrodes can be implemented according to the individual location and volume of the tumor. Most importantly, phantom studies demonstrate a fairly good homogeneity of heating along the electrodes and no critical dependence on insertion depth. Best homogeneity of specific absorption rate (SAR) was achieved with a spacing of less than 1 cm.

At present, these interstitial C-RF systems are being extensively tested on animal and human tumors in Nancy, France (Marchal et al. 1989), Rotterdam, The Netherlands (Deurloo et al. 1989; Visser et al. 1989), and Kyoto, Japan (Akuta et al. 1989).

Both interstitial R-RF and C-RF heating are compatible with thermocouple (TC) and fiberoptic (FO) thermometry. However, the generally applied thermometry measurements within the needles usually overestimate the actual heating profile within the implanted volume and are prone to artefacts; therefore, supplementary thermometry between the needle pairs and sufficient isolation is required to avoid possible electrical interference.

3.2 Microwave Technique

MW IHT utilizes radiative heating produced by high frequency power generators. The generators are connected to semirigid coaxial MW antennae, which radiate the electromagnetic energy into the absorbing tissue. The useful frequency range is 300–2450 MHz. Different antennae and generator designs have been developed, aiming for optimized individual antenna heating characteristics and power regulation. The initial antenna design developed at Dartmouth Medical Center (King et al. 1983) consisted of miniature, flexible coaxial cables operating at 915 MHz frequency with ≤1.6 mm diameter. Physically the antennae are characterized as electromagnetic monopole radiators consisting of an inner and outer conductor, with the inner conductor extending further than the outer conductor by a distance equal to one-quarter of the wavelength (in dielectric tissue). As the portion proximal to the termination of the outer conductor cable ("junction") also radiates similarly to the distal portion, the resulting power deposition pattern along the antenna direction is approximately ellipsoidal. However, the specific power deposition in the tissue depends upon various parameters: the frequency of operation, the dielectric properties of surrounding catheters (plastic coating) and tissue, as well as the insertion depths of the antennae (MECHLING and STROHBEHN 1986; CHAN et al. 1989; JAMES et al. 1989; STAUFFER et al. 1989).

Since the "active" length of the MW antennae as well as the 50% iso-SAR distribution is inversely related to the operating frequency, small lesions are better treated with higher frequencies (2450 MHz) while large lesions ≥6 cm diameter require a lower frequency (433 MHz) (MECHLING and STROHBEHN 1986). With appropriate power splitters and variable attenuators on each antenna feedline the radiated power levels can be varied to overcome the expected nonuniformities in power deposition.

Disadvantages of MW antennae include the limited longitudinal heating length, power heterogeneity at tissue interfaces, and the "cold spot" at the antenna tip, which can pose a major problem for more extensive tumors (STROHBEHN et al. 1984; TREMBLY 1985). The lack of "shielding" of normal tissues especially at antennae entry or exit points still has to be resolved, as high power reflections and consequently excessive heating can occur (TREMBLY 1985); hopefully, the development and use of special bolus material may help to overcome at least some of these problems. The main advantage of the MW technique is that conventional flexible brachytherapy afterloading plastic tubes can be employed to house the MW antennae, since insulating properties in tissue are not required for frequencies above 300 MHz.

A variety of improved MW antennae designs (incorporating multiple active sections, sleeves, chokes, helical coils) have been described in the literature, which promise extended heating length even to the antenna tip as well as improved and more hemogeneous SAR characteristics (SATOH and STAUFFER 1988; ROOS and HUGANDER 1988). The sophisticated interstitial microwave antenna array hyperthermia (IMAAH) system at Dartmouth Medical Center is designed to be phase-coherent for all antenna in order to maintain consistent and reproducible SAR distributions from the antenna arrays: the addition of phase shifting to each antenna is expected to enhance further the MW array performance by reducing the heterogeneity of power deposition in extensive MW antenna arrays (WONG et al. 1989).

Due to electrical interference only nonpertubing thermometry techniques can be utilized: nonperturbing thermistors measure temperatures every 15–20 s, when power deposition is interrupted for about 3 s, and FO thermometry devices like gallium arsenide (GAs) crystals or thermoluminescent materials. In contrast to the LCF technique with thermometry sensors being placed within the implanted needles, for most interstitial MW techniques invasive temperatures are usually underestimated, as long as they are measured "interstitially" between the antenna array; thermometry should be supplemented by integrated thermometry from within the MW antennae to document critical maximum temperatures, e.g., at the "junction" place of the implant (TURNER 1986; ASTRAHAN et al. 1988).

3.3 Hot Source Techniques

Passive dissipation of heat by thermal conduction is the basic mechanism of interstitial hot source hyperthermia techniques. There is neither a "driving force" on ions or molecules (as for MW techniques) nor an "active current flow" (as for RF techniques). Temperatures in the range of 45°–55°C within the heating sources are usually required to achieve sufficient heating of the tissue inbetween. In addition, all conductive heating techniques have to overcome the particular problem of heat loss due to local blood perfusion, which can change rapidly over time and with the spatial distribution of the implant (SONG 1984; REINHOLD and ENDRICH 1986).

3.3.1 Inductive Ferromagnetic Seed Technique

Unlike the previously described interstitial heating devices, ferromagnetic thermoseeds absorb power from an externally applied electromagnetic induction field in a contactless manner and do not require electrical or mechanical connection to the implant itself. Ferromagnetic (FM) seed hyperthermia utilizes magnetic induction forces preferentially to absorb energy in FM implants.

Several physical mechanisms have been applied including conductive (eddy current) heating of FM materials and hysteresis heating of ferro- or ferrimagnetic substances. In both instances heating is induced by means of thermal conduction rather than direct electromagnetic power absorption. This method was initially proposed by Burton for thermocoagulation of brain tissue (BURTON et al. 1966). It has been extensively tested in phantoms and in vivo experiments on various tumors (BREZOVICH 1987; BREZOVICH et al. 1984a, 1990; DESHUMUK et al. 1984; STAUFFER et al. 1984; SHIMM et al. 1989).

The attraction of FM material is its sudden decrease of magnetic permeability and loss of energy absorption when heated to sufficiently high temperatures. At a certain temperature, which is referred to as *transition or Curie point*, the FM material suddenly loses its ferromagnetism to become non- or paramagnetic; physically, as the Curie temperature is approached, a magnetic phase transition within the FM substance occurs which reduces its magnetic moment and the power absorption; hence at higher temperatures paramagnetic conditions are created within the previously ferromagnetic substances, which cease completely to absorb any energy by the induced magnetic field. FM materials become automatically thermally regulating near the predetermined Curie temperature whenever sufficiently intense magnetic fields are induced. Nowadays, appropriate metallurgical techniques allow the adjustment of the Curie point of specific alloys to any desired temperature value. Special biocompatible FM seeds have been developed, e.g., the nickel-copper (70.4 atom% nickel–29.6 atom% copper) or nickel-silicon seeds; they are fully biocompatible due to their electroplated coating consisting of a 55 atom% gold and 45 atom% silver alloy. These FM seeds or wires can be customised in such a way that they have their Curie transition near therapeutic temperatures (45°C–55°C).

The main advantage of this method is that the FM implant operates as an internal thermostat itself, which establishes an absolute temperature close to the Curie point within the implanted volume. The "intrinsic" automatic thermal regulation theoretically allows homogeneous heating of nonuniform perfused tissues without invasive thermometry.

Since direct heating of the whole body region by magnetic induction (eddy currents) has to be avoided, only low FM induction field frequencies between 0.1 and 2.0 MHz should be employed. The magnetic field is induced by a circuit consisting of an induction coil with a capacitor in parallel, driven by a RF generator at the resonant frequency. For clinical versatility the induction coil has to have a sufficiently large inner diameter and should be mobile enough to encompass any segment of the human body. Moreover, a certain extent of angulation in any direction is required to align the induction field orientation with the different axes of the implanted FM seeds. The amount of heating power which themoseeds are able to deliver depends mainly on blood perfusion and spacing between them. Theoretical and clinical experience indicates that for a typical 1-cm grid pattern 100–200 mW power per cm FM seed length should be able to heat adequately even highly perfused tumors (ATKINSON et al. 1984; BREZOVICH 1987). Parallel, equidistant, and close spacing of the FM seeds is strictly required. Deviations of ±15° away from the axis of the field may be acceptable.

The specific heating rate which is produced by a cylindrical-shaped ferromagnetic thermoseed depends upon various factors: (a) the radius of the individual thermoseed, (b) the specific magnetic permeability, (c) the intensity of the applied magnetic induction field, (d) the frequency of the applied magnetic induction field, and (e) the orientation of the FM seed within the magnetic induction field.

Typically an array of self-regulating FM seed is implanted with a density of one seed per cubic centimeter; the FM seed implant is used either as a permanent implant (together with percutaneous radiotherapy) in unresectable portions of tumors and areas of risk of local relapse or as a temporary implant (together with interstitial afterloading radiotherapy techniques). The method allows heating of any desired volume ≤ 12-cm diameter including deep-seated tumors without excessive heating of the surrounding normal tissues (SHIMM et al. 1989).

Extensive phantom and animal studies and first clinical experiences with FM seed hyperthermia have been done at the universities of Arizona (STAUFFER et al. 1984; FORSYTH et al. 1984; SHIMM et al. 1989), Alabama (BREZOVICH 1987; BREZOVICH et al. 1984a, b, 1990), Wisconsin (PALIWAL et al. 1989), and Japan (KOBAYASHI et al. 1986). Problems with seeds concern biocompatibility of copper and nickel material due to the porosity of the gold-silver coating layer, migration of implanted seeds, and imperfect temperature regulation (BREZOVICH 1987).

3.3.2 *Conductive Hot Water Perfusion Technique*

Recently another "hot source" heating technique has been developed at the University Clinic in Vienna (Model KHS9-W18, Otmar Handl GmbH, Vienna, Austria) which uses hot fluids (water) under constant high perfusion as a heating medium. As for the FM technique, only thermal conductivity governs the heat transfer within the implanted tissue (HANDL-ZELLER et al. 1987a,b; SCHREIER et al. 1990). Heat exchange occurs through liquid heated tubes or needles capable of providing an estimated heating rate of 150–200 $mW\,cm^{-1}\,°C^{-1}$. The applicators are designed to be compatible with standard brachytherapy equipment. The plastic or metallic applicators with a diameter of 1.6 mm and a wall thickness of 0.1 mm can be used both unidirectionally for through-and-through implants as well as bidirectionally for closed-end implants. The system consists of a water tank, a precision thermostat equipped with cooling and heating elements, pressure and suction pumps, a proportional integral differential (PID) device to survey water temperature, pressure, and flow, and two manifolds to distribute the water into the implant. The high-precision themostat controls the intraluminal water temperature between 43°C and 60°C stabilized to +/−0.1°C. The principle of this technique is simple, providing an "overcompensating perfusion system", which is capable of overriding normal and tumor tissue perfusion within the physiological range (SONG 1984; REINHOLD and ENDRICH 1986).

The advantages of this system include easy handling, simplicity of engineering, low cost, and the use of any thermometry technique, as no interference occurs like that with electromagnetic heating systems. The heating properties of HWP systems are strongly dependent on: (a) the spacing of the heated probes, (b) the specific flow rate of the hot medium (e.g., water), (c) the temperature of the hot medium, and (d) the physical properties of both hot medium as well as tissue (density, specific heat, dynamic viscosity, and conductivity).

Phantom and animal experiments demonstrate that, depending upon needle spacing and local perfusion characteristics, intraluminal temperatures of up to 50°C and a water flow of 2.5 $ml\,s^{-1}$ (with spacing 14 mm, perfusion rate 2.25 $kg\,m^{-3}\,s^{-1}$) are necessary to obtain a consistent homogeneous heating region at 42°C. Assuming the thermal conductivity of resting muscle tissue at 0.5 $W\,m^{-1}\,°C^{-1}$, thermal modelling reveals that a HWP system has about 30 to 40 times the required power in reserve. Nevertheless, regions of high local blood flow still comprise a significant problem even with 10-mm spacing between applicators. The use of water at even higher temperatures (≥50°C) or implementation of larger diameter tubes (1.9–2.3 mm) may alleviate these problems (SCHREIER et al. 1990).

3.3.3 *Conductive Laser Technique*

A very recent heating technique is derived from laser surgery technology. It utilizes implanted frosted-sapphire probes, which are illuminated via FO cables with laser light emitted from a Nd:YAG laser generator. The frosted coating of the probes diffuses the laser energy into the surrounding tissue, where the energy is absorbed in the layers immediately adjacent to the implant. Adjustment of the penetration characteristics is possible by varying the coating pattern of the probes and the laser wavelength and by manipulation of the pulse parameters. Thermal modelling in phantoms and first animal studies have just started to reveal the potential of this device (MILLIGAN and OVERHOLT 1989). Combined use with specific photosensitizing agents employed in photodynamic therapy should be possible in the near future.

3.4 Current Technical Status and Future Prospects

Interstitial radiotherapy and IHT implantation follows similar principles, as shown in Fig. 1. In comparison with external hyperthermia several advantages are clear for interstitial techniques: more precise definition of tumor size and tumor selection criteria, better homogeneity of tumor temperature distributions, better sparing of normal tissue, accessibility of deep tumors, use of multipoint invasive thermometry ("thermal mapping"), and estimation of local blood flow by thermal washout measurements. These advantages provide a strong rationale for further development and implementation of interstitial hyperthermia. Each modality has its own advantages and drawbacks (Table 1). The comparison between the various techniques led some investigators to believe that

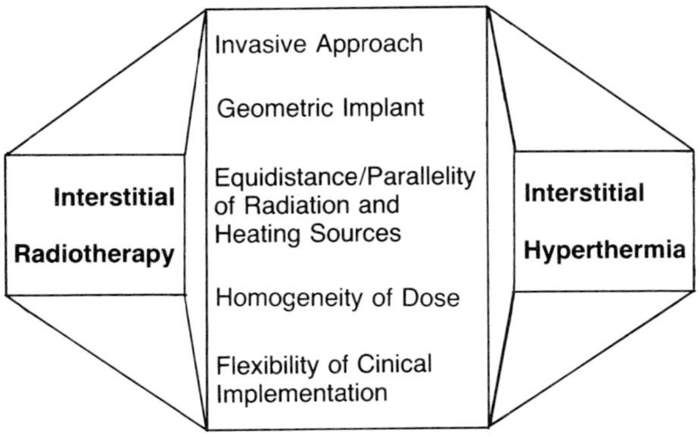

Fig. 1. Principles of interstitial thermoradiotherapy

the MW technique is superior to the LCF technique in some aspects of thermal performance characteristics (MECHLING and STROHBEHN 1986; PAULSEN et al. 1985), whereas other investigators favor the LCF technique with respect to the lesser variability of heating along the applicators and in complex implants (STAUFFER et al. 1989). Probably RF applicators are more advantageous for tumors, which can be implanted with an array of parallel, equal length, and equal distance metal needles (template technique), like those in the pelvis and in some areas of the head and neck and chest wall. MW antennae can be used for tumors which require nonparallel or flexible implants or are less accurate in terms of spacing and heating length, like those in the floor of the mouth, base of the tongue, or other head and neck lesions as well as chest wall and brain lesions. Hot source techniques, especially FM seeds, are useful in situations in which a high precision of implants can be achieved with few or no external cable connections, like thoracic, abdominal, and intracranial tumors.

The implantation guidelines for the use of IHT devices are more or less identical for all techniques:

- Close spacing of interstitial applicators (maximum 10–15 mm)
- Strict parallelism among all applicators (template technique)
- Equidistant positioning of all applicators
- Equidistant "active" heating length of all applicators
- Variable rearrangement of all applicators for adjustments
- Extensive thermometry within and outside of heating applicators including thermal mapping along the central axes of the implant

Several problems need to be resolved for the individual IHT techniques in the near future: (a) precise implant geometry for IHT and brachytherapy require good clinical expertise; (b) spacing between needles, antennae, or FM seeds has to meet specific physical characteristics; (c) heating patterns of applicator arrays need to correspond with interstitial radiation dose distribution; (d) invasive thermometry is limited to implanted probes, and mathematical modelling is required for 3-dimensional temperature calculations; (e) complex pretreatment planning and precise timing are required for these time- and personnel-consuming treatment procedures; (f) sophisticated, expensive technical equipment has to be utilized.

4 Clinical Experience

4.1 Treatment Rationale and Indications

From its initial stages of technical development, clinical interstitial thermoradiotherapy has been used in specialized institutions, and preliminary results from more than 500 patients have already been published. Most experience has accumulated in nonrandomized, phase I/II studies for palliative treatment of superficial or medium depth, primary advanced or local recurrent tumors, which had failed previous multimodality treatment approaches. Various treatment schedules have been reported: IHT has been used in combination with interstitial radiotherapy (IRT), either applying only one hyperthermia session before or after IRT, or applying two sessions of IHT before and after IRT. Only a few groups have combined IHT with percutaneous radiation therapy (PRT) or IHT and percutaneous hyperthermia (PHT) with PRT (TUPCHONG et al. 1988; YABUMOTO and SUYAMA 1984; INOUE et al. 1989).

Suitable situations for combining IHT and IRT are seen in tumors failing to achieve satisfactory local control when treated with conventional therapy: primary advanced and local recurrent tumors of the head and neck, N3 metastases, breast and chest wall lesions, gynecological, urogenital, anal canal, and rectal lesions (KAPP 1986; OVERGAARD 1989). Special techniques and treatment indications are given for brain (SALCMAN and SAMARAS 1983; ROBERTS et al. 1986; KOBAYASHI et al. 1986; SEEGENSCHMIEDT et al. 1987) and eye tumors (FINGER et al. 1986; ASTRAHAN et al. 1987). In abdominal surgery intraoperative radiotherapy and interstitial hyperthermia have been employed for pancreatic, bile duct, and colorectal tumors (COUGHLIN et al. 1985; MERRICK et al. 1988). Recently, intraluminal hyperthermia techniques were also investigated for application in esophageal, broncheal, and vaginal tumors as well as for benign prostate hyperplasia.

4.2 Clinical Results

Published data from interstitial hyperthermia trials so far differ in many details: type of hyperthermia technique employed (interstitial RF, MW, HS hyperthermia), interstitial and/or percutaneous radiation dose (20–60 Gy), number of IHT sessions (1–4), sequence of IHT and IRT (IHT before and/or after IRT), frationation between IHT sessions (48–120 h or weeks apart), therapeutic temperature levels (41°–45°C as minimum tumor temperature), and total treatment time (30–90 min). Despite these variations and considering the unfavorable patient selection, all clinical series have demonstrated encouraging results.

4.2.1 Interstitial Radiofrequency Hyperthermia

Table 2 summarizes the clinical experience with interstitial RF hyperthermia. In this collective experience of 299 patients treated with localized current field (LCF) and/or other interstitial RF techniques, a total of 169 patients (57%) achieved complete response (CR = 100% tumor regression), 65 patients (22%) partial response (PR = more than 50% tumor regression), and 57 patients (19%) no response (NR = less than 50% tumor regression) at initial treatment evaluation (after at least 1 month follow-up). Eight patients were not evaluable. The follow-up period ranged from 1–74 months. By combining complete and partial responders an overall response rate of 234/300 (78%) was noted for this treatment technique (VORA et al. 1982; MANNING et al. 1982; ARISTIZABAL and OLESON 1984; OLESON et al. 1984; YABUMOTO and SUYAMA 1984; COSSET et al. 1985, 1986; EMAMI et al. 1984, 1987; LINARES et al. 1986; GAUTHERIE et al. 1989b; VORA et al. 1989; YABUMOTO et al. 1989).

4.2.2 Interstitial Microwave Hyperthermia

Table 3 summarizes the clinical experience with interstitial MW hyperthermia, revealing a similar overall response rate in the collective experience of 215 patients treated with interstitial MW techniques. A total of 130 patients (60%) achieved CR, 61 patients (28%) PR, and 19 patients (9%) no change (NC) or progression (PD) as initial response. Five patients (3%) were not evaluable. The follow-up period varied between 1 and 48 months. The overall response rate (complete plus partial responders) was 191/215 (88%) (BICHER et al. 1984; STROHBEHN et al. 1984; PUTHAWALA et al. 1985; EMAMI et al. 1984, 1987; LAM et al. 1988; AOYAGI et al. 1989; INOUE et al. 1989) PETROVICH et al. 1989) SEEGENSCHMIEDT et al. 1989, 1990).

4.2.3 Interstitial Ferromagnetic Seed Hyperthermia

The universities of Arizona, Alabama, Wisconsin, and Kyoto, Japan have used thermally self-regulating FM seed hyperthermia clinically in the

Table 2. Clinical experience with interstitial radiofrequency hyperthermia

Reference	Year	Lesions	Clinical response				Toxicity (%)	Follow-up (months)
			CR (%)	PR (%)	NR (%)	Not evaluable		
VORA et al.	1982	16	69	6	19	1	6	1–3
OLESON et al.	1984	52	38	42	19	–	21	3–18
COSSET et al.	1985	29	66	14	–	6	25	>2
LINARES et al.	1986	10	30	70	–	–	40	short
EMAMI et al.	1987	9	56	33	11	–	25	6–48
GAUTHERIE et al.	1989	95	65	16	19	–	26	2
YABUMOTO et al.	1989	13	54	15	31	–	31	3–13
VORA et al.	1989	75	56	15	28	1	17	6–74
Total		299	57	22	19	8	22	

CR, complete response (100% tumor regression); PR, partial response (>50% tumor regression); NR, no response.

Table 3. Clinical experience with interstitial microwave hyperthermia

Reference	Year	Lesions	Clinical response				Toxicity (%)	Follow-up (months)
			CR (%)	PR (%)	NR (%)	Not evaluable		
Bicher et al.	1984	8	63	25	13	–	13	short
Strohbehn et al.	1984	6	50	33	17	–	17	short
Puthawala et al.	1985	43	74	2	–	–	21	>12
Emami et al.	1987	39	54	23	13	4	25	6–48
Aoyagi et al.	1989	10	20	40	40	–	20	–
Inoue et al.	1989	9	11	67	11	1	11	–
Petrovich et al.	1989	44	64	34	2	–	20	3–36
Own experience	1990	56	68	21	11	–	21	6–45
Total		215	60	28	9	5	21	

CR, complete response (100% tumor regression); PR, partial response (>50% tumor regression); NR, no response.

treatment of human tumors after extensive animal studies. Recently, the group from Arizona presented their initial results in 14 patients with primary advanced or recurrent tumors treated with FM seed hyperthermia (6 head and neck, 3 pelvic and 5 brain tumors). Safe and satisfactory heating was observed: 15/57 intratumoral points in the head and neck tumors, 17/48 in pelvic tumors, and 32/78 in brain tumors achieved time-averaged mean temperatures > 42°C. As relatively constant heating along the length of the implant occurs, the researchers believe this technique to be well suited for thermal modelling and estimation of blood perfusion. Due to the short follow-up, clinical data on tumor regression are still preliminary, but some patients are reported to be disease-free up to 18 months after diagnosis (Shimm et al. 1989).

4.2.4 Interstitial Hot Water Perfusion Hyperthermia

In a cooperative trial of the University Clinic for Radiotherapy and Radiobiology of Vienna and the Institute of Oncology in Ljubljana, the feasibility and safety of HWP hyperthermia have been examined in 18 patients with 23 lesions (14 head and neck tumors, 4 pelvic tumors, 3 tumors of the prostate, and 2 perineal recurrences of rectal carcinoma). Satisfactory heating and good patient tolerance were possible. In all, 4/23 (17%) tumors achieved CR and 17/23 (74%) PR. The complication rate was 22%; follow-up was 1–9 months (Handl-Zeller et al. 1987b; Schreier et al. 1990).

4.2.5 Personal Interstitial Microwave Hyperthermia Experience

In our own clinical experience at the Departments of Radiation Oncology of the University of Erlangen-Nürnberg and Hahnemann University, Philadelphia, in a cooperative phase I/II study between January 1986 and December 1989, 56 lesions (38 head and neck, 13 pelvic, and 5 other) in 55 patients (31 men, 24 women; 6–81 years old) received MW IHT combined with ^{192}Ir IRT. If possible, external radiation (ERT) was added. Mean dimensions of 52 lesions were $3 \times 4 \times 4.5\,cm^3$ with most tumor volumes ranging from 10 to $135\,cm^3$; four other lesions had extensive volumes $> 225\,cm^3$. IHT was applied immediately prior to and/or after low-dose ^{192}Ir IRT (20–30 Gy) for 1 h at 41°–44°C. ERT (40–50 Gy) was always delivered for AP (advanced primary) and LM (locally metastatic) lesions but was proportionally less for preirradiated LR (locally recurrent) lesions; the cumulative radiation dose did not exceed 110 Gy per site. IRT dose was 17–48 Gy (mean: 26.2 ± 8 Gy) at a dose rate of 25–70 cGyh (mean: 44 ± 12 cGy/h). Some 38 lesions received additional ERT (28–56 Gy, mean: 46 ± 8 Gy), and the total IRT plus ERT dose was 31–82 Gy (mean: 60 ± 18 Gy). Commercial hyperthermia systems (Lund/Buchler 4010 and Clini-Therm Mark VI/IX) were used. At 3 months FU 38 lesions (68%) showed CR, 12 (21%) PR, and 6 (11%) NC/PD. Long-term response of 40 lesions at 12 months FU revealed 35 patients (88%) with local control (LC) and 5 (12%) with in-field relaspses. Eight patients died before 12 months FU, 4 with LC and 4 with PD. In 16 lesions (29%)

acute side effects were observed, which resulted in 12 (21%) long-term complications with 3 lesions (7%) requiring plastic surgery.

Several prognostic factors were found: type of lesion, tumor volume, total radiation dose, quality of implant, and thermal parameters related to minimum temperature; such as minimum mean temperature T_{min} (mean) > 41°C (statistical trend) and quality of heating expressed as homogeneity of tumor temperatures, thermal quality, (TQ) 41°C > 75% (SEEGENSCHMIEDT et al. 1989, 1990).

4.3 Treatment Complications

In many reports IHT has been described as well tolerated with side effects similar to those observed with brachycurietherapy alone. However, in about 20%–25% of patients locoregional pain, discomfort, and systemic stress (tachycardia) can prevent effective heating performance. Patients treated under general anesthesia did not generally have more complications, but general anesthesia should be avoided if possible. In the published series the complication rate ranged between 6% and 40%, with a rate of 22% for RF hyperthermia (Table 2) and 21% for MW hyperthermia (Table 3), which is acceptable considering the frequent extensive pretreatment of these patients. In addition to minor complications (blisters and thermal burns), which heal within days or weeks, major complications (soft tissue necrosis and rapid tumor necrosis) have occasionally been observed. They can cause delayed healing over months, even with a concomitant complete response. Fistulae may occur in pelvic tumors (gynecological, urogenital, and rectal lesions), which sometimes require plastic surgery. Skin grafts are necessary for extensive skin defects.

4.4 Prognostic Treatment Factors

For external hyperthermia the minimum tumor temperature appears to be the most significant prognostic treatment factor (DEWHIRST et al. 1984) with best results above 43.0°C. EMAMI et al. (1987) have reviewed IHT trials with regard to the thermal quality of the hyperthermia treatment performance. Of 96 evaluable lesions with at least one satisfactory heating session, 77 (80%) achieved CR, whereas in 14 lesions with unsatisfactory heating, no CR was noted. Satisafactory heating was recorded when the entire gross tumor volume was included within the implant volume and the achieved temperature had reached a minimum of 42°C (EMAMI et al. 1987). The clinical use of other thermal parameters, e.g., the thermal dose concept, is still controversial.

In a similar way to hyperthermia equipment performance (HEP) rating (PAULSEN et al. 1985), we have established from our experience of 56 lesions a thermal quality score by assessing the rate of all measured tumor sensors yielding a time-averaged temperature > 41°C (TQ 41°C_{av}) from various temperature probes during a steady state treatment condition. We found a statistically significant higher CR rate for tumors which had achieved at least one treatment with TQ 41°C_{av} > 75% (SEEGENSCHMIEDT et al. 1989, 1990).

Tumor size has an impact on CR for several reasons: Larger lesions are more difficult to control, homogeneous implantation is more difficult, and homogeneous temperatures at the therapeutic level are more awkward to achieve in larger lesions (> 4 cm ∅) than in smaller ones, resulting often in poorer thermal quality (EMAMI et al. 1987).

A sufficient radiation dose appears to be another important parameter in conjunction with IHT treatments to achieve long-term local tumor control (COSSET et al. 1985; SEEGENSCHMIEDT et al. 1989, 1990). Some recurrences were recorded when low radiation doses in the range of only 20–30 Gy had been used. Thus, better results should be expected for primary advanced tumors with full-dose radiotherapy than for local recurrent lesions with compromised low-dose radiotherapy.

It is also important to recognize that CR is not an ideal criterion to evaluate the success of local hyperthermia within a palliative treatment concept because it does not account for recurrences outside the treated volume or death due to metastatic disease without recurrence of the tumor at the treated site. Therefore, LC appears to be the more appropriate measure to analyze long-term treatment success. It is believed that more than 75% of those patients exhibiting initially CR will have a long-term LC (PEREZ et al. 1987).

5 Conclusions

IHT follows the same principles and goals that apply for bracthycurietherapy: (a) optimal tumor dose including a sufficient margin around the lesion, (b) normal tissue sparing, (c) homogeneous dose distribution within the implant volume, and (d) better and long-term tumor control. However, optimal "heat dose' delivery is still hampered by technical and clinical insufficiencies of both treatment modalities such as unfavourable implant geometry, nonuniform heating patterns of antennae, or insufficient individual "power steering." Differences in vascular perfusion play another major role in thermal inhomgeneities. Many radiobiological questions still need to be answered. The future tasks of research are:

a) *Biological Tasks*
- Revealing the biomolecular mechanisms of thermal radiosensitization;
- Experiments using different fractionation, sequencing, and "heat doses";
- Prospective definition of a therapeutic heat dose ("thermal dose");

b) *Physics and Engineering Tasks*
- Improvement of individual applicator power steering;
- Homogeneous heating characteristics of heating probes;
- Development of noninvasive, 3-dimensional thermometry devices;

c) *Clinical Tasks*
- Definition of meaningful *palliative* treatment indications;
- Definition of meaningful *adjuvant* treatment indications;
- Optimization of planning, therapy, and quality assurance.

There is also a strong need for prospective randomized phase II/III trials to evaluate the various factors responsible for the observed therapeutic impact of combined thermoradiotherapy. Two studies have been initiated, the American RTOG 84-19 study (RTOG 1987) and the European ESHO 4-86 study (Overgaard 1987), which compare IRT alone versus combined IRT + IHT. The first study focuses on interstitial treatment of recurrent or persistent lesions and stratifies institution, histology, size of lesion, type of lesion, modality, and previous radiation dose as well as delivered dose to the study lesion. The second study concentrates on T1–T3 base of the tongue lesions eligible for primary treatment by external radiation plus brachytherapy. In the future, strong emphasis has to be put on strict quality assurance for IHT equipment and clinical implementation to control the broad spectrum of possible critical treatment parameters (Nussbaum 1984; Seegenschmiedt et al. 1989). These studies will provide more conclusive answers and further ideas for a broader adjuvant and palliative clinical use.

References

Akuta K, Hiraoka M, Jo S, Nishimura Y, Nagata Y, Masunaga S, Nohara H, Takahashi M, Abe M, Rhee JG, Song CW, Yamamoto I (1989) A newly developed interstitial applicator for RF current heating. In: Sugahara T, Saito M (eds) Hyperthermic Oncology 1988, vol 1. Taylor and Francis, London, pp 878–880

Aoyagi Y, Kanehira C, Kobori K, Hayakawa Y, Mochizuki S, Harada N (1989) Clinical experience with microwave interstitial hyperthermia. In: Sugahara T, Saito M (eds) Hyperthermic Oncology 1988, vol 1. Taylor and Francis, London, pp 601–603

Arcangeli G, Barni E, Cividalli A, Lovisolo G, Nervi C, Mauro F (1982) Hyperthermia by implantable applicators. In: Gautherie M (ed) Biomedical thermology. Liss, New York, pp 641–647

Aristizabal SA, Oleson JR (1984) Combined interstitial irradiation and localized current field hyperthermia: results and conclusions from clinical studies. Cancer Res [Suppl] 44: 4757s–4760s

Astrahan MA (1982) A localized current field hyperthermia system for use with 192 iridium interstitial implants. Med Phys 9: 419–424

Astrahan MA, Liggett P, Petrovich Z, Luxton G (1987) A 500 kHz localized current field hyperthermia system for use with ophthalmic plaque radiotherapy. Int J Hyperthermia 3: 423–432

Astrahan MA, Luxton G, Sapozink MD, Petrovich Z (1988) The accuracy of temperature measurements from within an interstitial microwave antenna. Int J Hyperthermia 4: 593–608

Atkinson WJ, Brezovich IA, Chakraborty DP (1984) Usable frequencies in hyperthermia with thermoseeds. IEEE Trans Biomed Eng 31: 70–75

Ben-Hur E, Elkind MM, Bronk BV (1974) Thermally enhanced radiosensitivity of cultered Chinese Hamster cells; inhibition of repair of sublethal damage and enhancement of lethal damage. Radiat Res 58: 38–51

Bicher HI, Moore Dw, Wolfstein RS (1984) A method for interstitial thermoradiotherapy. In: Overgaard J (ed) Hyperthermic Oncology 1984, vol 1. Taylor and Francis, London, pp 575–578

Brady LW, Seegenschmiedt MH (1987) Radiation therapy. JAMA 258: 2285–2287

Brezovich IA (1987) Ferromagnetics. In: Steeves RA, Paliwal BR (eds) Syllabus: a categorical course in radiation therapy. Hyperthermia. 73rd scientific assembly and annual meeting of the RSNA, Oak Brook, Illinois, RSNA, pp 117–126

Brezovich IA, Atkinson WJ, Lilly MB (1984a) Local hyperthermia with interstitial techniques. Cancer Res [Suppl] 44: 4752s–5756s

Brezovich IA, Takinson WJ, Chakraborty DP (1984b) Temperature distributions in tumor models heated by self-regulating nickel-copper alloy thermoseeds. Med Phys 11: 145–152

Brezovich IA, Lilly BM, Meredith RF, Weppelmann B, Henderson RA, Brawner W Jr, Salter MM (1990) Hyperthermia of pet animal tumours with self regulating ferromagnetic thermoseeds. Int J Hyperthermia 6: 117–130

Burton CV, Mozley JM, Walker AE, Braitman HE (1966) Induction thermocoagulation of the brain: a new neurosurgical tool. IEEE Trans Biomed Eng 13: 114–120

Chan KW, Chou CK, McDougall JA, Luk KH, Vora NL, Forell BW (1989) Changes in heating patterns of interstitial microwave antenna arrays at different insertion depths. Int J Hyperthermia 5: 499–508

Corry PM, Barlogie B (1982) Clinical application of high frequency methods for local hyperthermia. In: Nussbaum GH (ed) Physical aspects of hyperthermia. American Institute of Physics, New York, pp 307–322

Cosset JM, Brule JM, Salama AM, Damia E, Dutreix J (1982) Low-frequency (0.5 MHz) contact and interstitial techniques for clinical hyperthermia. In: Gautherie M (ed) Biomedical thermology. Liss, New York, pp 649–657

Cosset KM, Dutreix J, Dufour J, Janoray P, Damia E, Haie C (1984) Combined interstitial hyperthermia and brachytherapy: Institute Gustave Roussy technique and preliminary results. Int J Radiat Oncol Biol Phys 10: 307–312

Cosset JM, Dutreix J, Haie C, Gerbaulet A, Janoray P, Dewars JA (1985) Interstitial thermoradiotherapy: a technical and clinical study of 29 implantations performed at the Institute Gustave Roussy. Int J Hyperthermia 1: 3–13

Cosset JM, Dutreix J, Haie C, Mabire JP, Damia E (1986) Technical aspects of interstitial hyperthermia. In: Bruggmoser G, Hinkelbain W, Engelhardt R, Wannen-Macher M (eds) Locoregional high-frequency hyperthermia and temperature measurement. Springer, Berlin Heidelberg New York, pp 56–60 (Recent results in cancer research, vol 101)

Coughlin CT, Douple EB, Strohbehn JW, Eaton WL, Trembly BS, Wong TZ (1983) Interstitial hyperthermia in combination with brachytherapy. Radiology 148: 285–288

Coughlin CT, Wong TZ, Strohbehn JW, Colacchio TA, Sutton JE, Belch RZ, Douple EB (1985) Intraoperative interstitial microwave-induced hyperthermia and brachytherapy. Int J Radiat Oncol Biol Phys 11: 1673–1678

Deshmuk T, Damento M, Demer L, Forsyth K, De Young J, Dewhirst M, Cetas TC (1984) Feromagnetic alloys with curie temperatures near 50°C for use in hyperthermic therapy. In: Overgaard J (ed) Hyperthermic Oncology 1984, vol 1. Taylor and Francis, London, pp 571–574

Deurloo IKK, Visser AG, Ruifrok ACC, Lakeman RF, van Rhoon GC, Levendag PC (1989) Radiofrequency interstitial hyperthermia: a mulicentric program of quality assessment and clinical trials. In: Sugahara T, Saito M (eds) Hyperthermic Oncology 1988, vol 1. Taylor and Francis, London, pp 864–875

Dewhirst MW, Sim DA, Sapareto S, Connor WG (1984) The importance of minimum tumour temperature in determining early and long-term reponse of spontaneous pet animal tumours to heat and radiation. Cancer Res [Suppl] 44: 43–50

Doss JD, McCabe CW (1976) A technique for localized heating in tissue: an adjunct to tumor therapy. Med Instrum 10: 16–21

Emami B, Marks JE, Perez CA, Nussbaum GH, Leybovich L, von Gerichten D (1984) Interstitial thermoradiotherapy in the treatment of recurrent residual malignant tumors. Am J Clin Oncol 6: 699–704

Emami B, Perez CA, Leybovich L, Straube W, von Gerichten D (1987) Interstitial Thermoradiotherapy in the treatment of malignant tumours. Int J Hyperthermia 3: 107–118

Finger PT, Packer S, Kistner LM, Paglione R, Anderson LL, Svitra PP, Kim JH (1986) A thermoradiotherapy technique for choroidal melanoma. Endocuriether Hyperthermia Oncol 2: S-33–S-37

Forsyth K, Deshmuk R, De Young DW, Dewhirst MW, Cetas TC (1984) Recent clinical experience in pet animals with hyperthermic therapy in the head and neck region induced with inductively-heated ferromagnetic implants. In: Overgaard J (ed) Hyperthermic Oncology 1984, vol 1. Taylor and Francis, London, pp 599–602

Frazier OH, Corry PM (1984) Induction of hyperthermia using implanted electrodes. Cancer Res [Suppl] 44: 4864s–4866s

Gautherie M (1989) Interstitial hyperthermia: state of the art and prospects. In: Sugahara T, Sazito M (eds) Hyperthermic Oncology 1988, vol 2. Taylor and Francis, London, pp 63–68

Gautherie M, Cosset JM, Gerard JP, Horiot JC, Ardiet JM, El Akoum H, Alperovitch A (1989) Radiofrequency interstitial hyperthermia: a mulicentric program of quality assessment and clinical trials. In: Sugahara T, Saito M (eds) Hyperthermic Oncology 1988, vol 2. Taylor and Francis, London, pp 711–714

Gerner EW, Oval JH, Manning MR, Sim DA, Bowden GT, Hevezi J (1983) Dose rate dependence of heat radiosensitization. Int J Radiat Oncol Biol Phys 9: 1401–1404

Hall EJ (1985) The biological basis for endocurietherapy. Endocuriether Hyperthermia Oncol 1: 141–152

Handl-Zeller L, Kärcher KH, Schreier K, Handl O (1987a) Beitrag zur Optimierung interstitieller Hyperthermiesysteme. Strahlenther Onkol 163: 460–463

Handl-Zeller L, Kärcher KH, Lesnicar H, Budihna M, Schreier K (1987b) Newly developed liquid heated interstitial hyperthermia system. KHS-9/W18. Int J Hyperthermia 3: 567

Harisiadis L, Sung D, Kessaris N, Hall EJ (1978) Hyperthermia and low dose rate irradiation. Radiology 129: 195–198

Inoue T, Masaki N, Ozeki S, Ikeda H, Nishiyama K, Matayoshi Y, Kozuka T (1989) Clinical experience of interstitial hyperthermia combined with external radiation using MA-251 interstitial applicator. In: Sugahara T, Saito M (eds) Hyperthermic Oncology 1988, vol 1. Taylor and Francis, London, pp 598–600

James BJ, Strohbehn JW, Mechling JA, Trembly BS (1989) The effect of insertion depth on the theoretical SAR patterns of 915 MHz dipole antenna arrays for hyperthermia. Int J Hyperthermia 5: 733–747

Jones EL, Lyons BE, Douple EB, Dain BJ (1989) Thermal enhancement of low-dose irradiation in a murine tumour system. Int J Hyperthermia 5: 509–524

Joseph CD, Astrahan M, Lipsett J, Archambeau J, Forell B, George FW (1981) Interstitial hyperthermia and interstitial iridium-192 implantation: a technique and preliminary results. Int J Radiat Oncol Biol Phys 9: 827–833

Kapp DS (1986) Site and disease selection for hyperthermia clinical trials. Int J Hyperthermia 2: 139–156

King KWP, Trembly BS, Strohbehn JW (1983) The

electromagnetic field of an insulated antenna in a conducting or dielectric medium. IEEE Trans Biomed Eng 31: 574–583

Kobayashi T, Kida Y, Tanaka T, Kageyama N, Kobayashi H, Amemiya Y (1986) Magnetic induction hyperthermiam for brain tumors using ferromagnetic implant with low Curie temperature. J Neuroncol 4: 175–181

Lam K, Astrahan M, Langholz B, Jepson J (1988) Interstitial thermoradiotherapy for recurrent or persistent tumours. Int J Hyperthermia 4: 259–266

Le Bourgeois JP, Convert G, Dufour J (1978) An interstitial device for microwave hyperthermia of human tumors. In: Streffer C (ed) Cancer therapy by hyperthermia and radiation. Urban and Schwarzenberg, Baltimore, pp 122–124

Leeper DB (1984) Molecular and cellular mechanisms of hyperthermia alone or combined with other modalities. In: Overgaard J (ed) Hyperthermic Oncology 1984, vol 1. Taylor and Francis, London, pp 9–41

Lilly MB, Brezovich IA, Atkinson W, Charkraborty D, Durant JR, Ingram J, McElvein R (1983) Hyperthermia with implanted electrodes. In vitro and in vivo correlations. Int J Radiat Oncol Biol Phys 9: 373–382

Linares LA, Nort D, Brenner H, Shiu M, Ballon D, Anderson L, Alfieri A, Brennan M, Fuks Z, Hilaris B (1986) Interstitial hyperthermia and brachytherapy: a preliminary report. Endocuriether Hyperthermia Oncol 2: 39–44

Manning MR, Cetas TC, Miller RC, Oleson JR, Connor WG, Gerner EW (1982) Clinical hyperthermia: results of a phase I trial employing hyperthermia alone or in combination with external beam. Cancer 49: 205–216

Manning MR, Gerner EW (1983) Interstitial thermoradiotherapy. In: Storm FK (ed) Hyperthermia in cancer therapy. Hall, Boston, pp 467–477

Marchal C, Nadi M, Hofstetter S, Bey P, Pernot M, Prieur G (1989) Practical interstitial method of heating operating at 27.12 MHz. Int J Hyperthermia 5: 451–466

Mechling JA, Strohbehn JW (1986) A theoretical comparison of the temperature distributions produced by three interstitial hyperthermia systems. Int J Radiat Oncol Biol Phys 12: 2137–2148

Merrick HW, Milligan AJ, Greenblatt SH, Dobelbower RR (1988) Clinical experience with intraoperative interstitial hyperthermia and intraoperative radiation therapy. In: Abstracts of the 36th annual meeting of the Radiation Research Society, Philadelphia, Pennsylvania, April 16–21, 1988, p 45

Miller RC, Leith JT, Voemett RC, Gerner EW (1978) Effects of interstitial radiation therapy alone, or in combination with localized hyperthermia ona response of a mouse mammary tumour. Radiat Res 19: 175–180

Milligan AJ, Overholt BF (1989) Interstitial hyperthermia with a Nd:YAG. In: Abstracts of the 37th annual meeting of the Radiation Research Society, Seattle, Washington, March 18–23, 1989, p 48

Moorthy CR, Hahn EW, Kim JH, Feingold BS, Alifieri AA, Hilaris BS (1984) Improved response of a murine fibrosarcoma (Meth-A) to interstitial radiation when combined with hyperthermia. Int J Radiat Oncol Biol Phys 10: 2145–2148

Neyzari A, Cheung AY (1985) A review of brachyhyperthermia approaches for the treatment of cancer. Endocuriether Hyperthermia Oncol 1: 257–264

Nussbaum GH (1984) Quality assessment and assurance in clinical hyperthemia: Requirements and procedures. Cancer Res 44 (Suppl): 4811s–4817s

Oleson JR, Cetas TC (1982) Clinical hyperthermia with RF currents. In: Nussbaum GH (ed) Physical aspects of hyperthermia American Institute of Physics, New York, pp 280–366

Oleson JR, Manning MR, Sim DA, Heusinkveld M, Aristizibal SA, Cetas TC, Hevezi JC, Connor WG (1984) A review of the University of Arizona human clinical hyperthermia experience. Front Radiat Ther Oncol 18: 136–143

Overgaard J (1987) Hyperthermia as an adjuvant to radiotherapy. Review of the randomized multicenter studies of the European Society of Hyperthermic Oncology. Strahlenther Onkol 163: 453–457

Overgaard J (1989) The current and potential role of hyperthermia in radiotherapy. Int J Radiat Oncol Biol Phys 16: 535–549

Paliwal BR, Vandeby R Jr, Wakai R, Buechler D, Steeves RA, Shrivastava P, Partington B, Tompkins DT (1989) Studies of thermal distribution from ferromagnetic implants. In: Sugahara T, Saito M (eds) Hyperthermic oncology 1988, vol 1. Taylor and Francis, London, pp 854–856

Paulsen KD, Strohbehn JW, Lynch DR (1985) Comparative theoretical performance of two types of regional hyperthermia systems. Int J Radiat Oncol Biol Phys 11: 1659–1671

Perez CA, Emami B, Nussbaum G, Sapareto S (1987) Hyperthermia. In: Perez CA, Brady LW (eds) Principles and practice of radiation oncology. Lippincott, Philadelphia, pp 317–352

Petrovich Z, Langholz B, Lam K, Luxton G, Cohen D, Jepson J, Astrahan M (1989) Interstitial microwave hyperthermia combined with Iridium-192 radiotherapy for recurrent tumors. Am J Clin Oncol 12: 264–268

Prionas SD, Fessenden P, Kapp DS, Goffinet DR, Hahn GM (1989) Interstitial electrodes allowing longitudinal control of SAR distributions. In: Sugahara T, Saito M (eds) Hyperthermic Oncology 1988, vol 2. Taylor and Francis, London, pp 707–710

Puthwala AA, Syed AMN, Khalid MA, Rafie S, McNamara CS (1985) Interstitial hyperthermia for recurrent malignancies. Endocuriether Hyperthermia Oncol 1: 125–131

Reinhold HS, Endrich B (1986) Tumour microcirculation as a target for hyperthermia. Int J Hyperthermia 2: 117–137

Roberts DW, Coughlin CT, Wong TZ, Fratkin JD, Douple EB, Strohbehn JW (1986) Interstitial hyperthermia and iridium brachytherapy in treatment of malignant glioma. J Neurosurg 64: 581–587

Roos D, Hugander A (1988) Microwave interstitial applicators with improved longitudinal heating patterns. Int J Hyperthermia 4: 609–615

RTOG – Radiation Therapy Oncology Group (1987) Pre-meeting reports. Philadelphia, July 8–10, 1987, Vol 1. pp 217–218

Rude J, Luk KJ, Vora NL, Forell B, Findley D, Doggett S (1987) Multipoint feedback control of power: preliminary clinical results. In: Abstracts of the 35th annual meeting of the Radiation Research Society, Atlanta, February 21–26, 1987, p 6

Salcman M, Samaras GM (1983) Interstitial microwave hyperthermia for brain tumors. J Neurooncol 1: 225–236

Satoh T, Stauffer PR (1988) Implantable helical coil microwave antenna for interstitial hyperthemia. Int J Hyperthermia 4: 497–512

Schreier K, Budihna M, Lesnicar H, Handl-Zeller L, Hand

JW, Prior MV, Clegg ST, Brezovich IA (1990) Preliminary studies of interstitial hyperthermia using hot water. Int J Hyperthermia 6: 431–444

Seegenschmiedt MH, Sauer R (1989) Methoden und klinische Ergebnisse der interstitiellen Thermoradiotherapie. Strahlenther Onkol 165: 360–368

Seegenschmiedt MH, Brady LW, Karlsson UL, Black P, McCormack T (1987) A critical review of interstitial thermoradiotherapy for recurrent malignant astrocytoma: problems and promises. Int J Hyperthermia 3: 589

Seegenschmiedt MH, Sauer R, Herbst M, Thiel H-J, Fietkau R, Brady LW, Karlsson UL (1989). Interstitial hyperthermia for head and neck tumors: treatment planning and quality assurance (QA). In: Sugahara T, Saito M (eds) Hyperthermic Oncology 1988, vol 2. Taylor and Francis, London, pp 524–527

Seegenschmiedt MH, Sauer R, Fietkau R, Brady LW, Karlsson UL (1990) Primary advanced and local recurrent head & neck tumors: effective management with interstitial thermo-radiotherapy. Radiology 176: 267–274.

Shimm D, Stea B, Cetas TC, Buechler D, Carter L, Chen J, Dean S, Fletcher A, Guthkelch A, Haider S, Hodak J, Iacono R, Lutz W, Obeens E, Rossman K, Sinno R, Spetzler R, Cassady J (1989) Clinical results of interstitial hyperthermia using thermally regulating ferromagnetic seeds. In: Sugahara T, Saito M (eds) Hyperthermic Oncology 1988, vol 2. Taylor and Francis, London, pp 536–539

Song CW (1984) Effect of local hyperthermia on bloodflow and microenvironment. Cancer Res [Suppl] 44: 4721s–4730s

Stauffer PR, Cetas TC, Jones RC (1984) Magnetic induction heating of ferromagnetic implants for inducing localized hyperthermia in deep seated tumors. IEEE Trans Biomed Eng 31: 76–91

Stauffer PR, Sneed PK, Suen SA, Satoh T, Matsumoto K, Fike JR, Philips TL (1989) Comparative thermal dosimetry of interstitial microwave and radiofrequency-LCF hyperthermia. Int J Hyperthermia 5: 307–318

Strohbehn JW (1983) Temperature distributions from interstitial RF electrode hyperthermia systems: theoretical predictions. Int J Radiat Oncol Biol Phys 9: 1655–1667

Strohbehn JW, Douple EB, Coughlin CT (1984) Interstitial microwave antenna array systems for hyperthermia. Front Radiat Ther Oncol 18: 70–84

Trembly BS (1985) The effects of driving frequency and antenna length on power deposition within a microwave antenna array used for hyperthermia. IEEE Trans Biomed Eng 32: 152–157

Tupchong L, Nerlinger RE, Waterman FM (1988) Combined use of external beam radiation and interstitial heat for advanced recurrent head and neck carcinomas – a new approach. In: Abstracts of the 36th annual meeting of the Radiation Research Society, Philadelphia, April 16–21, 1988, p 42

Turner PF (1986) Interstitial equal-phased arrays for EM hyperthermia. IEEE Trans Biomed Eng 34: 572–578

Visser AG, Deurloo IKK, Levendag PC, Ruifrok ACC, Cornet B, van Rhoon GC (1989) An interstitial hyperthermia system at 27 MHz. Int J Hyperthermia 5: 265–276

Vora NL, Forell B, Joseph C, Lipsett JA, Archambeau J (1982) Interstitial implant with interstitial hyperthermia. Cancer 50: 2518–2523

Vora NL, Luk KH, Forell B, Findley DO, Lipsett JA, Pezner RD, Desai KR, Wong JYC, Hill B (1988) Intersitial local current field hyperthermia for advanced cancers of the cervix. Endocuriether Hyperthermia Oncol 4: 97–106

Vora NL, Forell B, Luk KH, Pezner RD, Desai KR, Lipsett JA, Wong JYA (1989) Interstitial thermoradiotherapy in recurrent and advanced carcinoma of malignant tumors: seven years experience. In: Sugahara T, Saito M (eds) Hyperthermic Oncology 1988, vol 1. Taylor and Francis, London, pp 588–590

Wong TZ, Ryan TP, Stohbehn JW, Jones KM (1989) A phase-coherent interstitial microwave antenna array hyperthermia system. In: Sugahara T, Saito M (eds) Hyperthermic Oncology 1988, vol 1. Taylor and Francis, London, pp 894–895

Yabumoto E, Suyama S (1984) Interstitial radiofrequency hyperthermia in combination with external beam radiotherapy. In: Overgaard (ed) Hyperthermic Oncology 1984, vol 1. Taylor and Francis, London, pp 579–582

Yabumoto E, Suyama S, Show K, Yamazaki T (1989) A phase I clinical trial of radiofrequency interstitial hyperthermia combined with external radiotherapy. In: Sugahara T, Saito M (eds) Hyperthermic Oncology, vol 1. Taylor and Francis, London, pp 591–593

9.2 Interstitial and Intraoperative Hyperthermia

CHRISTOPHER T. COUGHLIN

CONTENTS

1 Introduction 359
2 Method 359
3 Patients and Results 360
4 Intraoperative Implantations 362
References 363

1 Introduction

Over the past 15 years, interstitial techniques have become popular within the field of radiation oncology. Implantation techniques involve the uniform spacing and placement of a variety of radioactive materials directly within a tumor volume. With proper surgical assistance in carefully placing silastic catheters throughout a tumor volume, virtually any primary site can be implanted (HILARIS 1975). The energy of the isotope utilized for the implant governs the separation of the catheters and the number of catheters required for a uniform isodose distribution. Iridium 192 has become a popular isotope for implantation techniques within the USA. Iridium's 340-keV γ-ray emission dictates that a uniform dose distribution can be obtained by separating catheters by 0.5–1.0 cm. A 74.5-day half-life also enables the implant to be left in place for several days without significant decay of activity. Thus, iridium 192 is the basis for a very convenient system for implants.

Hyperthermia has also gained popularity in the past few years (STORM 1983). This modality involves the use of heat to kill tumor tissue selectively. If a tumor mass can be heated to between 42.5° and 45°C, a window of relatively selective cell-kill can be created for malignant tissue. When administered closely in time with irradiation, hyperthermia can be synergistic with the irradiation cell-kill effect (DEWEY et al. 1980).

CHRISTOPHER T. COUGHLIN, M.D., F.A.C.P., F.A.C.R., Professor of Radiation Oncology Dartmouth Hitchcock, Medical Center, 2 Maynard Street, Hanover, NH 03756, USA

2 Method

It has been difficult to develop systems adequate to administer heat for the delivery of hyperthermia. Over the past 10 years an interstitial microwave system has been developed at the Dartmouth-Hitchcock Medical Center in cooperation with the Thayer School of Engineering at Dartmouth (STROHBEHN et al. 1986). Microwave antennas have been designed of a small enough diameter to fit through the standard interstitial catheters used for the delivery of brachytherapy. These antennas can deliver heat to a given tumor volume based on the geometry of the implant. The spacing of the implantation catheters is based on the iridium dosimetry. Optimal uniformity of the iridium dose conforms very closely to that necessary to heat a tumor volume evenly. The antenna is designed to heat a 0.5 to 1.0-cm radius easily when used in an array. Creating a uniform irradiation dose throughout a tumor volume usually requires more catheters thatn are necessary to deliver a uniform hyperthermia treatment. Conduits used for iridium can serve as the additional catheters necessary for the thermometry required for hyperthermia. Another useful feature for the delivery of hyperthermia has been the development of a computerized feedback control system (STROHBEHN et al. 1985) which allows the delivery of power through each microwave antenna to be monitored and controlled in an instantaneous feedback loop. This makes it possible to regulate continuously the power deposition within the tumor volume and therefore to maintain nearly constant temperatures throughout the tumor mass over time. Blood flow varies as a function of heat and time in a nonhomogeneous fashion throughout a mass, leading to much closer monitoring and control of the hyperthermia treatment.

A more extensive description of the technical aspects of interstitial hyperthermia has been published elsewhere (Strohbehn et al. 1984). The Dartmouth Technique involves the use of microwave antennas that can be inserted through interstitial catheters with a 2.2-mm outer diameter and a 1.8-mm inner diameter which allows placement of the antenna and 4 Luxtron fiberoptic probes. Radiofrequency antennas or thermomagnetic seeds have not been used at our institution. A microwave interstitial system most commonly utilizes a 915-MHz power source. Power sources of 433 and 2450 MHz are also available to accomodate both larger and smaller tumor diameters, respectively. Each power source has an optimal length of useful heating. For the 2450-MHz system, it is approximately 2 cm in length; for the 915-MHz system, it is approximately 5 cm in length; and for the 433-MHz system, it is 8–10 cm along the axis of the anetnna.

Table 2. Intended protocol at optimal dosage day after implantation

Heat:	43°C for 60 min
Iridium:	6000 cGy at 5.0 cm from plane of implant over 4–7 days
Heat: (Second treatment)	43°C for 60 min
Plus external beam irradiation if clinically indicated	

3 Patients and Results

As of January 1988, 58 patients have been implanted with 113 interstitial microwave-induced hyperthermia treatments administered to a variety of tumor sites. The majority of these patients were treated with a 915-MHz power source and involved tumor thicknesses in the 4–6-cm range. The standard implant rules for iridium dosimetry were utilized to accommodate the entire tumor volume (Hilaris 1975). A breakdown of these primary sites are given in Table 1. As can be seen from this distribution, virtually any primary site can be approached with interstitial techniques. Table 2 gives a breakdown of the intended sequence and optimal tumor dose.

The degree of difficulty for an implant varies, depending upon the primary site. For brain tissues, the technique involves close neurosurgical cooperation and has been described fully elsewhere (Roberts et al. 1987). Catheters are inserted using a stereotactic procedure with a Lexell stereotactic headframe and computerized tomography (CT) images from a GE 9800 CT scanner. The algorithms from the CT scan are converted into stereotactic coordinates. The neurosurgeon and radiation oncologist jointly decide the number and orientation of the catheters. Once the catheters are in place, microwave antennas can be inserted along with the appropriate thermometry devices. The goal is to achieve 43°C for 60 min to a 5-mm margin around the ring-enhancing volume based upon the CT scan. This often necessitates the use of higher temperatures in the center of the tumor volume (Fig. 1). Every effort is made to

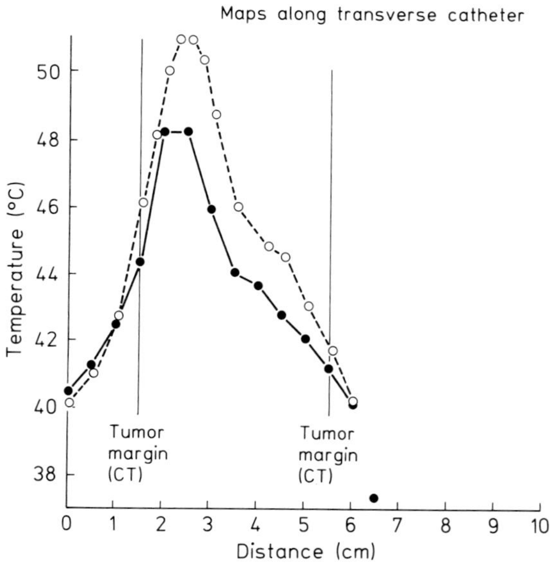

Fig. 1. Thermal map across a brain tumor with margins indicated. *Crosses* represent a temperature sweep at 20 min, and *Circles* indicate a sweep at 40 min during a hyperthermia session. Peak temperatures within the tumor reached 51°C while distal margin temperatures did not reach the therapeutic threshold of 42°C. Also, note the change in temperatures with time

Table 1. Patients studied arranged according to tumor site

Site	Brain	Head and neck	Lung	Gastrointestinal	Gynecologic	Superficial
	21	18	11	1	15	5

keep normal brain tissue at temperatures lower than 42°C. During each heat treatment, there is continuous monitoring of the EEG, visual evoked potentials, and frequent neurological examination. After hyperthermia treatment, the microwave antennas and thermometry devices are removed, and the catheters are loaded with iridium 192. Seeds of 1.0 mCi are utilized, the number being governed by the iridium dosimetry. The goal is to deliver 6000 cGy to a 5-mm margin around the ring-enhancing tumor volume, which usually takes 4–7 days. Once this dose is delivered, a second hyperthermia treatment can be given. Upon completion of the second heat treatment, all of the catheters are removed. If the patient has not received prior irradiation, a full course of external beam treatment is given with the intent of delivering 6000 cGy to the tumor volume over 6 weeks, using a shrinking field technique. This approach has been well-tolerated: Eighteen patients have been treated in this manner with acceptable side effects. Four patients have had a transient increase in their paresis, and four patients have had a permanent increase in their local neurological defect. One patient required emergency decompression toward the end of his iridium treatment. Further refinements in the dosimetry as well as in the total dosages of both iridium and hyperthermia are in progress.

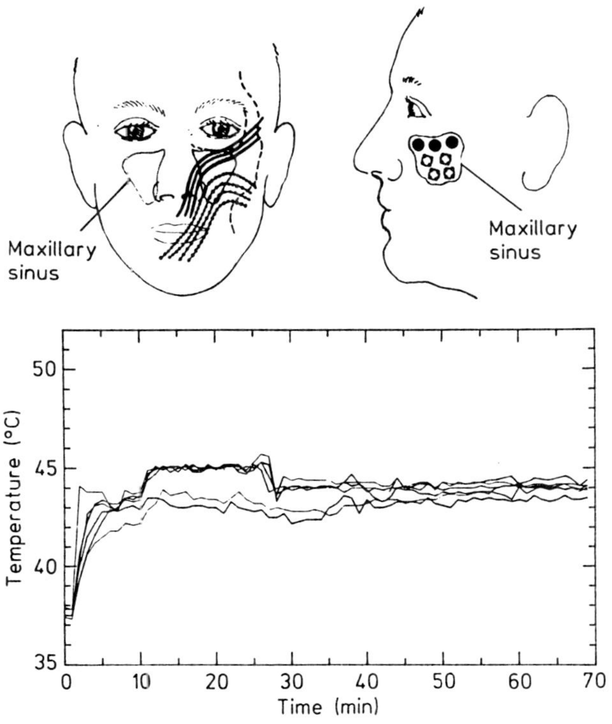

Fig. 2. Schema of a maxillary sinus implant. The thermal map is plotted as temperature versus time for each of the six microwave antennas. The seventh catheter was used for thermometry. Clustering of temperature data in the therapeutic range over the duration of the hyperthermia session is seen

Ten patients with advanced head and neck carcinomas were implanted. Primary sites have included the maxillary antrum, base of the tongue, buccal mucosa, and salivary gland. Standard implantation techniques are performed under general anesthesia with the assistance of a head and neck surgeon (Coughlin et al. 1985). The number of catheters necessary to encompass the tumor volume is dictated by the iridium dosimetry but usually requires 5–10-mm separation. Once the implant is complete and the patient is stabilized, microwave antennas can be inserted through the catheters. The goal is to heat the tumor volume to 42.5°C for 1 h. One constraint in the head and neck region is that bone pain can be temperature limiting. Sedation is often required during the hyperthermia session. Pain is directly related to the amount of power input necessary and subsides immediately upon reduction of the power to each anetnna. An example is given in Fig. 2. Iridium is administered with a target goal of 4000–6000 cGy to a 5-mm radius around the tumor volume over 4–7 days. Once this is completed, a second hyperthermia treatment is delivered. The catheters are then removed, and patients can be given external beam radiation therapy is they have not been heavily pretreated. This technique allows excellent palliation, and there are four long-term survivors.

Biliary cancers represent another unique technical challenge. These tumors are treated with intracavitary techniques that have been previously described (Ryan et al. 1987; Coughlin et al. 1987). After documentation of an obstructing biliary carcinoma, percutaneous drainage is accomplished with a standard French ring biliary drainage catheter. Through this catheter, a microwave antenna and a thermometry device can be placed in direct opposition to the site of obstruction. It is not possible to measure temperature in a raidal fashion, but proximal and distal margins as well as maximum temperatures within the catheter can be recorded (Fig. 3). Doses similar to those for other cancers were given, with the goal of 43°C for 1 h at the tumor margins. These treatments have bracketed iridium irradiation with doses that have varied between 2000 and 6000 cGy, calculated at 1 cm from the catheter in a radial fashion. A full

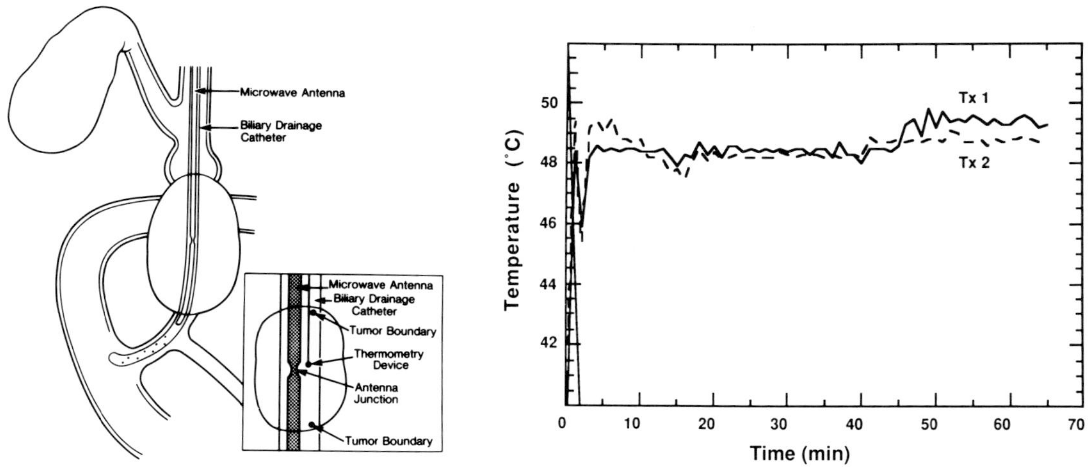

Fig. 3. Schema of a biliary implant with a thermal representation of both treatment sessions at the antenna junction. Temperatures well within the therapeutic range can be easily maintained throughout the course of each session

course of external beam radiation is delivered, if possible, after completion of the interstitial technique. Thirteen patients have been treated in this fashion, with seven responses sufficient for removal of their biliary draining catheters. These treatments were administered is an entirely asymptomatic fashion and with no subsequent difficulties from bleeding or infection.

A similar technical approach was used to deliver doses to gastrointestinal tract, gynecological lung, and superficial cancers. Once again, these implants are usually performed in the operating room with the assistance of an attending surgeon. Catheters can be exteriorized from any primary site and function as conduits through which microwave antennas, thermometry devices, and iridium can be administered.

4 Intraoperative Implantations

It became clear after a number of attempts at intraoperative implantations that one needed to develop a more efficient delivery system. To that end, a Phillips 305 orthovoltage X-ray machine was obtained for use in an intraoperative setting. The technique for radiation treatment was similar to that developed by Rich (1986) and Piontek et al. (1986) at the New England Deaconess Hospital. Once this delivery system was functioning, it became clear that intraoperative hyperthermia was also quite feasible. The two areas of major interest have been colorectal carcinoma and pancreatic cancer. These diseases are difficult to control locally and are amenable to direct surgical exposure. Irradiation doses are governed by the volume of tumor to be treated: 1250 cGy for microscopic residual disease, 1500 cGy for less than 2-cm residual disease, and 1750 cGy for greater than 2-cm residual disease. A variety of cone diameters and bevel angles is available to meet the clinical need.

The hyperthermia system available is based upon the use of ultrasound. Transducers with a variety of diameters are available, as described previously (Ryan et al. 1986; Coughlin 1987). The transducer diameters vary from 4 to 10 cm. The transducer of the appropriate diameter can be placed directly on a tumor once it has been properly exposed. Thermometry devices can be directly inserted into the tumor mass and record continuously. The power to the transducer is regulated through a feedback control system from the thermometry devices. The goal is to obtain a constant temperature throughout the tumor volume for as long as necessary, ideally, 42.5°C at the tumor boundary is to be maintained for 1 h. To date, 14 patients have been treated, 7 with pancreatic carcinoma and 7 with colorectal carcinoma. Figure 4 shows a representative thermal mapping. Great care must be taken to avoid heating of the small bowel, large bowel, or ureter. Work is being conducted to assess the tolerance of small nerves to these hyperthermia and irradiation doses. Large blood vessels, muscle, connective tissue, and bone do not appear to be readily

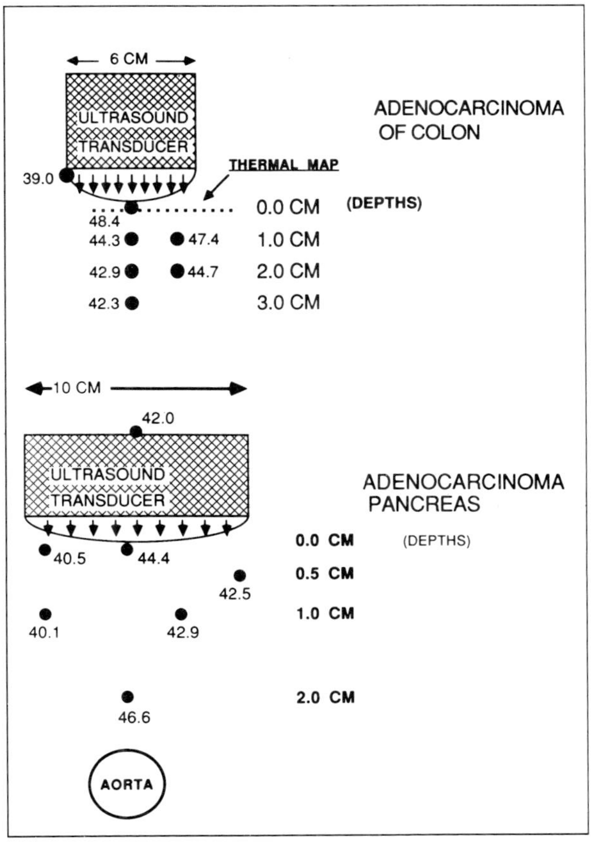

Fig. 4. Schema of transducer placement of thermal maps for an adenocarcinoma of colon and of pancreas. This indicates therapeutic temperatures directly measured at various depths below the transducer

damaged by this approach. It is difficult to assess tumor shrinkage for these deep-seated tumors, as this involves only indirect evaluation. Follow-up CT scans as well as physical examination are used to check the tumor response. External beam radiation is also utilized after the patient has recovered from the surgical procedure. The impression arising is that intraoperative hyperthermia is technically feasible, and as these techniques become refined, they can be integrated into a full treatment program for these difficult cancers.

References

Coughlin CT (1987) Intraoperative irradiation and hyperthermia, 8th International Congress on Radiation Research, Edinburgh

Coughlin CT, et al. (1985) Implantation of maxillary antrum for delivery of iridium brachytherapy and microwave-induced hyperthermia. Radiation Research Society, Los Angeles

Coughlin CT, et al. (1987) Treatment of obstructing biliary cancers with interstitial microwave-induced hyperthermia and iridium brachytherapy. European Society for Hyperthermic Oncology, Cardiff

Dewey WC, et al. (1980) Cell biology of hyperthermia and radiation, Raven, New York (Radiation biology in cancer research)

Hilaris BS (1975) Handbook of interstitial brachytherapy. Publishing Sciences, Acton

Piontek RW, et al. (1986) Design and dosimetric properties of an intraoperative radiation therapy system using orthovoltage x-ray unit. Int J Radiat Oncol Biol Phys 12: 255–259

Rich TA (1986) Intraoperative radiotherapy. Radiother Oncol 6: 207–221

Roberts DW, et al. (1987) Iridium-192 brachytherapy in combination with interstitial microwave-induced hyperthermia for malignant glioma. Appl Neurophysiol 50: 281–291

Ryan TA, et al. (1986) Temperature analysis of ultrasound-induced hyperthermia in patients with superficial tumors. 34th Annual Meeting of the Radiation Research Society, Las Vegas

Ryan T, et al. (1987) SAR evaluation of interstitial clinical heating for biliary obstruction due to cancer. 13th Northeast Bioengineering Conference, Phildelphia

Storm FK (1983) Hyperthermia in cancer therapy. Hall, Boston

Strohbehn JW, et al. (1984) Frontiers in radiation therapy oncology, vol 18. Karger, Basel

Strohbehn JW, et al. (1984) Hyperthermia and cancer therapy: a review of biomedical engineering contributions and challenges. IEEE Trans Biomed Eng BME 31: 779–787

Strohbehn JW, Mechling JA (1986) Interstitial techniques for clinical hyperthermia. In: Hand JW, James JR (eds) Physical techniques for clinical hyperthermia. Somerset, England, Research Studies, pp 201–287

9.3 Interstitial Thermoradiotherapy in the Treatment of Malignant Tumors

BAHMAN EMAMI and CARLOS A. PEREZ

CONTENTS

1 Introduction 365
2 Method of Interstitial Hyperthermia 365
2.1 Resistive Radiofrequency Heating 365
2.2 Radiative Microwave Heating 365
2.3 Inductively Heated Ferromagnetic Seeds 366
2.4 Experience at Our Institution 366
3 Results 366
4 Complications 367
5 Discussion 367
References 369

1 Introduction

The history of using heat to treat diseases (tumors) dates back as far as 400 B.C. Recent interest, however, in using heat to treat human malignancies is based on biological and clinical research of the past 2 decades. Most of the current experience with hyperthermia is with external local hyperthermia and superficial tumors (PEREZ and EMAMI 1989). Significant problems exist with this type of heating such as: The inability of external applicators to achieve the correct temperature within the target volume and a lack of adequate depth of penetration. Interstitial hyperthermia appears to be a suitable alternative in certain clinical situations. It has several advantages (EMAMI et al. 1987): (a) More uniform heating within the target volume, (b) more accurate and comprehensive temperature measurement, (c) there is probably better sparing of normal tissue due to the fact that the elements are implanted within the tumor only, and (d) suitable for both deep and superficial tumors.

BAHMAN EMAMI, M.D., Professor of Radiology, chief, Hyperthermia Section, CARLOS A. PEREZ, M.D., Professor of Radiology, Director Radiation Oncology Center, Mallinckrodt Institute of Radiology, Washington University Medical School, 4939 Audubon Avenue, St. Louis, MO-63110, USA

2 Method of Interstitial Hyperthermia

The following methods are used for interstitial hyperthermia in clinics: resistive radiofrequency (LCF), radiative microwave (microwave antennas), inductively heated ferromagnetic seeds, and other methods (hot water tubes).

2.1 Resistive Radiofrequency Heating

With the LCF technique, the therapist encompasses the target volume (to be heated) with rows of parallel needles (electrodes), and heat is produced by currents driven between electrically connected arrays of these needles. From the clinical point of view, LCF can generally provide a relatively homogeneous temperature distribution within the slabs of tissue encompassed by the implant. Potential problems and possible solutions when using this technique in the clinic are as follows: (a) Patient discomfort (if implant left in situ for several days for more than one session of heating), (b) potential for distortion of implant geometry (parallelsim) if implant is left in situ for a few days, (c) incompatibility with afterloading interstitial radiotherapy, and (d) possible heating of normal tissue overlying the tumor volume. To solve the last problem, one can insulate the section of needles passing through normal tissue. In this case a precise pretreatment planning is mandatory.

2.2 Radiative Microwave Heating

With this technique the therapist utilizes radiative microwave antennas which are usually placed within the preimplanted catheters in the tumor. Each antenna heats a specific volume within the tumor. Given that many needles are usually used to implant a given tumor volume, the interaction and interrelationships between these antennas are important. This method has some advantages for clinical use including flexibility, adaptability in combination with brachytherapy, comfort to the patient, and ease of use for the physician. It also has some disadvantages: (a) Fixed length of heated volume (potential solution is to heat various compartments of tumors at different sessions); (b) unsatisfactory temperature distribution in very superficial tumors, which can be avoided by the creation of complementary tissue equivalent over the tumor; and (c) unheated area ("dead space") at the tip of the antenna. With the new design of antenna this problem has improved but has not been completely solved (SATHIASEELAN et al. 1990).

As with the LCF technique, precise spacing is very important, and an inappropriate implant technique may result in cold or hot spots.

2.3 Inductively Heated Ferromagnetic Seeds

This is a passive technique in which ferromagnetic seeds are implanted within the tumor, and the heating is induced with an external device. The major difference between this type of heating and the two previous techniques is that in the latter the power is deposited within the tissue whereas with ferromagnetic seeds the power is deposited in the seed and the heating of the tissue is a function of heat conduction within the tissue. Several potential problems with this technique need to be overcome, including long curie point, low biocompatability, seed migration, and need for longitudinal implantation of the seeds.

Our experience is only with the first two techniques; we have not utilized ferromagnetic seeds in our clinic.

2.4 Experience at Our Institution

We have used both LCF and microwave techniques in the thermoradiotherapy of recurrent tumors. We believe there should be no difference between the two techniques when an appropriate selection is made based on tumor size and site and when the technique is properly executed. Usually, the choice between these methods is heavily influenced by the institutional preference and the availability of technical expertise.

We have analyzed the results of 48 recurrent and/or persistent tumors in 46 patients treated by interstitial thermoradiotherapy (EMAMI et al. 1987). There were 27 men and 19 women. The youngest patient was 37 years old and the oldest, 83 years old. The primary tumor sites were head and neck 29 lesions, breast 6 lesions, pelvis 7 lesions, and other areas 6 lesions. Using the average dimension method $\left(\frac{A+B+C}{3}\right)$, there were 13 lesions less than 4 cm, 32 lesions 4–10 cm, and 3 lesions of 10 cm in average dimension. All these tumors had persisted after prior courses of conventional treatments.

Interstitial radiation was delivered with ^{192}Ir interstitial afterloading technique except in nine patients in whom on external course of radiation therapy was carried out. The total dose varied from 2000 to 6000 cGy, depending on the previous dose of radiation to the site.

Hyperthermia was administered either with the LCF technique (19 patients) or by microwave antenna (39 lesions). Patients treated with the former received one hyperthermia session followed by a course of brachytherapy; in patients treated with the latter there were two hyperthermia sessions, one before and one immediately after a course of interstitial radiation. With both techniques, the aim was to achieve a minimal measured tumor temperature of 42.5°C for 60 min. To achieve this goal we often had to accept higher temperatures within the target volume.

3 Results

Four patients were lost to follow-up. Of the remaining 44 patients available for evaluation of the response, there were 26 (59%) complete responders and 12 partial responders. The follow-up varied from 6 to 36 months. These results are in general agreement with those in other published series. A summary of the reported series in the literature is given in Table 1. Of the total of 286 lesions, 181 (63.5%) showed a complete response and 32 (24%), a partial response. There was no statistically significant difference between the complete response rate achieved with the LCF technique (52%) versus the microwave technique (88.2%). The apparent difference between the results with the two techniques is due to the technical difficulties in the earlier series with LCF techniques (YABUMOTO and Suyama 1984; OLESON

Table 1. Summary of Clinical results of interstitial thermoradiotherapy

Microwave				Radiofrequency			
Author	No. of evaluable patients	CR	PR	Author	No. of evaluable patients	CR	PR
Bicher et al. (1984)	8	5	2	Cosset et al. (1985)	23	19	4
Emami et al. (1987)	35	21	9	Emami et al. (1987)	9	5	3
Petrovich et al. (1990)	44	28	15	Linares et al. (1986)	10	5	5
Puthawala et al. (1985)	43	37	6	Oleson et al. (1984)	52	20	22
Strohbehn et al. (1984)	6	3	2	Vora et al. (1982)	16	11	1
Total	136	120 (88.2%)	27 (19.8%)		117	61 (52.1%)	37 (31.6%)

CR, complete response; PR, partial response (over 50% regression of tumor volume)

et al. 1984). With the proper LCF technique, Cosset et al. (1985) achieved 19 of 23 (82.6%) complete responses, which is comparable to the reported series using the microwave technique.

The term "complete response" and "tumor control" have often been used synonymously in reporting the results from hyperthermia trials. The fact is that not all complete responses will eventually result in long-term tumor control. Three institutions have reported information on the long-term tumor control with interstitial thermoradiotherapy. Of 82 patients with initial complete response from these three institutions, 68 patients (82.9%) experienced long-term tumor control. This is similar to the results reported from patients treated with external local hyperthermia as reported by Perez and Emami (1989).

Of 32 patients with recurrent head and neck cancers who were treated with interstitial thermoradiotherapy, 27 had at least one satisfactory hyperthermia session. From these, 18 patients achieved a complete response (66.6%) and 5 (18.5%), a partial response. Fourteen of 18 lesions initially showing a complete response achieved long-term tumor control with a tumor control/complete response ratio of 77.7%. Of 6 patients with recurrent breast carcinoma, 5 achieved a complete response.

The results of this series were analyzed as a function of heating quality. Arbitrarily, we have defined satisfactory hyperthermia when the entire tumor volume is encompassed within the implant volume and when the minimal tumor temperature of 42.5°C is achieved within measured points. Following these criteria, of the 48 lesions treated, 37 patients had at least one satisfactory hyperthermia session, and 11 did not. The complete response rate was 26 of the 37 with satisfactory hyperthermia sessions (70.2%) and 0 of 11 with unsatisfactory hyperthermia sessions. The difference between these two groups is statistically significant. Similar results have been reported by Cossett et al. (1985) and Puthawala et al. (1985).

4 Complications

Almost all patients treated in this series had recurrent or persistent tumors and had undergone unsuccessfully prior conventional treatments including radiotherapy. Therefore, in many the course of interstitial hyperthermia was a desperate salvage attempt. Of 44 patients available for analysis, there were 8 severe complications and 4 additional mild to moderate complications. This is in general agreement with reports from other institutions (Table 2). The overall complication rate is about 30% with approximately 10% severe complication rate.

5 Discussion

The goal of any local regional therapy is to achieve uncomplicated local regional control of the tumor. A complete response rate of 60%–80% and a partial response rate of 20%–30% appear to be very encouraging. More importantly, the majority of these complete responders (83%) evidence long-term local control. At first glance, the complication rate appears to be high but considering the type of tumors treated with interstitial thermo-

Table 2. Summary of complications with interstitial thermoradiotherapy

Microwaves				Radiofrequency			
Author	No. of patients	All complications	Severe complications	Author	No. of patients	All complications	Severe complications
Emami et al. (1987)	35	9 (25.7%)	6 (17.1%)	Cosset et al. (1985)	23	12 (52%)	4 (17.4%)
Petrovich et al. (1990)	44	10 (22.7%)	2 (8.7%)	Emami et al. (1987)	9	3 (33%)	2 (22%)
Puthawala et al. (1985)	43	9 (21%)	5 (11%)	Linares et al. (1986)	10	4 (40%)	1 (10%)
				Oleson et al. (1984)	52	10 (19%)	6 (11%)
				Vora et al. (1982)	16	3 (19%)	1 (6%)

radiotherapy trials, an approximately 12% severe complication rate is acceptable.

One of the important aspects of interstitial thermoradiotherapy is its adaptability to various clinical situations. Interstitial thermoradiotherapy has been utilized in the treatment of recurrent brain tumors by Sneed et al. (1986) Roberts (1988), and Salcman and Samaras (1983). These preliminary trials have established the feasibility of this technique in the treatment of brain tumors, but the numbers are too small as yet for the assessment of the response rate. Interstitial thermoradiotherapy has also been employed in the treatment of recurrent head and neck tumors. From our experience and that of others (Puthawala et al. 1986), the overall complete response rate of 66%–81% has been achieved. Over 80% of these complete responses involve long-term tumor control. Reported complication rates range from 2% to 12%. Interstitial thermoradiotherapy has also been utilized in the treatment of recurrent gynecological tumors. Vora et al. (1988) used it in the treatment of 14 patients with stages III and IV cervical cancer. An initial complete response of the primary tumor was achieved in 8 patients and partial response, in 1 patient. Five of 8 complete responders were alive without any evidence of tumor on follow-up ranging from 6 to 47 months. This technique has also been used in the treatment of primary advanced breast carcinoma (Vora et al. 1986). Of 11 patients in their series, there were 10 complete responders (8 with long-term tumor control) and 1 partial responder.

Interstitial hyperthermia has also been used in combination with intraoperative interstitial hyperthermia in the treatment of deep-seated tumors (Goldson et al. 1987). These initial trials have established the feasibility of the intraoperative use of this technique, but the numbers are too small for any conclusion with regard to therapeutic advantage.

Recurrent or persistent tumors have been treated with interstitial radiation (brachytherapy) alone. Syed et al. (1977) and Emami and Marks (1983) have reported on the results of using interstitial radiotherapy alone for these lesions. Accumulative complete response rate in the two series were 31 of 54 patients (57.4%). This compares favorably with some of the results reported with interstitial thermoradiotherapy. There is no controlled randomized study published comparing both treatment regimens. The Radiation Therapy Oncology Group is currently studying this specific question. No results are available yet.

There are important prognostic factors which should be considered in both the design of clinical trials as well as the analysis of the therapeutic outcome in interstitial thermoradiotherapy: Tumor related (site, size); treatment related (temperature such as minimum and maximum, time and number of hyperthermia sessions); other modalities (radiation, chemotherapy); irradiation (dose and fractionation); and most importantly the quality of the hyperthermia sessions. As reviewed before, even with very simplistic criteria for the quality of hyperthermia sessions, there was a 70% complete response rate in patients with at least one satisfactory hyperthermia session versus 0% for patients without any good hyperthermia sessions. Therefore, it is extremely important that the above criteria be included in future clinical trials. The physical characteristics of hyperthermia systems and their relationship to the clinical situation at hand should be fully appreciated. Finally, the application of strict quality assurance criteria for interstitial radiotherapy and interstitial hyperthermia cannot be overemphasized.

References

Bicher HI, Wolfstein RS, Fingerhut AG, Frey HS, Lewinsky BS (1984) An effective fractionation regime for interstitial thermoradiotherapy – preliminary clinical results. In: Overgaard J (ed) Hyperthermic oncology 1984, vol 1. Taylor and Francis, London pp 575–578

Cossett JM, Dutreix J, Haie C, Gerbaulet A, Janoray P, Dewar JA (1985) Interstitial thermoradiotherapy: a technical and clinical study of 29 implantations performed at the Institut Gustave-Roussy. Int J Hyperthermia 1: 3–13

Emami B, Marks JE (1983) Retreatment of recurrent carcinoma of the head and neck by afterloading interstitial 192Ir implants. Laryngoscope 93: 1345–1347

Emami B, Perez CA, Leybovich L, Staube W, von Gerichten D (1987) Interstitial thermoradiotherapy in treatment of malignant tumors. Int J Hyperthermia 3: 107–118

Goldson AL, Smyles JM, Ashayeri E, Dewitty R, Nibhanupudy JR, King G (1987) Simultaneous intraoperative radiation therapy and intraoperative interstitial hyperthermia for unresectable adenocarcinoma of the pancreas. Endocuriether Hyperthermia Oncol 3: 201–208

Linares LA, Nori D, Brenner H, Shiu M, Ballon D, Anderson L, Alfieri A, Brennan M, Fucs Z, Hilaris B (1986) Interstitial hyperthermia and brachytherapy: a preliminary report. Endocuriether Hyperthermia Oncol 2: S39–S44

Oleson JR, Sim DA, Manning MR (1984) Analysis of prognostic variables in hyperthermia treatment of 161 patients. Int J Radiat Oncol Biol Phys 10: 2231–2239

Perez CA, Emami B (1989) Clinical trials with local (external and interstitial) irradiation and hyperthermia. Current and future perspectives. Radiol Clin North Am 27: 525

Petrovich Z, et al. (1990) Interstitial thermoradiotherapy for recurrent head and neck cancers. Am J Otolaryngol (in press)

Puthawala AA, Nisar AM, Sheikh Khalid MA, Rafie S, McNamara S (1985) Interstitial hyperthermia for recurrent malignancies. Endocuriether Hyperthermia Oncol 1: 125–131

Puthawala AA, Syed AMN, Sheikh KMA, Syed R (1986) Thermoendocurietherapy for recurrent and/or persistent head and neck cancers. Int J Radiat Oncol Biol Phys 12: 110

Roberts DW (1988) Interstitial hyperthermia of the brain. In: Paliwal BR, Hetzel FW, Dewhirst MW (eds) Biological, physical and clinical aspects of hyperthermia. American Association of Physicists in Medicine, New York, pp 300–314 (Medical physics monograph, vol 16)

Salcman M, Samaras GM (1983) Interstitial microwave hyperthermia for brain tumors: results of a phase-I clinical trial. J Neurooncol 1: 225–236

Sathiaseelan V, Leybovich L, Emami B, Stauffer P, Straube W (1990) Performance characteristics of improved microwave interstitial antennas for local hyperthermia. Int J Hyperthermia (in press)

Sneed PK, Matsumoto K, Stauffer P, Fike JR, Smith V, Gutin PH (1986) Interstitial microwave hyperthermia in a canine brain model. Int J Radiat Oncol Biol Phys 12: 1887–1897

Strohbehn JW, Douple EB, Coughlin CT (1984) Interstitial microwave antenna array systems for hyperthermia. Front Radiat Ther Oncol 18: 70–74

Syed AMN, Feder BH, George FW (1977) Afterloading interstitial implant in the treatment of oral cavity and oropharyngeal cancers. Radiol Clin (Basel) 46: 390–397

Vora N, Forell B, Joseph C, Lipsett J, Archambeau JO (1982) Interstitial implant with interstitial hyperthermia. Cancer 50: 2518–2523

Vora N, Shaw S, Forell B, Desai K, Archambeau J, Penzer R, Lipsett J, Covell J (1986) Primary radiation combined with hyperthermia for advanced (stage III–IV) and inflammatory carcinoma of breast. Endocuriether Hyperthermia Oncol 2: 101–106

Vora NL, Luk KH, Forell B, Findley DO, Lipsett JA, Pezner RD, Desai KR, Wong JYC, Hill B (1988) Interstitial local current field hyperthermia for advanced cancers of the cervix. Endocuriether Hyperthermia Oncol 4: 97–106

Yambumoto E, Suyama S (1984) Interstitial radiofrequency hyperthermia in combination with external beam radiotherapy. In: Hyperthermia oncology 1984, vol 1. Taylor and Francis, London

9.4 Experiences with Interstitial Hyperthermia as a Sole Treatment Modality or Combined with Radiotherapy

TAKEHIRO INOUE, NORIE MASAKI, SHUZI OZEKI, HIROSHI IKEDA, and TAKAHIRO KOZUKA

CONTENTS

1 Introduction 371
2 Materials and Methods 371
3 Results 372
3.1 Case Report 373
4 Discussion 373
References 374

1 Introduction

Several clinical studies have revealed that a combination therapy of radiation and hyperthermia is more effective than radiation alone. Interstitial hyperthermia is an improvement over external hyperthermia, as interstitial radiation therapy is over external irradiation. However, interstitial irradiation is not always recommended to treat large tumors. The purpose of this report is to evaluate the effectiveness of interstitial microwave hyperthermia in combination with external radiation therapy for advanced large tumors.

2 Materials and Methods

From October 1986 through December 1988, a total of 10 large tumors were treated with interstitial hyperthermia. The patients' characteristics are given in Table 1. There were 6 men and 4 women, with an age ranged of 35–66 years. Seven patients with cervical lymph nodes involved with metastatic disease, two with involved inguinal or femoral nodes and one with primary skin cancer of the face were treated. Seven tumors were squamous cell carcinomas, and the others included one each of malignant melanoma, basal cell carcinoma, and large cell carcinoma. Eight tumors were untreated, and two were recurrent tumors previously treated with conventional radiotherapy. All patients had extensive disease and large tumors (size of 4 cm or more in diameter).

Interstitial hyperthermia was delivered using a MA-251 antenna with 630-MHz microwave through multiple coaxial antennae. Figure 1 shows the MA-251 interstitial antenna with multiplexer of microwave made by the BSD Corporation (Salt Lake City, Utah, USA). The thermistor is mounted within the interstitial antennae 19 mm from its tip. The blind-ended Teflon catheters used to guide the interstitial antennae were implanted into the tumor under local anesthesia. Two to six guide catheters were arranged in single or double plane with 1–2 cm separation. The interstitial antennae were inserted in these catheters. Additional catheters for thermometry purposes were implanted as well. One catheter near the center of the tumor volume was allocated for the Bowman thermistor. This probe was considered to be the principal temperature control point. Therapeutic temperature was defined as at least 42°C. The temperature of the antenna was monitored during treatment using the indwelling thermistor.

It was attempted to maintain the temperature of the tumor at 42°C for 30–45 min. Interstitial hyperthermia was applied once a week, 1–2 h after radiation therapy. Two patients had three applications of interstitial hyperthermia, four patients had two, and four patients had one. The implanted guide catheters were removed after each treatment.

External radiation was applied in nine patients using cobalt 60 or 10-MV X-ray. The remaining patient had previously been irradiated with 58 Gy over 6 weeks, 1 year and 4 months before and received instead two treatments of interstitial hyperthermia alone without radiation therapy. Total doses of 50–60 Gy with conventional fractionation were delivered to four previously untreated pa-

TAKEHIRO INOUE, M.D., NORIE MASAKI, M.D., SHUZI OZEKI, M.S., HIROSHI IKEDA, M.D., TAKAHIRO KOZUKA, M.D., Department of Radiology, Osaka University Medical School 1-1-50 Fukushima, Fukushima-ku, Osaka, 553, Japan

Table 1. Details of patients treated by interstitial hyperthermia combined with external radiation

	Age	Sex	Tumor site	Tumor size (mm)	Primary site	Histology
1	43	M	Inguinal	150 × 100	Foot	Malignant melanoma
2	55	M	Face	64 × 60	Face	Basal cell carcinoma
3	66	F	Femoral	48 × 35	Vulva	Squamous cell carcinoma
4	43	M	Upper Neck	57 × 30	Nasopharynx	Squamous cell carcinoma
5	45	M	Upper Neck	43 × 37	Hypopharynx	Squamous cell carcinoma
6	58	F	Upper Neck	90 × 80	Hypopharynx	Squamous cell carcinima
7	58	F	Lower Neck	60 × 50	Lung	Squamous cell carcinoma
8	61	M	Upper Neck	60 × 50	Unknown	Squamous cell carcinoma
9	35	M	Lower Neck[a]	70 × 68	Unknown	Large cell carcinoma
10	56	F	Upper Neck[a]	70 × 50	Tongue	Squamous cell carcinoma

[a] Previously treated by irradiation.

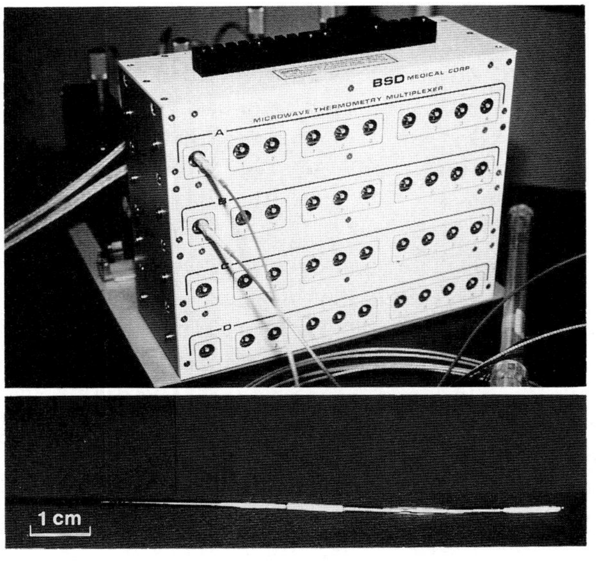

Fig. 1. Multiplexer of microwave and MA-251 interstitial antenna

tients and 30–40 Gy to two patients with a poor general condition and to one previously irradiated with 70 Gy over 5 days with interstitial radiation for tongue cancer, 8 months before. One patient was treated with 40 Gy of external radiation and twice with interstitial hyperthermia, but she refused further treatment. Another one was treated with one session of 6 Gy of external radiation and once with interstitial hyperthermia, but no further treatment was considered because of the poor general condition due to advanced disease.

Tumor response was evaluated 4 weeks after the end of radiation therapy. A complete response (CR) was defined as complete disappearance of the tumor, a partial response (PR) as a 50% or greater reduction, and no change (NC) as less than 50% reduction. Patients were followed up for at least 2 years or until death.

3 Results

Eighteen treatments of interstitial hyperthermia were applied in 10 patients. Five were given in a single plane arrangement using 2 or 3 antennae, and 13 in a double plane using 4 or 6 antennae. For 12 of the 18 heatings, a temperature in the tumor of 42°–45°C was obtained (Table 2). In double plane application, 11 of 13 treatments (85%) achieved 42°–45°C. However, in the single plane arrangement only 1 of 5 reached 42.5°C.

A patient (no. 7) experienced a small blister at the puncture site of the guide catheter on the next day after the second heating, but it healed within 3 weeks without complications. There were no severe complications such as skin burn or soft tissue necrosis.

Treatment resulst are shown in Table 3. Of 10 lesions, one showed CR, seven PR, one NC, and the remaining one was not evaluable. A patient with metastatic femoral lymph nodes (no. 3) achieved CR within 2 weeks after 50 Gy of radiation combined with 2 applications of interstitial hyperthermia and maintained CR for 4 months. She died 8 months after treatment due to primary and nodal recurrence. Three of seven tumors evaluated as PR were surgically removed after combination therapy of interstitial hyperthermia and radiation. One of these patients (involving basal cell carcinoma of the face) has been controlled and well for 2 years and 1 month, and

Table 2. Tumor temperature with interstitial hyperthermia

	No. of antennae	<41.9°C	42°C–45°	Total
Single plane	2	3	1	4
	3	1	–	1
Double plane	4	2	6	8
	6	–	5	5
Total		6	12	18

another two died 1 year after treatment due to recurrence of the primary tumor. The patient (no. 9) treated with interstitial hyperthermia alone showed only NC, and the tumor regrew within 3 months. The remaining patient (no. 1) was not evaluable because he died within 2 weeks after treatment due to hemorrhage from intraabdominal metastasis.

3.1 Case Report

A 58-year-old woman (no. 6) who had an inoperable bulky neck node metastasis (maximum diameter: 9 cm) from hypopharyngeal carcinoma was treated with interstitial hyperthermia and external radiation (Fig. 2). Two treatments of interstitial hyperthermia with six antennae in a double plane were applied and combined with a total dose of 40 Gy of external radiation. The tumor markedly decreased in size to less than 4.5 cm in diameter and was evaluated as PR. She refused any further treatment, but the tumor remained this size for 3 months. She died 6 months after hyperthermia with locoregional regrowth and bony metastases.

4 Discussion

The superiority of combination therapy with radiation and hyperthermia to radiation alone has been clinically shown (Arcangeli et al. 1988; Valdangi et al. 1988). Furthermore, combined therapy of interstitial hyperthermia and interstitial radiation therapy was reported to have a higher complete response and local control rate than external hyperthermia combined with external radiation therapy. Complete response rates of 83% and 69% were reported by Cosset et al. (1985) and Emami et al. (1984), respectively. Puthawala et al. (1985) obtained a 74% local control rate. However, effective application of interstitial irradiation is difficult for large tumors in the head and neck region. We tried to use interstitial hyperthermia combined with external irradiation instead of interstitial irradiation for advanced and large tumors.

The interstitial antennae used in this study are capable of heating a cylindrical volume of tissue with a length of about 4 cm and a diameter of 1 cm (Lum et al. 1988).

To achieve effective heating for large tumors, at least 4 antennae are needed and should be arranged in a double plane. According to our experience, only one of 5 treatments using 2 or 3 antennae in a single plane obtained effective heating. We used 4 antennae in four patients and 6 in three patients. For interstitial antennae it is better to place as many catheters as possible.

In this study, the rapid response of the lesion was observed within a week or two when once a week interstitial hyperthermia was added to con-

Table 3. Tumor response and prognosis after interstitial hyperthermia combined with external radiation

	Inter. HT	Ext. RT	Tumor response	Op.	Prognosis	
1	1	6 Gy	NE	–	3 weeks	DTNM
2	1	60	PR	+	2 years 1 month	NED
3	2	50	CR	–	8 months	DTN
4	3	60	PR	+	1 year	DT
5	1	60	PR	+	1 year	DTM
6	2	40	PR	–	6 months	DTNM
7	2	30	PR	–	1 months	DTNM
8	3	40	PR	–	4 months	DTNM
9	2	–	NC	–	8 months	DN
10	1	38	PR	–	2 months	DN

Inter. HT, interstitial hyperthermia; Ext. RT, external radiotherapy; Op., surgical removal after combination therapy; NED, no evidence of disease; DTNM, death with primary tumor, nodal, and distant metastasis, respectively; NC, no change; PR, partial response; CR, complete response; NE, not evaluable.

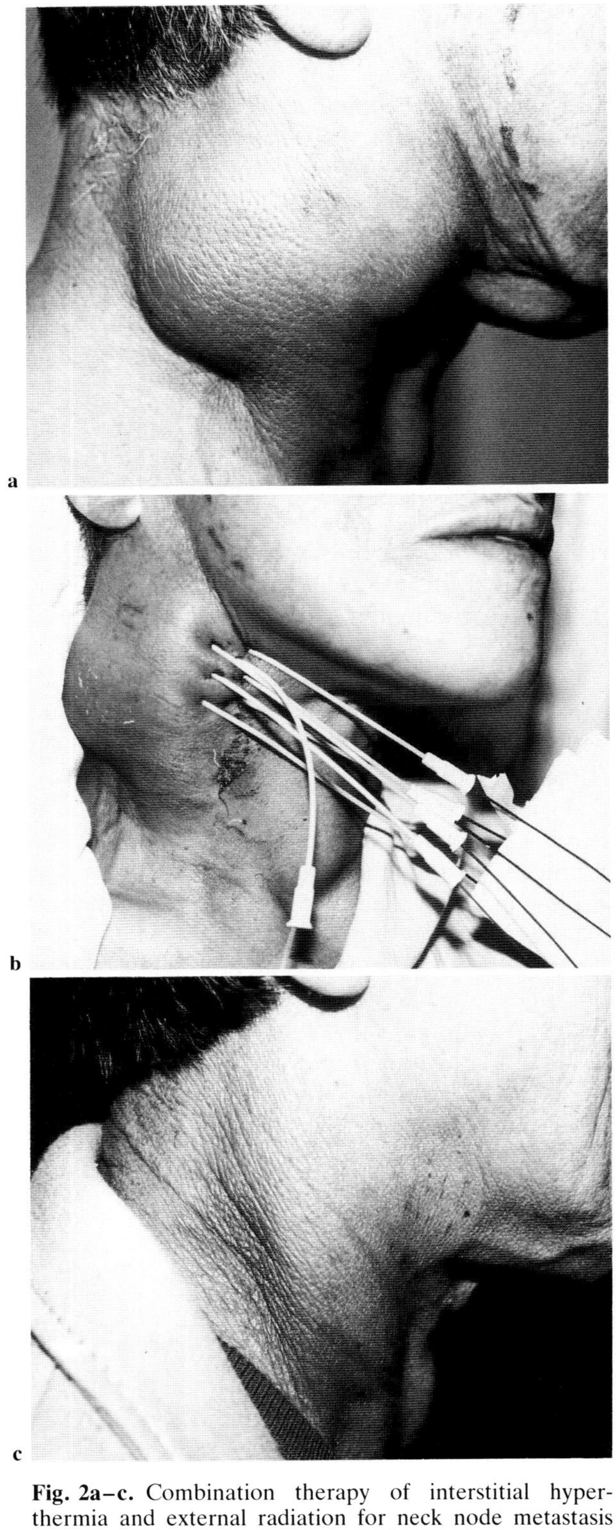

Fig. 2a–c. Combination therapy of interstitial hyperthermia and external radiation for neck node metastasis from hypopharynx carcinoma (no. 6). **a** Before treatment, **b** during interstitial hyperthermia applying 6 antennae and one thermistor, and **c** partial response 4 weeks after combination therapy of 2 applications of interstitial hyperthermia and 40 Gy of external radiation

ventional radiotherapy. There was some difficulty in applying interstitial hyperthermia more than twice in combination with 20 Gy of radiation. Radiation therapy alone with a dose of 20–40 Gy was given thereafter. This combined treatment was tolerable without major complications.

Four tumors were treated with 50–60 Gy of external radiation and interstitial hyperthermia. One showed CR, and three evidenced PR and were completely surgically removed. However, four tumors which received 30–40 Gy of radiation combined with 1–3 interstitial hyperthermia treatments showed PR and regrew within 6 months after the end of treatment. Although the patient number of this study is too small to evaluate the role of interstitial hyperthermia, 2 or 3 times of interstitial hyperthermia combined with 50–60 Gy of external radiation were safely applied and seemed to lead to a higher response rate than external radiation alone.

References

Arcangeli G, Overgaard J, Gonzalez DG, Shrivastava PN (1988) Hyperthermia trials. Int J Radiat Oncol Biol Phys 14: s93–s109

Cosset JM, Dutreix J, Haie C, Gerbaulet A, Janoray P, Dewar JA (1985) Interstitial thermoradiotherapy: a technical and clinical study of 29 implantations performed at the Institute Gustave-Roussy. Int J Hyperthermia 1: 3–13

Emami B, Marks JE, Perez CA, Nussbaum GH, Leybovich L, von Gerichten D (1984) Interstitial thermoradiotherapy in the treatment of recurrent/residual malignant tumors. Am J Clin Oncol 7: 699–704

Lum K, Astrahan M, Langholz B, Jepson J, Cohen D, Luxton G, Petrovich Z (1988) Interstitial thermoradiotherapy for recurrent or persistent tumors. Int J Hyperthermia 4: 259–266

Puthawala AA, Syed N, Sheikh KMA, Rafie S, McNamara S (1985) Interstitial hyperthermia for recurrent malignancies. Endocuriether Hyperthermia Oncol 1: 125–131

Valdagni R, Amichetti M, Pani G (1988) Radical radiation alone versus radical radiation plus microwave hyperthermia for N3 (TNM-UICC) neck nodes: A prospective randomized clinical trial. Int J Radiat Oncol Biol Phys 15: 13–24

Subject Index

abdominoperineal rectotomy
anal canal cancer 230
afterloading
high, low dose rate remote afterloading 38, 39
hypofractionated, HDR regime, dermal tolerance 15
^{192}Iridium, breast cancer, boosting the tumor bed 201
manual
advantages, disadvantages 157
head and neck tumors 153–157
afterloading devices
low, high dose rate
comparison, advantages, disadvantages 292, 307, 311–317
radioisotopes 333, 334
remote, Buchler-, Gamma-Med-, Selectron-principles 276, 277
remote-controlled, high dose rate
Decatron using ^{192}Ir sources 320
remote-controlled, high, low dose rate
applicators 333, 335
dose distribution, radiographic documentation 312, 313, 319
gynecological tumors 333–339
afterloading plate methode
cervix carcinoma
high dose rate 298, 299
late results 302, 303
stage related dosage 301
afterloading system
^{137}Cs sources 26
intracavitary, ^{60}Co 28
afterloading techniques
anorectal obturators, vaginal tubes 39
brachytherapy, ^{192}Ir needles, wires, nylon-, teflon tubes 153
brain, permanent and temporary interstitial irradiation 70, 81–84
breast cancer, interstitial boost irradiation 191–198
^{137}Cs, ^{60}Co, ^{192}Ir, intracavitary therapy, gynecological tumors 271, 272, 285–290, 291–297
^{137}Cs, dose distribution, female pelvis 334
dose calculation 39, 40, 41
endocurietherapy with ^{192}Ir needles and wires 153–157
gynecological tumors 271, 272, 276, 280, 285–290, 291–297, 307–310, 311–317, 329–331, 333–339
head and neck tumors, interstitial brachytherapy 145, 146
high dose rate, conventional brachytherapy, comparison 329–331, 333–339
high and low dose rate 38, 39, 70
critical comparison 53–57
^{192}Iridium, breast cancer 179, 187
^{192}Iridium contact irradiation, brachycurietherapie, brain tumors 70
manual, remote
low, high dose, gynecological tumors 276–283
radiation protection of staff 191, 291, 319
tongue carcinoma 172
uniplanar, biplanar template, Paris system, applicator sets 38, 39
afterloading therapy
applicators 335
applicator position, organs of risk 287, 289, 312
cervix carcinoma 279–283, 289, 298, 299, 302–304, 308, 309, 322, 323, 329
isodose configuration 330
doses, comparison of measurement and calculation, organs at risk 289, 290
dosimetry 285–290
fractionated, malignant inoperable brain, tumors 70, 85–88
high dose rate
clinical results, endometrium carcinoma 327, 328
conventional brachytherapy, comparison 329–331
primary treatment lethality 320
^{226}Ra therapy, comparison of results 311–317, 336
intracranial tumors 75
organs at risk, comparison of measured and calculated doses 289, 290
radioisotopes used, characteristics 285
source specification, check of source position 285
survival rates, gynecological tumors 295, 298–305
air kerma rate constant
absorbed dose, definition 32
γ-ray brachytherapy 193
American Joint Committee on Cancer (AJCC)
oral cavity, oropharynx carcinomas, tumor classification 131, 132
anal canal cancer
abdominoperineal resection 211, 212, 215, 229–232
chemotherapy 212, 215, 216, 221–227
classification, WHO 207, 216
clinical results 209–213, 215–219, 221–227, 229–232
combination therapy: external irradiation, chemotherapy, ^{192}Iridium implantation 215–219, 223, 224, 231
CT diagnosis 207
incidence 207
^{192}Iridium implantation, afterloading technique 153, 154, 210, 215–219, 221–227
lymphatic spread 207
multimodal therapy, indications 229
pelvic metastases 213
prognostic factors 207, 211
radiochemotherapy, combined, Lyon experience 209–213
radiotherapy versus radiotherapy plus chemotherapy, results 212
sexual practice, sexual transmitted diseases 208
sphincter preservation 212, 232
split-course treatment external beam therapy, ^{192}Iridium-, ^{60}Co implants 210
surgery, diagnosis, clinical results 217, 222, 229–232
thermoradiotherapy 350
TNM system 207, 216
anatomy
tongue 169, 170
angiogram
brain, arteriovenous malformation, follow-up, cure rate by megavoltage therapy 95
applicators
low, high dose afterloading 333, 335
arteriovenous malformations
brain, external stereotactic focal irradiation 91–93
with linear accelerator 95–99

astrocytoma
 brachycurietherapie 73
 survival times, postoperative, interstitial therapy 68, 72
^{198}Au
 brain tumors, stereotactic interstitial irradiation 68
 implantation, tongue carcinoma 169-175
 marker seeds, inadequate retropubic implantation 259
 physical properties 28
 reference air kerma rate, strength of source 31
 seeds, interstitial therapy, gynecological tumors 276
 specific dose factor, brain 82, 83
 tongue carcinoma, interstitial therapy, technique, results 37, 38, 169-175
autoradiography
 loaded isotope sources, correct, incorrect source assembly 197
 quality control 31

β-irradiation, choroidal melanoma, ^{106}Ru/^{106}Rh plaques, long term results 105, 110-120
bilary tract carcinoma
 interstitial hyperthermia, technique 361, 362
biopsy
 anal canal cancer, residual tumor 231
 choroidal melanoma 114, 115
 prostatic cancer 245, 248, 258, 265, 266
 stereotactic, brain tumors 69
bladder
 dosimetry, intracavitary irradiation of cervix carcinoma 279-283, 308
 organ at risk, dose calculation, afterloading therapy 287, 289, 290
 radiation protection, brachytherapy, cervix cancer 301, 308
 radiation reactions, brachytherapy 279, 315, 325, 326, 328
 shape 279, 280
bone marrow
 dose rate dependence, normal tissue tolerance 20
 dose-rate effect, brachytherapy, radiobiology 11, 12, 14
brachycurietherapy
 brain, permanent and temporary interstitial irradiation, experimental dose effects 81-84
 brain tumors
  clinical results 65, 70
  complications 69, 71, 73
  computer program 64
  CT scans 64
  dose specification 63
  dosimetry 65
  fractionated afterloading therapy 85-88
  gliomas, low grade, nonresectable (WHO I, II) 71, 72
  implantation techniques 62, 63, 67, 70
  ^{192}Iridium contact irradiation, see afterloading technique
  patient selection criteria 63
  pituitary adenomas, survival 71
  results of treatment 65, 70
  selection of radionuclides 62
 breast cancer, Paris system 179-185
 clinical results 86
 complications 69, 71, 81, 82, 83, 85, 87, 97
 external stereotactic focal irradiation 89-94
 mesencephalic brain stem, pons region tumors 73
 parasellar tumors, indications, clinical results 71
 pituitary adenomas, mortality, complications 71
 pituitary tumors, indications, clinical results 71
 skull base tumors 71
 tumors
  diencephalic region 72
  midline structures 72
  pineal region 73
  ventricular system 72
brachytherapy
 advanced head and neck cancer, Erlangen experience 159-167
 afterloading, low, high dose rate 333-339
 afterloading techniques 153-157
  Erlangen experience 159-167
 anal canal cancer 207-232
 bladder, radiation reactions 279, 315, 328
 bone marrow, dose rate effect, radiobiology 11, 12
 brain tumors 61-99
 breast cancer 179-204
 cervix carcinoma 279, 299, 307-310, 311-317, 319-331
 choroidal melanoma with ^{106}Ru/^{106}Rh plaques 103-111, 113-127
 combination with external beam therapy, radiobiology 4, 5
 combination with teletherapy, techniques, critical organs 45
 conventional, high dose rate afterloading, comparison 329-331
 dose rate effects 3, 4, 9, 11
  experimental tumors, radiobiology 16, 17
 dose response, dose rate and volumes, linear quadratic model 45
 endometrium carcinoma 307-310
 γ-ray-, air kerma rate constant 193
 gynecological tumors 271-283, 285-290, 333-339
 head and neck tumors 131, 145-152, 159-167
 history 35, 37
 instruments 36, 192
 interstitial, head and neck tumors 145-152, 159-167
 intestinal mucosa, dose-rate effect, radiobiology 12
 isoeffect comparison of high and low dose rate 56
 linear quadratic model, dose rate 45
 low, middle, high dose rates, gynecological tumors 337
 lung, dose-rate effect, radiobiology 13, 14
 oral cavity, oropharynx carcinoma 144
 oral mucosa, dose-rate effect, radiobiology 13
 perioperative, breast cancer 182
 physical fundamentals 25-33
 prostatic cancer, technique, results 236, 237, 239, 245-256, 257-262, 275
 radionuclides used, energy 26
 results, oral cavity lesions 134
 rigid guide needle technique, lip carcinoma 134
 ^{106}Ru/^{106}Rh plaques, choroidal melanoma 103-111, 113-127
 splite-course, breast cancer 182, 185
 time-dose fractionation (TDF) 5
brachytherapy techniques
 afterloading techniques, high, low dose rate 38, 39
 chemical modifiers 46
 classic presciption rules 39
 dose calculation 43, 44
 dose, time, volume dependent response 45
 experimental and future techniques 46
 "free spaced" interstitial permanent application 41
 gynecologic application rules 39, 40
 high and low dose rate, clinical comparison 53-57
 hyperthermia 46
 instrumentarium 36
 interstitial, intracavitary applicators 37
 intracavitary ^{226}Ra application, Stockholm system (Radiumhemmet), Paris system (Curie Foundation) 37
 isodoses 44
 molds, external applicators 36
 monoclonal antibodies carrier 46
 new isotopes and sources 46
 nomographs 41
 orthogonal and stereo shift projection 42
 radioactive sealed, unsealed sources 35
 radiography 42

reference points, isodoses 44
source localization, three dimensional reconstruction 42
teletherapy combination 45
template and multiplanar temporary application 41
tomography 42

brain
arteriovenous fistula, external stereotactic focal irradiation 91-93
with linear accelerator 95-99
experimental dose effects 81
implantation techniques, brachycurietherapy 62, 63, 64
interstitial irradiation, permanent and temporary, experimental dose effects 81-84

brain tumors
afterloading therapy, fractionated 85-88
astrocytoma, survival times, postoperative, interstitial therapy 68, 69, 71, 72
^{198}Au stereotactic interstitial irradiation 68
brachycurietherapie
basic principles 61-66
preimplantation planning, CT, MRI 64
^{60}Co curietherapie 67
^{60}Co stereotactic interstitial irradiation 68
^{157}Cs, intracranial curietherapie 67
diencephalic region, interstitial stereotactic irradiation, ^{192}Iridium, ^{125}Iodine 72, 73
ependymoma, survival times, postoperative, interstitial therapie 68
fractionated afterloading therapy 85-88
glioblastoma, survival times, postoperative, interstitial therapy 68, 70
glioma
nonresectable, low grade, imaging stereotaxic implantation of radionuclides 71, 72
radiosensitivity 81
survival time, tumor dose, fractionated high dose rate 70, 71
histologic examination, biopsy 68
^{125}I stereotactic interstitial irradiation 68, 71, 72
imaging stereotaxic implantation of radionuclides 67-79
implantation techniques
clinical results 65
selection of radionuclides 62, 63
surgical procedures 64
target volume 63
interstitial hyperthermia, clinical results 360, 361, 368
interstitial thermoradiotherapy 350
^{125}Iodine, interstitial curietherapie, stereotactic 67, 68, 72
^{192}Iridium, curietherapie 68, 69, 72
meningioma, survival times, interstitial therapy 68
mesencephalic brainstem, pons region, imaging stereotaxic implantation of ^{192}Iridium, ^{125}Iodine 72, 73
metastases, solitary, external stereotactic focal irradiation 69, 75, 89-94
midline structures, imaging stereotactic implantation of ^{192}Iridium followed by ^{125}Iodine 72, 73, 75
MRI, definition of thrue extent 62
oligodendroglioma, survival times, postoperative, interstitial therapy 68
^{32}P stereotactic interstitial irradiation 68
parasellar, imaging stereotaxic curietherapy 74
parasellar, pituitary interstitial therapy, indications, clinical results 71
pituitary adenomas, imaging stereotactic radioisotope implantation 70, 71, 74
pons region, interstitial stereotactic irradiation, ^{192}Iridium, ^{125}Iodine 72, 73
postoperative survival times, interstitial therapy 68
radionuclides, stereotactic interstitial irradiation procedures 68
skull base, brachycurie therapy 71
^{182}Ta stereotactic interstitial irradiation 68
thermal map, interstitial hyperthermia 360, 361
thermoradiotherapy, results 350, 366, 368
ventricular system, imaging stereotactic implantation of radionuclides 72, 73
^{90}Y stereotactic interstital irradiation 68

breast cancer
additional tumor foci 199, 200
afterloading techniques 191-198
^{192}Iridium 179, 187, 201
brachycurietherapy 179-185
clinical results 182-184
intraoperative, lumpectomy 182
local recurrence, incidence, therapy 182, 183
breast preservation, interstitial therapy 190
chemotherapy 183, 184
clinical results, interstitial boost irradiation 189, 190, 201-203
dose distribution, ^{192}Iridium implantion 195
dose-response relationship
as function of tumor volume 200
primary radiotherapy 201
ductal carcinoma in situ 199
exposure-free station, ^{192}Iridium sources preparation 195
interstitial boost irradiation
dosimetry 189
indications 187
results 189
technique 187, 191, 194
interstitial therapy, dosimetry 180
^{192}Iridium implantation, dose distribution 195
^{192}Iridium seeds, implantation techniques 188, 189
^{192}Iridium wires, interstitial boost irradiation 191
lumpectomy
interstitial therapy, dosimetry 181
intraoperative brachytherapy 182
mastectomy, local recurrence, brachytherapy 185
microscopic extension of tumor 188
prognostic factors 199
quadrantectomy, ^{192}Iridium afterloading 201
radiotherapy, primary, dose-response relationship 201
recurrence, interstitial boost irradiation 190
recurrence rates
^{192}Iridium afterloading 201
prognostic factors 199
recurrence, salvage lumpectomy 184, 185
salvage mastectomy 183
target volume, definition 146, 180, 188, 199
theoretical considerations 199-201
thermoradiotherapy 350
tumor size, dosimetry 181
tumorectomy, external irradiation, results 187, 190

buccal mucosa, carcinoma, interstitial therapy 134, 136, 137, 147

carcinoma
anal canal
abdominoperineal resection 211, 212, 215, 229-232
chemotherapy 212, 215, 216, 221-227
classification, WHO 207, 216
clinical results 209-213, 215-219, 221-227, 229-232
combination therapy: external irradiation, chemotherapy, ^{192}Iridium implantation 215-219, 223, 224, 231
CT diagnosis 207
incidence 207
^{192}Iridium implantation, afterloading technique 153, 154, 210, 215-219, 221-227
lymphatic spread 207
pelvic metastases 213
prognostic factors 207, 211

carcinoma, anal canal
  radiochemotherapy, combined, Lyon experience 209-213
  radiotherapy versus radiotherapy plus chemotherapy, results 212
  sexual practice, sexual transmitted diseases 208
  sphincter preservation 212
  split-course treatment, external beam irradiation, ^{192}Iridium-, ^{60}Co implantation 210
  surgery, diagnosis, clinical results 217, 222, 229-232
  TNM system 207, 216
 biliary tract, interstitial hyperthermia, technique 361, 362
 breast cancer
  additional tumor foci 199
  afterloading techniques 179, 187, 191-198
  brachycurietherapy, clinical results 179-185
  brachycurietherapy, local recurrence, treatment 182, 183
  breast preservation, interstitial therapy 190
  chemotherapy 183, 184
  clinical results, interstitial boost irradiation 189, 190, 201
  dose distribution, ^{192}Iridium implantation 195
  ductal carcinoma in situ 199
  exposure-free station, ^{192}Iridium sources preparation 195
  interstitial boost irradiation, dosimetry 189
  interstitial boost irradiation, techniques 187, 191, 194
  intraoperative brachycurietherapy, lumpectomy 182
  ^{192}Iridium implantation 188, 189, 191, 195
  lumpectomy 181, 182
  mastectomy, local recurrence, brachytherapy 183, 185
  microscopic tumor extension 188
  prognostic factors 199
  recurrence, therapy 184, 185, 190, 199
  target volume, definition 146, 180, 188, 199
 buccal mucosa
  interstitial therapy, results 134, 135
  plastic tube technique (Pernot technique) 136
 cervix
  afterloading plate method, high dose rate 298, 299
  afterloading therapy, results 308, 309, 317, 337
  afterloading therapy, CT, portio to rectum and bladder distances 289
  bladder dosimetry, intracavitary irradiation 279-283
  brachytherapy, complications 293, 302, 339
  clinical results, low, high dose rate irradiation 279, 302, 304, 308, 317, 322, 329, 337
  external beam therapy 271, 321, 336
  high dose rate afterloading, prospective clinical trial 307-310
  high dose rate afterloading, ^{226}Ra therapy, comparison 311-317
  interstitial therapy, dosimetry 275-283
  intracavitary irradiation, bladder-, rectum dosimetry 279-283
  isodose configuration 330
  low and high dose rate brachytherapy, comparison of doses, reference points 338
  megavolt therapy 271, 314
  one- and three way applicators, radiography 287, 289
  physical properties of radionuclides 272
  physics, afterloading plate method 300, 301
  postoperative intravaginal treatment regime 321
  ^{226}Radium treatment techniques, results 271, 272, 273, 311
  side effects, low and high dose rate therapy 302, 315, 339
  standard afterloading treatment 336
  survival rates, afterloading therapy 295, 298-305, 336, 337
  telecobalt therapy 307
  three channel Alth applicators, ^{60}Co, ^{137}Cs 300
  uterine, vaginal applicators, Stockholm-, Paris-, Manchester-, Munich techniques 272, 273, 308
 colon, thermal map, interstitial hyperthermia 363
 endometrium
  afterloading therapy, results 275, 303, 308, 309, 314, 327
  high dose rate afterloading, prospective clinical trial 307-310
  high dose rate afterloading, ^{226}Radium therapy, comparison 311-317
  intracavitary therapy, results 275, 303
  ^{192}Iridium wire tube packing 302
  metastatic spread 327
  packing method, Heyman, Stockholm, results 307, 308
  postoperative intravaginal treatment, results 327
  primary intracavitary treatment 321, 322
  radiotherapy with Heyman capsules, Ries eggs, ^{60}Co pearls 275
  survival rates 275, 303, 307-309, 314, 327
  "umbrella" applicator, low, high dose rate afterloading 333, 336
 gynecological, see gynecological tumors
 head and neck
  advanced, brachytherapy, Erlangen experience 159-167
  afterloading brachytherapy 153-157, 159-167
  brachytherapy, general cave of patients, before and after 148
  brachytherapy, principles, techniques 146, 147
  brachytherapy, results of treatment 148, 150
  interstitial thermoradiotherapy, results 350, 360, 361
 hypopharynx, hyperthermia, external beam irradiation, results 372
 larynx, brachytherapy, Erlangen experience 159
 lips
  interstitial therapy, results 133, 134
  Erlangen experience 159
 lung, hyperthermia
  clinical results 360
  external beam irradiation, results 272
 mammary
  brachycurietherapie 179-185
  interstitial boost irradiation, indications, technique, results 187-190
 mouth, interstitial therapy 138
 nasopharynx
  hyperthermia, external beam irradiation, results 272
 nasopharynx, interstitial brachytherapy 146
  Erlangen experience 159
 neck, recurrent, two plan implantation, brachytherapy 147
 oral cavity, oropharynx, interstitial therapy, clinical results 133-144
 pancreas, thermal map, interstitial hyperthermia 363
 prostatic
  ^{198}Au marker seeds, inadequate retropubic implantation 259
  biopsy 245, 246, 258, 265, 266
  bone metastases 246, 249
  brachytherapy 236, 237, 239, 245-256, 257-267
  chemotherapy 265
  classification 235, 236, 241, 245, 257
  clinical results 237, 238, 241-244, 250, 251, 253, 254, 263-267
  complications 239, 242, 246, 248, 252
  CT cross sections 235, 237, 239, 243, 264

diagnosis 245, 246, 258, 265, 266
dosimetry, interstitial therapy 237, 248
external beam irradiation, results 238, 239, 247, 248, 249, 253, 264
^{198}gold marker seeds, inadequate retropubic implantation 259
histology 246, 249, 250
hormonal therapy 243, 248
incidence 235
indications, transperineal sonographically guided ^{125}Iodine seed implantation 261, 264
interstitial brachytherapy 235, 236, 239, 241-244, 245-256, 263-267
^{125}Iodine implantation, Mick applicator 235, 236, 239, 241, 248-254, 258-262
lymph node involment 236, 246, 258, 266
lymphadenectomy, histologic grade, survival times 250, 258, 266
lymphangiography 246
metastases 235, 236, 246
morbidity, treatment related 242, 246, 247, 251
orchiectomy, hormonal therapy 248
prognostic factors 235
prostatectomy, radical, results 238, 245, 264, 266
recurrence rates 250
sonography 235, 239, 243, 257-262
surgery, lymphadenectomy, histologic grade, survival times 250, 258, 266
surgery, orchiectomy 248
surgery, radical prostatectomy, results 238, 245, 264, 266
survival times 237, 238, 239, 247, 250, 251, 263-267
TNM system 235, 236, 241, 245, 266
treatment results 237, 238, 247, 250, 251, 263-267
ultrasound guided ^{125}Iodine implantation 257-262, 264
volumetry, sonography 258, 264
rectum, interstitial thermoradiotherapy 350, 351
salivary glands, interstitial hyperthermia 361
soft palate, interstitial therapy, results 135
tongue
^{198}Au implantation, technique, results 169-175
interstitial brachytherapy, results 135, 136, 142, 147
interstitial hyperthermie 361, 372
tonsillar region, interstitial therapy 142
uvula, endocurietherapie by guide gutter technique 140
vaginal
applicators, afterloading, low, high dose rate 272, 273, 276, 333, 335
interstitial, intracavitary therapy, techniques 275
intravaginal postoperative treatment regime 321, 323
radiation reactions 302
vaginal stump, secondary carcinoma of vagina 327
vallecular region, curietherapy, plastic tube implantation 143
vulva, hyperthermia, external beam irradiation, results 272
cell
biology, hyperthermic cytotoxicity, radiosensitization 344
repair, recovery, time factor, normal healthy tissue, tumor 329
surviving fraction curves, changing the dose per fraction of high dose rate 55
toxicity, hyperthermic radiosensitization 344
cell cycle
radiosensitivity 9
cell damage
sublethal, repair during CLDRI 9
cell progression, proliferation
dose rate effect 9, 10, 45
cell survival curves
see dose survival curves
cells
anoxic, experimental tumors, oxygen enhancement ratio 16
hypoxic tumor-, radioresistance 315, 337
"incomplete repair" model of THAMES, mathematical analysis 19
repair capacity, tissue tolerance, mathematical analysis 19
repair half times, experimental tumor cell systems 19
repair kinetics, dose rate, brachytherapy, combination with teletherapy 45
theoretical generation curves of sublethal damage, three half times of repair 18
central nervous system
dose-rate effect, radiobiology 13, 14
cervix carcinoma
afterloading plate method, high dose rate 298, 299
afterloading therapy
CT, minimal distances from portio to rectum and bladder 289
results 308, 309, 317, 337
bladder dosimetry, intracavitary irradiation 279-283
brachytherapy, complications 293, 302, 339
clinical results, high, low dose rate irradiation 302, 304, 308, 309, 317, 322, 323, 329, 337
external beam therapy 271, 321, 336
high dose rate afterloading
prospective clinical trial 307-310
^{226}Radium therapy, comparison 311-317
high dose rate afterloading techniques, clinical results 279, 280, 291-297, 298-305, 333-339
interstitial therapy, dosimetry 275-283
intracavitary irradiation
bladder dosimetry 279-283
history 291
primary, Stockholm regime 320, 321
isodose configuration, high dose rate afterloading therapy 330
low and high dose rate brachytherapy, comparison of doses in the reference points 338
megavolt therapy 271, 314
one- and three way applicators, radiography 287, 289
physical properties of radionuclides 272
physics, afterloading plate method 300, 301
postoperative intravaginal treatment regime 321
^{226}Radium treatment techniques 271, 272, 273, 311
side effects, low and high dose rate therapy 302, 315, 339
standard afterloading treatment 336
survival rates, afterloading therapy, radium therapy 295, 298-305, 336, 337
telecobalt pendulum therapy 307
three channel Alth applicators, ^{60}Co, ^{137}Cs 300
uterine, vaginal applicators, Stockholm-, Paris-, Manchester-, Munich-techniques 272, 273, 308
chemotherapy
anal canal carcinoma 212, 215-219, 221-227
brachytherapy 46
breast cancer, before and after 183
breast cancer, telebeam ^{60}Co irradiation, clinical results 184
prostatic cancer 265
radiochemotherapy anal canal cancer, results 209-213
chest wall
tumors, interstitial thermoradiotherapy 350
choroidal melanoma
β-irradiation with ^{106}Ru/^{106}Rh plaques, long term results 103-111, 113-127
before and after brachytherapy with ^{106}Ru/^{106}Rh plaque 106, 107

cervix carcinoma
brachytherapy 103-111, 113-122
clinical results 104-111, 117-127
complications of therapy, radiogenic 108, 109, 110, 119, 120
cryosurgical treatment 116
current approaches to diagnosis and management 114, 115
external beam therapy 114
eye enucleation 109, 113, 115
after β-therapy 109, 110
interstitial thermoradiotherapy 346
^{125}Iodine eye plaque therapy, results 119, 346
laser photocoagulation 113, 115
light coagulation, additional 108, 113, 115
metastases, therapy, results 104, 114, 117
pre-, posttreatment tumor dimension 116
radiotherapy, repeated 108
RTOG study 118
side effects, radiogenic 108, 109, 110
survival rates 109
therapeutic modalities, discussion 113
therapy, comparison: Eye enucleation, radiocobalt eye-plaque therapy 115
thermoradiotherapy, interstitial, ^{125}I plaque treatment 346
TNM system 104
tumor size 108
Xenon coagulation 108
classification
anal canal cancer, WHO 207, 216
cervical cancer 308
choroidal melanoma 104
endometrial cancer 308
^{192}Iridium-, ^{125}Iodine at Erlangen treated patients 160
oral cavity, oropharynx carcinoma 131, 132
prostatic cancer
after lymphadenectomy 258, 266
Memorial Sloan-Rettering Cancer Center 235, 236, 241, 245
tongue carcinoma 170, 171
clinical results
afterloading therapy, cervix-, endometrium carcinoma 303, 304, 308, 309, 313, 314, 322, 323, 329, 337
anal canal cancer 209-213, 215-219, 221-219, 221-227, 229-232
brain, arteriovenous malformations, external stereotactic focal irradiation 92, 93, 95, 97
brain tumors
afterloading therapy 86
postoperative survival times, interstitial therapy 68
breast cancer
brachycurietherapy 182-184
European Group of Curietherapists 190
interstitial boost irradiation 189
^{292}Iridium afterloading 201, 202
primary chemotherapy, telebeam ^{60}Co irradiation 184
cervix carcinoma 302, 304, 308, 309, 317, 322, 323, 329, 337
choroidal melanoma
eye enucleation versus radiocobalt eye plaque therapy, comparison of results 115
pre-, posttreatment dimensions 116-120
^{106}Ru/^{106}Rh plaque curietherapy 104-111, 117-127
eye tumors, malignant 116-120
gynecological tumors
annual report on results of treatment, Deutsche Gesellschaft für Gynäkologie, Halle 295
high dose rate irradiation 303, 304, 311-317, 337
head and neck tumors, brachytherapy, interstitial 147, 148, 150
interstitial brachytherapy
^{125}Iodine implantation 162
oral cavity-, oropharynx carcinoma 135, 136, 147
interstitial hyperthermia 351, 352, 360, 361, 366, 367, 372-374
interstitial radiofrequency hyperthermia 351
interstitial thermoradiotherapy 351-354, 366, 367, 372, 373
intraoperative hyperthermia 359-363
mouth floor carcinoma, interstitial therapy 135, 136, 147
oral cavity, oropharynx carcinoma, curietherapy 135, 136, 147
prostatic cancer 237, 238, 241-244, 250, 260, 261, 263-267
external beam therapy 247, 248
external beam therapy, comparison with implantation techniques 253, 254
^{226}Radium therapy, conventional 311-317
soft palate carcinoma, curietherapy 135, 136, 149
thermoradiotherapy 351
tongue carcinoma
^{198}Au interstitial therapy 174, 175
curietherapy 135, 136, 142, 147
tonsillar region carcinoma, interstitial therapy 135, 136, 147
^{60}Co
abdominal irradiation, dose effect curves 13
afterloading plate method, cervix carcinoma 298
anal canal cancer, treatment 210
biaxial telecobalt pendulum therapy, cervix carcinoma 307
brain tumors, stereotactic interstitial irradiation 68
choroidal, retinal tumors, brachytherapy 36
curietherapy, techniques 67
episcleral applicators, brachytherapy techniques 117
eye-plaque therapy, choroidal melanoma, eye enucleation, comparison of results 115-120
high dose rate afterloading devices, gynecologic tumor therapy 292, 333, 334
intracavitary high dose afterloading, gynecological tumors 285, 337
intracavitary therapy, gynecological tumors 271, 272
needles, interstitial therapy 25
pearl technique, therapy of endometrium carcinoma 275
physical properties 28, 272
plaques, choroidal melanoma treatment 117, 118
reference air kerma rate, strength of source 31
rotation therapy, cervix cancer, survival rates 337
three channel Alth applicator, cervix cancer 300
colon
carcinoma, thermal map, interstitial hyperthermia 363
complications
afterloading high dose rate therapy, organs at risk, gynecologic tumors 288, 293, 295, 296, 308, 315, 317, 325, 326
afterloading plate method, cervix cancer 302
afterloading therapy, fractionated, brain tumors 85, 87
anal canal cancer treatment 209
biopsy, CT-stereotactic, midbrain/brainstem/pons region tumors 69, 73
brain, arteriovenous malformation, external focal irradiation 97
breast cancer, ^{192}Iridium afterloading 202, 203
cervix carcinoma, low and high dose rate therapy 339
choroidal melanoma treatment 108, 109, 110, 119, 120
head and neck tumors 150
interstitial brachytherapy, lip cancer 134, 135, 138, 150
interstitial irradiation
anal canal cancer 209
brain, chronic vasogenic edema 81, 83
brain, radionecrosis 81, 82, 83
breast cancer 202, 203
interstitial thermoradiotherapy 367
mandibular, dental, interstitial therapy, orocavity-, oropharynx carcinoma 133, 138, 150

- pelvic lymphadenectomy, ^{125}Iodine seed implantation 261
- prostatic cancer 239, 242, 246, 248, 252, 261
- ^{226}Ra therapy, conventional 315
- rectum, high dose rate afterloading 293, 315, 328
- soft tissue necrosis, interstitial therapy 165, 166
- stereotactic biopsy, implantation of radioisotopes 69
- stereotactic ^{192}Ir curietherapy, pituitary adenomas 71
- thermoradiotherapy 353, 367, 368

computer programs
- brain, external stereotactic focal irradiation 90, 91
- brain tumors, brachytherapy 64
- dose calculation, prostatic cancer 237

conjunctiva
- melanoma, therapy, results 117

^{137}Cs
- afterloading technique, dose distribution, female pelvis 26, 334
- intracavitary afterloading therapy, gynecological tumors 285, 333, 334
- intracavitary therapy, gynecological tumors 271, 272
- low dose rate afterloading devices, gynecologic tumor therapy 292
- physical properties 26, 272
- reference air kerma rate, strength of source 31
- remote afterloading technique, plate method, cervix carcinoma 298, 299
- three cannel Alth applicator, cervix cancer 300

^{137}Cs gamma irradiation
- dose survival curves 9
  - experimental rhabdomyosarcoma 16
- standard Fletcher application, isodoses 4

CT
- afterloading therapy, cervix carcinoma, minimal distances from portio to rectum and bladder 289
- anal canal cancer, early diagnosis 207
- brain, arteriovenous fistula 91
- brain tumors
  - contrast enhanced region 64
  - verification of CT scans 64, 65
- cerebral metastasis, before and after stereotactic irradiation 91
- choroidal melanoma 114
- dosimetry, interstitial therapy, oral cavity-, oropharynx carcinoma (Paris system) 133, 139
- guided stereotactic implantation techniques 61, 73
- head and neck tumors, treatment planning 145, 146
- interstitial hyperthermia, stereotactic headframe of Lexell 360
- neck metastases 143
- pelvic lymph node involvement, prostatic cancer 241, 243, 264
- prostatic cancer 237, 241
- stereotactic ^{192}Ir, ^{125}I brachycurietherapy 73, 74

curietherapy
- Européen Groupe de Curiethérapie, brain tumors, treatment results 135
- European Group of Curietherapists, breast cancer, experiences 190
- see brachytherapy, brachycurietherapy

death
- after β-irradiation, choroidal melanoma 109
- anal canal cancer 211
- metastases, choroidal melanoma 105

definition
- absorbed dose, Gray (Gy) units 32
- air kerma rate constant 32
- "average life", Ta 30
- miligram radium equivalent (mg Ra eq), radioisotopes 30
- ^{226}Radium therapy, classic, high, medium, low dose rates 313
- reference air kerma rate, radioisotopes 30, 31
- target volume, tumor therapy 146

dental prophylaxis
- interstitial therapy, oral cavity, oropharynx carcinoma 133

diagnosis
- choroidal melanoma 114
- prostatic cancer 245
- tongue carcinoma 171

dose
- absorbed
  - afterloading therapy 40, 41
  - definition, calculation in a medium 32, 33
  - organs at risk, intracavitary afterloading, calculation, measurement 288
- alpha/beta ratio, dose rate relative effectiveness 4, 29
- brain, arteriovenous malformations, external stereotactic focal irradiation 91, 98
- calculation
  - afterloading therapy, graphic methods, measurement, comparison 289
  - brachytherapy 43, 44
  - brain tumors, brachycurietherapy 63
  - ^{125}I seeds, interstitial therapy 41
  - interstitial therapy, oral cavity-, oropharynx carcinoma (Paris system) 133, 139
- calculation, prostatic cancer 237
  - ^{125}Iodine seed implantation 248
- ^{60}Co pearls, therapy of endometrium carcinoma 275
- critical dose-rate, V-79 cells, HeLa cells, differences 10, 11
- distribution
  - applicators in situ, biplane radiographs 312, 313
  - ^{198}Au seeds, interstitial therapy 172, 173
  - female pelvis, ^{226}Ra conventional brachytherapy, ^{137}Cs remote, controlled afterloading techniques 334, 337
  - homogeneity, brachytherapy, prostatic cancer 260
  - interstitial source localization, radiography 42
  - ^{192}Iridium sources, interstitial therapy 195
  - reference points, brachytherapy, gynecological tumors 291, 292
  - threedimensional, source localization, radiographic methods 42, 43
- distribution calculation, point, line, radioactive sources 43
- dose rate calculation, "free spaced" interstitial permanent application 41
- dose rate effect
  - afterloading techniques 39, 41
  - bone marrow 11, 12
  - cell progression, proliferation 5, 7, 9, 10
  - central nervous system 13, 14
  - correction, interstitial boost irradiation 197
  - experimental tumor cell systems, repair half times 16, 17, 19
  - fractionated irradiation 11, 17
  - intestinal mucosa 12, 13
  - isoeffect comparison of high and low dose rate 56
  - lung, carcinoma, radiobiology 17
  - lung, fibrosis 13, 14
  - mathematic formulations 17-22
  - mouse lip, percutaneous irradiation 13
  - oral mucosa 13, 17
  - shin 15, 16
  - spinal cord 14, 15
  - dose rate factors
  - afterloading techniques, gynecological tumor therapy 294
  - isoeffect doses low and high dose rate afterloading 293, 294
- dose rates
  - low, middle and high-, fall, brachytherapy, gynecological tumors 337
  - ^{226}Radium treatment techniques, cervix carcinoma, Hamburg-, Houston methods 274
  - ^{226}Radium treatment techniques, cervix, carcinoma, Heidelberg method 333-339

dose rates
^{226}Radium treatment techniques, cervix carcinoma, Paris method 272
^{226}Radium treatment techniques, cervix carcinoma, Stockholm method 273
dose volume analysis, intracavitary therapy 45
equivalent-, low to high dose rate conversion 338
experimental dose effects, brain, permanent and temporary interstitial irradiation 81-84
extrapolated tolerance dose (ETD) 4, 5
extrapolated tumor dose, relative effectiveness 5
high dose rate
brachycurietherapy, ^{192}Iridium afterloading 70, 191
interstitial boost irradiation technique 191
high dose rate afterloading
cervix carcinoma 279, 292, 337
gynecological tumors, complications 293
gynecological tumors, dose distribution, reference points 291, 292
high, low dose rate afterloading techniques 39, 191
homogenity, intracavitary therapy 37
interstitial boost irradiation, breast cancer (quality assurance program) 193
interstitial therapy
carcinoma, buccal mucosa 134, 136
carcinoma, tongue 136, 142
inverse dose rate effect, HeLa cells 10
LD_{50}, dose rate effect studies, lung 13
$LD_{50/30}$, bone marrow stem cells 12
linear quadratic model, repair kinetics, skin 3-5, 9, 15, 18
low dose irradiation, theoretical generation curves of sublethal damage 18
low dose rate therapy, radiobiology 5, 7, 9
nominal standard-(NSD), time dose fractionation (TDF) 5
prescription, interstitial temporary, permanent applications 41, 42
profiles, collimated narrow beams 90
radiation, operator during endocurietherapy 157
radioisotopes, dose distribution around sources 32
response curve, mouse lung damage 14
single, total, afterloading, high dose rate technique, radiobiology 292
skin tolerance dose, radiobiology 15, 17
specification
biologic effects 5
brachycurietherapy, brain tumors 63
specific dose factor, brain, γ-emitters, comparison 82
survival curves
^{137}Cs gamma irradiation C_3H cells 9
experimental tumors, radiobiology 16
isoeffective doses, jejunal crypt stem cells 13
mammalian cells 8
target-, head and neck tumors 146
tolerance-, repair half times, α/β ratio, mathematical analysis 19
tolerance dose for acut skin response 17
total isoeffective doses, mathematical formulation 17
tumor doses, brain tumors, fractionated high dose rate brachycurietherapy 70
dosimetry
afterloading therapy 41, 193, 285-290, 292, 293
brain tumors 86
breast cancer 193, 195
biologic, teletherapy, SEITZ and WINTZ 45
bladder, intracavitary irradiation, cervix carcinoma 279-283, 285-290, 308
brachycurietherapy, brain tumors 65
brachytherapy, oral cavity-, oropharynx cancer 133, 139
breast cancer
interstitial boost irradiation, Paris system 189, 193, 195
interstitial therapy 180, 193, 195
clinical, interstitial boost irradiation, breast cancer 195
dose rate correction, interstitial boost irradiation 197
see dose rate effect
^{192}Iridium sources, well-type ionisation chamber 193
Paris system, reference dose 189, 193, 195
prostatic cancer
Hahnemann University 248
New York system 236
radioisotopes, activity 27, 28, 36, 41, 45, 65, 86, 133, 139, 180, 189, 193, 195
ring-type linear activity meter 194

endocurietherapy
guide gutter technique
carcinoma of palate, uvula 140
single, double wires 155, 156
^{192}Ir hypodermic needles, chema 156
^{192}Ir needles, wires, afterloading techniques 153-157
^{192}Ir, radiation protection 156, 157
^{192}Ir, Reverdin's needle 155
radiation dose to the operator 157
endometrium carcinoma
afterloading therapy, results 308, 309, 314, 327
clinical results of therapy 275, 303, 307-309, 314, 327
high dose rate afterloading
prospective clinical trial 307-310
^{226}Radium therapy, comparison 311-317
intracavitary therapy 275
treatment results 303
^{192}Iridium wire tube packing 302
metastatic spread 327
packing method, Heyman, Stockholm, results 307, 327
postoperative intravaginal treatment, results 327
primary intracavitary treatment 321, 322
radiotherapy with Heyman capsules, Ries eggs, ^{60}Co pearls 275
survival rates 275, 303, 307-309, 314, 327
"umbrella" applicator, low, high dose rate afterloading 333, 336
ependymoma
survival times, postoperative, interstitial therapy 68, 69
errors
dose calculation, afterloading therapy 289
experimental dose effects brain, γ-emitters, permanent low activity, low dose rate, temporary, high activity, high dose rate implants, comparison 83
brain, interstitial irradiation 81-84
external beam therapy
anal canal cancer 210, 215-219, 223, 224, 231
Bragg peak effect, choroidal melanoma 114
breast cancer 182
buccal mucosa carcinoma 134, 135, 136
cervix carcinoma 321, 336
combination with brachytherapy, radiobiology 4, 5
dose rate effects 3, 4
gynecological tumors 271, 314, 336
head and neck tumors 145, 151
interstitial hyperthermia, combination therapy, results 372, 374
mathematical dose formulations 16-22
oral cavity, oropharynx carcinoma 133, 135

palatotonsillar carcinoma 142
prostatic cancer, results 238, 239, 245, 247, 248, 253, 254, 264
radiobiology 3-6
time dose fractionation (TDF) 5
eye
enucleation after β-therapy, choroidal melanoma 109, 110, 113
enucleation, choroidal melanoma 115, 116
versus radiocobalt eye-plaque therapy, comparison of results 115
metastases, incidence 114
primary malignant tumors, American Cancer Society, incidence 114
tumors
malignant, therapy, results 117
thermoradiotherapy 350
eye application
^{106}Ru/^{106}Rh, choroidal melanoma 103, 104

fistulae
bladder, rectum, vagina, afterloading therapy, cervical cancer 302, 308, 328
pelvic tumors, following interstitial thermoradiotherapy 353
urethra-rectal, after ^{125}Iodine seed implantation, prostatic cancer 261
fluoroscopy
cervix carcinoma, bladder dosimetry 281
Foley catheter
ballon, bladder dosimetry 282, 283

Geiger-Müller counter
radioisotopes, homogeneity test 31, 32
glioblastoma
survival times, postoperative, interstitial therapy 68
glioma
brachycurietherapy, survival time, tumor dose, fractionated high dose rate 70
nonresectable
external stereotactic focal irradiation 89-94
WHO, I and II low grade, stereotactic radionuclide implantation 67-79
^{198}gold
interstitial therapy
gynecologic tumors 276
prostatic cancer 253
marker seeds, inadequate retropubic implantation 253
physical properties 28
radon seeds, choroidal melanoma, interstitial therapy 116
seeds, interstitial therapy, tongue carcinoma 169-175
gynecological tumors
afterloading devices, low and high dose rates 292
gynecological tumors
afterloading techniques 271, 272, 276, 279-283, 285-290, 291-297
tumor control 39, 40
afterloading therapy
high dose rate, conventional brachytherapy, comparison 319-331
high dose rate, ^{192}Iridium, dosimetry, localization 285-290
high, low dose rate 333-339
annual report on results of treatment, Deutsche Gesellschaft für Gynäkologie, Halle 295
bladder dosimetry, technical aspects 279-283
brachytherapy
comparison with high dose rate afterloading 329-331
complications 288, 293
dose distribution, reference points 291, 292
methods, future developments 277
radiation protection of staff 291
techniques, results 39, 40, 279-283, 285-290, 291-297
cervix carcinoma
high dose rate afterloading, techniques 279, 280, 291
high dose rate afterloading techniques, clinical trial 307-310
high dose rate irradiation, clinical results 298-305
megavolt therapy 271
one - and three way applicators 287
^{226}Radium treatment techniques 271, 272, 273, 274
clinical results of therapy 295, 298-305, 308, 309, 313, 314, 323, 324
dose rate fall, low, middle, high dose rates brachytherapy, isodoses 337
endometrium carcinoma
high dose afterloading therapy, prospective trial 307-310
intracavitary radiotherapy 275, 333-339
external beam therapy 271, 314, 336
high dose rate afterloading
prospective clinical trial 307-310
^{226}Ra therapy, comparison 311-317
high dose rate afterloading techniques
dose rate factors, evaluation 294
instruments 335, 336
high, low dose rate afterloading 333-339
interstitial remote afterloading 277-290, 291-317, 319-331
interstitial therapy, needles, wires, seeds 276
interventional radiation therapy, techniques 271-278
intracavitary application of radioisotopes, dose specification 45
intracavitary techniques, physical properties of sources 271, 272, 285-290
^{192}Iridium afterloading devices 292
^{192}Iridium afterloading, hight dose rate, dosimetry, localization 285-290
low and hight dose rate afterloading devices 292
megavolt therapy 271, 322
^{226}Radium treatment techniques, Paris-, Stockholm, Manchester-, Munich-, Hamburg-, Houston methods 271-274
survival rates, afterloading therapy 295, 298-305, 324, 325, 337
vaginal carcinoma, interstitial, intracavitary therapy 275

head and neck tumors
advanced, brachytherapy, Erlangen experience 159-167
afterloading, brachytherapy 153-157, 159-167
brachytherapy, interstitial 145-152
general care of patients, before and after 148
planning, evaluation 145
principles, techniques 146, 147
results of treatment 148, 150
interstitial hyperthermia
clinical results 350, 360, 361, 366, 368
external beam irradiation, results 372, 374
morbidity, acute, late 150
radiotherapy 129-167
thermoradiotherapy 350
histology
prostatic cancer 246, 249, 250
history
anal canal cancer, therapeutic situation 230
brachytherapy 35
cancer therapy with radium 25
clinical radiobiology 7
dosimetry
biologic, teletherapy, SEITZ and WINTZ 45
radioisotopes 36
interstitial brachytherapy 25, 145
^{198}Au seeds 169
breast cancer 179
interstitial hyperthermia 343
intracavitary brachytherapy, gynecological tumors 291, 298
intracavitary therapy, ^{226}Ra application, Stockholm system (Radiumhemmet), Paris system (Curie Foundation) 37
^{125}Iodine implantation, prostatic cancer 263

history
  radiobiology 7
  radioisotopes
    activity units 28
    interstitial, intracavitary therapy 25
  ^{226}Radium treatment techniques, gynecology 271, 272, 291
  thermoradiotherapy 365
  uveal tract tumors, diagnosis, treatment 114
hyperthermia
  interstitial
    advantages 365
    brachytherapy techniques 46
    capacitive radiofrequency technique 446
    chemotherapy, malignant tumors 371
    computerized feedback control system 359
    conductive hot water perfusion, technique 349
    conductive laser technique 349
    current technical status, future prospects 349, 350
    external beam irradiation, combinated therapy, results 372–374
    hot source techniques 347
    inductive ferromagnetic seed technique 347, 348
    intraoperative, techniques, results 357, 358, 362, 363
    microwave technique 346, 347, 359, 360
    radiofrequency-, clinical experience 351
    radiosensitizing, chemosensitizing agents 343
    standard implantion techniques 361
    techniques, survey 344, 345
    techniques, thermoradiotherapy, clinical experience 343–357
    therapeutic treshold temperature of 42 °C 360, 361, 373
    thermoradiotherapy, malignant tumors 365–370
    tumor therapy as sole treatment modality or with radiotherapy 371–374
hypopharynx
  carcinoma, hyperthermia, external beam irradiation, results 372
hysterography
  endometrium carcinoma 275

^{125}I
  brain tumors, stereotactic interstial irradiation 68
  interstitial therapy, techniques 37
  seeds, dose calculation, interstitial therapy 41
imaging stereotactic implantation of radionuclides
  intracranial tumors 67–79
implantation technique
  ^{198}Au seeds, tongue carcinoma 173
  ^{192}Iridium seeds, breast cancer 188, 189
incidence
  anal canal carcinoma 207
  prostatic cancer 235
indications
  ^{198}Au, interstitial radiotherapy 172
  brachytherapy, oral cavity-, oropharynx carcinoma 133
  breast cancer
    interstitial boost irradiation 187
  breast cancer
    interstitial therapy 182
  CT-stereotactic ^{192}Ir-, ^{125}I brachycurietherapy 73
  plastic tube technique, hairpin implant, tongue carcinoma 137
  radiotherapy, pituitary adenoma 71
  transperineal, sonographically guided ^{125}Iodine seed implantation, prostate cancer 261, 264
instruments
  afterloading applicators, low, high dose rate 36, 192, 335
  teflon carrier tubes with seeds and spacers, prostatic cancer therapy 259
interstitial boost irradiation
  breast cancer, technique, dosage 187–190, 191–198
    complications 202, 203
interstitial hyperthermia
  see hyperthermia, interstitial
interstitial irradiation
  brain, permanent and temporary, experimental dose effects 81–84
interstitial therapy
  afterloading techniques 85–88, 153–157
  anal canal cancer 209–213, 230
    combined with external beam therapy 210, 215–219
  astrocytoma, ^{192}Iridium implantation 73
  ^{198}Au, implantation, tongue carcinoma 169–175
  brain
    experimental dose effects 81–84
    radionecrosis 82, 83
  brain tumors, brachycurietherapy
    basic principles 61–66
    clinical results 65
    dosimetry 65
    gliomas 75, 76
    implantation techniques 62, 63, 64
    presurgical CT scans 64
    selection of radionuclides 62
    thalamus tumors 72
  buccal mucosa, carcinoma, results 134, 135
  cervix carcinoma 275
  classic systems, Paris-, New York model 146
  complications 69, 71, 73, 81, 83, 85, 87, 133, 135, 138
  curietherapy, techniques, radioisotopes, stereotactic biopsy, therapy 69, 70–79
  dose volume analysis 45
  "free spaced" interstitial permanent source application, dose rate calculation 41
  guide gutter-, loop techniques 140, 146, 147, 155
  hairpin implantation 137, 138, 140
  head and neck tumors 145–152
    Erlangen experience 159–167
  high and low dose implants, critical comparison 53–57
  history 25
  ^{125}I seeds, dose calculation 41
  instruments 36, 192, 276
  lip carcinoma 133, 134, 135
  lymph node matastases 157
  mouth carcinoma 135, 138, 140
  neck metastases, CT 135, 143
  needles, wires, seeds, gynecological tumors 276
  oligodendroglioma, survival times 68
  oral cavity, oropharynx carcinoma (Paris technique) 133–144
  palate-, uvula carcinoma guide gutter technique 140
  parasellar, pituitary tumors, indications, clinical results 70, 71
  plastic tube technique (Pernot technique), palatotonsillar implants 136, 137, 141, 143
  prostate, sonography, dose distribution 43
  prostatic cancer 245–256, 257–262, 263–267
    Freiburg experiences 241–244, 250, 252
  radioactive sources, physical properties 25–30
  radionuclides, number of stereotactic interstitial irradiation procedures 68
  reference points, reference isodose 44, 45
  Reverdin's needles, retromandibular, cervical lymphadenopathy 143
  skull base tumors 71
  target volumen 180
  techniques, permanent, temporary application of radioisotopes 37, 38
  therapeutic concept 160
  tongue, base, carcinoma 135, 142
  tongue, oral, carcinoma 135, 136, 137, 142
  tonsillar carcinoma 142
  vaginal carcinoma 275
intraoperative hyperthermia

interstitial, techniques, clinical results 362, 363
intestinal mucosa
dose rate effect, radiobiology 12, 13
intracavitary therapy
advantages 316
cervix carcinoma, bladder dosimetry 279-283
dose volume analysis 45
endometrium carcinoma 275
gynecologic, dose specification 45
physical properties of sources 25-30
radionuclides used, characteristics 285
^{226}Radium application, Stockholm system (Radiumhemmet), Paris system (Curie Foundation) 37
techniques, instrumentation 37, 38
see brain tumors
intracranial malignant tumors
indications, CT stereotactic ^{192}Ir-, ^{125}I brachycurietherapy 73, 75
intrauterine application
Heyman capsules, Ries eggs, ^{60}Co pearls, endometrium carcinoma 275
^{226}Radium, Paris-, Stockholm-, Manchester-, Munich-, Hamburg-, Houston methods 271-274
"umbrella" applicator, endometrium carcinoma 333, 336
^{125}Iodine
implantation, ultrasound guided, prostatic carcinoma 257-262, 263-267
implants
prostatic cancer, Mick applicator 236, 239, 241, 248-254, 263-267
temporary, permanent, head and neck tumors, therapeutic concept 146, 160
interstitial irradiation, brain, permanent implants, experimental dose effects 81-84
interstitial therapy, survival times 162
physical properties 29
plaques, choroidal melanoma treatment 117, 346
seed therapy
physical and radiobiological data 264
ultrasonically guided 264
seeds
implantation technique, brain tumors, target volume 63
implantation, tongue carcinoma 172
interstitial therapy, gynecological tumors 276
iridectomy, iridocyclectomy
choroidal melanoma 114, 116
^{192}Iridium
afterloading technique, breast cancer 179, 181, 191
afterloading therapy, physical measurements, dosimetry, localization 285-290, 333, 334
brain tumors, stereotactic interstitial irradiation 68
curietherapy, lip carcinoma, results 134, 135
dose distribution calculation, Monte Carlo method 43
high dose rate afterloading devices, gynecologic tumor therapy 292
implantation
afterloading technique, anal canal cancer 153, 154, 210, 215-219, 223, 224, 231
astrocytoma 73
superimposed dose distribution, radiographs 196
implants
permanent, temporary, Erlangen experiences 159-167
temporary, head and neck tumors 146
interstitial therapy, survival times 163
irradiation, interstitial hyperthermia, combination therapy 361, 362
needles, wires, endocurietherapy
afterloading techniques 153-157
head and neck tumors 153-157
radiation protection 156
needles, wires, seeds, interstitial therapy, gynecological tumors 276
nylon-, teflon tubes, wires, loading, schema 153
physical properties 27
plaques, choroidal melanoma treatment 117, 118
point source, exposure rate in air, in a lucite phantom, comparison 287
reference air kerma rate, strength of source 31
seeds, target volume, implantation techniques, brain tumors 63
source preparation, exposure-free station 194, 195
sources, dosimetry, well-type ionisation chamber 193
wire tube packing, endometrium carcinoma 302
wires
automatic encapsulator 154
breast cancer, interstitial boost irradiation 191
constructional details 27
iris, melanoma, therapy, results 117
isodose distribution
^{198}Au seed model 173
prostatic cancer, control 260
isodoses
configuration, afterloading therapy, cervix carcinoma 330, 334
curves, bladder dosimetry 282
dose rate fall in gynecological low, middle and high dose rate brachytherapy 337
female pelvis, ^{226}Ra conventional brachytherapy, ^{137}Cs afterloading technique 334
isocentric stereoradiographs, interstial boost irradiation 191, 192
reference-, interstitial therapy, breast cancer 180, 193

jejunum
dose rate dependence, experimental curves 20
epithelium, dose rate effect, radiobiology 12
tolerance curves, normalized to their acute ED_{50} values 21

larynx
carcinoma, brachytherapy, Erlangen experiences 159
leukaemic cells
dose rate effect, total body irradiation 11
linear accelerator
brain, arteriovenous malformations, stereotactic focal irradiation 95-99
parametrium irradiation, cervix carcinoma 314
linear energy transfer (LET)
dose suvival curves, mammalian cells 8
lips
carcinoma
interstitial therapy, results 133, 134, 135, 149, 157
triangular implant pattern, curietherapy, reference isodose 134
lumpectomy, breast cancer
dosimetry 181
intraoperative brachycurietherapy 182
lung
carcinoma
hyperthermia, clinical results 360
hyperthermia, external irradiation, results 372
Lewis, dose rate dependence of radiosensibility 17
dose rate effect, radiobiology 13, 14, 20
pneumonitis, dose response curves, thoracic irradiation 14
tolerance curves, normalized to their acute ED_{50} values 21
tumors, interstitial hyperthermia, clinical results 360, 372
lymph nodes
dissection
neck irradiation, oral cavity-, oropharynx carcinoma 133
prostatic cancer 239
metastases,
interstitial therapy, Erlangen experiences 157

lymph nodes
prostatic cancer 236, 246
metastases, tongue carcinoma 170, 171
radical neck dissection 172
regional, anal canal cancer 207
lymphadenectomy
prostatic cancer
histological grade, survival times 250, 252, 258
indications 246, 252
lymphadenopathy
anal canal cancer 207
breast cancer, ductal carcinoma in situ 119
cervical, retromandibular, plastic tube implantation, Reverdin's needles 143
prostatic cancer 236
lymphangiography
prostatic cancer 246
lymphatic spread
anal canal cancer 207

mammary carcinoma
see breast cancer
mastectomy
breast cancer, chemotherapy 184
mathematical formulations
afterloading techniques 41
brachytherapy
dose calculation 43, 44
nomographs 41
dose rate effect 17–22
isoeffect comparison of high and low dose rate 56
radioactive sources, dose distribution calculations interstitial therapy 44, 45
reference air kerma rate, radioisotopes, calculation 285, 286
maxillary carcinoma
interstitial hyperthermia, thermal map 361
megavolt therapy
cervix carcinoma 271, 322
melanoma
choroidal, see choroidal melanoma
meningioma
survival times, postoperative, interstitial therapy 68
metastases
anal canal cancer 213
bone, prostatic cancer 242, 246, 249
brain
before and after stereotactic irradiation, CT 91
external stereotactic focal irradiation 89–94
choroidal melanoma, therapy, results 104, 114, 117
endometrial carcinoma 327
eye, incidence 114
N classification, oral cavity, oropharynx carcinoma according UICC, AJCC 131, 132
neck
CT 143
hyperthermia, external radiotherapy 374
interstitial therapy 143, 147, 161
prostatic carcinoma 236, 242, 246, 249
thermoradiotherapy 350
tongue carcinoma, survival rates 170, 171, 174, 175
metastatic dissemination
prostatic cancer, transurethral resection 246
Monte Carlo method
dose distribution calculations, interstitial therapy 43, 44
morbidity
acute, late, head and neck tumors 150
prostatic cancer, treatment related 242, 246, 247, 251
mouth
carcinoma, interstitial therapy 134, 135, 136, 161
MRI
brain tumor, definition of true extent 62
chorioral melanoma, diagnosis 114
head and neck tumors, tumor volume 146
mucosa
buccal, carcinoma
interstitial brachytherapy 134
interstitial hyperthermia 361
plastic tube implantation technique 136, 137
results of curietherapy 134, 135
intestinal, dose rate effect, radiobiology 12, 13, 20
late reactions, portio, cervix, vagina, afterloading plate method 302
oral, dose rate effect mathematical formulation 19, 20
rectal, bleeding, teleangiectasia, anal canal cancer treatment 210

nasopharynx
carcinoma, hyperthermia, external beam irradiation, results 372
carcinoma, interstitial brachytherapy 146
neck
carcinoma, recurrent, two plan implantation, brachytherapy 147
dissection
residual tumor 161
tongue carcinoma 172
metastases
interstitial therapy 143, 147
5-year survival rates, tongue carcinoma 174, 175
tumor, implants in position, localization radiograph 161
nylon-, teflontubes
endocurietherapy, afterloading techniques 153, 154

oligodendroglioma
survival times, postoperative, interstitial therapy 68
oral cavity, oropharynx carcinoma
afterloading brachytherapy 153–157
base of tongue, interstitial therapy 142
brachytherapy 144
Erlangen experiences 159–167
buccal mucosa,
interstitial therapy, results 134, 135
plastic tube technique (Pernot technique) 136
classification 131, 132
interstitial therapy (Paris technique) 133–144
lips, interstitial, therapy 133, 135
localization, treatment results 149, 159
mouth floor, interstitial therapy 138, 140, 147, 149
neck metastases, interstitial therapy 143
oral tongue, interstitial therapy 136
recurrent disease, ^{192}Iridium implantation therapy 163
soft palate, curietherapy, results 135
TNM system, results of therapy 131, 132, 135
tongue carcinoma
curietherapy, results 136, 137, 142, 145, 147
hair pin implant, reference isodose 137
tonsillar region, interstitial therapy 141, 142, 147, 149
uvula carcinoma, endocurietherapy, guide gatter technique 140
vallecular region carcinoma, curietherapy, plastic tube implantation 143
oral mucosa
dose rate effect, radiobiology, mathematical formulation 12, 13, 17
organs of risk
absorbed dose calculation, methods 288, 289
applicator position, intracavitary afterloading therapy 287, 312, 313
osteonecrosis
head and neck tumors, brachytherapy 150, 162

^{32}P, brain tumors, stereotactic interstitial irradiation 68
pancreas
tumors, thermal map, interstitial hyperthermia 363
parametrium
irradiation, linear accelerator, cervix carcinoma 314
pathology
brain, radionecrosis, after interstitial irradiation 81, 82

breast cancer, additional tumor foci 199, 200
pelvis
female
dose reference points, intracavitary brachytherapy 291
frontal view, dose distribution, ^{226}Ra conventional brachytherapy, ^{137}Cs remote controlled afterloading techniques 334
sagittal plane, applicator position control 288
percutaneous irradiation
dose rate effect, oral mucosa 13
see external beam therapy
perineum
anal canal cancer
abdominoperineal resection 211
^{60}Co-, ^{192}Iridium implantation therapy 210
implants, ^{192}Ir needles, wires, afterloading brachytherapy 153, 154
transperineal implantation, ^{125}Iodine seeds, prostatic cancer 258, 260, 264
pharynx
carcinoma
brachytherapy, treatment, results 149, 161
see oral cavity-, oropharynx carcinoma
vallecular region, plastic tube implantation 143
physical fundamentals
afterloading, high and low dose rate techniques 53, 54
radioisotopes 25-33
physical measurements
intracavitary high dose rate afterloading therapy 285-290
physical properties of sources
absorbed dose, definition, calculation 32
activity, SI units 27, 28
air kerma rate constant 32
^{198}Au 28, 271, 272
^{198}Au seeds 172, 259
"average life", T_a 30
brachytherapy 35, 46
^{137}Caesium 26, 271, 272
calibration 32
^{60}Co 28, 271, 272
contamination tests 31
dose distribution around sources 32
energy 26
gold 198 28, 271, 272
half-life time 28
homogeneity test 31, 36
identification 31
intracavitary, interstitial therapy 25, 271, 272, 286
intracavitary techniques, gynecological tumors 271, 272
^{125}Iodine 29, 259
seed therapy 264
^{192}Iridium 27, 271, 272
radioisotopes, brachytherapy 35, 271, 272
^{226}Radium 29, 272
reference air kerma rate 32
^{106}Ruthenium 30
SI units for activity 28
source strength, specification 30
physics
3-channel applicator, brachytherapy, cervix cancer 300, 301
pineal region
tumors, brachycurietherapy 73
pituitary
chromophob, eosinophilic adenoma, brachycurietherapy 71
pneumonitis
dose response curve, mouse lung damage 14
prognosis
anal canal cancer 207, 211
breast cancer 299
choroidal melanoma 116, 117
interstitial hyperthermia, combined with external radiation, tumor response 373
prostatic cancer 235
tongue carcinoma 171
prognostic factors
anal canal cancer 207, 211
interstitial thermoradiotherapy 353, 367, 368
prostate
needle implantation, technique 237
sonography, interstitial therapy, dose distribution 43
prostatectomy
prostatic cancer, techniques, indications 246, 247, 264
prostatic cancer
^{198}Au marker seeds, inadequate retropubic implantation 251
biopsy, effect on survival 245, 246, 258, 265, 266
bone metastases 246, 249
brachytherapy 236, 237, 239, 245-256, 257-267
chemotherapy 265
classification 235, 236, 241, 245, 257
clinical results 237, 238, 241-244, 250, 251, 253, 254, 263-267
complications of treatment 239, 242, 246, 248, 252
CT cross sections 235, 237, 239, 243, 264
diagnosis 245, 246, 258, 265, 266
dosimetry, interstitial therapy 237, 248
external beam irradiation, results 238, 239, 247, 248, 249, 253, 264
^{198}gold marker seeds, inadequate retropubic implantation 259
histologic pattern 246, 249, 250
hormonal therapy 243, 248
incidence 235
indications, transperineal, sonographically guided ^{125}Iodine seed implantation 261, 264
interstitial brachytherapy 235, 236, 239, 241-244, 245-256, 263-267
^{125}Iodine implantation, Mick applicator 235, 236, 239, 241, 248-254, 257-262
lymph node
involment 236, 246, 258, 266
lymph node
dissection 239, 266
lymphadenectomy, histologic grade, survival times 250, 258, 266
lymphadenopathy 236, 238, 257
lymphangiography 246
metastases, lymph nodes, pelvic 235, 236
metastatic dissemination, transurethral resection 246
morbidity, treatment related 242, 246, 247, 251
orchiectomy, hormonal therapy 248
prognostic factors 235
prostatectomy, radical 238, 245, 264, 266
recurrence rates 250
sonography
^{123}Iodine implantation 257-262
transrectal 235, 239, 243, 258
survival times 237, 238, 239, 250, 251
TNM system, Memorial Sloan-Kettering Cancer Center 235, 236, 241, 245, 266
transurethral resection, metastatic dissemination 246
treatment alternatives, indications, advantages, disadvantages, complications 264
treatment results 237, 238, 247, 263-267
ultrasound guided ^{125}Iodine implantation 257-262, 264
volumetry, sonography 258, 264
prostatic irradiation
morbidity, treatment related 247
proton beam therapy
brain, arteriovenous malformations 93
protons
radiotherapy, choroidal melanoma 120

quality control
breast cancer, interstitial boost irradiation 193
clinical radioisotope sources 31, 32

radiation exposure
interstitial implantation of ^{198}Au seeds 174
medical personnel, brachytherapy, prostatic cancer 260, 266

radiation hazard
  staff protection, remote controlled afterloading therapy 191
radiation protection
  ^{198}Au seeds, implantation techniques 174
  interstitial boost irradiation 191
  medical staff, manual afterloading treatment 291, 319
  radioactive wires 156, 157
  rectum, bladder, brachytherapy, cervix cancer 300, 301
radiation reactions
  bladder, following brachytherapy 279
radiobiology
  afterloading, high dose rate, single and total doses 292
  air kerma rate constant, absorbed dose, definition, calculation 32
  ^{198}Au seeds, interstitial therapy 169
  biologic dose response, late effects 5
  bone marrow, dose rate dependence of normal tissue tolerance 20
  brachytherapy
    cervix cancer 3-6, 301, 302
    combination with external beam therapy 4, 5
  brain, permanent and temporary interstitial irradiation, experimental dose effects 81-84
  cell cycle, radiosensitivity 9
  cell progression, proliferation, dose rate effect 9, 10
  cell proliferation, intestinal mucosa 12
  cell repair, recovery, time factor, norma tissue, tumor tissue 329
  cell surviving curves, changing of dose per fraction, high dose rate 55
  ^{60}Co radiation, dose-rate effects 13
  critical dose-rate, V-79 cells, HeLa cells differences 10, 11
  ^{137}Cs gamma irradiation, dose survival curves 9
  dermal tolerance doses 15
  dose rate dependence of radiosensitivity, experimental tumor cells 17
  dose rate effect
    bone marrow 11, 12
    central nervous system 14, 15
    fractionated irradiation 11
    intestinal mucosa 12, 13
    leukaemic cells 11
    lung 13
    oral mucosa 13
    relative effectiveness (RE) 4
    skin, tolerance dose 15, 16
  dose response curves, mouse lung damage 14
  dose specification, biologic effects 5
  dose survival curves
    ^{137}Cs gamma irradiation 9
    experimental tumors, hypoxic cells 16
    HeLa cells, inversion of dose-rate effect 10
    mammalian cells 8
  effect on tumor, single, total dose, hight dose rate afterloading 292, 336
  equating CLDRI and HDR fractionation 21
  equivalent dose, low to high dose rate conversion 338
  experimental tumor cell systems, repair half times 29
  extended linear quadratic formalism, mathematical formulations 18
  external beam therapy 3-6
    combination with brachytherapy 4, 5
    extrapolated tolerance dose (ETD) 4, 5
    fractionated radiotherapy lung, pneumonitis 14
    fundamentals 7-24
    Gy for Gy, biologic effects 3
    high and low dose rate afterloading 54
    history 7
    hyperthermia, complementary interaction of irradiation 344
    hypoxic cells, radiosensitivity 7
    hypoxic tumor cells, radioresistance 315, 337
    incomplete repair model, mouse pneumonitis 14
    intracellular repair during CLDRI 12
    intrinsic cell sensitivity 11
    inverse dose rate effect, HeLa cells 10
    ^{125}Iodine seed therapy 264
    isodoses, standard Fletcher caesium application 4
    jejunal crypts, dose rate dependence, experimental curves 20
    jejunum, tolerance curves, normalized to their acute LD_{50} values 21
  late radiation effects 5
    LDRI, biological effectiveness, skin 15
    linear energy transfer (LET) 8
    linear quadratic dose relationship, oral mucosa 13
    linear quadratic model
    dose rate, brachytherapy with teletherapy 45
  linear quadratic model
    repair kinetics, skin 3-5, 9, 15, 18
    linear quadratic survival curves, mathematical deduction 18
    low dose rate therapy, biological effectiveness 5, 7, 8, 9
    lung
    carcinoma, radiosensibility, dose rate dependence 17
    dose rate dependence of normal tissue tolerance 20
    pneumonitis, dose response curves 14
  lung tolerance curves, normalized to their acute ED_{50} values 21
  mathematical formulations 17-22
  nominal standard dose 5
  oxygen enhancement ratio, high dose rate 337
  physics, fundamentals 25-33
  pneumonitis, dose-response curves 14
  radiation damage, kinetic studies 9
  radiobiological calculations, linear quadratic model, Paris system 191
  radioisotopes, physical properties 25
  radiosensitivity
    hypoxic cells 7
    thermoradiotherapy, biomolecular mechanisms 354
  ^{226}Radium therapy, classic 311, 312
  reference linear kerma rate, definition, calculation 32, 33
  "relative effectiveness" (RE), definition 21
  "shoulder", dose survival curves, sublethal damage repair (SLDR) 8, 21, 22
  skin, tolerance curves normalized to their acute LD_{50} values 21
  spinal cord radiosensitivity 14, 15
  spinal cord, tolerance curves, normalizad to their acute ED_{50} values 21
  sublethal radiation damage, repair during CLDRI 7, 9
  theoretical generation curves of sublethal cell damage 18
  thermoradiotherapy
    biomolecular mechanisms 354
    complementary interaction of heat and radiation 344
  time dose fractionation (TDF) 5
  tolerance dose
    acute, chronic irradiation, experimental curves 20, 21
    extrapolated 4, 5, 19, 20
    repair half times, α/β ratio, mathematical analysis 19, 20
  tumor cell systems, repair half times 19
  tumor dose, extrapotated, relation effectiveness 4, 5, 16
  X radiation, dose rate effects 13, 16
radiochemotherapy
  combined, anal canal cancer, Lyon experiences 209-213
radiography
  applicator position
    dose distribution 312, 313
    organs of risk, afterloading therapy 287
  brachytherapy techniques 42
  breast cancer, isocentric stereoradiographs 191, 192

cervix carcinoma, bladder dosimetry 281
check of radioactive source position 287
^{192}Iridium implantation, superimposed dose distribution 195, 196
methods for optimize seed distribution, prostatic cancer 251, 265
one-, three way applicators for treatment of cervix carcinoma 287, 289
tonsillar carcinoma, radioactive implants 162, 163

radioisotopes
absorbed dose 32
activity 27
air kerma rate constant 32
applicators, quality control 31, 32
^{198}Au, physical properties 172
"average life", T_a 30
brachytherapy, physical properties 35, 46
calibration 32
decay mode 25-30
disintegration, α-, β-, γ-emitters 25-30
dose calculation in a medium 32
dose distribution around sources 32
dosimetry 27, 28, 36, 41, 45, 65, 86, 133, 139, 180, 189, 193, 195
energy, α-, β-, γ-radiation 26
half life time 28
high dose rate afterloading, physical parameters 286, 333, 334
homogeneity test 31, 32
identification, quality control 31
intracavitary afterloading, characteristics 285
intracavitary techniques, physical properties 271, 272
leakage, surface contamination tests 31
linear activimeter, homogeneity test 31, 32
linear applicator, radiography, control of position 287
loaded sources, autoradiography, correct, incorrect source assembly 197
milligram radium equivalent (mg Ra eq), definition 30
production, disintegration 25
radiation protection 156, 157, 174, 191
radioactive plaques, choroidal melanoma treatment 117
reference air kerma rate, calculation 285
selection
implantation techniques, brain tumors 62, 63, 68
number of stereotactic interstitial irradiation procedures 68
source calibration, lucite phantom 286
source localization, three-dimensional reconstruction, radiographic methods 42, 43
source preparation
exposure free station 194, 195
isoeffect evaluation, nomogram 197
sources
autoradiography, correct, incorrect source assembly 197
point-, line-, dose distribution calculations 43
specification of source strength 30
surface contamination, swab test 31

radionecrosis
anal canal 209, 210
brain, γ-emitters, permanent low activity, low dose rate; temporary, high activity, high dose rate implants, comparison 83
soft tissue
interstitial brachytherapy 165, 166
interstitial brachytherapy lip, cancer 134, 135, 138

radionuclides
see radioisotopes

radioresistance
hypoxic tumor cells 315, 337

radiosurgery
brain, arteriovenous malformations 92, 93

radiosensitivity
brain, normal parenchyma 61
brain glioma 81
cell cycle 9
choroidal melanoma 116
dose rate dependence, experimental tumors, radiobiology 17, 20
hypoxic cells 7, 16, 17
increased, hyperthermia 344
intrinsic cellular, low dose-rate, division delay, correlation 11

radiosensitization
thermal, biomolecular mechanisms 354

radiosurgery
oral cavity, oropharynx carcinoma 133

radiotherapy
afterloading, high and low dose techniques, critical comparison 53-57
^{198}Au interstitial therapy, technique, tongue carcinoma 172
^{198}Au seeds, intratumoral irradiation 169
brachytherapy
combination with external beam therapy 45
physical fundamentals 25-33
radiobiology 3-6
brain, arteriovenous fistula, external stereotactic focal irradiation 91-93
with linear accelerator 95-99
brain tumor metastases, external stereotactic focal irradiation 89-94
brain tumors, brachycurietherapy 61-99
breast cancer
cosmetic results 183
primary irradiation 202, 203
choriordal melanoma, ^{106}Ru/^{106}Rh plaques 103-111, 113-127
combination with interstitial hyperthermia 350
external stereotactic focal irradiation 89-94
fractionated, late skin response, dermal tolerance 15
head and neck tumors 129-167
hyperthermia
interstitial, combination effect 350
tumor response, prognosis 373
imaging stereotactic implantation of radionuclides 67-79
interstitial irradiation, brain, experimental dose effects 81-84
interstitial thermoradiotherapy, techniques, results 343-357, 359-363, 365-374
interventional, gynecological tumors 271-278
intracavitary, interstitial therapy, radioisotopes, physical properties 25
intraoperative, thermoradiotherapy 350
isoeffect comparison of high and low dose rate 56
percutaneous irradiation, radiobiology 13
photon treatment, linear accelerator 97
physics, fundamentales 25-33
pituitary adenomas, indications, results 71
postoperative, brain tumors 85
primary, dose-response relationship, breast cancer 201
role in the management of malignant eye tumors 113
stereotactic single high dose irradiation 92
teletherapy, oropharyngeal mucosa, dose rate 13
see thermoradiotherapy
thermoradiotherapy, interstitial, clinical experience, techniques 343-357
tongue carcinoma, survival rates 171, 172, 174, 175

^{226}Radium
brachytherapy
cervix cancer, survival rates 337
dose distribution, female pelvis 334
curietherapy, lip carcinoma, results 134, 135

^{226}Radium
interstitial therapy
gynecological tumors 271
techniques 37
intracavitary afterloading therapy, gynecological tumors 285, 311, 337
intracavitary application, Stockholm system (Radiumhemmet), Paris system (Curie Foundation) 37
intrauterine application 272, 273, 274, 311
needles, head and neck tumors, interstitial therapy 145
needles, wires, seeds, interstitial therapy, gynecological tumors 276
physical properties 29
therapy, conventional
complications 315
high, medium, low dose rate, definition 313
high, medium, low dose rates, results of treatment 314
radiobiology 311, 312
vs. high dose rate afterloading, survival rates 314
therapy, interstitial, intracavitary, gynecological tumors 271
treatment techniques, gynecological tumors, Paris-, Stockholm-, Manchester-, Munich-, Hamburg-, Houston methods 271–274
rectum
carcinoma, interstitial thermoradiotherapy 350
organs at risk, dose calculation, afterloading therapy 287, 289, 290, 308
radiation exposure, reduction, possibilities 300
side effects, low, high dose rate afterloading 292, 293, 302, 308, 315, 325, 328
reference air kerma rate
measurement, experimental set-up 286
radioisotopes, definition, calculation 30, 31, 32, 285
reference points
brachytherapy, gynecological tumors 291, 292
retinoblastoma
incidence 114
therapy, results 117
rhabdomyosarcoma
dose rate dependence of radiosensitivity, radiobiology 16, 17
^{106}Ru/^{106}Rh plaque therapy
choroidal melanoma 103–111
^{106}Ruthenium
applicators, brachytherapy 36
physical properties 30

salvage brachytherapy
breast cancer, local recurrence 184, 185
sexual practice
anal canal cancer 208
salivary glands
carcinoma, interstitial hyperthermia 361
skin
defects following interstitial thermoradiotherapy 353
dose rate effect, radiobiology 15, 16
epilation dose, dose rate dependence, experimental curves 20
tolerance curves, normalized to their acute ED_{50} values 21
tolerance dose for acute response, mathematical formulation 17
soft tissue necrosis
interstitial therapy 165, 166
lip cancer 134, 135, 138
sonography
choroidal melanoma 114
interstitial therapy, prostate, dose distribution 43
prostatic cancer
prostate mapping, volumetry 258
seed distribution 243
transrectal 235, 239, 258
transvesical, transrectal, afterloading therapy, dose calculation 289
tumor volume, head and neck tumors 146
ultrasound-controlled implantation technique 243, 257–262, 264
source localization
three dimensional reconstruction, radiographic methods 42, 43
sphincter ani
involvement, anal canal cancer 207
preservation, anal canal cancer 212, 232
spinal cord
sensitivity to LDRI 14, 15
tolerance curves, normalized to their acute ED_{50} values 21
staging
see classification, TNM system
stereo shift projection
source localization, threedimensional reconstruction 42, 43
^{90}Strontium
applicators, plane, concave 36
surgery
abdominal, intraoperative thermoradiotherapy 350, 351
anal canal cancer, diagnosis, treatment results 229–232
lymphadenectomy, prostatic cancer, histological grade, survival times 250, 252
orchidectomy, prostatic cancer 248
prostatectomy, prostatic cancer 245, 246, 247
survival rates
breast cancer, interstitial boost irradiation 190
cells, see dose survival curves
cervical carcinoma, afterloading brachytherapy, radium therapy 295, 298–305, 309, 314, 324, 325, 337
choroidal melanoma, eye enucleation, radiocobalt eye-plaque therapy, comparison 115, 118
^{137}Cs gamma irradiation, C_3H cells
endometrium carcinoma 309, 327
experimental tumors, radiobiology 16
^{125}Iodine-, ^{192}Iridium implantation therapy, external irradiation 162, 163
isoeffective doses, jejunal crypt stem cells 13
patients with choroidal melanoma treated by ^{106}Ru/^{106}Rh β-ray applicator 109
prostatic cancer
effect of biopsy 245, 246, 250, 251
Kaplan-Meier method 237, 238, 250, 251
tongue carcinoma, ^{198}Au seeds, implantation 174, 175

^{182}Tantalum
brain tumors, stereotactic interstitial irradiation 68
wires
dose distribution calculation, threedimensional 43
interstitial therapy 25
target volume
^{198}Au seed model 173
brachycurietherapy, breast cancer 180, 199
interstitial boost irradiation, breast cancer 188, 199
interstitial thermoradiotherapy 343, 365
three dimensional, Alth applicators, cervix cancer 301, 302
tumor therapy, definition 146
teletherapy
cervix cancer
doses in the reference points 338
telecobalt pendulum therapy 307
combination with brachytherapy, techniques 45
see external beam therapy, external stereotactic focal irradiation
oral mucosa, radiobiology 13
therapy
anal canal cancer
combined radiochemotherapy 209–213
external irradiation, chemotherapy, ^{192}Iridium implantation 215–219, 223, 224, 231
open questions 231
organ-sparing treatment 207
see brachytherapy, teletherapy, interstitial therapy
brachytherapy
chemotherapy 46

radiobiology 3-6
techniques 35-51
choroidal melanoma 103-111, 113-127
comparison of results: Eye enucleation versus radiocobalt eye-plaque therapy 115
external beam therapy, radiobiology 3-6
hormonal, prostatic cancer 243, 248
intracavitary, interstitial, physical properties of radioactive sources, techniques 25-30, 37
lowe dose rate, radiobiology 5, 7, 9
multimodal, anal canal carcinoma 229
radionecrosis, anal canal 210
results see clinical results
tongue carcinoma 171, 172
thermal map
interstitial hyperthermia, tumor therapy 360, 361, 363
thermoradiotherapy
interstitial
American RTOG 84-89 study, clinical experience 354
anal canal carcinoma 350, 351
biological considerations 343, 344
biomolecular mechanisms of cell radiosensitization 354
brain tumors 350, 368
capacitive radiofrequency technique 346, 365
choroidal melanoma 346
clinical results 351-354, 366, 367, 372, 373
complications 353, 367, 368
complimentary interaction of heat and irradiation, cell biology 344
complimentary interaction of heat and irradiation, clinical experiences 372, 373
Curie point, loss of ferromagnetism 348
European ESHO 4-86 study, clinical experience 354
ferromagnetic technique 345, 347-349
ferromagnetic technique, clinical experience 351, 352
head and neck tumors 350, 360, 361, 366, 372
heating techniques 343, 365
hot water perfusion technique 345, 349
hot water perfusion technique, clinical experience 252
hyperthermic cytotoxicity, radiosensitization 344
indications 350, 351
inductive ferromagnetic seed technique 347, 348, 366
intraoperative radiotherapy, abdominal surgery 350, 351
microwave technique 345-347, 365, 366, 372
microwave technique, clinical experience 351
multipoint invasive thermometry ("thermal mapping") 349
principles 350
prognostic factors 353, 367, 368
radiofrequency, clinical experience 351
radiofrequency, technique 345, 346
rectum carcinoma 350, 351
resistive local current field technique 344
results 351-354, 366, 367, 372, 373
self-regulating ferromagnetic seed technique 345, 347-349
specific absorption rate 346, 348
target volume 343, 365
techniques, survey 344, 345
thermometry 346, 359
TNM system
anal canal cancer 207, 216
choroidal melanoma 104
^{192}Iridium-, ^{125}Iodine at Erlangen treated patients 160
Manual for Staging Cancer, American Joint Committee on Cancer 236, 241
oral cavity, oropharynx carcinoma, clinical results of therapy 131, 132, 135
prostatic cancer
after lymphadenectomy 258, 266
Memorial Sloan-Kettering Cancer Center 235, 236, 241, 245
tongue carcinoma 170, 171
tolerance dose
acute, chronic irradiation, experimental curves 20
repair half times, α/β ratio, mathematical analysis 4, 5, 9
tongue
anatomy, topography 169, 170
carcinoma
hairpin implant, reference isodose 137
hyperthermia, external beam irradiation, results 361, 372
interstitial brachytherapy, ^{198}Au 169-175
loop implantation technique, brachytherapy 147
lymph node metastases, localization 171
prognosis 171
5-year survival rates 174, 175
lymphatic capillary flow 170
tonsil
carcinoma, interstitial brachytherapy, results 135, 136, 142, 147, 149, 161
radioactive implants, localization radiograph 161, 162
tonsillar region
carcinoma, interstitial therapy 142
topography
tongue 169, 170
total body irradiation
leukaemic cells, dose-rate effect 11
treatment results
see clinical results, survival rates
tumor
additional foci, breast cancer 199
bed boosting, ^{192}Iridium afterloading 201
brachytherapy, radiobiology 3-6
cell biology, thermoradiotherapy, hyperthermic cytotoxicity, radiosensitization 344
cell system
hypoxic, radioresistance 315
repair half times 19
classification, see TNM system
control, afterloading techniques 39
differentiation, survival, prostatic cancer, relations 242, 243
dimension, pre-, posttreatment, choroidal melanoma 116
dose
lip carcinoma, interstitial therapy 133, 139
prostatic cancer, ^{125}Iodine seed therapy 263
dose rate effects, radiobiology 4
extension, microscopic, breast cancer 188
external beam therapy, radiobiology 3-6
extrapolated dose, relative effectiveness 5
hyperthermia, thermoradiotherapy, therapeutic, treshold temperature of 42 °C, thermal maps 360, 361
isodoses, standard Fletcher cesium application 4
localization, prostatic cancer 240
minimum temperature, prognostic factor, thermoradiotherapy 353
necrosis after initial thermoradiotherapy 353
regression after interstitial therapy, prostatic cancer 266
response, hyperthermia with external radiotherapy, prognosis 373
size, choroidal melanoma 108
target volume, interstitial hyperthermia 343, 365
temperature distribution, thermoradiotherapy 349, 360
therapy, interstitial thermoradiotherapy, indications 344, 345, 350
tissue, normal tissue, cell repair, recovery time factor 319
treatment duration, mathematical formulations 17
volume
dose-response relationship, breast cancer 200
head and neck tumors, sonography, CT, MRI 146
interstitial hyperthermia 359

tumors
anal canal carcinoma
abdominoperineal resection 211, 212, 215, 229-232
chemotherapy 212, 215, 216, 221-227
classification, WHO 207, 216
combination therapy: external irradiation, chemotherapy, ^{192}Iridium implantation 215-219, 223, 224, 231
CT diagnosis 207
incidence 207
^{192}Iridium implantation, afterloading technique 153, 154, 210, 215-219, 221-227
lymphatic spread 207
metastases 213
prognostic factors 207, 211
radiochemotherapy, Lyon experience 209-213
radiotherapy versus radiotherapy plus chemotherapy, results 212
sexual practice, sexual transmitted disease 208
sphincter preservation 212
split-course therapy, external beam therapy, ^{192}Iridium-, ^{60}Co implantation 210
surgery, diagnos, clinical results 217, 222, 229-232
thermoradiotherapy 350
TNM system 207, 216
brain
afterloading therapy, fractionated 85-88
astrocytoma, survival times, postoperative interstitional therapy 68, 69, 71, 72
^{198}Au stereotactic interstitial irradiation 68
brachycurietherapy, basic principles 61-66
^{60}Co curietherapy 67
^{60}Co stereotactic interstitial irradiation 68
^{157}Cs intracranial curietherapy 67
diencephalic region, interstitial stereotactic irradiation, ^{192}Iridium, ^{125}Iodine 72, 73
ependymoma, survival times 68
fractionated afterloading therapy 85-88
glioblastoma, survival times, postoperative, interstitial therapy 68
glioblastoma, survival times, tumor doses 70
glioma, radiosensitivity 81
glioma, imaging stereotaxic implantation of radionuclides 71, 72
histologic examination, biopsy 68
^{125}I stereotactic interstitial implantation 68, 71, 72
imaging stereotactic implantation of radionuclides 67-79
implantation techniques, clinical results 65
implantation techniques, surgical procedure 64
implantation techniques, target volume 63
interstitial hyperthermia, results 360, 361
interstitial thermoradiotherapy 350
^{125}Iodine, interstitial, stereotactic curietherapy 67, 68, 72
^{192}Iridium curietherapy 68, 69, 72
meningioma, survival times, interstitial therapy 68, 71, 72
mesencephalic brain stem, imaging stereotactic implantation of ^{192}Iridium, ^{125}Iodine 72, 73
metastases, external stereotactic focal irradiation 69, 75, 89-94
midline structures, stereotactic implantation of ^{192}Iridium, ^{125}Iodine 72, 73, 75
MRI, definition of thrue extent 62
oligodendroglioma, survival times, postoperative interstitial therapy 68
^{32}P stereotactic interstitial irradiation 68
parasellar, imaging stereotactic curietherapy 74
pituitary adenomas, stereotactic radioisotope implantation 70, 71, 74
pons region, interstitial stereotactic irradiation, ^{192}Iridium, ^{125}Iodine 72, 73
^{182}Ta stereotactic interstitial irradiation 68
thermal map, interstitial hyperthermia 360
ventricular system, imaging stereotactic implantation of radionuclides 72, 73
^{90}Y stereotactic interstitial irradiation 68
breast cancer
interstitial boost irradiation 187-290
thermoradiotherapy 250, 366, 367
buccal mucosa carcinoma
interstitial brachytherapy 134
interstitial hyperthermia 361
plastic tube implantation technique 136, 137
results of curietherapy 134, 135
cervical cancer
afterloading plate method, high dose rate 298, 299
afterloading therapy, results 302-304, 308, 309, 317, 322, 323, 329, 337
bladder dosimetry, intracavitary irradiation 279-283
brachytherapy, complications 293, 302, 339
classification 308
external beam therapy 321, 336
high dose rate afterloading, prospective clinical trial 307-310
high dose rate afterloading, ^{226}Radium therapy, comparison 311-317
high dose rate techniques, clinical results 279, 280, 291-297, 298-305, 333-339
interstitial therapy, dosimetry 275-283
intracavitary therapy, Stockholm regime 320, 321
isodose configuration, high dose rate afterloading therapy 330
low and high dose rate brachytherapy, comparison of doses in the reference points 338
megavolt therapy 271, 314
one- and three way applicators, radiographs 287, 289
physical properties of radioisotopes 272
physics, afterloading plate method 300, 301
postoperative intravaginal treatment regime 321
^{226}Radium therapy, techniques 271-273, 311
side effects, low and high dose rate therapy 302, 315, 339
standard afterloading techniques 336
survival rates, afterloading therapy 295, 298-305, 336, 337
telecobalt pendulum therapy 307
three channel Alth applicators, ^{60}Co, ^{137}Cs 300
TNM system 308
uterine, vaginal applicators, Stockholm-, Paris-, Manchester-, Munich-techniques 272, 273
eye, thermoradiotherapy 350
choroidal melanoma, brachytherapy techniques 36
with ^{125}I plaques 346
with ^{106}Ru/^{106}Rh plaques, long term results 103-111, 113-122
choroidal melanoma, interstitial thermoradiotherapy 346
endometrium carcinoma
afterloading therapy, results 308, 309, 314, 327
clinical results of therapy 275, 303, 307-309, 314, 327
high dose rate afterloading, prospective clinical trial 307-310
high dose rate afterloading, ^{226}Radium therapy 311-317
intracavitary therapy, results 275, 303

^{192}Iridium wire tube packing 302
metastatic pread 327
packing method, Heyman, Stockholm, results 307, 327
postoperative intravaginal treatment results 327
primary intracavitary therapy 321, 322
survival rates 275, 303, 307–309, 314, 327
"umbrella" applicator, low, high dose rate afterloading 333, 336
experimental, dose rate effect, radiobiology 16, 17
eye
choroidal melanoma 113–122, 123–127
enucleation after β-therapy 109, 110, 113
metastases, incidence 114
primary malignant, American Cancer Society 114
retinoblastoma, brachytherapy techniques 36, 114
^{106}Ru/^{106}Rh applicators 103, 104
therapeutic results 117, 123
thermoradiotherapy 350
gynecological, see gynecological tumors
head and neck
advanced, brachytherapy, Erlangen experience 159–167
interstitial brachytherapy 145–152
interstitial hyperthermia, clinical results 360, 361, 368, 372
interstitial thermoradiotherapy 350
hypopharynx, hyperthermia, external beam irradiation, results 372, 374
implantation techniques, selection of radionuclides 62, 63
larynx, brachytherapy, Erlangen experiments 159, 161
lip carcinoma, curietherapy, results 134, 135, 161
lung, interstitial hyperthermia, clinical results 360, 372
lymph node metastases, interstitial therapy 135, 143, 147, 157, 161
melanoma, choroidal, brachytherapy with ^{106}Ru/^{106}Rh plaques, long term results 103–111
mouth floor, interstitial therapy 135, 138, 140, 161
nasopharynx
hyperthermia, external beam irradiation, results 372
interstitial brachytherapy 146
oral cavity
oropharynx carcinoma, American Committee on Cancer (AJCC), tumor classification 131, 132
oropharynx carcinoma, interstitial therapy, results 133–144
pancreas, thermal map, interstitial hyperthermia 363
parametrium, afterloading techniques 40
parasellar, pituitary, interstitial therapy, indications, clinical results 71
perineum, ^{192}Ir implantation, afterloading technique 153, 154
pituitary adenomas, imaging stereotactic implantation of radionuclides 70, 71, 74
poorly vasculized, radiosensitivity 344
prostatic cancer
^{198}Au marker seeds, inadequate retropubic implantation 259
biopsy 245, 246, 258, 265, 266
bone metastases 246, 249
brachytherapy 236, 237, 239, 245–256, 257–267
chemotherapy 265
classification 235, 236, 241, 245, 257
clinical results 237, 238, 241–244, 250–254, 263–267
complications 239, 242, 246, 248, 252
CT cross sections 235, 237, 239, 243, 264
diagnosis 245, 246, 258, 265, 266
dosimetry, interstitial therapy 237, 248
external beam irradiation, results 238, 239, 247, 248, 249, 253, 264
histology 246, 249, 250
hormonal therapy 243, 248
indications, transperineal sonographically guided ^{125}Iodine seed implantation 261, 264
interstitial brachytherapy 235, 236, 239, 241–244, 245–256, 263–267
^{125}Iodine implantation, Mick applicator 235, 236, 239, 241, 248–254, 258–262
lymphadenectomy, histologic grade, survival times 250, 258, 266
lymphadenopathy 236, 246, 258, 266
lymphangiography 246
metastases 235, 236, 246
morbidity, treatment related 242, 246, 247, 251
orchiectomy, hormonal therapy 248
prognostic factors 235
radical prostatectomy, results 238, 245, 264, 266
recurrence rates 250
risk of local tumor progression 259
sonography 235, 239, 243, 257–262
surgery 238, 245, 248, 250, 258, 266
survival times 237, 238, 239, 250, 251
TNM system 235, 236, 241, 245, 266
treatment results 237, 238, 247, 250, 251, 263–267
volumetry, sonography 258, 264
radiotherapy with Heyman capsules, Ries eggs, ^{60}Co pearls 275
retinal, brachytherapy techniques 36, 114
skull base, brachycurietherapy 71
soft palate carcinoma, interstitial therapy results 135
thermoradiotherapy, clinical experience 351, 352, 360, 361, 366, 367, 372
tongue
^{198}Au implantation 169–175
hyperthermia, external beam irradiation, results 372
tongue carcinoma, interstitial therapy, results 135, 135, 141, 142
uvula carcinoma, endocurietherapy, guide gutter technique 140
vaginal
applicators, afterloading, low, high dose rate 272, 273, 276, 333, 335
interstitial therapy, techniques 275
intravaginal postoperative treatment regime 321, 323
radiation reactions 302
vaginal stump, secondary carcinoma of vagina 327
vallecular region carcinoma, curietherapy, plastic tube implantation 143
vulva, hyperthermia, external irradiation, results 372

"umbrella" applicator
low, high dose rate afterloading therapy, endometrium carcinoma 333, 336
ureter
fibrosis, complication of afterloading plate method 302, 328
organ at risk, high dose afterloading 291
stenosis, low, high dose rate afterloading 328
uterine, vaginal applicators
cervix carcinoma, Stockholm-, Paris-, Manchester-, Munich techniques 272, 273
uterus
endometrium carcinoma, see endometrium carcinoma
intrauterine irradiation, "Umbrella" applicator, low, high dose rate afterloading 333, 336
uvea melanoma, see choroidal melanoma

uveal tract
malignant tumors, controversy regarding best approach to diagnosis and treatment 114
uvula carcinoma endocurietherapy, guide gutter technique 140
treatment results 149

vagina
applicators, afterloading, low, high dose rate 272, 273, 276, 333, 335
carcinoma, interstitial, intracavitary treatment techniques 275
intravaginal postoperative treatment regime, cervix carcinoma 321, 323
radiation reactions, brachytherapy, cervix cancer 302
stump, secondary carcinoma of vagina 327
vulva carcinoma, hyperthermia, external beam irradiation, results 372

X rays
abdominal irradiation, dose effect curves 13
dose survival curves, rhabdomyosarcoma 16

^{90}Y, brain tumors, stereotactic interstitial irradiation 68

List of Contributors

GERHART ALTH, Prim., Prof. Dr.
Allgemeines Krankenhaus der Stadt Wien-Lainz
Abteilung Strahlentherapie
Wolkersbergenstr. 1
1130 Wien, Austria

HEINER ANNWEILER, Dr.
Universitätsklinikum der Gesamthochschule Essen
Radiologische Klinik und Poliklinik
Hufelandstr. 55
4300 Essen 1, Germany

H. W. ANTON, Dr.
Radiologische Universitätsklinik
Abteilung Gynäkologisch-geburtshilfliche Radiologie
Voßstr. 9
6900 Heidelberg, Germany

J. M. ARDIET, Dr.
Centre Léon Bérard
Departement de Radiotherapie
28, rue Laënnec
69008 Lyon, France

JAMES J. AUGSBURGER, M.D.
Oncology Service
Wills Eye Hospital
Philadelphia, PA 19107, USA

KURT BAIER, Dipl.-Phys.
Strahlenabteilung der Universitäts-Frauenklinik
Josef-Schneider-Str. 4
8700 Würzburg, Germany

FRANÇOIS BAILLET, Prof. Dr.
Centre des Tumeurs
Hôpital Salpêtrière
47–83, Boulevard de l'Hôpital
75634 Paris, France

H. BARTELINK, M.D., Ph.D.
Chairman
Netherlands Cancer Institute
Antoni van Leeuwenhoekziekenhuis
Plesmanlaan 12
1066 CX Amsterdam, The Netherlands

J. H. BORGER, M.D.
Netherlands Cancer Institute
Antoni van Leeuwenhoekziekenhuis
Plesmanlaan 12
1066 CX Amsterdam, The Netherlands

ANNE MARIE BOROFSKY, M.D.
Department of Radiation Oncology and Nuclear Medicine
Hahnemann University
Mail Stop 200; Broad and Vine
Philadelphia, PA 19102-1192, USA

LUTHER W. BRADY, M.D.
Professor and Head
Department of Radiation Oncology and Nuclear Medicine
Hahnemann University
School of Medicine
Mail Stop 200; Broad and Vine
Philadelphia, PA 19102-1192, USA

M. BUSCH, Prof. Dr.
Universitätsklinikum der Gesamthochschule Essen
Radiologische Klinik und Poliklinik
Hufelandstr. 55
4300 Essen 1, Germany

H. BUSSE, Prof. Dr.
Universitäts-Augenklinik
Domagkstr. 15
4400 Münster, Germany

DANIEL CHASSAGNE, Dr.
Département de Carcinologie
Service de Radiothérapie
Hôpital Henri Mondor
51, Avenue de Maréchal de Lattre de Tassigny
94010 Créteil, France

J. L. CHASSARD, Dr.
Centre Léon Bérard
Département de Radiotherapie
28, rue Laënnec
69008 Lyon, France

CHRISTOPHER T. COUGHLIN, M.D.
Professor and Chairman of Radiation Oncology
Dartmouth Hitchcock Medical Center
2 Maynard Street
Hanover, NH 03756, USA

JOHN L. DAY, Ph.D.
Department of Radiation Oncology and Nuclear Medicine
School of Medicine
Hahnemann University
Mail Stop 200; Broad and Vine
Philadelphia, PA 19102-1192, USA

JÜRGEN DUNST, Dr.
Strahlentherapeutische Klinik und Poliklinik der Universität Erlangen-Nürnberg
Universitätsstr. 27
8520 Erlangen, Germany

BAHMAN EMAMI, M.D.
Professor of Radiology, Chief
Hyperthermia Section
Radiation Oncology Center
Mallinckrodt Institute of Radiology
Washington University Medical School
4939 Audubon Avenue
St. Louis, MO 63110, USA

RITA ENGENHART, Dr.
Abteilung Klinische Radiologie
Radiologische Klinik der Universität Heidelberg
Im Neuenheimer Feld 400
6900 Heidelberg, Germany

HELMUT ERNST, Prof. Dr.
Abteilung für Radiologie und Strahlentherapie
Universitätsklinikum Steglitz
Freie Universität Berlin
Hindenburgdamm 30
1000 Berlin 45, Germany

WILLIAM A. FAIR, M.D.
Department of Surgery
Memorial Sloan Kettering Cancer Center
New York, NY 10021, USA

RAINER FIETKAU, Dr.
Strahlentherapeutische Klinik der Universität Erlangen-Nürnberg
Universitätsstr. 27
8520 Erlangen, Germany

ZVI FUKS, M.D.
Department of Radiation Oncology
Memorial Sloan Kettering Cancer Center
New York, NY 10021, USA

FRANE PAUL GALL, Prof. Dr.
Chirurgische Klinik
Maximiliansplatz
8520 Erlangen, Germany

J. P. Gerard, Dr.
Département de Radiotherapie
Hôpital Lyon Sud
69310 Pierre Bénite, France

Bruce J. Gerbi, Ph.D.
Department of Therapeutic Radiology
Radiation Oncology
University of Minnesota Hospitals and Clinics
Harvard Street at East River Road
Minneapolis, MN 55455, USA

Felix H. Glaser, Prof. Dr. Sc.
Klinik und Poliklinik für Radiologie
Medizinische Akademie Erfurt
Nordhäuser Str. 74
5010 Erfurt, Germany

Gerhard Grabenbauer, Dr.
Strahlentherapeutische Klinik und Poliklinik der Universiät Erlangen-Nürnberg
Universitätsstr. 27
8520 Erlangen, Germany

Laval Grimard, Dr.
Ottawa Regional Cancer Centre
Ottawa, Ontario, Canada

Josef Hammer, Dr.
Institut für Radiotherapie
Krankenhaus der Barmherzigen Schwestern
Seilerstätte 4
4010 Linz, Austria

Hans-Peter Heilmann, Prof. Dr.
Hermann-Holthusen-Institut für Radiotherapie
Allgemeines Krankenhaus St. Georg
Lohmühlenstr. 5
2000 Hamburg 1, Germany

Basil S. Hilaris, M.D.
Professor and Chairman
New York Medical College
Department of Radiation Medicine
Valhalla, NY 10595, USA

Hiroshi Ikeda, M.D.
Department of Radiology
Osaka University Medical School
1-1-50 Fukushima, Fukushima-ku
Osaka, 553, Japan

Takehiro Inoue, M.D.
Department of Radiology
Osaka University Medical School
1-1-50 Fukushima, Fukushima-ku
Osaka, 553, Japan

H. Junkermann, Dr.
Radiologische Universitätsklinik
Abteilung Gynäkologisch-geburtshilfliche Radiologie
Voßstr. 9
6900 Heidelberg, Germany

Ulf L. Karlsson, M.D.
Department of Radiation Oncology and Nuclear Medicine
School of Medicine
Hahnemann University
Mail Stop 200; Broad and Vine
Philadelphia, PA 19102-1192, USA

Bernhard N. Kimmig, Priv.-Doz. Dr.
Abteilung Klinische Radiologie
Radiologische Klinik der Universität Heidelberg
Im Neuenheimer Feld 400
6900 Heidelberg, Germany

Karl-Heinz Kloetzer, Dr. Sc.
Klinik für Radiologie der Friedrich-Schiller-Universität
Bachstr. 18
6900 Jena, Germany

Dieter Kob, Prof. Dr.
Klinik für Radiologie der Friedrich-Schiller-Universität
Bachstr. 18
6900 Jena, Germany

Christopher Koprowski, M.D.
Department of Radiation Oncology and Nuclaer Medicine
Hahnemann University
Mail Stop 200; Broad and Vine
Philadelphia, PA 19102-1192, USA

Takahiro Kozuka, M.D., Professor
Department of Radiology
Osaka University Medical School
1-1-50 Fukushima, Fukushima-ku
Osaka, 553, Japan

Johann C. Kummermehr, Dr.
Institut für Strahlenbiologie
Gesellschaft für Strahlen- und Umweltforschung (GSF)
Ingolstädter Landstr. 1
8042 Neuherberg, Germany

Klaus Kuphal, Dr.
Albert-Ludwigs-Universität
Radiologische Klinik
Abteilung Strahlentherapie
Hugstetter Str. 55
7800 Freiburg, Germany

Chung K. Lee, M.D.
Department of Therapeutic Radiology
Radiation Oncology
University of Minnesota Hospitals and Clinics
Harvard Street at East River Road
Minneapolis, MN 55455, USA

Seymour H. Levitt, M.D., Professor
Department of Therapeutic Radiology
Radiation Oncology
University of Minnesota Hospitals and Clinics
Harvard Street at East River Road
Minneapolis, MN 55455, USA

David A. Lightfoot, M.A.
Department of Radiation Oncology and Nuclear Medicine
Hahnemann University
Mail Stop 200, Broad and Vine
Philadelphia, PA 19102-1192, USA

P. K. Lommatzsch, Prof. Dr.
Augenklinik der Universität
Liebigstr. 14
7010 Leipzig, Germany

Klaus Lutz, Dr.
Urologische Klinik,
Katharinenhospital
Kriegsbergstr. 60
7000 Stuttgart 10, Germany

Hans-Bruno Makoski, Prof. Dr.
Strahlenklinik-Radioonkologie-Nuklearmedizin
Städtische Kliniken Duisburg
Zu den Rehwiesen 9/Kalkweg
4100 Duisburg 1, Germany

Ginette Marinello, Ph.D.
Département de Carcinologie
Centre Hôpitalo Universitaire Henri Mondor, 51, Avenue de Maréchal de Lattre de Tassigny
94010 Créteil, France

Arnold M. Markoe, M.D.
Department of Radiation Oncology and Nuclear Medicine
School of Medicine
Hahnemann University
Mail Stop 200; Broad and Vine
Philadelphia, PA 19102-1192, USA

Norie Masaki, M.D.
Department of Radiology
Osaka University Medical School
1-1-50 Fukushima, Fukushima-ku
Osaka, 553, Japan

Robert E. Maxwell, M.D.
Department of Neurosurgery
University of Minnesota Hospitals and Clinics
Harvard Street at East River Road
Minneapolis, MN 55455, USA

Jean Jaques Mazeron, Dr.
Département de Carcinologie
Service de Radiothérapie
Hôpital Henri Mondor
51, Avenue du Maréchal de Lattre de Tassigny
94010 Créteil, France

J. F. Montbarbon, Dr.
Centre Léon Bérard
Départment de Radiotherapie
28, rue Laënnec
69008 Lyon, France

Chitti R. Moorthy, M.D.
New York Medical College
Department of Radiation Medicine
Valhalla, NY 10595, USA

Reinhold G. Müller, Priv.-Doz. Dr.
Institut für Radiologie
Krankenhausstr. 12
8520 Erlangen, Germany

Rolf-P. Müller, Prof. Dr.
Klinik und Poliklinik für Strahlentherapie der Universität zu Köln
Joseph-Stelzmann-Str. 9
5000 Köln 41, Germany

Fritz Mundinger, Prof. Dr.
Ärztlicher Direktor a.D. der Abteilung Stereotaxie und Neuronuklearmedizin der Universität Freiburg
St. Josefs-Krankenhaus
Hermann-Herder-Str. 1
7800 Freiburg, Germany

Dattatreyudu Nori, M.D.,
Chairman, Department of Radiation Oncology
56-45 Main Street
Flushing, NY 11355, USA

Colin G. Orton, Ph.D.
Director, Medical Physics
Gershenson Radiation Oncology Center
Harper-Grace Hospitals and Wayne State University
3990 John R. Street
Detroit, MI 48201, USA

Christoph B. Ostertag, Prof. Dr.
Abteilung Stereotaktische Neurochirurgie
Neurochirurgische Universitäsklinik
7800 Freiburg, Germany

Shuzi Ozeki, M.S.
Department of Radiology
Osaka University Medical School
1-1-50 Fukushima, Fukushima-ku
Osaka, 553, Japan

J. Papillon, Dr.
Professeur á la Faculté
Radiologiste des Hôpitaux
12, Quai Général-Sarrail
69006 Lyon, France

Carlos A. Perez, M.D.
Professor of Radiology
Radiation Oncology Center
Mallinckrodt Institute of Radiology
Washington University Medical School
St. Louis, MO 63110, USA

Bernard Pierquin, Prof. Dr.
Département de Carcinologie
Service de Radiothérapie
Hôpital Henri Mondor
51, Avenue de Maréchal de Lattre Tassigny
94010 Créteil, France

Gudrun Pipard, M.D.
Division of Radiotherapy
University Hospital of Geneva
21, rue Alcide Jentzer
1211 Geneva, Switzerland
Address for Correspondence:
23, rue St. Martin
74160 Julien en Genevois, France

Roger A. Potish, M.D.
Associate Professor
Departments of Therapeutic Radiology and of Obstetrics and Gynecology
University of Minnesota Hospitals and Clinics
Harvard Street at East River Road
Minneapolis, MN 55455, USA

R. Pötter, Priv.-Doz. Dr.
Strahlentherapeutische Klinik der Universität Münster
4400 Münster, Germany

Markus Riccabona, Dr.
Urologische Abteilung
Krankenhaus der Barmherzigen Schwestern Seilerstätte 4
4010 Linz, Austria

Karsten Rotte, Prof. Dr.
Strahlenabteilung der Universitäts-Frauenklinik
Josef-Schneider-Str. 4
8700 Würzburg, Germany

Rolf Sauer, Prof. Dr.
Strahlentherapeutische Klinik der Universität Erlangen-Nürnberg
Universitätsstr. 27
8520 Erlangen, Germany

P. Schlag, Prof. Dr.
Leiter der Sektion Chirurgische Onkologie
Chirurgische Universitätsklinik
Im Neuenheimer Feld 110
6900 Heidelberg, Germany

Gerhard Schlegel, Dr.
Radiologische Klinik
Katharinenhospital Stuttgart
Kriegsbergstr. 60
7000 Stuttgart 10, Germany

M. Heinrich Seegenschmiedt, Dr.
Strahlentherapeutische Klinik und Poliklinik der Universität Erlangen-Nürnberg
Universitätsstr. 27
8520 Erlangen, Germany

Jerry A. Shields, M.D.
Oncology Service
Wills Eye Hospital
Philadelphia, PA 19107, USA

Joachim Slanina, Prof. Dr.
Albert-Ludwigs-Universität
Radiologische Klinik
Abteilung Strahlentherapie
Hugstetter Str. 55
7800 Freiburg, Germany

H. Sommerkamp, Prof. Dr.
Abteilung Urologie
Universitätsklinikum
Hugstetter Str. 55
7800 Freiburg, Germany

Anne Sparenberg, Dr.
Abteilung Radiologie und Strahlentherapie
Universitätsklinikum Steglitz der Freien Universität Berlin
Abteilung Radiologie und Strahlentherapie
Hindenburgdamm 30
1000 Berlin 45, Germany

Wolfgang Spitzer, Priv.-Doz. Dr.
Klinik und Poliklinik für Kieferchirurgie
Glückstr. 11
8520 Erlangen, Germany

Volker Sturm, Prof. Dr.
Neurochirugische Klinik der Universität Köln
Abteilung Stereotaxie und funktionelle Neurochirurgie
Joseph-Stelzmann-Str. 9
5000 Köln, Germany

Anca E. Tchelebi, M.D.
New York Medical College
Department of Radiation Medicine
Valhalla, NY 10595, USA

Norfried Thesen, Dr.
Klinik und Poliklinik für Strahlentherapie der Universiät Köln
Joseph-Stelzmann-Str. 9
5000 Köln 41, Germany

A. J. SUBANDONO TJOKROWARDOJO, M.D.
Dr. Daniel den Hoed Cancer Center and Rotterdam Radio-Therapeutic Institute
Groene Hilledijk 301
3075 EA Rotterdam, The Netherlands

P. TOURAINE-ROMESTAING, Dr.
Département de Radiothérapie
Hôpital Lyon Sud
69310 Pierre Bénite, France

KLAUS-RÜDIGER TROTT, Prof. Dr.
Department of Radiation Biology
Medical College of St. Bartholomew's Hospital
Charterhouse Square
London EC1M 6BQ, UK

PETER C. VERAGUTH, Prof. Dr.
Universitätsklinik für Strahlentherapie Inselspital
3010 Bern, Switzerland

A. G. VISSER, Ph.D.
Dr. Daniel den Hoed Cancer Center and Rotterdam Radio-Therapeutic Institute
Groene Hilledijk 301
3075 EA Rotterdam, The Netherlands

D. VON FOURNIER, Prof. Dr.
Radiologische Universitätsklinik
Abteilung Gynäkologisch-geburtshilfliche Radiologie
Voßstr. 9
6900 Heidelberg, Germany

MICHAEL WANNENMACHER, Prof. Dr.
Radiologische Universitätsklinik
Abteilung Klinische Radiologie
(Schwerpunkt Strahlentherapie)
Im Neuenheimer Feld 400
6900 Heidelberg, Germany

MANFRED WEIDENBECHER, Prof. Dr.
Klinik und Poliklinik für Hals-Nasen-Ohrenkranke
Waldstr. 1
8520 Erlangen, Germany

KLAUS WEIGEL, Dr.
Abteilung Neurochirurgie
Universitätsklinikum Steglitz
Freie Universität Berlin
Hindenburgdamm 30
1000 Berlin 45, Germany

WILLET F. WHITMORE, M.D., Professor
Department of Surgery
Memorial Sloan-Kettering Cancer Center
New York, NY 10021, USA

FRANK WILSON, Dr.
Departement de Carcinologie
Service de Radiothérapie
Hôpital Henri Mandor
51, Avenue de Maréchal de Lattre de Tassigny
94010 Créteil, France

G. WOLF, Dipl. Phys.
Radiologische Universitätsklinik
Abteilung Gynäkologisch-geburtshilfliche Radiologie
Voßstr. 9
6900 Heidelberg, Germany

N. WOLF, Dr.
Chirurgische Klinik der Universität Erlangen-Nürnberg
Maximiliansplatz
8520 Erlangen, Germany

REGINALD WOODLEIGH, M.S.
Department of Radiation Oncology and Nuclear Medicine
School of Medicine
Hahnemann University
Mail Stop 200; Road and Vine
Philadelphia, PA 19102-1192, USA

MR Medical Radiology

Diagnostic Imaging and Radiation Oncology

Series Editors:
L. W. Brady, M. W. Donner, H.-P. Heilmann, F. Heuck

This series recognizes the demand for an international state-of-the-art account of the developments reflecting the progress in the radiological sciences. Each volume conveys an overall picture of a topical theme so that it can be used as a reference work without taking recourse to other volumes.

The contents of the volumes concentrate on new and accepted developments in a manner appropriate for review by physicians engaged in the practice of radiology.

Springer-Verlag
Berlin Heidelberg New York London Paris Tokyo Hong Kong Barcelona

C. W. Scarantino (Ed.)

Lung Cancer

Diagnostic Procedures and Therapeutic Management with Special Reference to Radiotherapy

1985. XI, 173 pp. 42 figs. Hardcover.
ISBN 3-540-13176-0

H. R. Withers, University of California, Los Angeles, CA; **L. J. Peters,** University of Texas, Houston, TX (Eds.)

Innovations in Radiation Oncology

1987. XVII, 329 pp. 111 figs. Hardcover.
ISBN 3-540-17818-X

G. E. Laramore, University of Washington, Seattle, WA (Ed.)

Radiation Therapy of Head and Neck Cancer

1989. XII, 237 pp. 123 figs. Hardcover.
ISBN 3-540-19360-X

J.H. Anderson, The Johns Hopkins University, Baltimore, MD (Ed.)

Innovations in Diagnostic Radiology

1989. XIII, 213 pp. 144 figs. some in color. Hardcover. ISBN 3-540-19093-7

R. R. Dobelbower Jr., Toledo, OH (Ed.)

Gastrointestinal Cancer

Radiation Therapy

1990. XV, 301 pp. 76 figs. 90 tabs. Hardcover.
ISBN 3-540-50505-9

E. Scherer, C.Streffer, University of Essen; **K.-R. Trott,** London (Eds.)

Radiation Exposure and Occupational Risks

1990. XI, 150 pp. 32 figs. 55 tabs. Hardcover.
ISBN 3-540-51174-1

S. E. Order, The Johns Hopkins University, Baltimore, MD; **S. S. Donaldson,** Stanford University, Stanford, CA

Radiation Therapy of Benign Diseases

A Clinical Guide

1990. VIII, 214 pp. 103 tabs. Hardcover.
ISBN 3-540-50901-1